# *Advances in*
# ORGANOMETALLIC CHEMISTRY

## VOLUME 43

# Advances in Organometallic Chemistry

EDITED BY

ROBERT WEST

DEPARTMENT OF CHEMISTRY
UNIVERSITY OF WISCONSIN
MADISON, WISCONSIN

ANTHONY F. HILL

DEPARTMENT OF CHEMISTRY
IMPERIAL COLLEGE OF SCIENCE,
TECHNOLOGY, AND MEDICINE
LONDON, ENGLAND

FOUNDING EDITOR

F. GORDON A. STONE

VOLUME 43

ACADEMIC PRESS
San Diego    London    Boston    New York
Sydney    Tokyo    Toronto

Academic Press
*a division of Harcourt Brace & Company*
525 B Street, Suite 1900, San Diego, California 92101-4495, USA
http://www.apnet.com

Academic Press Limited
24-28 Oval Road, London NW1 7DX, UK
http://www.hbuk.co.uk/ap/

International Standard Book Number: 0-12-031143-7

PRINTED IN THE UNITED STATES OF AMERICA
98   99   00   01   02   03   QW   9   8   7   6   5   4   3   2   1

# Contents

## Silylenes Coordinated to Lewis Bases

### JOHANNES BELZNER and HEIKO IHMELS

## Chemistry of Ruthenium–Carbide Clusters Ru$_5$C(CO)$_{15}$ and Ru$_6$C(CO)$_{17}$

### PAUL J. DYSON

# Transition Metal Heteroaldehyde and Heteroketone Complexes

## HELMUT FISCHER, RÜDIGER STUMPF, and GERHARD ROTH

# Recent Progress in Transition Metal-Catalyzed Reactions of Silicon, Germanium, and Tin

## JENNIFER A. REICHL and DONALD H. BERRY

# Organometallic Compounds of the Heavier Alkali Metals

## J. DAVID SMITH

# Organometallic Complexes in Nonlinear Optics II: Third-Order Nonlinearities and Optical Limiting Studies

## IAN R. WHITTALL, ANDREW M. McDONAGH, MARK G. HUMPHREY, and MAREK SAMOC

# Contributors

*Numbers in parentheses indicate the pages on which the authors' contributions begin.*

JOHANNES BELZNER (1), Institut für Organische Chemie der Georg-August-Universität Göttingen, D-37077 Göttingen, Germany

DONALD H. BERRY (197), Department of Chemistry and Laboratory for Research on the Structure of Matter, University of Pennsylvania, Philadelphia, Pennsylvania 19104

PAUL J. DYSON (43), Centre for Chemical Synthesis, Department of Chemistry, Imperial College of Science, Technology, and Medicine, South Kensington, London SW7 2AY, United Kingdom

HELMUT FISCHER (125), Fakultät für Chemie, Universität Konstanz, Fach M727, D-78457 Konstanz, Germany

MARK G. HUMPHREY (349), Department of Chemistry, Australian National University, Canberra, ACT 0200, Australia

HEIKO IHMELS (1), Institut für Organische Chemie der Universität Würzburg, D-97074 Würzburg, Germany

ANDREW M. MCDONAGH (349), Department of Chemistry, Australian National University, Canberra, ACT 0200, Australia

JENNIFER A. REICHL, Department of Chemistry and Laboratory for Research on the Structure of Matter, University of Pennsylvania, Philadelphia, Pennsylvania 19104

GERHARD ROTH (125), Fakultät für Chemie, Universität Konstanz, Fach M727, D-78457 Konstanz, Germany

MAREK SAMOC (349), Australian Photonics Cooperative Research Centre, Laser Physics Centre, Research School of Physical Sciences and Engineering, Australian National University, Canberra, ACT 0200, Australia

J. DAVID SMITH (267), School of Chemistry, Physics, and Environmental Science, University of Sussex, Brighton BN1 9QJ, United Kingdom

RÜDIGER STUMPF (125), Fakultät für Chemie, Universität Konstanz, Fach M727, D-78457 Konstanz, Germany

IAN R. WHITTALL (349), Department of Chemistry, Australian National University, Canberra, ACT 0200, Australia

# Preface

This 43rd volume of *Advances in Organometallic Chemistry* marks the transition of Professor F. Gordon A. Stone, F.R.S., C.B.E., to the status of Founding Editor of the series that he instigated and coedited for 34 years.

Organometallic chemistry has grown dramatically over the past four decades. Research in this vibrant field is by nature dendritic; each year has brought new divergences and subdisciplines. It is therefore easy to forget how nascent the field was when this series began in 1964, prefaced by the words: "Organometallic chemistry now seems well on its way toward establishing its identity as an important domain of science . . . We hope that chemists with a general interest in this field will, turning these pages, be able to find recent information about many of the most active areas in organometallic chemistry." In the interim, Gordon's breadth of interest and insight has allowed him to identify emergent fields and rising scientists. Forty-three volumes later, over 350 invited authors have contributed to this goal of providing an author list that reads like an organometallic chemist's "Who's Who."

The enormous impact of Gordon Stone's research in organometallic chemistry continues to receive wide recognition. His long list of honorary doctorates, prestigious awards and commendations, and visiting professorships and the establishment of the "Gordon Stone Lectures" attest to the central role he has played in the development of our science. Moreover, he has been a champion of the support of fundamental research through his considerable influence on science policy on both sides of the Atlantic.

On this occasion, it is particularly appropriate to consider Gordon's contributions to scholarship and scientific publishing. He has served in an editorial or advisory role to every key journal concerned with our science. With E. W. Abel and G. Wilkinson, he conceived of and coedited the preeminent and essential reference works in our field, *Comprehensive Organometallic Chemistry,* editions 1 and 2.

We are pleased to congratulate Gordon on these achievements and to welcome him in his new role as Founding Editor. His continued advice and enthusiasm for the series will be invaluable. We both wish him all the best in his continuing researches.

*Robert West*
*Anthony F. Hill*

# Silylenes Coordinated to Lewis Bases

## JOHANNES BELZNER

Institut für Organische Chemie der Georg-August-Universität Göttingen
D-37077 Göttingen, Germany

## HEIKO IHMELS

Institut für Organische Chemie der Universität Würzburg
D-97074 Würzburg, Germany

## I

## INTRODUCTION

Silylenes are, in a certain sense, the younger but bigger siblings of carbenes, and their chemistry has been reviewed several times.[1] In 1978, Gaspar asked the question: "Are we to conclude from our present knowledge that nature has designed silylenes to humbly mimic carbenes, or is the situation different?"[1a] Today, we know that silylenes and carbenes have a number of properties in common, but also show significant differences.

One of these differences is the fact that $H_2Si$: has a singlet ground state with an empty p orbital and a doubly occupied $\sigma$ orbital, which is in contrast to the well-known triplet ground state of $H_2C$:. Moreover, calculations show that, with the exception of silylenes bearing electropositive substituents such as $Li(H)Si$:, $Li_2Si$:, and $HBe(H)Si$:, all silylenes are singlet ground state species and that the singlet–triplet gap often is bigger than that in the corresponding carbenes.[2] Due to the availability of the unoccupied p-orbital, silylenes are prone to react with Lewis bases, which may donate their free lone pair into the empty orbital (Scheme 1). This donor–acceptor interaction results in the formation of a silylene complex. In the last two

1

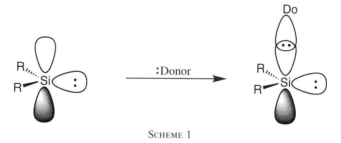

SCHEME 1

decades, much interest has been focused on this new class of complexes, and a considerable number of investigations, either of theoretical or of experimental nature, have been performed to obtain better insight into the chemistry of Lewis base–silylene complexes. This review gives a brief summary of the literature that documents the efforts to explore this new field of organosilicon chemistry.

Scanning the literature one notices that silylene–Lewis base complexes are drawn either using an arrow, which points from the heteroatom toward the silicon center, thus indicating a dative bond (structure **A**), or as a 1,2-dipole with a covalent bond (structure **B**). Because the theoretical results concerning the extent of charge separation in silylene–Lewis base complexes or silaylides are contradictory (Section II,A) we will use both grapical descriptions of these compounds alternatively and interchangeably.

**A**                    **B**

## II

## THEORETICAL STUDIES OF SILYLENE–LEWIS BASE COMPLEXES

Only few theoretical studies have been devoted exclusively to the coordination of Lewis bases to silylenes. Most calculational evidence for the formation of silylene–Lewis base complexes was obtained from the computational investigation of the insertion reaction of $H_2Si$: into various H–X $\sigma$ bonds, where X is an heteroatom center possessing one or more free electron pairs. As can be seen from Fig. 1 these exothermic reactions proceed via initial formation of a donor–acceptor complex and subsequent rearrangement through a three-membered cyclic transition state to yield the eventual insertion product. The kinetic stability of the initial silylene–Lewis base complex depends on the depth of this local minimum, i.e., on the complexation energy $\Delta E_{compl}$ as well as on the height of the energy barrier for the rearrangement ($\Delta E_{rearr}$).

In the following, we will present the results of semiempirical and *ab initio* calculations addressing the formation and subsequent reactions of silylene–Lewis base complexes. The review is organized according to the nature of the coordinating base; i.e., we will start with Lewis bases of group 14 and go through the periodic table to group 17 Lewis bases.

## A. *Complexes with Carbon Nucleophiles*

The only carbon nucleophile whose coordination to silylenes was investigated theoretically is carbon monoxide. The main interest centers on the question whether this nucleophile forms a Lewis acid–Lewis base complex

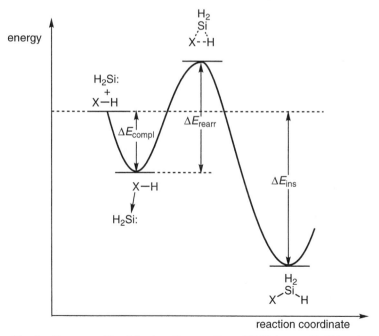

FIG. 1. Energy profile of the insertion reaction of $H_2Si$: into an X–H bond.

**1** or a silaketene **2** with silylenes. Using semiempirical methods such as MNDO or AM1, Arrington et al.[3] localized a planar silaketene structure as minimum for the $Me_2SiCO$ system. The Si–C bond lengths of 162.5 pm (MNDO) and 164.4 pm (AM1) are clearly in the range of Si=C double bonds as expected for a silaketene. In contrast, at the HF/3-21G level a $Me_2Si: \leftarrow CO$ complex exhibiting a pyramidal silicon center was shown to be 84 kJ mol$^{-1}$ more stable than the planar silaketene structure.[3] The Si–C bond length of 289.1 pm and the C–Si–CO bond angle of 89° are in good agreement with a complex in which a free electron pair at the carbon is donated into the empty p orbital at the silicon center. Ab initio calculations by Hamilton and Schaefer[4] at considerably higher levels localized a weakly bound nonplanar structure **1** (Si–C: 193.8 pm, H–Si–C: 88.6°) as the equilibrium structure of $H_2SiCO$. The planar ketene-like structure **2**, which was calculated to be 77.5 kJ mol$^{-1}$ higher in energy using the CCSD method and a polarized triple-$\zeta$ basis set, represents at this level of theory the transition state for the inversion of the silicon center.[4] Similar results were obtained for the $Me_2SiCO$ system.[4] Finally, based on ab initio energies obtained at the MP2/3-21G*//3-21G* level of theory, an alternative structure **3**, which features a CO molecule bridging two silylene units, was suggested by Zakharov and Zhidomirov[5] for the adduct of $R_2Si:$ (R = H, Me) with carbon monoxide.

### B. Complexes with Group 15 Nucleophiles

The reaction of $H_2Si:$ with $NH_3$ was investigated theoretically for the first time by Raghavachari et al.[6a] at the MP4SDTQ/6-31G**//6-31G* level. A staggered complex **4** is formed by interaction between the lone pair of $NH_3$ and the empty p orbital of singlet $H_2Si:$. The length of the dative Si $\leftarrow$ N bond is 208.9 pm; i.e., it is considerably longer than the calculated Si–N bond length of silylamine (172 pm). Complex **4** is located in a fairly deep minimum as evidenced by the binding energy of 105.1 kJ mol$^{-1}$ as well as the energy barrier of 160.4 kJ mol$^{-1}$ for the rearrangement of **4**, which would yield the insertion product **5** (Scheme 2). A more recent study by Conlin et al.[7] estimated the bond dissociation enthalpy of **4** as 97 $\pm$ 10 kJ mol$^{-1}$ at the G2 level of theory. The fact that the Mulliken charges of the $H_2Si:$ and $NH_3$ units of **4** are equal within $\pm0.20$ was interpreted by the authors as evidence against an ylid structure of complex **4**. In contrast to these results, Schoeller et al.[8] found that, based on Löwdin populations, electron density is transferred significantly from the $NH_3$ unit to the $H_2Si$ moiety in **4**, and accordingly described **4** as a zwitterionic ylide species. CIS/6-31+G* calculations suggested the $S_0 \rightarrow S_1$ transition to be shifted

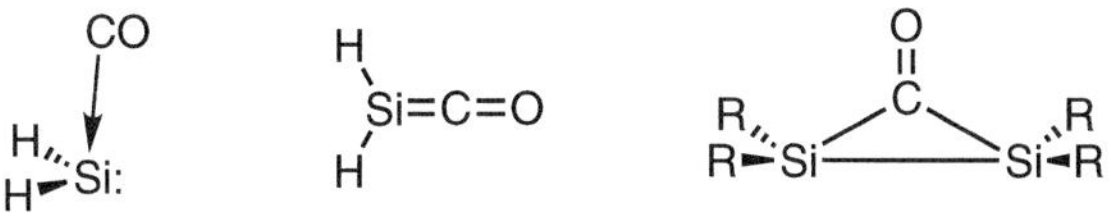

**1**    **2**    **3**

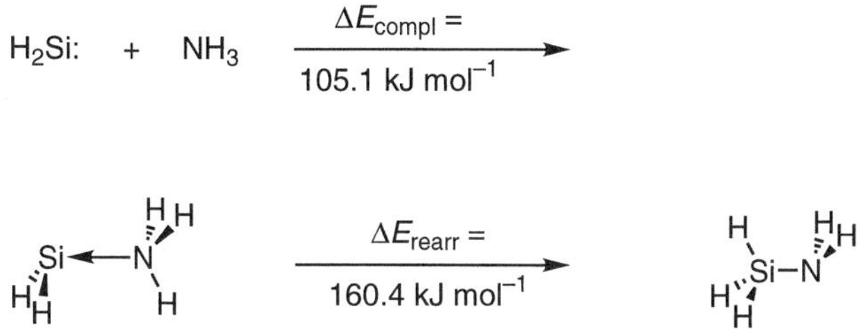

Scheme 2

from 485 in free $H_2Si$: to 301 nm in complex **4**;[7] this theoretical result is in good agreement with the hypsochromic shift, which is experimentally observed for matrix-isolated silylene–Lewis bases complexes (Section III).

The phosphine–silylene complex **6** is less stable than **4** as evidenced by a lower binding energy of 73.3 kJ mol$^{-1}$ as well as a lowered activation energy of 82.4 kJ mol$^{-1}$ for the rearrangement of **6** into **7** at the MP4SDTQ/6-31G**//6-31G* level (Scheme 3).[6a] A similar decrease in binding energy was found by Schoeller *et al.*[8a] comparing the ammonia complex **4** ($\Delta E_{compl}$ = 103.0 kJ mol$^{-1}$) with phosphine complex **6** ($\Delta E_{compl}$ =

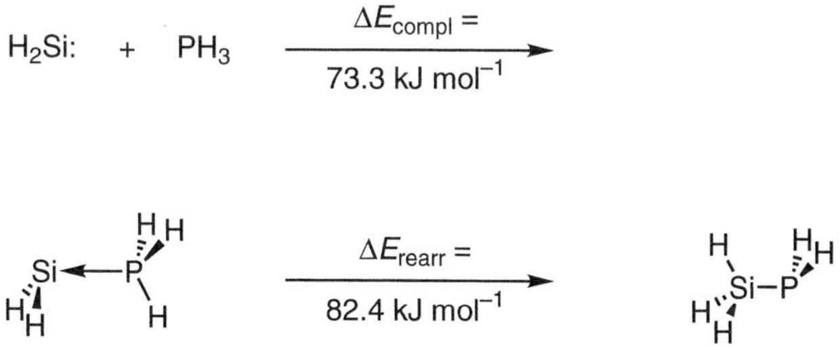

**6**    **7**

Scheme 3

87.5 kJ mol$^{-1}$), when an appreciably higher level of theory was used. In addition, these authors investigated the influence of substituents at phosphorus and silicon in **6** on the binding energy.[8a] It was found that the binding energy is increased by the introduction of electron-withdrawing substituents (Cl and F) at phosphorus in **6** as well as of electron-donating SiH$_3$ groups at silicon. These results suggest, in line with the calculated Löwdin populations, that complex **6** has appreciable zwitterionic character.

The same authors also studied the coordination of heavier pnictogen–hydrogen compounds XH$_3$ to H$_2$Si:. The X–Si distances in these complexes correspond to slightly elongated single bonds. In general, the binding energy decreases going from NH$_3$ to BiH$_3$ as Lewis base (Table I).

The addition of two molecules of NH$_3$ is calculated to result in the formation of a $C_{2v}$ symmetrical staggered structure **8** (Scheme 4).[7,8a] The Si–N bond length of 236.7 pm in **8**, which was found at the G2 level, is significantly longer than in the monoadduct **4** (208.9 pm).[7] The bond dissociation energy is only 7 ± 20 kJ mol$^{-1}$ at the G2 level, and thus **8** is expected to dissociate easily into **4** and NH$_3$;[7] similar results were obtained by Schoeller *et al.*[8a] In contrast, the $C_{2v}$ symmetrical bisadducts of PH$_3$, AsH$_3$, and SbH$_3$ with H$_2$Si: are transition states, whereas the corresponding bisadduct of BiH$_3$ is an energy minimum again.[8a]

### C. Complexes with Chalcogen Nucleophiles

Silylenes were reported to react rapidly with molecular oxygen,[9] and a silanone $O$-oxide **9,** which, in a very formal sense, may also be viewed as a sily-

TABLE I

CALCULATED BOND LENGTHS $d_{Si-X}$ (pm), BINDING ENERGIES $\Delta E_{compl}$, AND ACTIVATION ENERGIES $\Delta E_{rearr}$ (kJ mol$^{-1}$, ZPE INCLUDED) OF ADDUCTS H$_2$Si ← XH$_3$ (X = N, P, As, Sb, Bi)

| Donor | $d_{Si-X}$ | $\Delta E_{compl}$ | $\Delta E_{rearr}$ |
|---|---|---|---|
| NH$_3$ | 208.9,[a,b] 206.9[c] | 105.1,[a] 97 ± 10,[b] 103.0[c] | 160.4[a] |
| PH$_3$ | 240.4,[a] 239,[d] 241.7,[e] 235.2[c] | 73.3,[a] 95.4,[d] 87.5[c] | 82.4[a] |
|  |  |  | 101.9[d] |
| AsH$_3$ | 248.5[c] | 68.2[c] | — |
| SbH$_3$ | 264.8[c] | 71.6[c] | — |
| BiH$_3$ | 277.2[c] | 47.3[c] | — |

[a] Ref. 6a: MP4SDTQ/6-31G**//6-31G*.
[b] Ref. 7: G2.
[c] Ref. 8a: MP2/CEP-31g(2d,p)//MP2/CEP-31g(2d,p).
[d] Ref. 6b: MP3/6-31G*//3-21G*.
[e] Ref. 8b: CEPA-1.

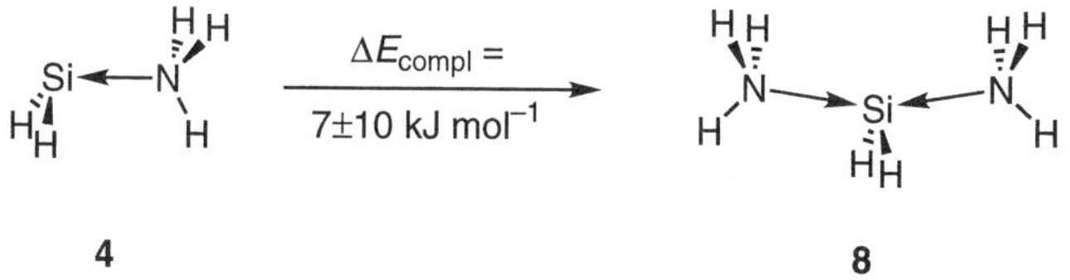

SCHEME 4

lene–oxygen adduct, or a siladioxirane **10** was discussed as an alternative product structure.[10-12] According to HF/6-31G* calculations, triplet **9** is a minimum at the potential energy surface.[10] The Si–O bond length is 170.8 pm, which is only slightly longer than the experimentally measured length of the covalent Si–O bond in $Me_3SiOOSiMe_3$ (168 pm).[13] No stable silanone $O$-oxide or silyene–oxygen complex structure could be localized at the singlet potential energy surface at the HF/6-31G* level. However, $C_1$ symmetrical singlet **9**, revealing considerable 1,3-diradical character, was found to be minimum when electron correlation was included (MP2/6-31G*).[11] As expected for a 1,3-diradical species **9**, the Si–O bond length of 174.8 pm is in the range of covalent Si–O bonds. According to these calculations, **9** rearranges easily to siladioxirane **10**. This process is highly exothermic (267.1 kJ mol$^{-1}$) and proceeds via a transition state whose energy is only 27.2 kJ mol$^{-1}$ higher than that of **9**.[11,12] Thus, singlet silanone $O$-oxide is expected to be kinetically unstable with regard to its isomerization to siladioxirane.

The reaction of $H_2Si$: with $H_2O$ was actually the first silylene insertion reaction to be investigated by theoretical methods. Raghavachari *et al.*[14] calcu-

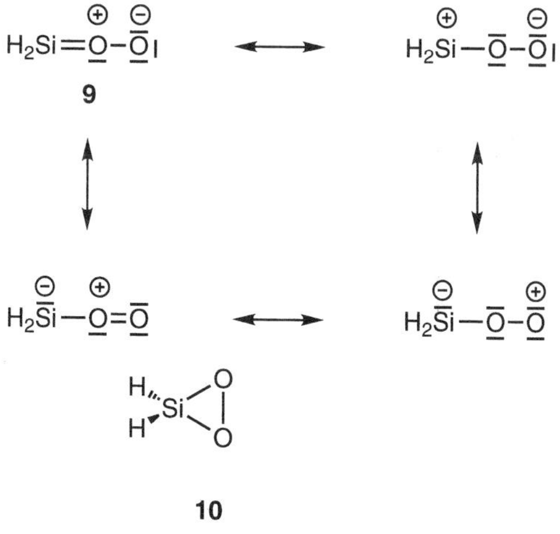

**10**

lated a complexation energy of 50.7 kJ mol$^{-1}$ at the MP4SDTQ/6-31G**//3-21G* level for the silylene–water adduct **11** (Scheme 5). The silicon lone pair of $C_1$ symmetrical **11** is nearly orthogonal to the free electron pair at oxygen. In contrast to silanone $O$-oxide **9** (see earlier), the Si–O bond length of **11** (196 pm) is significantly longer than a covalent Si–O single bond (163 pm in silanole), thus characterizing **11** as a weakly bound silylene–water complex. A 1,2-hydrogen shift transforms **11** into silanol **12,** which is 242.8 kJ mol$^{-1}$ more stable than **11**; the activation energy of this rearrangement is 98.0 kJ mol$^{-1}$. More recent calculations[6a,8a,15–17] using more elaborate *ab initio* levels basically confirm these early results (Table II). Similar results were obtained for the coordination of dimethylsilylene (**13**) to $H_2O$ (Table II).[16]

Reaction of $H_2Si$: with silanol[14] and methanol[18] proceeds in a manner qualitatively analogous to the reaction with water. The complexation energy of silanol to $H_2Si$: is 54.0 kJ mol$^{-1}$ at the MP4SDTQ/6-31G**//3-21G* level.[14] The subsequent insertion reaction into the O–H bond proceeds via a transition state 74.1 kJ mol$^{-1}$ higher in energy than the initially formed silylene–silanol complex. The corresponding values for the reaction with methanol are 83.3 kJ mol$^{-1}$ ($\Delta E_{compl}$) and 85.8 kJ mol$^{-1}$ ($\Delta E_{rearr}$) at the MP4//6-311G(2df,p)//MP2(full)/6-31G* level; i.e., the overall energy barrier for the insertion of $H_2Si$: into the O–H bond of methanol is vanishingly small.[18]

The insertion of $H_2Si$: into the C–O bond of furan was calculated using the semiempirical PM3 method to involve the initial formation of a silylene–furan complex ($\Delta E_{compl} = 57$ kJ mol$^{-1}$). However, keeping in mind that the *ab initio* calculated activation energy for the rearrangement of the methanol–silylene complex to the corresponding C–O insertion product is 169 kJ mol$^{-1}$,[18] the height of the activation energy barrier ($\Delta E_{rearr} = 1$ kJ mol$^{-1}$) seems to be severely underestimated with the PM3 method.[19] Similarly questionable results of semiempirical calculations were obtained for the reaction of dichlorosilylene with furan.[19]

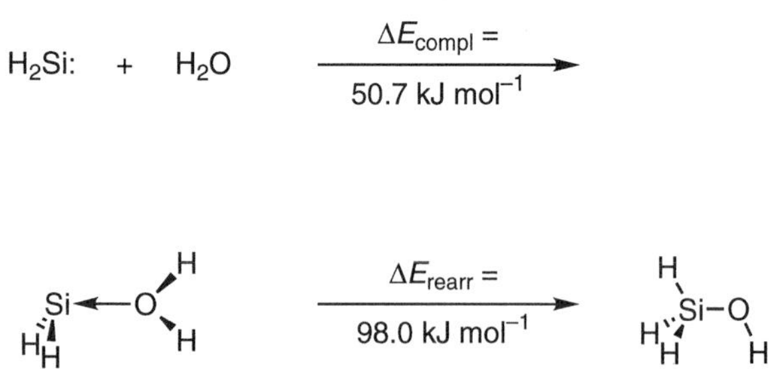

SCHEME 5

TABLE II

COMPLEXATION ($\Delta E_{compl}$), ACTIVATION ($\Delta E_{rearr}$), AND INSERTION ENERGIES ($\Delta E_{ins}$) (kJ mol$^{-1}$, ZPE INCLUDED) CALCULATED FOR THE REACTION OF $R_2Si$: (R = H, Me) WITH $H_2O$; BOND LENGTHS $d_{Si-O}$ (pm) OF COMPLEX $H_2Si \leftarrow OH_2$ (11)

| R | $\Delta E_{compl}$ | $\Delta E_{rearr}$ | $\Delta E_{ins}$ | $d_{Si-O}$ | Ref. |
|---|---|---|---|---|---|
| H | 50.7 | 98.0 | 293.6 | 197 | 14[a] |
| H | 55.7 | 92.1 | 293.1 | 212.9 | 6a[b] |
| H | 48.6 | 92.5 | 289.7 | 209.8 | 16[c] |
| H | 48.6 | — | 290.2 | 208.9 | 17[d] |
| H | 29 | 39.5 | 289 | 200 | 15[e] |
| H | 55.3 | — | — | 213.2 | 8a[f] |
| Me | 52.8 | 65.3 | 323.7 | 212.7 | 16[g] |

[a] MP4SDTQ/6-31G**//3-21G*.
[b] MP4SDTQ/6-31G**//6-31G*.
[c] QCISD(T)/6-311+G(3df,3pd)//MP2/6-31G(d,p).
[d] G2.
[e] BAC-MP4 calculated at MP4/6-311G(2df,p//MP2/6-31G(d).
[f] MP2/CEP-31g(2d,p)//MP2/CEP-31g(2d,p).
[g] MP2/6-31G*//MP2/6-31G*.

Not surprising, the insertion of silylene into the S–H bond of hydrogen sulfide shows qualitatively the same features as the reaction with water. However, the complexation energy (36.0 kJ mol$^{-1}$) and the activation energy (56.1 kJ mol$^{-1}$) of the rearrangement are lower at the MP4SDTQ/6-31G**// 3-21G* level.[14] $\Delta E_{compl}$ increases to 55.3 kJ mol$^{-1}$ when calculated at the MP2/CEP-31g(2d,p)//MP2/CEP-31g(2d,p) level.[8a]

The comprehensive theoretical study of Schoeller et al.[8a] is the only one that includes the complexes of silylene with the heavier hydrogen chalcogenides such as $H_2Se$ and $H_2Te$. As was found for the complexes of silylenes with group 15 donor centers (see earlier), the Si–X distances of these compounds are slightly longer than the corresponding covalent single bond. In contrast to the silylene–pnictogen complexes, the complexation energy varies to a considerably less extent on changing the heteroatom center of the Lewis base to the heavier chalcogens (Table III).

D. *Complexes with Halogen Nucleophiles*

Clark and Schleyer[20] were the first to report theoretical evidence for the complexation of $H_2Si$: by halogen donors when investigating the structural isomers of the silylenoid $H_2SiLiF$. A silylene–lithium fluoride complex 14 was found to be only 5.4 kJ mol$^{-1}$ less stable at the MP2/6-21G//3-21G

TABLE III

CALCULATED BOND LENGTHS $d_{Si-X}$ (pm) AND BINDING
ENERGIES $\Delta E_{compl}$ (kJ mol$^{-1}$, ZPE INCLUDED) OF ADDUCTS
$H_2Si \leftarrow XH_2$ (X = O, S, Se, Te)[8a]

| Donor | $d_{Si-X}$ | $\Delta E_{compl}$ |
|---|---|---|
| OH$_2$ | 213.2 | 55.3 |
| SH$_2$ | 246.6 | 51.9 |
| SeH$_2$ | 259.9 | 52.3 |
| TeH$_2$ | 273.6 | 61.5 |

level than the metallosilicenium ion **15,** the structure of which is analogous to that of carbenoids. As expected for a dative bond, the Si–F distance in **14** (181.0 pm) is appreciably elongated in comparison to the Si–F bond in $H_2SiF_2$, which was determined to be 157.7 pm.[21]

A loose complex **16,** in which, unlike in lithium fluoride–complex **14,** the F–H bond bisects the H–Si–H angle of the silylene unit, is predicted to be formed by the interaction of $H_2Si$: with hydrogen fluoride. Here the length of the Si–F bond was determined to be 238 pm at the HF/6-31G* level. The complexation energy at the MP4SDTQ/6-31G**//6-31G* level is 29.3 kJ mol$^{-1}$, and the rearrangement of **16** to the H–F insertion product has an activation energy of only 40.6 kJ mol$^{-1}$. Thus, this complex is not a likely candidate to be observed experimentally.[6a]

Analogously, the fluorosilane–silylene complex **17** is characterized by its long Si–F distance (238.1 pm) as well as by the low complexation ($\Delta E_{compl}$ = 22.9 kJ mol$^{-1}$) and activation energy ($\Delta E_{rearr}$ = 39.6 kJ mol$^{-1}$), which were calculated at the MP4/6-31G*//6-31G* level. Complex **17** appears to be a transient intermediate, the experimental detection of which would be a considerable task.[22]

The complex stability decreases further down the periodic table. The complex **18** resides in a fairly shallow minimum as shown by a complexation energy of 4.2 kJ mol$^{-1}$ and an activation energy of the rearrangement of 32.2 kJ mol$^{-1}$.[6a] Again, semiempirical methods such as AM1 and PM3 overestimate the energy of the complex formation of $H_2Si$: and dichlorosilylene with hydrogen chloride appreciably, whereas the activation energy of the rearrangement is underestimated.[19]

In contrast to the coordination of hydrogen chloride, the complexation energy of the $C_s$ symmetrical silylene–bromoform adduct **19,** in which the silylene is coordinated to a bromine center, is calculated to be 97.6 kJ mol$^{-1}$ using the PM3 method.[23] However, keeping in mind the inaccuracy of PM3 results concerning the complexation energies of other silylene–

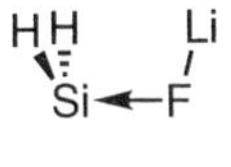

14 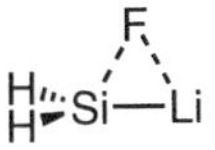          15 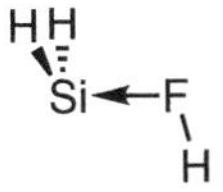          16

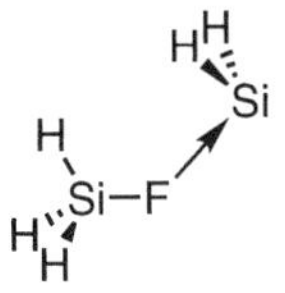

17 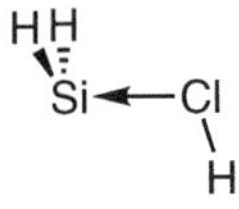          18 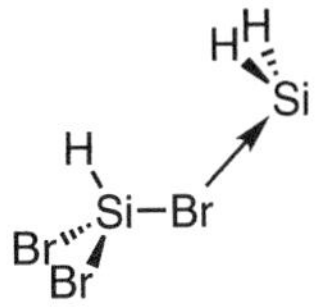          19

Lewis base complexes (see earlier), this unexpectedly high value is questionable.

III

## SPECTROSCOPIC PROPERTIES OF SILYLENE–LEWIS BASE ADDUCTS

Silylenes are short-lived intermediates, and their detection requires fast methods such as ultraviolet (UV)[24] or laser-induced fluorescence spectroscopy.[25] The characteristic absorption maxima in the UV–visible spectra of these species, which are assigned to n → p transitions of electrons at the silicon atom, were used as a fingerprint to prove the occurrence of silylenes in matrices or solution. In addition, these transient species, which under normal conditions are too short lived to be observed by a slow detection method such as infrared (IR) spectroscopy, can be isolated in inert hydrocarbon or noble gas matrices, thus allowing the accurate measurement of their IR spectra.

The first silylene–Lewis base complex to be detected by spectroscopic methods was **11,** which was generated by cocondensation of silicon and water in an argon matrix and identified by its IR spectrum.[26] Only a few other IR spectroscopic investigations of the reaction of silylenes with Lewis bases have been reported to date. The intense IR band at 1084 cm$^{-1}$ which was observed when Mes$_2$Si: (**20a**) was generated in an oxygen matrix at 16 K was associated with the formation of a silanone $O$-oxide **9.**[10] Isotopic labeling experiments as well as comparison with the calculated IR spectrum of parent H$_2$SiO$_2$ support this interpretation. The reaction of

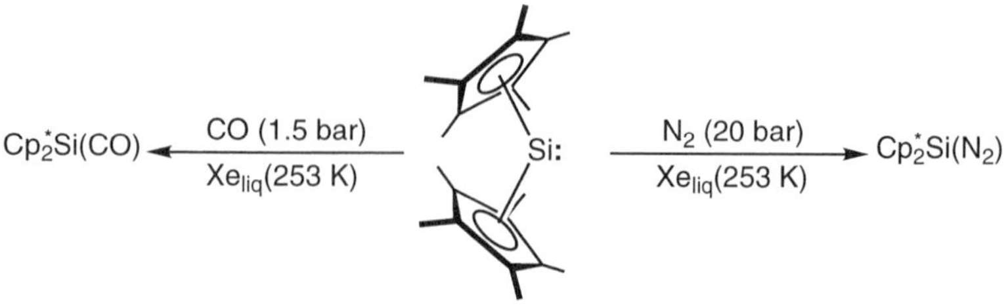

$$Cp_2^*Si(CO) \xleftarrow[Xe_{liq}(253\ K)]{CO\ (1.5\ bar)} \quad Si: \quad \xrightarrow[Xe_{liq}(253\ K)]{N_2\ (20\ bar)} Cp_2^*Si(N_2)$$

22                              21                              23

SCHEME 6

dimethylsilylene (**13**) with carbon monoxide in an argon matrix was followed by Arrington et al.[3] by means of IR spectroscopy. An intense band at 1962 cm$^{-1}$, which is shifted to a lower wavenumber than free carbon monoxide by 187 cm$^{-1}$, was assumed to be indicative of the formation of an adduct between the nucleophile and the silylene **13**.[3] However, comparison of the experimental IR frequencies with those calculated for adduct **1** and silaketene **2** did not allow an unambiguous distinction between these alternative structures. Moreover, it was argued by Zakharov et al.[5] that the experimental IR frequencies are also in good agreement with a bridged structure. Finally, Tacke et al.[27] observed a new IR absorption of low intensity at 2065 cm$^{-1}$, when a saturated solution of **21** in liquid xenon was subjected to a pressure of 1.5 bar carbon monoxide (Scheme 6). This band is shifted to lower wavenumbers in comparison to free carbon monoxide (2143 cm$^{-1}$) and was assigned to the formation of complex **22** between **21** and CO. Similar IR spectroscopic experiments with dinitrogen as dopant indicated the formation of **23**. The structures of **22** and **23** are unknown; however, it must be assumed that the coordination of the cyclopentadienyl ligand is $<\eta^5$ in order to accommodate a further Lewis base such as CO or N$_2$ as ligand in the coordination sphere of silicon.

By far more is known about the UV spectra of silylene–Lewis base complexes in matrices as well as in solution. To observe silylene–Lewis base complexes in matrices, an appropriate silylene precursor, in most cases a cyclic or acyclic oligosilane, is photolyzed at low temperature in an inert hydrocarbon or noble gas matrix, which is doped with small amounts (2–5%) of the respective Lewis base. In a few cases, 2-methyltetrahydrofuran (2-MeTHF), which serves as matrix material as well as a Lewis base, is used. When a rigid matrix such as a 3-methylpentane (3-MP) or 2-MeTHF matrix is used, the formation of the silylene complexes normally requires the matrix to be annealed in order to allow the molecules to diffuse through the softening matrix. In contrast, when the silylene is generated in an *a priori* soft matrix, e.g., using a mixture of 3-MP and isopentane (IP), in the presence of the Lewis base, the free silylene cannot be detected in most cases. Direct formation of the silylene complex is instead observed.

As seen from Table IV, the feature most characteristic of the formation of a matrix silylene–Lewis base complex is the hypsochromic shift of the visible absorption band of the silylene on complexation. At this point it is worthwhile to mention that as early as 1982 it was reported[28] that the absorption maxima of methyl-substituted silylenes in a nitrogen matrix are shifted to significantly higher energies than those of the same silylenes in an inert argon matrix. However, the cause of the shift difference was not addressed in this publication.

Not surprising, an intramolecularly coordinated silylene **25** exhibits a comparable shift[29]: When tetrasilane **24** was photolyzed in a 3-MP matrix at 77 K, a band at $\lambda_{max}$ = 478 nm was detected. Because the noncoordinated silylene **26** absorbs at $\lambda_{max}$ = 662 nm, this hypsochromic shift was taken as evidence for the formation of silaylide **25** (Scheme 7).

Table IV also shows that the $\lambda_{max}$ values of the complexes are influenced even more by the substituents at silicon than by the nature of the Lewis base. The smallest blue shift on complexation occurs with silylenes, which are substituted by electron-donating groups such as aryloxy groups. This may be explained by intramolecular stabilization of the silylene by back-bonding of the heteroatom substituent, thus reducing the electrophilicity of the silicon center and weakening the silylene–donor interaction.[30]

The only exception to these uniform spectroscopic properties of silylene–Lewis base complexes, i.e., the hypsochromic shift on complexation, is silacarbonyl ylide **27,** the absorption maximum of which is shifted to wavelengths longer than that of free dimesitylsilylene (**20a**);[31] an explanation of this observation is yet to come.

Whereas UV spectroscopic data are available for base complexes of a variety of matrix-isolated silylenes (Table IV), solution data are reported almost exclusively for complexes of dimethylsilylene (**13**) (Table V). Again, the complexation of this silylene is accompanied by a significant blue shift of the absorption maximum. Probably due to "matrix effects" of unknown nature, the $\lambda_{max}$ values of dimethylsilylene complexes in solution are shifted to wavelengths shorter than those of the corresponding matrix-isolated complexes.

Nuclear magnetic resonance (NMR) spectroscopic data have been reported by Takeda et al.[32] for silylene–isocyanide complexes **28–30**, which are stable at room temperature due to the bulky substituents. The $^1J_{SiC}$ coupling constants of 38.6 Hz (**28**), 22.1 Hz (**29**), and 1.0 Hz (**30**) are considerably smaller than that of a Si–C single bond and indicate a weak Si–C interaction as expected for a dative bond.

TABLE IV

ABSORPTION MAXIMA $\lambda_{max}$ (nm) OF SELECTED, MATRIX-ISOLATED SILYLENES AND
SILYLENE–BASE COMPLEXES

| $R^1R^2Si$: | Matrix[a] | $\lambda_{max}$ ($R^1R^2Si$:) | Donor | $\lambda_{max}$ (complex) | Ref. |
|---|---|---|---|---|---|
| $Me_2Si$: | $N_2$ | 450 | CO | 342 | 3 |
| $Me_2Si$: | 3-MP | 454 | CO | 354 | 56 |
| $Mes(t\text{-}Bu)Si$: | 3-MP | 505 | CO | 328 | 56 |
| $Mes_2Si$: | 3-MP | 580 | CO | 338 | 56 |
| $Mes(OAr)Si$:[b] | 3-MP | 398 | CO | 345 | 56 |
| $Me_2Si$: | 3-MP | 450 | $Et_3N$ | 287 | 30 |
| $Mes(t\text{-}Bu)Si$: | 3-MP | 505 | $Et_3N$ | 345 | 30 |
| $Mes_2Si$: | 3-MP | 580 | $Et_3N$ | 350 | 30 |
| $Mes(OAr)Si$:[b] | 3-MP | 395 | $Et_3N$ | 348 | 30 |
| $Mes_2Si$: | IP/3-MP (4:1) | 573 | $N$-Methylpyrrolidine | 324 | 53 |
| $Me_2Si$: | 3-MP | 450 | $Bu_3P$ | 287 | 30 |
| $Mes(t\text{-}Bu)Si$: | 3-MP | 505 | $Bu_3P$ | 345 | 30 |
| $Mes_2Si$: | 3-MP | 580 | $Bu_3P$ | 325 | 30 |
| $Mes(OAr)Si$:[b] | 3-MP | 395 | $Bu_3P$ | 346 | 30 |
| $Mes_2Si$: | IP/3-MP (4:1) | 573 | $Bu_3P$ | 338 | 53 |
| $Mes(t\text{-}Bu)Si$: | 3-MP | 505 | $i$-PrOH | 333 | 30 |
| $Mes_2Si$: | 3-MP | 580 | $i$-PrOH | 325 | 30 |
| $Mes(OAr)Si$:[b] | 3-MP | 395 | $i$-PrOH | 320 | 30 |
| $Me_2Si$: | 3-MP | 450 | $Et_2O$ | 299 | 30 |
| $Mes(t\text{-}Bu)Si$: | 3-MP | 505 | $Et_2O$ | 348 | 30 |
| $Mes_2Si$: | 3-MP | 580 | $Et_2O$ | 320 | 30 |
| $Mes(OAr)Si$:[b] | 3-MP | 395 | $Et_2O$ | 332 | 30 |
| $Me_2Si$: | 3-MP | 450 | THF | 280 | 30 |
| $Mes(t\text{-}Bu)Si$: | 3-MP | 505 | THF | 350 | 30 |
| $Mes_2Si$: | 3-MP | 580 | THF | 328 | 30 |
| $Mes(OAr)Si$:[b] | 3-MP | 395 | THF | 330 | 30 |
| $Mes_2Si$: | IP/3-MP (4:1) | 573 | THF | 320 | 53 |
| | 3-MP | 505 | 2-MeTHF | 390 | 54 |
| | 3-MP | 475 | 2-MeTHF | 345 | 54 |
| $Mes_2Si$: | IP/3-MP (6:4) | 573 | | 610 | 31 |
| $Me_2Si$: | 3-MP | 450 | $t\text{-}Bu_2S$ | 322 | 30 |
| $Mes(t\text{-}Bu)Si$: | 3-MP | 505 | $t\text{-}Bu_2S$ | 368 | 30 |

TABLE IV (*continued*)

| $R^1R^2Si$: | Matrix[a] | $\lambda_{max}$ ($R^1R^2Si$:) | Donor | $\lambda_{max}$ (complex) | Ref. |
|---|---|---|---|---|---|
| Mes$_2$Si: | 3-MP | 580 | $t$-Bu$_2$S | 316 | 30 |
| Mes(OAr)Si:[b] | 3-MP | 395 | $t$-Bu$_2$S | 350 | 30 |
| Mes$_2$Si: | IP/3-MP (4:1) | 573 | Thiolane | 315 | 53 |
| Mes$_2$Si: | IP/3-MP (6:4) | 573 | 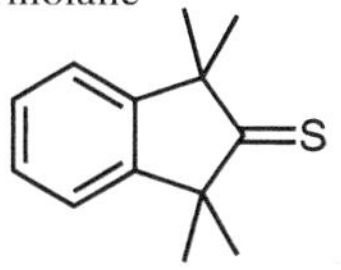 | 485 | |

[a] 3-MP, 3-methylpentane; IP, isopentane.
[b] OAr, 2,6-diisopropylphenoxy.

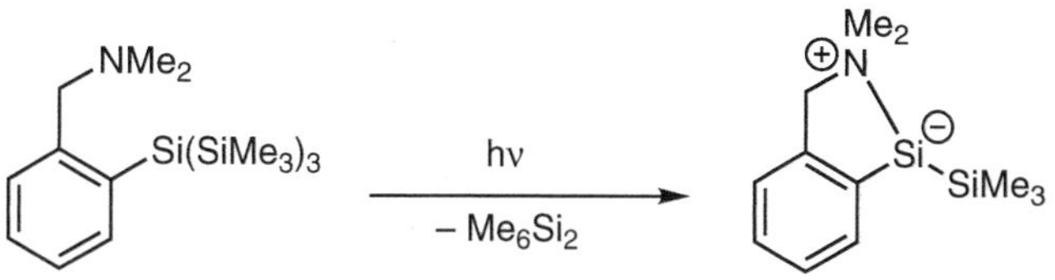

**24**        **25**

SCHEME 7

TABLE V

ABSORPTION MAXIMA $\lambda_{max}$ (nm) OF SELECTED SILYLENES AND SILYLENE–BASE COMPLEXES IN SOLUTION

| $R^1R^2Si$: | Solvent | $\lambda_{max}$ ($R^1R^2Si$:) | Donor | $\lambda_{max}$ (complex) | Ref. |
|---|---|---|---|---|---|
| Me$_2$Si: | Cyclohexane | 465 | Et$_3$N | <270 | 33 |
| Me$_2$Si: | Cyclohexane | 465 | H$_3$CCN | 340 | 33 |
| Me$_2$Si: | Cyclohexane | 470 | $N$-Methylpyrrolidine | 285 | 34 |
| Me$_2$Si: | Cyclohexane | 470 | Bu$_3$P | 310 | 34 |
| Me$_2$Si: | Cyclohexane | 465 | Et$_2$O | 305 | 33 |
| Me$_2$Si: | Cyclohexane | 465 | THF | 310 | 33,55 |
| Me$_2$Si: | Cyclohexane | 470 | THF | 310 | 34 |
| Me$_2$Si: | Cyclohexane | 470 | Thiolane | 335 | 34 |
| Me$_2$Si: | Hexane | Not observed | HCBr$_3$ | 338 | 23 |

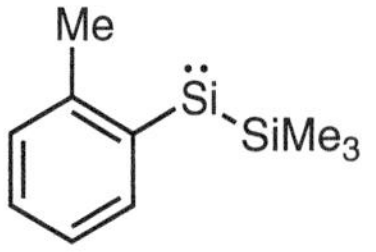

**26**

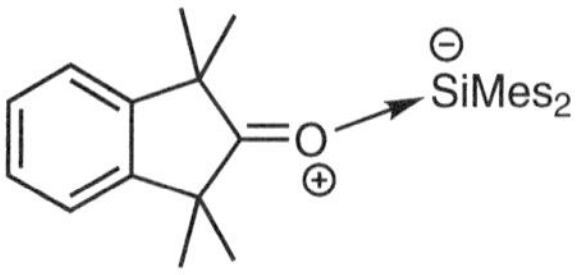

**27**

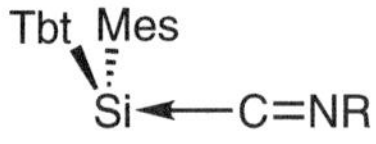

Tbt Mes

Si ◄—C=NR

**28, 29, 30**

Tbt = 2,4,6-[(Me$_3$Si)$_2$CH]$_3$C$_6$H$_2$
Mes = 2,4,6-Me$_3$C$_6$H$_2$
**28**: R = 2,4,6-(*i*-Pr)$_3$C$_6$H$_2$
**29**: R = Tbt
**30**: R = 2,4,6-(*t*-Bu)$_3$C$_6$H$_2$

## IV

## KINETIC STUDIES

The laser flash photolysis technique allows the observation of rapid reactions and determines the lifetime of reactive intermediates. Thus, it is not surprising that this method was used extensively to obtain information about the formation, stability, and reactivity of silylene–Lewis base complexes.

The rate of complex formation of dimethylsilylene with a variety of Lewis bases was found to be close to the diffusion limit in cyclohexane at room temperature.[33,34] The results of Yamaji et al.[34] indicate that the rate of this reaction is governed not so much by electronic factors such as the HOMO energy of the Lewis base as by steric hindrance around the heteroatom center. In contrast, Baggott et al.[35] found a satisfying correlation between rate constants and ionization energies of the nucleophile for the reaction of dimethylsilylene with various oxygen-containing substrates in the gas phase.

Levin et al.[33] recorded the decay of the UV-absorption maxima of a variety of dimethylsilylene–Lewis base complexes in cyclohexane at room temperature and found that the lifetimes of these complexes are significantly longer than that of noncoordinated dimethylsilylene (Table VI). Gillette et al.[36] estimated the lifetime of complex **31d** to be 1 min at −135°C in 2-MeTHF solution.

The reactivity of silylene–Lewis base complexes toward a variety of silylene traps such as olefins, alcohols, and silanes is significantly reduced, as can be seen from Table VII.

TABLE VI

LIFETIMES ($10^{-6}$ sec) OF "FREE" DIMETHYLSILYENE AND
DIMETHYLSILYLENE–LEWIS BASE COMPLEXES IN CYCLOHEXANE
AT 295 K[33]

| Donor | Lifetime[a] |
|-------|-------------|
| None | 0.15–1.2 |
| Tetrahydrofuran | 12.5 |
| Diethyl ether | 4.0 |
| 1,4-Dioxane | 4.3 |
| Triethylamine | 29 |
| 1,4-Diazabicyclo[2.2.2]octane | 62 |
| Acetonitrile | 2.3 |

[a] ±15%.

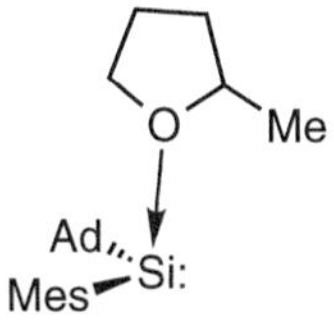

**31d**

According to the reactivity–selectivity principle, reduced reactivity of a compound is expected to be accompanied by increased selectivity of its reactions. Exactly this was observed in early experiments by Steele and Weber[37] when the insertion reaction of dimethylsilylene (**13**) into the O–H bond of alcohols was performed in different solvents: In a more polar solvent such as THF the reaction of the silylene with the

TABLE VII

BIMOLECULAR RATE CONSTANTS [$M^{-1}$ s$^{-1}$] FOR THE TRAPPING OF "FREE" AND THF-
COMPLEXED DIMETHYLSILYLENE (**13**) IN CYCLOHEXANE AT 295 K[33]

| Trapping agent | Rate constant ("free" silylene) | Rate constant (complexed silylene) |
|----------------|--------------------------------|-----------------------------------|
| EtO–H | $9.2 \times 10^9$ | $1.9 \times 10^8$ |
| t-BuO–H | $10.2 \times 10^9$ | $4.4 \times 10^7$ |
| $H_2C{=}CH(CH_2)_3CH_3$ | $7.3 \times 10^9$ | $2.4 \times 10^6$ |
| $(n\text{-Pr})_3Si{-}H$ | $2.9 \times 10^9$ | $8.6 \times 10^5$ |

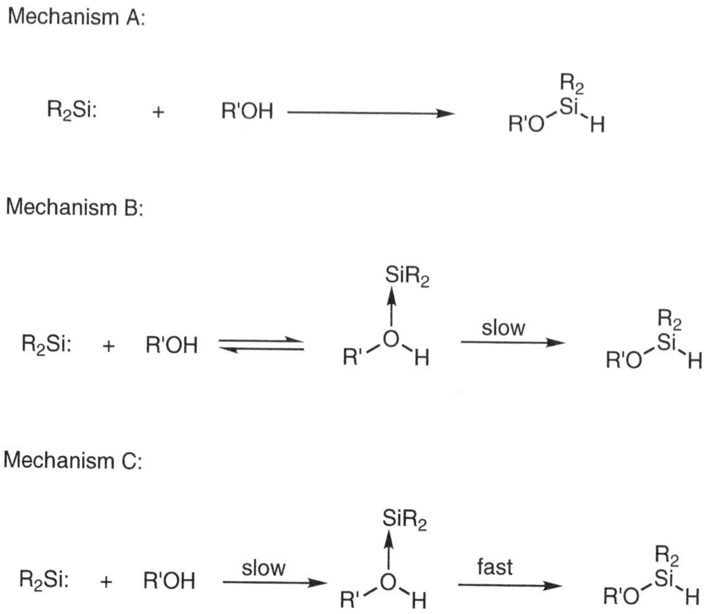

Mechanism A:

$$R_2Si: \quad + \quad R'OH \quad \longrightarrow \quad R'O{\overset{R_2}{\underset{}{\diagdown Si \diagup}}} H$$

Mechanism B:

$$R_2Si: \quad + \quad R'OH \quad \rightleftharpoons \quad R'{\overset{\overset{\displaystyle SiR_2}{\uparrow}}{\diagdown O \diagdown}} H \quad \xrightarrow{\text{slow}} \quad R'O{\overset{R_2}{\underset{}{\diagdown Si \diagup}}} H$$

Mechanism C:

$$R_2Si: \quad + \quad R'OH \quad \xrightarrow{\text{slow}} \quad R'{\overset{\overset{\displaystyle SiR_2}{\uparrow}}{\diagdown O \diagdown}} H \quad \xrightarrow{\text{fast}} \quad R'O{\overset{R_2}{\underset{}{\diagdown Si \diagup}}} H$$

SCHEME 8

sterically less shielded alcohol is more favored than in cyclohexane. This solvent-modified selectivity was interpreted in terms of initial formation of a more stable and hence more selectively reacting silylene–Lewis base complex. Furthermore, it was shown by the same authors that the preference of silylene **13** for O–H insertion over Si–H insertion is higher in THF than in noncoordinating solvents such as $n$-decane or cyclohexane.[38] These results, which were obtained by product analysis of competition experiments, were corroborated later by direct determination of absolute rate constants using laser flash photolysis[33]: The ratio $k_{O-H}/k_{Si-H}$ of the insertion rate of dimethylsilylene (**13**) into EtO–H and $(n\text{-Pr})_3$Si–H increases from 3.2 in pure cyclohexane to 221 in cylohexane in the presence of THF.

Kinetic isotope effects (KIE) of silylene insertion reactions were determined in solution as well as in the gas phase. Steele and Weber[39] found KIEs between 1.8 and 2.3 for the insertion of silylene **13** into various alcohols; no significant variation of these values was observed with a change of solvent from cyclohexane to THF. These results are consistent with two mechanisms (Scheme 8), both of them proceeding via a nonlinear or triangular transition state. Mechanism B, which is favored over mechanism A by the authors, is in agreement with theoretical results, which, in general,

show that the activation energy for the rearrangement of the silylene–
alcohol complex is higher than the complexation energy (see Section II,C).
In contrast, a very small KIE was determined for the insertion of **13** into
methanol in the gas phase, and it was concluded that this reaction proceeds
via initial, rate-determining complex formation, followed by rapid hydrogen
transfer (Scheme 8, mechanism C).[35] No doubt, further studies need to
be done to solve these puzzling discrepancies between solution and gas
phase experiments.

# V

## DONOR–ACCEPTOR INTERACTIONS OF SILYLENES IN
## ORGANOMETALLIC REACTIONS

In preparative organosilicon chemistry, complexes or ylides of silylenes
either are generated intentionally as target molecules, in most cases to
further investigate their properties, or they are found coincidentally as
intermediates in silylene reactions. Thus, in most cases the synthesis of
complexes or ylides of silylenes, whether isolable or unstable, relates
strongly to the methods to generate silylenes. Basically, transient or stable
silylenes are generated photolytically or thermally from appropriate precursors. In thermal reactions, the silylene precursor may be simply heated
or, in some cases, the addition of another reagent is required to get the
target silylene.

This section presents reactions that proceed via the formation of silylene–
Lewis base complexes. The material is organized according to the method
used for silylene generation.

### A. Irradiation of Oligosilanes in the Presence of Lewis Bases

It is well established in organosilicon chemistry that oligosilanes extrude
silylenes on irradiation.[9,24,40] The most frequently used precursor for this

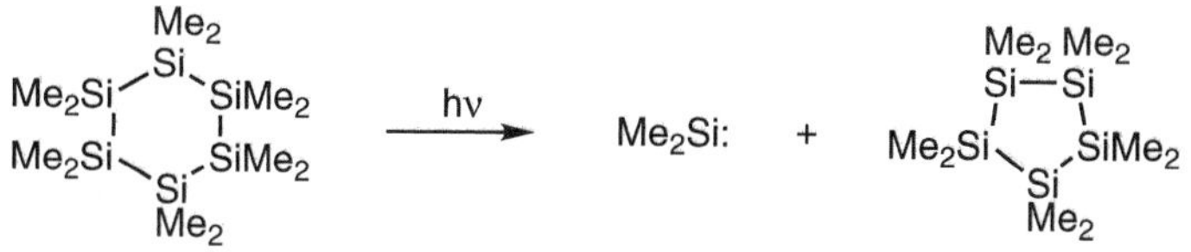

$$\underset{\textbf{32}}{\overset{\begin{array}{c}\text{Me}_2\\\text{Si}\end{array}}{\overset{\text{Me}_2\text{Si}\diagup\diagdown\text{SiMe}_2}{\underset{\text{Me}_2\text{Si}\diagdown\underset{\text{Si}}{\diagup}\text{SiMe}_2}{\underset{\text{Me}_2}{}}}} \quad \xrightarrow{h\nu} \quad \underset{\textbf{13}}{\text{Me}_2\text{Si:}} \quad + \quad \underset{\textbf{33}}{\overset{\begin{array}{c}\text{Me}_2\ \text{Me}_2\\\text{Si}-\text{Si}\end{array}}{\overset{\text{Me}_2\text{Si}\diagdown\underset{\text{Si}}{\diagup}\text{SiMe}_2}{\underset{\text{Me}_2}{}}}}$$

SCHEME 9

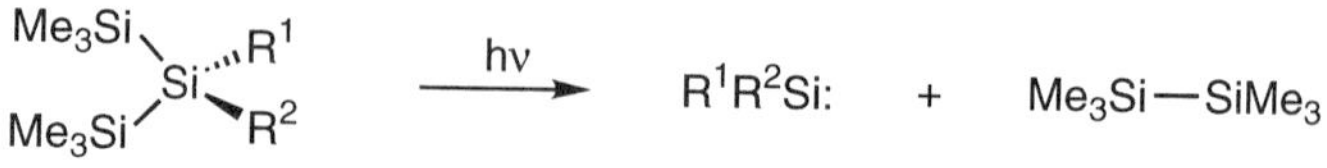

$$\text{Me}_3\text{Si} \diagdown \underset{\text{Me}_3\text{Si} \diagup}{\text{Si}} \diagup_{\displaystyle \sim \text{R}^1}^{\displaystyle \sim \text{R}^2} \quad \xrightarrow{\text{h}\nu} \quad \text{R}^1\text{R}^2\text{Si:} \quad + \quad \text{Me}_3\text{Si}-\text{SiMe}_3$$

**34**　　　　　　　　　　　　**35**　　　　　　**36**

SCHEME 10

methodology is dodecamethylcyclohexasilane (**32**) because it is easily available and can be handled without any precautions. Photolysis of **32** leads to the formation of dimethylsilylene (**13**) and decamethylcyclopentasilane (**33**) (Scheme 9).[41] However, a considerable number of other oligosilanes have been used to generate silylenes.[24,40] In most cases, where $R^1$ and/or $R^2$ is an aryl substituent, a trisilane of the general structure **34** is irradiated so that the only by-product of this reaction is hexamethyldisilane (**36**), which does not disturb further investigations (Scheme 10).

The photochemical approach using **32** as precursor was employed by Ando *et al.*,[42] who investigated the reaction of dimethylsilylene (**13**) with adamantanone (**37**) and norbornanone. Similar investigations have been performed with thermally generated dimethylsilylene (**13**)[43]; however, for convenience, the photochemical pathway was preferred in further studies. Generation of **13** by irradiating **32** in the presence of **37** yielded the products **40** and **41** (Scheme 11).[42] This result was rationalized by the initial attack of the silylene on the carbonyl double bond to give the corresponding siloxirane **38,** which is assumed to be in an equilibrium with carbonyl ylide **39**. The latter further dimerizes to **41** or reacts with **37** to form **40**. Interestingly, the possibility of the direct formation of the ylide **39** by the reaction of the silylene with the ketone was not discussed. The proposed intermediacy of siloxiranes in the course of the addition of silylenes to ketones was later confirmed by the isolation of a stable siloxirane **43** from the reaction of a photochemically generated, bulkily substituted silylene **20a** with the sterically demanding ketone **42** (Scheme 12).[44] It was shown that the siloxirane **43** converts back photochemically to silylene **20a,** which can be intercepted by trapping reagents such as triethylsilane and methanol.[45] Ketone **42** is formed as a by-product, which undergoes decarbonylation on further irradiation. Again, a carbonyl ylide **44** was proposed as the reactive intermediate in the reaction of the siloxirane. The existence of **44** was later proven by matrix isolation studies of siloxirane **43**.[31] Irradiation of **43** at low temperatures with $\lambda = 254$ nm light resulted in the formation of the intensely blue-colored ylide **44** that was detected by UV spectroscopy ($\lambda_{\max} = 610$ nm). Electron spin resonance experiments showed that a biradical, which had been proposed in early studies by Ando *et al.*,[43] is not formed. The most interesting result of these experiments, however, was

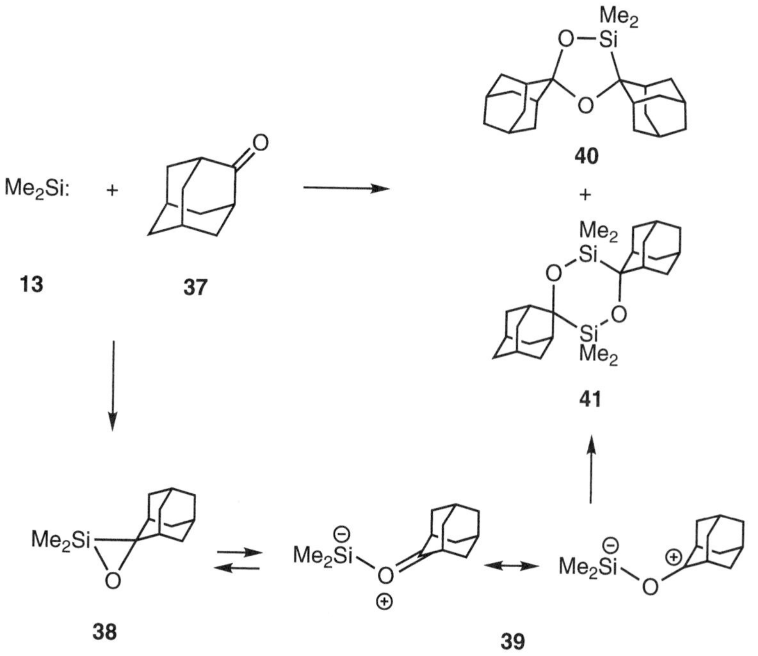

**SCHEME** 11

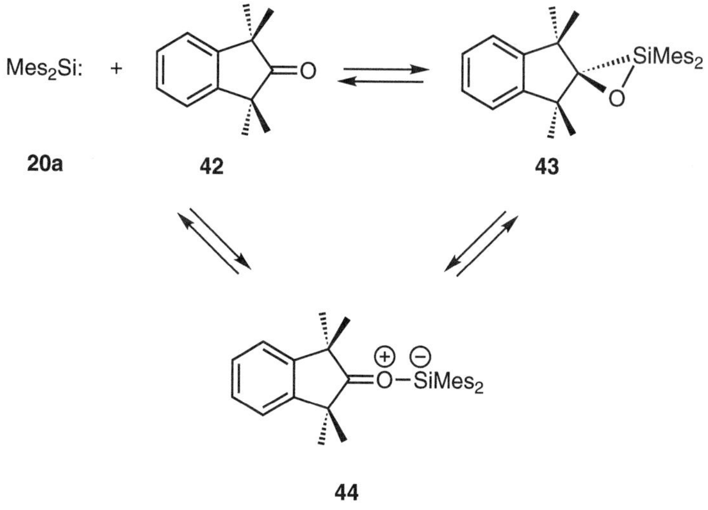

**SCHEME** 12

the observation that the same absorption band indicative of the formation of **44** appears during low-temperature generation of silylene **20a** by photolysis of the corresponding trisilane in the presence of ketone **42** after careful annealing of the matrix without intermediate formation of siloxirane **43**. With this result, the question whether an ylide or a siloxirane is the actual initially formed intermediate of a silylene reaction with ketones still remains unanswered.

The photolysis of dodecamethylcyclohexasilane (**32**) was also utilized in the work of Steel and Weber[37] (Scheme 13). In these studies, the insertion reaction of dimethylsilylene (**13**) into the O–H bond of alcohols was investigated in several solvents. The authors found a decreasing reactivity of dimethylsilylene (**13**) toward alcohols in tetrahydrofuran and diethyl ether compared to the results in hydrocarbons, and an increased selectivity of the silylene was observed. In competition reactions of the silylene with a mixture of different alcohols it was demonstrated that in etheral solvents the silylene exhibits a preference for the less hindered reagent. These results were rationalized by the formation of a silylene–ether complex **45** that reduces the reactivity of the silylene significantly, thus—according to the reactivity–selectivity concept—increasing the selectivity. In related studies, Weber *et al.* investigated the reactivity of photolytically generated **13** toward oxetanes[46] and vinyl epoxides.[47] With oxetane as the substrate, products **50** arising from the insertion of the silylene into the cyclic ether as well as **51**, which is formed by silylene insertion and subsequent ring opening, were observed (Scheme 14).[46] Vinyloxirane reacts with **13** to yield oxasilacyclohexene **47** and oligomerization products **48** and **49** of the corresponding silanone (Scheme 14).[47] These reactions and analogous reactions of substituted oxetanes and vinyl epoxides are assumed to proceed via ylides **52** as reaction intermediates. In comparable studies, Goure and Barton[48] also investigated the reaction of dimethylsilylene (**13**) with cyclooctene oxide (**53**) and found that the photolysis of **32** in the presence of this epoxide yielded cyclic siloxanes **57a–d** and cycloctene (**55**) (Scheme 15). These products are assumed to result from the reaction of the intermediate dimethylsilanone (**56**), and in fact, in the presence of hexamethylcyclotrisiloxane ($D_3$), the trapping product of silanone **56**

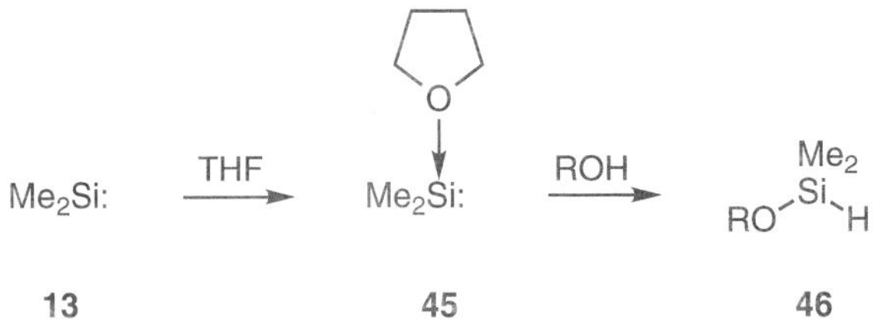

SCHEME 13

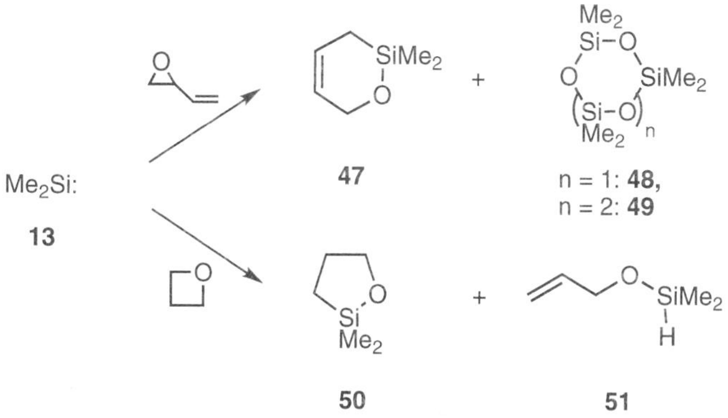

SCHEME 14

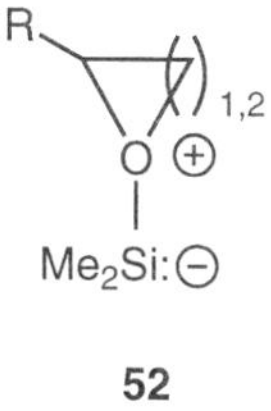

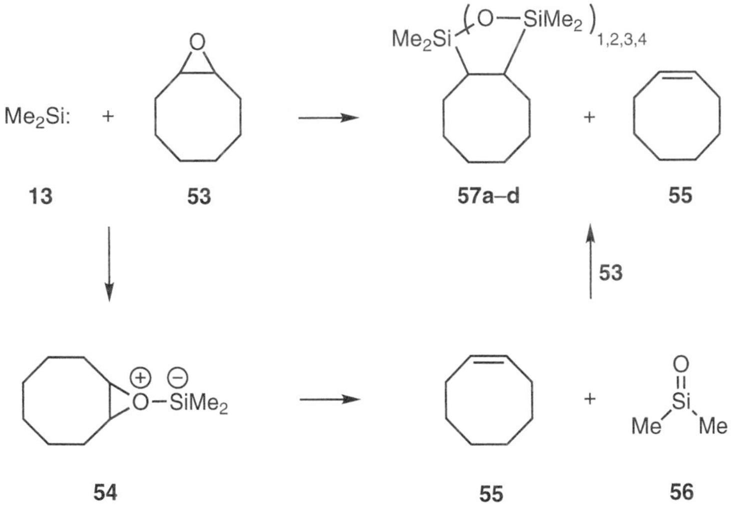

SCHEME 15

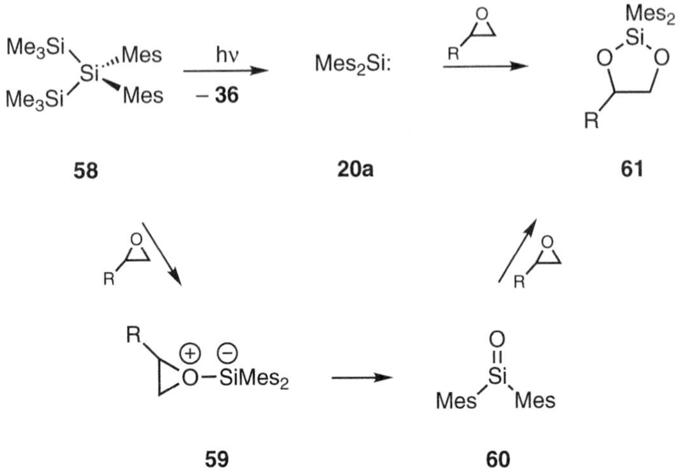

**58**                          **20a**                          **61**

**59**                          **60**

R = H, Me, Et, Ph

SCHEME 16

was isolated. One possible reaction pathway to the silanone that was discussed involves the formation of ylide **54** resulting from the reaction of silylene **13** with the epoxide. Product **54** subsequently dissociates to highly reactive **56** and cyclooctene (**55**). However, it was suggested by the authors that ylide **54** alternatively could transfer the silanone moiety in a concerted pathway without any involvement of free silanones. During the reaction of epoxides with dimesitylsilylene (**20a**), which was generated by photolysis of trisilane **58**, dioxasilacyclopentane **61** is formed among a terminal alkene, hexamethyldisilane, and various oligomers of silanone **60** (Scheme 16).[49] Similar to the reaction mechanism proposed earlier by Goure and Barton,[48] ylide **59** and silanone **60** are assumed to be reactive intermediates in this reaction.

When **32** is photolyzed in the presence of methyl allyl ether (**62**),[50,51] the generated silylene **13** initially coordinates to the ether oxygen to give **63**, which subsequently rearranges to allylsilane **64** (Scheme 17), a reaction

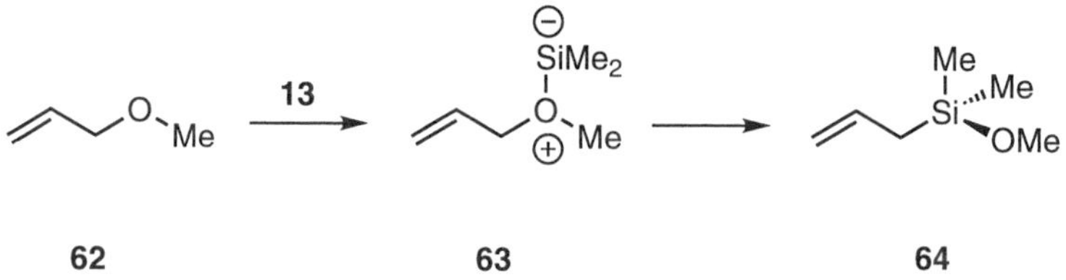

**62**                    **63**                    **64**

SCHEME 17

pattern that was also discovered for the reaction of **13** with other allylic ethers and thio ethers.[52] The same experimental protocol was used to investigate the kinetic isotope effect of silylene reactions with alcohols and silanes (Section IV).[39]

Trisilane **58** was irradiated in a soft hydrocarbon matrix that was doped with Lewis bases such as cyclic amines, tributylphosphine, tetrahydrofuran, and tetrahydrothiophene, and it was shown for the first time by UV spectroscopy (Table IV) that the generated dimesitylsilylene (**20a**) interacts coordinatively with the aforementioned dopants.[53] Warming of the matrix resulted, as shown in Scheme 18 for the complex between **20a** and *N*-methylpyrrolidine, in the formation of tetramesityldisilene (**66a**) and products probably arising from the reaction of the silylene or the labile silaylide **65** with the donor molecule. Similarly, Gillette *et al.*[36] generated silylenes **20a–d** in a rigid 3-MP matrix at low temperature that were detected by their characteristic UV absorptions. When the hydrocarbon matrix was doped with 2-MeTHF, a new blue-shifted band at $\lambda_{max} = 350$ nm was formed on annealing of the matrix. Further warming resulted in the formation of disilenes, a reaction that was also observed in warming of hydrocarbon matrices containing only these silylenes in the absence of 2-MeTHF. On photolysis of **20d** in a pure 2-MeTHF matrix, however, the 350-nm band was detected in addition to the silylene absorption already before annealing of the matrix, and on careful warming the silylene band decreased with concomitant growth of the new band. Further warming yielded a product of a reaction of the silylene with the matrix. Competition experiments with alcohols in various solvents such as pentane, diethyl ether, and 2-MeTHF confirmed the enhanced selectivity in donating solvents that Steele and Weber[37] had reported earlier. All of these results made it evident that complexes **31a–d** formed in these matrices (Scheme 19), and the absorption band at $\lambda = 350$ nm was ascribed to these complexes (Table IV). Furthermore, performing the photolysis of **20d** in 2-MeTHF at $-135°$C and determining the decay rate of the absorption band of **31d** showed that, even at this "higher" temperature, the silylene–ether complex has a lifetime of

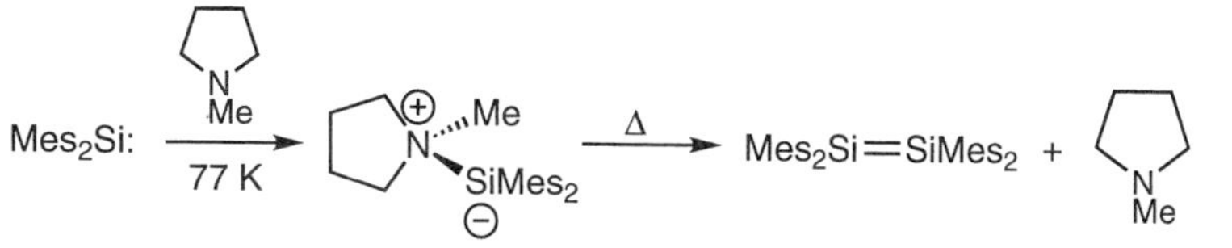

SCHEME 18

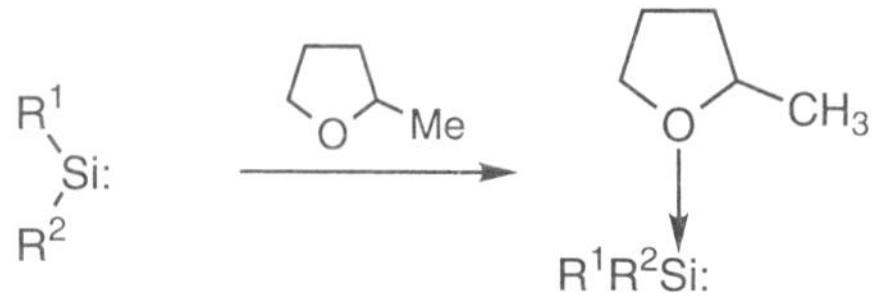

**20a–d**                                    **31a–d**

**a**: $R^1 = R^2 = $ Mes; **b**: $R^1 = t$-Bu, $R^2 = $ Mes; **c**: $R^1 = $ Mes,
$R^2 = 2,6\text{-}(i\text{-Pr})_2OC_6H_3$; **d**: $R^1 = $ Mes, $R^2 = $ 1-adamantyl

SCHEME 19

1 min under these conditions. Later, these matrix studies were extended to other donor molecules such as tertiary amines, sulfides, and phosphines,[30] and the investigated silylenes formed complexes with these donor molecules in matrix. As has been found for dimethylsilylene (**13**),[39] even alcohols form complexes with the investigated, sterically more shielded silylenes, and complex formation was found to depend on the steric bulk of the alcohol. Whereas no alcohol complexes of silylenes **20a–d** could be observed by means of UV spectroscopy with $t$-BuOH and EtOH as matrix material, the corresponding complexes with $i$-PrOH and $n$-BuOH were identified by their characteristic blue-shifted absorption maxima when the alcohol matrix was annealed. In recent, related studies, isolation of the first vinyl-substituted silylene at low temperature was achieved by irradiation of the trisilane **68** in a 3-MP matrix.[54] When the photolysis was performed in 2-MeTHF, a higher coordinated divalent silicon species **69** was detected by UV spectroscopy (Scheme 20).

Levin et al.[55] observed, in line with the qualitative results of Gillette et al.,[36] that less bulkily substituted silylene **13** is stabilized when coordinated by Lewis bases. Dimethylsilylene **13** was generated photochemically from **32** in cyclohexane solutions at 296 K, and it was shown by means of UV spectroscopy that the lifetime of this silylene is 0.4 $\mu$s under the conditions

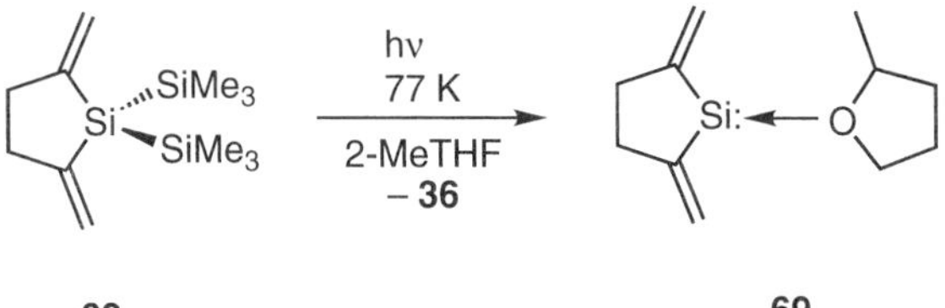

**68**                                          **69**

SCHEME 20

employed. Performing this reaction in the presence of donating reagents such as tetrahydrofuran, diethyl ether, triethylamine, and acetonitrile led to the formation of the silylene–Lewis base complexes, which were identified by their blue-shifted absorption band and showed a considerably longer lifetime (see Table VI) than the uncoordinated silylene. Subsequently, the reaction rates of these complexes toward alcohols, alkenes, and silanes were determined by the same authors,[33] and the so far solely qualitative observation that silylene complexes are less reactive than the corresponding noncoordinated silylene was confirmed. The authors explained the reduced reactivity by the limited availability of the vacant 3p orbital of the silylene–Lewis base complex.

At the same time, Arrington et al.[3] exploited the low-temperature methodology further by generating dimethylsilylene (**13**) by photolysis of dodecamethylcyclohexasilane (**32**) or from dimethyldiazidosilane (**70**) in the presence of carbon monoxide in a 3-MP matrix. The reaction was monitored by UV and IR spectroscopy. However, it was not possible to determine unambiguously from comparison of experimental and calculated IR data whether the complex **71a** or the silaketene **72a** had been formed. Pearsall and West[56] obtained similar results while investigating sterically, besides **13,** more hindered silylenes **20b–d** that reacted with carbon monoxide in hydrocarbon matrices. The fact that disilenes were formed on warming of the matrix provided evidence that complexes **71b–d** are the actual intermediate instead of silaketenes **72b–d,** which are unlikely to react to disilenes on warming. The analogous complex **71a** was also favored by Arrington et al.[3] as the likely intermediate of the reaction of **13** with carbon monoxide.

Tris(trimethyl)phenylsilane **73** extrudes phenyl(trimethylsilyl)silylene **74** on irradiation (Scheme 21).[57] When the photoreaction is carried out in the presence of alkyl chlorides, which are rather weak Lewis bases, the products **75** of a formal insertion of the silylene in the C–Cl bond were observed. This contrasts with the former observation that chlorobenzene does not interact with silylenes.[53] With sec-butylchloride as the substrate, chlorodisilane **76** was found as an additional product, which was the only silicon-containing product in the reaction when **74** was reacted with tert-butylchloride. The initial step in the reaction is assumed to be the attractive interaction of the silylene with the chlorine atom to form zwitterion **77.** Subsequent migration of the alkyl moiety gives the corresponding silylchlorides **75,** whereas with sec-butylchloride and tert-butylchloride a proton in the $\alpha$-position to the halogen-bearing carbon center is removed and finally disilane **76** is formed. Similar alkylchloride–silylene complexes were suggested by Oka and Nakao[58] to be formed on reaction of **73** with a variety of methyl chlorides.

Corriu et al.[29] generated a transient silylene by the irradiation of trisilane **24** (Scheme 22). Photolysis in the presence of 2,3-dimethyl-1,3-butadiene

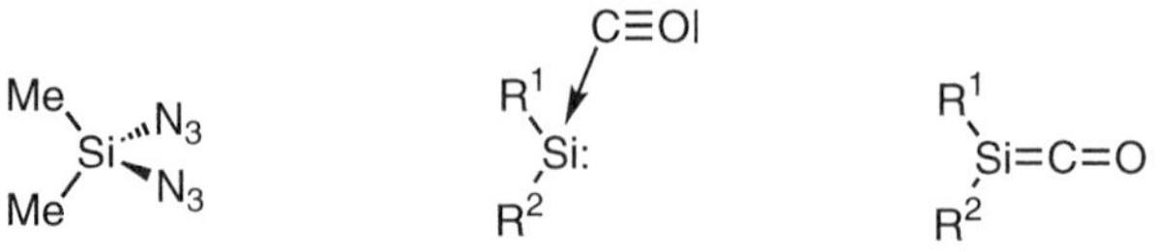

**70**          **71a-d**          **72a-d**

**a**: $R^1 = R^2 = Me$; **b**: $R^1 = R^2 = Mes$; **c**: $R^1 = t\text{-Bu}$, $R^2 = Mes$;
**d**: $R^1 = Mes$, $R^2 = 2,6\text{-}(i\text{-Pr})_2OC_6H_3$

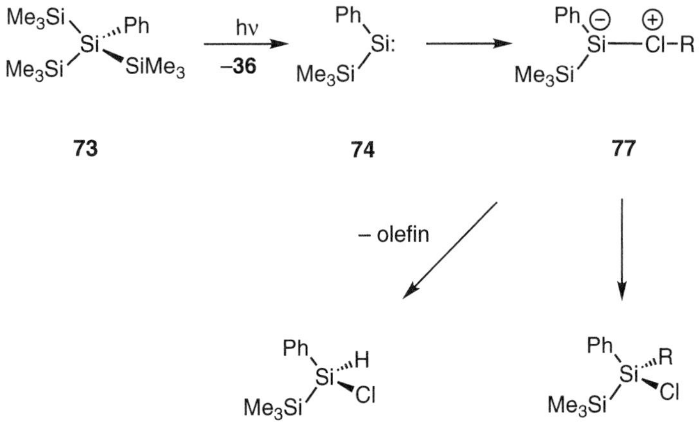

**73**                **74**                **77**

– olefin

**76**                          **75**

$R = n\text{-}C_8H_{17}$, $s\text{-Bu}$, $t\text{-Bu}$

Scheme 21

and triethylsilane, respectively, gave silacyclopentene **79** and disilane **80** as silylene-trapping products. Because of the chelating substituent on the silicon, the silylene is assumed to be intramolecularly coordinated, thus forming a sila-ammonium ylide **25**. Unfortunately, the dimethylamino-methyl substituent is a chromophore itself, so that the photoinduced cleavage of the benzyl–nitrogen bond is a disturbing side reaction under these conditions, giving rise to the formation of tolyl-substituted silane **78**. Irradiation of **24** in matrices at low temperatures resulted in the formation of the silylene-ylide **25** among the corresponding disilene and cleavage products that could be identified by UV spectroscopy (Section III).

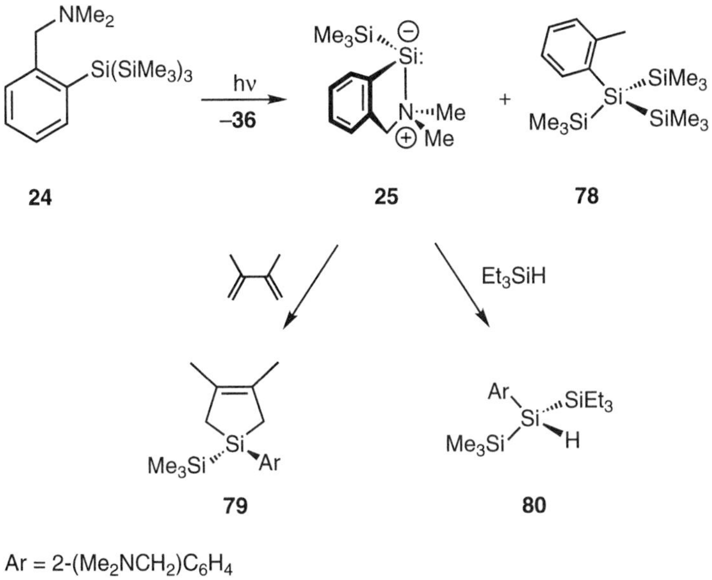

Ar = 2-(Me$_2$NCH$_2$)C$_6$H$_4$

SCHEME 22

## B. *Miscellaneous Photochemical Pathways*

The donor–acceptor complex **82** is suggested to be formed on photolysis of the silylene–iron complex **81** as a precursor compound, which already features a higher coordinated silicon center (Scheme 23).[59,60] Silylene complex **82** reacted with alcohols, alkenes, and alkynes to give the expected trapping products **83–85**. The authors exploited this concept further by introducing chelating substituents on the silicon center to facilitate an entropically more favored intramolecular coordination.[61] For example, the iron–silylene complex **86** extrudes photolytically the intramolecularly coordinated silylene **87,** which can be intercepted by diphenylacetylene, 2,3-dimethylbutadiene, and alcohols to give the silylene-trapping products **88–90** (Scheme 24).[61]

## C. *Thermally and Chemically Generated Silylenes*

Kummer and Köster[62] were the first to mention a silylene complex in organometallic synthesis. Treatment of hexachlorodisilane with 2,2'-bipyridyl resulted in a stable 1 : 2 adduct, which was interpreted as a dichlorosilylene coordinated to two bipyridyl molecules. However, because of the restricted possibilities at that time, no further experimental or spectroscopic

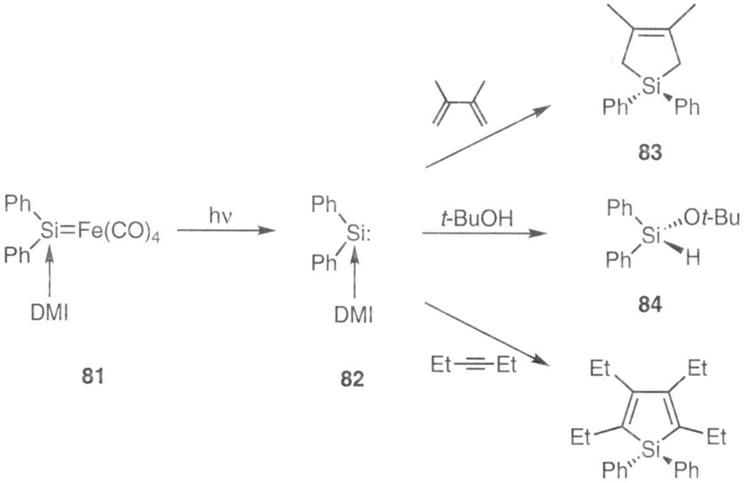


81 82 83 84 85

DMI = *N,N*-dimethylimidazolidinone

SCHEME 23

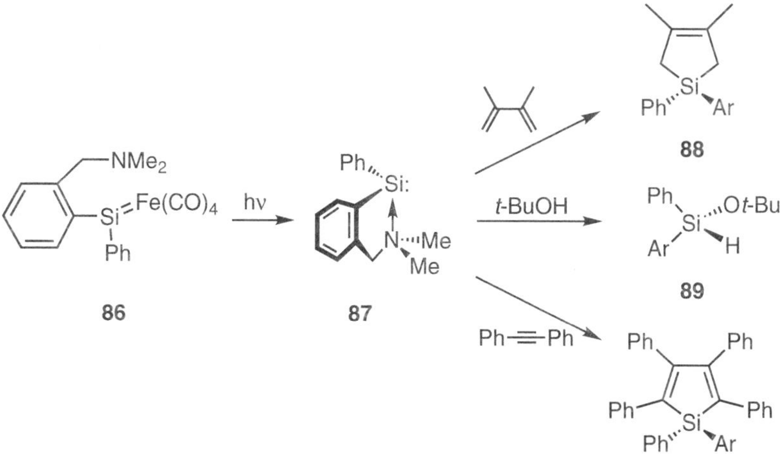


86 87 88 89 90

Ar = 2-$(Me_2NCH_2)C_6H_4$

SCHEME 24

evidence was presented for the postulated structure. To date, some groups have established that several silylenes react with 2,2′-bipyridyl to give covalent reaction products instead of complexes.[63,64] Thus, it should be noted that in the aforementioned experiment of Kummer and Köster,[62] a product other than a silylene complex may be formed. Base-induced disproportionation of chlorodisilanes was further investigated by Herzog *et al.*[65] Chlorodisilanes were activated by the addition of 1-methylimidazole under homogeneous conditions or heterogeneously with bis(dimethylamido)-phosphoric acid connected to a silicate carrier.[66] In all cases, insertion products of one or more silylenes into Si–Cl bonds of chlorosilanes were found. For the reaction of disilane **91** the silylene formation was explained by an initial donor–acceptor interaction between the chlorodisilane and the nitrogen atom of the Lewis base to give a penta-coordinated silicon compound **92** (Scheme 25), which subsequently undergoes disproportionation to chloromethylsilylene **93** and methyltrichlorosilane. Silylene **93,** which is supposed to be higher coordinated, reacts with starting material via insertion into the Si–Cl bond to give trisilane **94.** The subsequent insertion of silylene **93** into **94** takes place regioselectively in the 2-position, forming 2-silyltrisilane **95.** The authors explained this selectivity by the higher acceptor strength of the silicon in the 2-position, but did not discuss whether the coordination of the silylene, which possibly decreases its electrophilicity, may cause this selectivity as well.

Silylenes were shown for the first time by Seyferth and Annarelli[67] to be generated in the course of a thermally induced [1+2]-cycloreversion of silacyclopropanes. This methodology was used to generate transient dimeth-

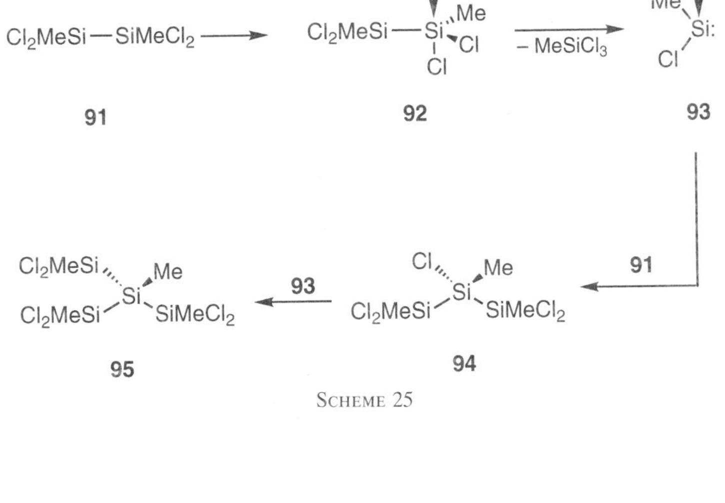

SCHEME 25

ylsilylene (**13**) by thermolysis of silacyclopropane **96** (Scheme 26). Thermolysis of **96** in the presence of triphenylphosphine and ketones yielded the six-membered rings **100** as reaction products along with the recovered phosphine.[68] Yields and reaction times were improved by using dimethylphenylphosphine instead of triphenylphosphine. The phosphine remained unchanged in the reaction mixture, and the reaction products were not formed in the absence of the phosphine, so the phosphine is assumed to act as a catalyst. Thus, the initial product in the thermal reaction of the silacyclopropane with the phosphine is supposed to be the ylide **97,** formed either via the formation of an uncoordinated silylene **13** and subsequent complexation or directly, without involvement of "free" **13,** in a concerted reaction from silacyclopropane **96.** The addition of **97** to the ketone yields betaine **98.** Resembling the reaction of the Corey ylide[69] with ketones, an unstable siloxirane **99** is formed and dimerizes to the observed products. With these experiments, Seyferth *et al.* demonstrated for the first time that the interaction of a silylene with a Lewis base significantly changes its chemical reactivity.

The [1+2] cycloreversion approach to silylenes was also used by Berry

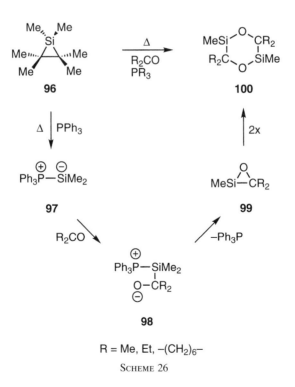

R = Me, Et, –(CH$_2$)$_6$–

SCHEME 26

and Jiang,[70] who investigated the reaction of silylene **13** with tantalum hydride **101** using hexamethylsilacyclopropane **96** as the precursor (Scheme 27). In first experiments, it was discovered that in addition to **102,** which is the expected insertion product of the **13** in the Ta–H bond of hydride **101,** disilacyclobutane **103,** which is known to be the insertion product of **13** into the silacyclopropane **96,**[71] was formed as a by-product in high yields. The authors showed, however, that the addition of four equivalents of trimethylphosphine decreased the amount of this by-product significantly, and they reasoned that an intermediate complex between the phosphine and the silylene is responsible for the enhanced selectivity of the silylene. This assumption was further confirmed by the observation that the addition of other Lewis bases such as triethylamine and tetrahydrofuran also results in the suppression of disilacyclobutane formation.

A different synthetic pathway to a higher coordinated silylene, leading to the first X-ray analysis of the solid-state structure, was presented by Karsch *et al.*[72] Treatment of a mixture of hexachlorodisilane and lithium phosphinosilylmethanide with lithium naphthalide led to complex **104.** The solid-state structure of this product shows that the two equatorial phosphorus atoms are covalently bound to the silicon center and the two axial phosphines act as donating ligands. The third position of the trigonal plane is occupied by the lone pair of the silylene. Thus, the silicon atom has a electron sextet and this compound can be regarded as a higher coordinated silylene. However, no experiments were performed to confirm that this compound actually undergoes typical silylene reactions.

Belzner[64] discovered a mild access to an intramolecularly coordinated silylene by the thermolysis of cyclotrisilanes **105** at moderate temperatures (40–60°C). It was furthermore shown[73] that cyclotrisilane **105** and silylene

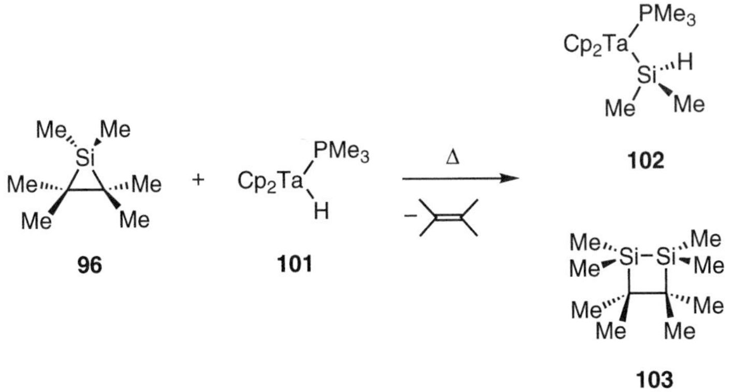

SCHEME 27

**104**

**106** are in equilibrium, presumably via the corresponding disilene. Cyclotrisilane **105** transfers all of its silylene subunits **106** quantitatively to a variety of reagents such as alkenes,[73] alkynes,[74,75] ketones,[76] isocyanates,[77] and nitriles[78] to yield the silylene-trapping products **107–111** (Scheme 28). The dimethylaminomethyl substituent is assumed to facilitate the ring cleavage of **105** by intramolecular coordination of the amino group to the extruded silylene, thus stabilizing this reactive intermediate. Such a coordinated silylene is supposed to exhibit decreased electrophilicity due to the blocked p orbitals.[79] It was demonstrated by competition experiments that **106** reacts faster with electron-poor than with electron-rich triple bonds.[79] This nucleophilic behavior of silylene **106** was rationalized by intramolecular blocking of the electrophilic p orbitals via the electron-donating amine so that the doubly occupied $\sigma$ orbital with its nucleophilic properties determines the reactivity of the silylene.

An interesting case of a divalent silicon species is represented by the stable decamethylsilicocene **21,** introduced by Jutzi,[80] which reacts like a silylene toward several trapping reagents.[80] When **21** was dissolved in liquid xenon at 253 K under CO pressure, Tacke *et al.*[27] found that a new compound, whose composition was assumed to be a silylene–carbon monoxide complex **22** of unknown structure (Scheme 6), was detected by IR spectroscopy. Considering that the starting material **21** is already highly coordinated, it was suspected that each Cp* unit in **22** and **23** is coordinated to the silicon center by less than five carbon atoms. Upon warming and evaporation of xenon, silylene **21** was regained. The same experiment was performed under nitrogen pressure, and a complex between the silylene and the nitrogen was formed, but complex formation was found to be a minor process. The formation of a higher equilibrium concentration of the same complex was achieved by performing the reaction in neat liquid nitrogen at 88 K.

Corriu *et al.*[29] presented several attempts to generate intramolecularly coordinated silylenes either chemically or photochemically. One approach used to generate a silylene employs the metal-mediated reductive dehalogenation of dihalosilanes **112** and **115,** respectively. Trapping experiments

with several butadienes gave, in some cases, the silylene reaction products
**113, 114,** and **116** in low to moderate yields (Schemes 29 and 30). However,
the authors mentioned that a silylenoid rather than a silylene in these
reactions may be the reactive intermediate. This is in agreement with observations of Boudjouk and Gaspar, who investigated different approaches to
silylenes.[81] Dehalogenation of dihalosilanes led to silylenoids as reactive
intermediates, which showed a selectivity toward trapping reagents that
differs significantly from the selectivity of the "free" silylenes formed in
the thermolysis or photolysis of silacyclopropanes. A similar approach made
use of the reductive dehalgenation of difluorosilane **117**.[82] Again, a transient

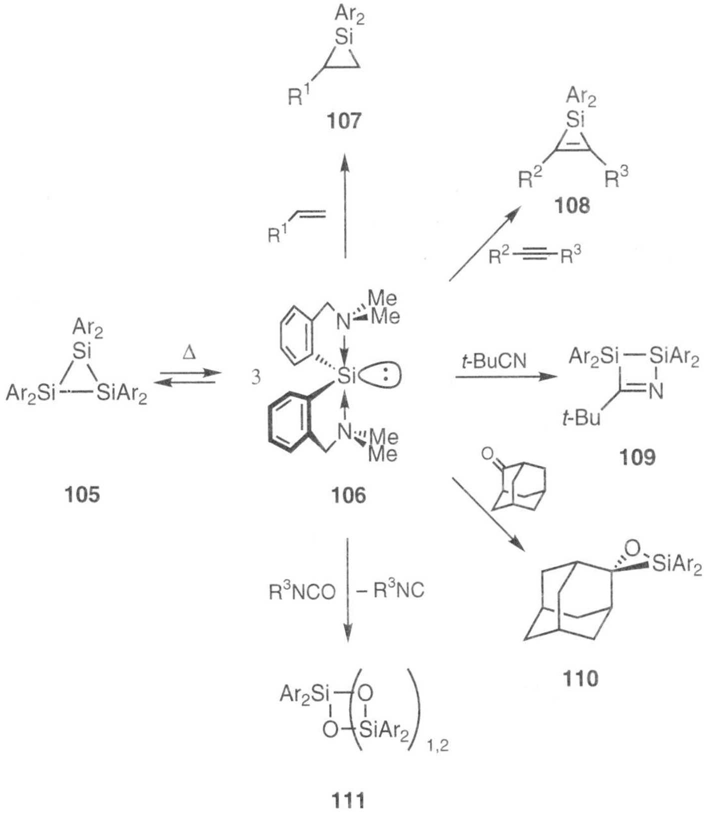

Ar = 2-(Me$_2$NCH$_2$)C$_6$H$_4$; R$^1$ = $n$-Pr, $n$-Bu, SiMe$_3$; R$^2$- R$^3$ = (CH$_2$)$_6$;
R$^2$ = Ph, R$^3$ = Ph; R$^2$ = $n$-Pr, R$^3$ = H; R$^2$ = Me$_3$Si, R$^3$ = H; R$^2$ = Ph, R$^3$ = H

Scheme 28

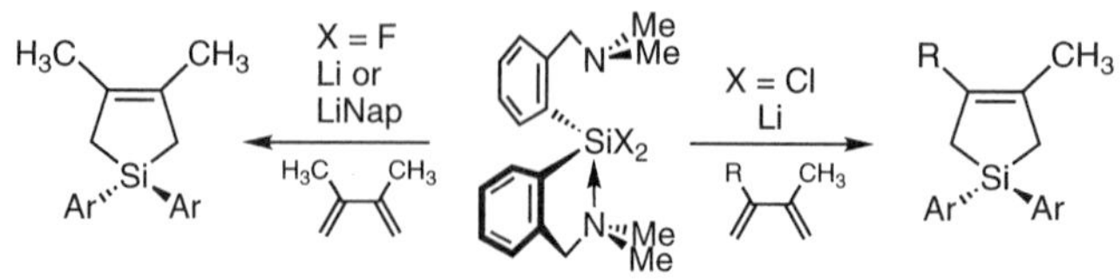

Ar = 2-(Me₂NCH₂)C₆H₄; R = H, Me

Scheme 29

silylene that was assumed to be intramolecularly coordinated was intercepted by 2,3-dimethyl-1,3-butadiene.

Tamao et al.[83] found that a higher coordinated silylene **119** can be formed from penta-coordinated silane **118** (Scheme 31). Warming a solution of **118** in toluene or dimethylformamide in the presence of diphenylacetylene or 2,3-dimethyl-1,3-butadiene resulted in the formation of silylene-trapping products **120** and **121**. Interestingly, no 1:1 reaction product between the silylene and the acetylene was isolated. Thus, it must be concluded that the insertion of silylene **119** into a Si–C bond of initially formed silacyclopropene is faster than the addition to the triple bond of the acetylene so that the silacyclopropene cannot be isolated under the reaction conditions.

The sterically crowded, kinetically stable disilene **122** acts on heating as a source of silylene **123** that exhibits the usual reactivity of silylenes, as

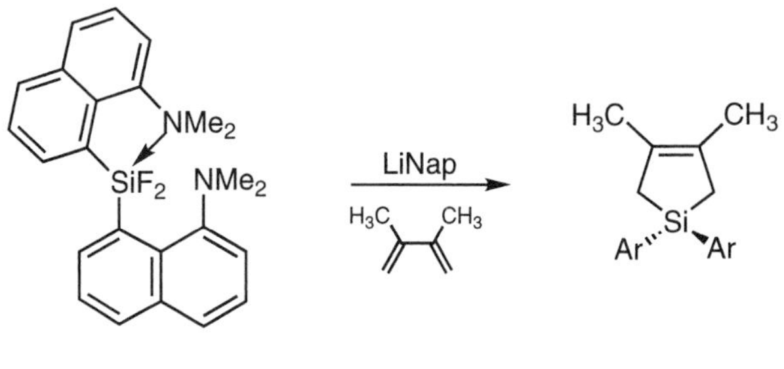

Ar = 8-(Me₂N)C₁₀H₆

Scheme 30

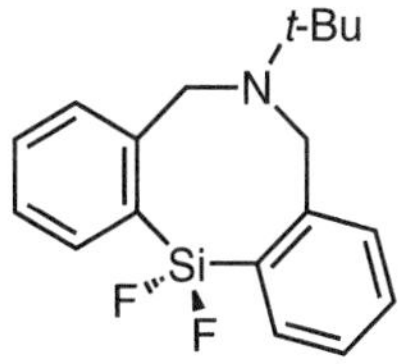

**117**

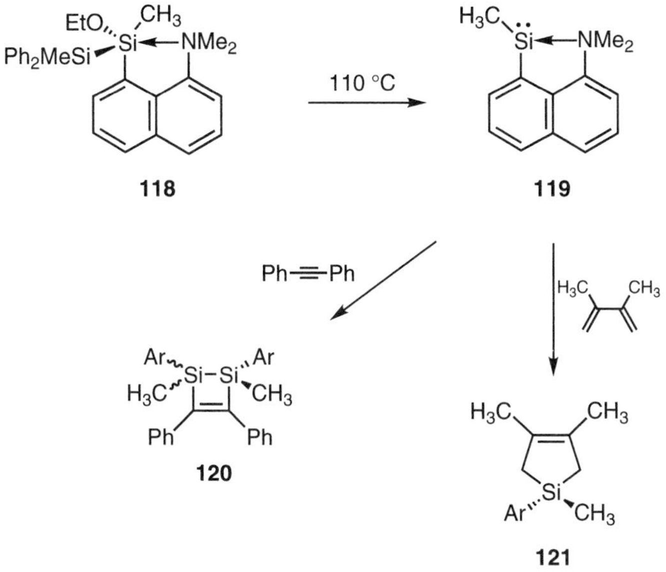

**118**                                                    **119**

**120**

**121**

Ar = 8-(Me$_2$N)C$_{10}$H$_6$

SCHEME 31

was shown by trapping experiments.[32] When silylene **123** is generated in the presence of bulky isocyanides, deeply colored products are formed almost quantitatively and are thermally stable up to 60°C (Scheme 32). Investigation of NMR spectra, especially the $^{29}$Si–$^{13}$C-coupling constants, confirms that there is only a weak interaction between the silicon and the isocyanide carbon so that the structure of the products is best described as a silylene–Lewis base complex **28–30** (see Section III). Further investigations of the chemical reactivity of **28–30** showed that silylene **123** can be transferred from **28–30** to various reagents such as triethylsilane, 2,3-dimethyl-1,3-butadiene, and methanol to give the silylene-trapping products

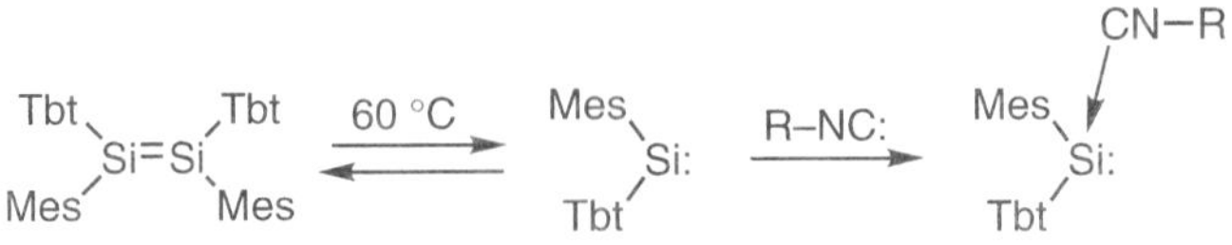


**122** **123** **28–30**

**28**: R = 2,4,6-($i$-Pr)$_3$C$_6$H$_2$; **29**: R = Tbt; **30**: R = 2,4,6-($t$-Bu)$_3$C$_6$H$_2$;
Mes = 2,4,6-Me$_3$C$_6$H$_2$; Tbt = 2,4,6-[(Me$_3$Si)$_2$CH]C$_6$H$_2$

SCHEME 32

**124–127** (Scheme 33). In the presence of methanol, **128a–c** were formed as the products resulting from protonation of the complex and subsequent addition of the methoxide ion. Weidenbruch *et al.*[84] also investigated the reactions of a bulky silylene with isocyanides, but no formation of a Lewis-base analogous to **28–30** was observed.

Schäfer and Weidenbruch[85] showed that transition metals catalyze the extrusion of silylenes from cyclotrisilanes, and this methodology was used to

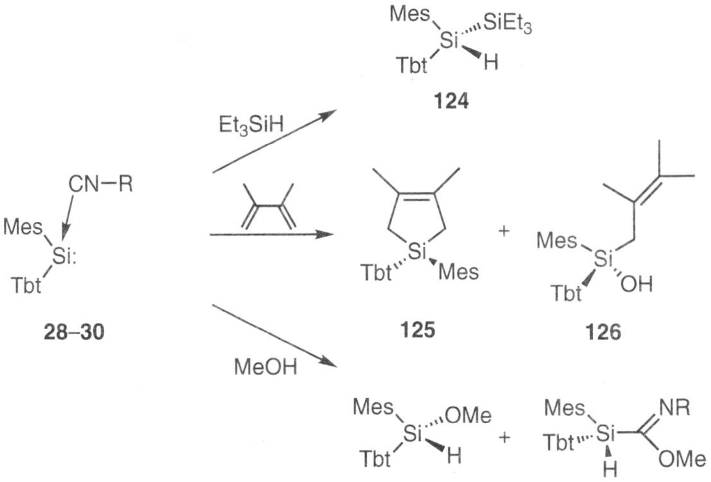


**124** **28–30** **125** **126** **127** **128a-c**

**28**: R = 2,4,6-($i$-Pr)$_3$C$_6$H$_2$; **29**: R = Tbt; **30**: R = 2,4,6-($t$-Bu)$_3$C$_6$H$_2$;
Mes = 2,4,6-Me$_3$C$_6$H$_2$; Tbt = 2,4,6-[(Me$_3$Si)$_2$CH]C$_6$H$_2$

SCHEME 33

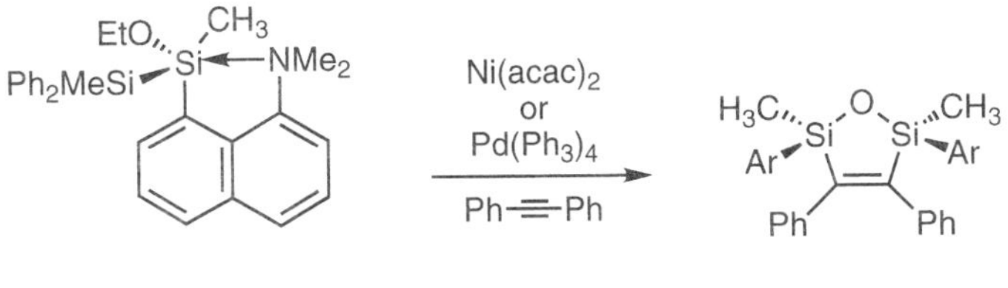

**129**

**130**

Ar = 8-(Me$_2$N)C$_{10}$H$_6$

Scheme 34

generate an intramolecularly coordinated silylene from an acyclic trisilane.[86] Trisilane **129,** which bears a chelating aminonaphthyl substituent, is cleaved by catalytic amounts of Ni(acac)$_2$ or Pd(PPh$_3$)$_4$. In the presence of diphenyl-acetylene, the cyclic disiloxane **130** was isolated as a reaction product (Scheme 34). Although the reaction mechanism is not quite clear, it is

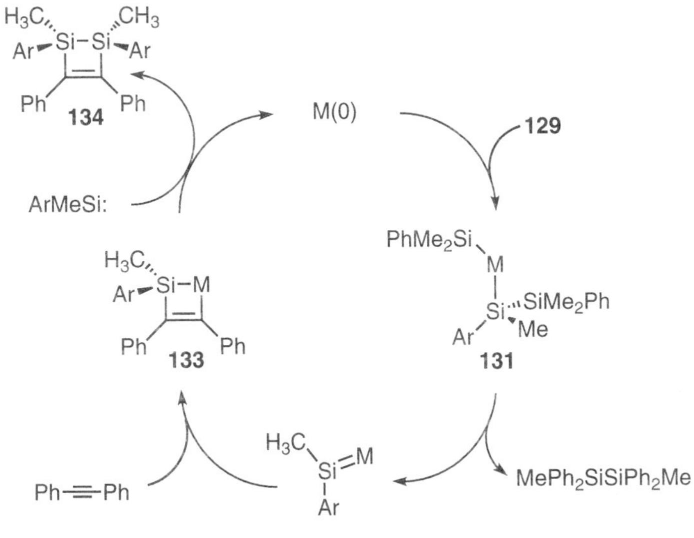

Ar = 8-(Me$_2$N)C$_{10}$H$_6$; M = Pd, Ni

Scheme 35

assumed that the catalytic cycle is initiated by the oxidative addition of the trisilane to the metal center and subsequent elimination of a disilane to give a metal–silylene complex **132** (Scheme 35). This silylene complex may extrude the higher coordinated silylene, which adds to the acetylene and eventually yields 1,2-disilacyclobutene **134** via the corresponding silacyclopropene. Alternatively, the acetylene may react, as shown in Scheme 35, within the coordination sphere of the complex to form pallada-silacyclobutene **133**. Further insertion of a silylene followed by reductive elimination finally yields **134,** which is oxidized to **130** under the workup conditions employed.

ACKNOWLEDGMENTS

Financial support from the Fonds der Chemischen Industrie, the Deutsche Forschungsgemeinschaft, and the Bundesministerium für Bildung, Wissenschaft, Forschung und Technologie is gratefully acknowledged by the authors.

REFERENCES

(1) (a) Gaspar, P. P. In *Reactive Intermediates*; Jones, M.; Moss, R. A., Jr., Eds.; Wiley: New York, 1978, Vol. 1, pp. 229–277; (b) Gaspar, P. P. In *Reactive Intermediates*; Jones, M.; Moss, R. A., Jr., Eds.; Wiley: New York, 1981, Vol. 2, pp. 335–385; (c) Gaspar, P. P. In *Reactive Intermediates*; Jones, M.; Moss, R. A., Jr., Eds.; Wiley: New York, 1985, Vol. 3, pp. 333–427; (d) Almond, M. *Short-Lived Molecules*, Ellis Horwood: New York, 1990, pp. 93–120; (e) Safarik, I.; Sandhu, V.; Lown, E. M.; Strausz, O. P.; Bell, T. N. *Res. Chem. Intermed.* **1990**, *14*, 105.

(2) Luke, B. T.; Pople, J. A.; Krogh-Jespersen, M.-B.; Apeloig, Y.; Karni, M.; Chandrasekhar, J.; Schleyer, P. v. R. *J. Am. Chem. Soc.* **1986**, *108*, 270.

(3) Arrington, C. A.; Petty, J. T.; Payne, S. E.; Haskins, W. C. K. *J. Am. Chem. Soc.* **1988**, *110*, 6240.

(4) Hamilton, T. P.; Schaefer, H. F., III, *J. Chem. Phys.* **1989**, *90*, 1031.

(5) Zakharov, I. I.; Zhidomirov, G. M., *React. Kinet. Catal. Lett.* **1990**, *41*, 59.

(6) (a) Raghavachari, K.; Chandrasekhar, J.; Gordon, M. S.; Dykema, K. J. *J. Am. Chem. Soc.* **1984**, *106*, 5853; (b) Dykema, K. J.; Truong, T. N.; Gordon, M. S. *J. Am. Chem. Soc.* **1985**, *107*, 4535.

(7) Conlin, R. T.; Laakso, D.; Marshall, P. *Organometallics* **1994**, *13*, 838.

(8) (a) Schoeller, W. W.; Schneider, R. *Chem. Ber./Recueil* **1997**, *130*, 1013; (b) Schoeller, W. W.; Busch, T. *Chem. Ber.* **1992**, *125*, 1319.

(9) Gaspar, P. P.; Holten, D.; Konieczny, S.; Corey, J. Y. *Acc. Chem. Res.* **1987**, *20*, 329.

(10) Akasaka, T.; Nagase, S.; Yabe, A.; Ando, W. *J. Am. Chem. Soc.* **1988**, *110*, 6270.

(11) Nagase, S.; Kudo, T.; Akasaka, T.; Ando, W. *Chem. Phys. Lett.* **1989**, *163*, 23.

(12) Patyk, A.; Sander, W.; Gauss, J.; Cremer, D. *Angew. Chem.* **1989**, *101*, 920. *Angew. Chem. Int. Ed. Engl.* **1989**, *28*, 898.

(13) Käss, D.; Oberhammer, H.; Brandes, D.; Blaschette, H. *J. Mol. Struct.* **1977**, *40*, 65.

(14) Raghavachari, K.; Chandrasekhar, J.; Frisch, M. J. *J. Am. Chem. Soc.* **1982**, *104*, 3779.

(15) Zachariah, M. R.; Tsang, W. *J. Phys. Chem.* **1995**, *99*, 5308.

(16) Su, S.; Gordon, M. S. *Chem. Phys. Lett.* **1993**, *204*, 306.

(17) Lucas, D. J.; Curtiss, L. A.; Pople, P. J. *J. Chem. Phys.* **1993**, *99*, 6697.

(18) Lee, S. Y.; Boo, B. H. *Theochem J. Mol. Struct.* **1996**, *366*, 79.

(19) Abronin, I. A.; Avdyukhina, N. A.; Chernyshev, E. A. *Russ. Chem. Bull.* **1994**, *43*, 751.

(20) Clark, T.; Schleyer, P. v. R. *J. Organomet. Chem.* **1980**, *191*, 347.

(21) Laurie, V. W. *J. Chem. Phys.* **1957**, *26*, 1359.

(22) Schlegel, H. B.; Sosa, C. *J. Phys. Chem.* **1985**, *89*, 537.

(23) Taraban, M. B.; Plyusnin, V. F.; Volkova, O. S.; Grivin, V. P.; Leshina, T. V.; Lee, V. Y.; Faustov, V. I.; Egorov, M. P.; Nefedov, O. M. *J. Phys. Chem.* **1995**, *99*, 14719.

(24) See, e.g., Michalczyk, M. J.; Fink, M. J.; De Young, D. J.; Carlson, C. W.; Welsh, K. M.; West, R.; Michl, J. *Silicon, Germanium, Tin and Lead Compounds,* **1986**, *9*, 75, and references cited therein.

(25) See, e.g., Harper, W. W.; Ferall, E. A.; Hilliard, R. K.; Stogner, S. M.; Grev, R. S.; Clouthier, D. J. *J. Am. Chem. Soc.* **1997**, *119*, 8361 and references cited therein.

(26) Ismael, Z. K.; Hauge, R. H.; Fredin, L.; Kauffman, J. W.; Margrave, J. L. *J. Chem. Phys.* **1982**, *77*, 1617.

(27) Tacke, M.; Klein, C.; Stufkens, D. J.; Oskam, A.; Jutzi, P.; Bunte, E. A. *Z. Anorg. Allg. Chem.* **1993**, *619*, 865.

(28) Reisenauer, H. P.; Mihm, G.; Maier, G. *Angew. Chem.* **1982**, *94*, 864; *Angew. Chem. Int. Ed. Engl.* **1982**, *21*, 854.

(29) Corriu, R.; Lanneau, G., Priou, C.; Soulairol, F.; Auner, N.; Probst, R.; Conlin, R.; Tan, C. *J. Organomet. Chem.* **1994**, *466*, 55.

(30) Gillette, G. R.; Noren, G. H.; West, R. *Organometallics* **1989**, *8*, 487.

(31) Ando, W.; Hagiwara, K.; Sekiguchi, A. *Organometallics* **1987**, *6*, 2270.

(32) Takeda, N.; Suzuki, H.; Tokitoh, N.; Okazaki, R.; Nagase, S. *J. Am. Chem. Soc.* **1997**, *119*, 1456.

(33) Levin, G.; Das, P. K.; Bilgrien, C.; Lee, C. L. *Organometallics* **1989**, *8*, 1206.

(34) Yamaji, M.; Hamanishi, K.; Takahashi, T.; Shizuka, H. *J. Photochem. Photobiol. A: Chem.* **1994**, *81*, 1.

(35) Baggott, J. E.; Blitz, M. A.; Frey, H. M.; Lightfoot, P. D.; Walsh, R. *J. Chem. Kinet.* **1992**, *24*, 127.

(36) Gillette, G. R.; Noren, G. H.; West, R. *Organometallics* **1987**, *6*, 2617.

(37) Steele, K. P.; Weber, W. P. *J. Am. Chem. Soc.* **1980**, *102*, 6095.

(38) Steele, K. P.; Tzeng, D.; Weber, W. P. *J. Organomet. Chem.* **1982**, *231*, 291.

(39) Steele, K. P.; Weber, W. P. *Inorg. Chem.* **1981**, *20*, 1302.

(40) Steinmetz; M. G. *Chem. Rev.* **1995**, *95*, 1527.

(41) Ishikawa, M.; Kumada, M. *J. Organomet. Chem.* **1972**, *42*, 325.

(42) Ando, W.; Ikeno, M.; Sekiguchi, A. *J. Am. Chem. Soc.* **1978**, *100*, 3613.

(43) Ando, W.; Ikeno, M.; Sekiguchi, A. *J. Am. Chem. Soc.* **1977**, *99*, 6447.

(44) Ando, W.; Hamada, Y.; Sekiguchi, A. *Tetrahedron Lett.* **1982**, *23*, 5323.

(45) Ando, W.; Hamada, Y.; Sekiguchi, A. *J. Chem. Soc. Chem. Commun.* **1983**, 952.

(46) Gu, T.-Y. Y.; Weber, W. P. *J. Am. Chem. Soc.* **1980**, *102*, 1641.

(47) Tzeng, D.; Weber, W. P. *J. Am. Chem. Soc.* **1980**, *102*, 1451.

(48) Goure, W. F.; Barton, T. J. *J. Organomet. Chem.* **1980**, *199*, 33.

(49) Ando, W.; Ikeno, M.; Hamada, Y. *J. Chem. Soc. Chem. Commun.* **1981**, 621.

(50) Tzeng, D.; Weber, W. P. *J. Org. Chem.* **1981**, *46*, 693.

(51) Tortorelli, V. J.; Jones Jr., M. *J. Chem. Soc. Chem. Commun.* **1980**, 785.

(52) Chihi, A.; Weber, W. P. *Inorg. Chem.* **1981**, *20*, 2822.

(53) Ando, W.; Sekiguchi, A.; Hagiwara, K.; Sakakibara, A.; Yoshida, H. *Organometallics* **1988**, *7*, 558.

(54) Kira, M.; Maruyama, T.; Sakurai, H. *Tetrahedron Lett.* **1992**, *33*, 243.

(55) Levin, G.; Das, P. K.; Lee, C. L. *Organometallics* **1988**, *7*, 1231.
(56) Pearsall, M.-A.; West, R. *J. Am. Chem. Soc.* **1988**, *110*, 7228.
(57) Ishikawa, M.; Nakagawa, K.-I.; Katayama, S.; Kumada, M. *J. Organomet. Chem.* **1981**, *216*, C48.
(58) Oka, K.; Nakao, R. *J. Organomet. Chem.* **1990**, *390*, 7.
(59) Corriu, R. J. P.; Lanneau, G. F.; Chauhan, B. P. S. *Organometallics* **1993**, *12*, 2001.
(60) Corriu, R. J. P.; Chauhan, B. P. S.; Lanneau, G. F. *Organometallics* **1995**, *14*, 1646.
(61) Chauhan, B. P. S.; Corriu, R. J. P.; Lanneau, G. F.; Priou, C.; Auner, N.; Handwerker, H.; Herdtweck, E. *Organometallics* **1995**, *14*, 1657.
(62) Kummer, D.; Köster, H. *Angew. Chem.* **1969**, *81*, 897; *Angew. Chem. Int. Ed. Engl.* **1969**, *8*, 878.
(63) (a) Weidenbruch, M.; Schäfer, A.; Marsmann H. *J. Organomet. Chem.* **1988**, *354*, C12; (b) Weidenbruch, M.; Lesch, A.; Marsmann, H. *J. Organomet. Chem.* **1990**, *385*, C47; (c) Jutzi, P., personal communication.
(64) Belzner, J. *J. Organomet. Chem.* **1992**, *430*, C51.
(65) Herzog, U.; Richter, R.; Brendler, E.; Roewer, G. *J. Organomet. Chem.* **1996**, *507*, 221.
(66) Richter, R.; Schulze, N.; Roewer, G.; Albrecht, J. *J. Prakt. Chem.* **1997**, *339*, 145.
(67) Seyferth, D.; Annarelli, D. C. *J. Am. Chem. Soc.* **1975**, *97*, 7162.
(68) Seyferth, D.: Lim, T. F. O. *J. Am. Chem. Soc.* **1978**, *100*, 7074.
(69) Melvin, L. S.; Trost, B. M. *Sulfur Ylides,* Academic Press: New York, 1975 and references cited therein.
(70) Berry, D. H.; Jiang, Q. *J. Am. Chem. Soc.* **1987**, *109*, 6210.
(71) Seyferth, D.; Annarelli, D. C.; Vick, S. C.; Duncan, D. P. *J. Organomet. Chem.* **1980**, *201*, 179.
(72) Karsch, H. H.; Keller, U.; Gamper, S.; Müller, G. *Angew. Chem.* **1990**, *102*, 297; *Angew. Chem. Int. Ed. Engl.* **1990**, *29*, 295.
(73) Belzner, J.; Ihmels, H.; Kneisel, B. O.; Gould, R. O.; Herbst-Irmer, R. *Organometallics* **1995**, *14*, 305.
(74) Belzner, J.; Ihmels, H. *Tetrahedron Lett.* **1993**, 6541.
(75) Belzner, J.; Ihmels, H.; Kneisel, B. O.; Herbst-Irmer, R. *J. Chem. Soc. Chem. Commun.* **1994**, 1989.
(76) Belzner, J.; Ihmels, H.; Pauletto, L. *J. Org. Chem.* **1996**, *61*, 3315.
(77) Belzner, J.; Ihmels, H.; Kneisel, B. O.; Herbst-Irmer, R. *Chem. Ber.* **1996**, *129*, 125.
(78) Belzner, J.; Ihmels, H.; Noltemeyer, M. *Tetrahedron Lett.* **1995**, *45*, 8187.
(79) Belzner, J. In *Organosilicon Chemistry III: From Molecules to Materials*; Auner, N.; Weis, J., Eds.; VCH: Weinheim, 1997, pp. 58–64.
(80) Jutzi, P. In *Frontiers of Organosilicon Chemistry*; Bassindale, A. R.; Gaspar; P. P., Eds.; Royal Society of Chemistry: Cambridge, 1991, pp. 307–318.
(81) (a) Boudjouk, P.; Samaraweera, U.; Sooriyakumaran, R.; Chrisciel, J.; Anderson, K. R. *Angew. Chem.* **1988**, 1406; *Angew. Chem. Int. Ed. Engl.* **1990**, *27*, 1355; (b) Pae, D. H.; Xiao, M.; Chiang, M. Y.; Gaspar, P. P. *J. Am. Chem. Soc.* **1991**, *113*, 1281.
(82) Carré, F. H.; Corriu, R. J. P.; Lanneau, G. F.; Merle, P.; Soulairol, F.; Yao, J. *Organometallics* **1997**, *16*, 3878.
(83) Tamao, K.; Nagata, K.; Asahara, M.; Kawachi, A.; Ito, Y.; Shiro, M. *J. Am. Chem. Soc.* **1995**, *117*, 11592.
(84) (a) Weidenbruch, M.; Brand-Roth, B.; Pohl, S.; Saak, W. *Angew. Chem.* **1990**, *102*, 93; *Angew. Chem. Int. Ed. Engl.* **1990**, *29*, 90; (b) Weidenbruch, M.; Brand-Roth, B.; Pohl, S.; Saak, W. *Polyhedron* **1991**, *10*, 1147.
(85) Schäfer, A.; Weidenbruch, M. *J. Organomet. Chem.* **1985**, *282*, 305.
(86) Tamao, K.; Tarao, Y.; Nakagawa, Y.; Nagata, K.; Ito, Y. *Organometallics* **1993**, *12*, 1113.

ADVANCES IN ORGANOMETALLIC CHEMISTRY, VOL. 43

# Chemistry of Ruthenium–Carbide Clusters Ru₅C(CO)₁₅ and Ru₆C(CO)₁₇

### PAUL J. DYSON

*Center for Chemical Synthesis*
*Department of Chemistry*
*Imperial College of Science, Technology, and Medicine*
*South Kensington, London SW7 2AY*
*United Kingdom*

# I

## INTRODUCTION

The chemistry of high nuclearity carbonyl clusters has been studied in considerable detail over the last few decades and many of the unique properties that they possess are now well understood. A large number of reviews and books are available that are concerned with these intriguing compounds and a few key texts are referenced.[1-6] The systematic study of specific clusters has led to an accumulation of knowledge concerning their chemical and physical properties. For example, the trinuclear clusters, $M_3(CO)_{12}$ (M = Fe, Ru, and Os) have been subjected to extensive reactivity studies and hundreds of derivatives of these clusters have been reported in the literature. Among the most intensely studied group of high nuclearity carbonyl clusters are those containing interstitial main group atoms and two general review articles have been devoted to these compounds.[7,8] More specifically, the ruthenium–carbide clusters $Ru_5C(CO)_{15}$ **1** and $Ru_6C(CO)_{17}$ **2** have been studied extensively and this reviews their chemistry. Because **2** provided the first example of a cluster to contain a completely encapsulated carbide atom, much of the early work focused on studying the properties of this atom. Reactivity studies of **1** and **2** range from simple ligand substitution reactions to those in which the cluster core undergoes polyhedral transformations or those in which heteronuclear clusters are prepared. It has been found that certain organic ligands exhibit stereochemical nonrigid behavior on the nuclear magnetic resonance (NMR) time scale when coordinated to these clusters and migration of ligands over the cluster surface has even been observed.

# II

## SYNTHESIS AND CHARACTERIZATION OF $Ru_5C(CO)_{15}$ AND $Ru_6C(CO)_{17}$

### A. Synthesis

The hexaruthenium cluster $Ru_6C(CO)_{17}$ **2** was first isolated as a by-product from the thermolysis of $Ru_3(CO)_{12}$ **3** in arene solvents, e.g., toluene, $C_6H_5Me$; m-xylene, $C_6H_4Me_2$-1,3; and mesitylene, $C_6H_3Me_3$-1,3,5.[9,10] It was formed together with the related arene-coordinated clusters $Ru_6C(CO)_{14}(\eta$-arene) (arene = $C_6H_5Me$, $C_6H_4Me_2$-1,3 and $C_6H_3Me_3$-1,3,5) in which the arene replaces three carbonyl groups on one of the ruthenium atoms. In the initial reports these clusters were characterized by mass spectrometry, infrared (IR), and $^1H$-NMR spectroscopy, and the single crystal

X-ray structure of the mesitylene derivative was reported shortly afterward.[11] This represented the second structurally characterized cluster containing an interstitial atom [the structure of $Fe_5C(CO)_{15}$ having already been established][12] and the first example of a cluster with a completely encapsulated carbide atom. At the time that the synthesis of **2** was first reported, another paper described the synthesis of a cluster also obtained from **3** when heated to 150°C in either benzene or cyclohexane. Based on an estimation of the mass of this compound from a differential vapor pressure measurement, the authors suggested that this compound corresponded to $Ru_6(CO)_{18}$.[13] It was subsequently noted from a comparison of $\nu_{CO}$ IR data and a structural determination that this compound was in fact **2**.

A number of other synthetic routes to **2** have also been developed. For example, the vacuum pyrolysis of $Ru_3(CO)_{12}$ **3** affords **2** in 5% yield.[14] Higher yielding routes to **2** include the autoclave reaction of **3** under an atmosphere of ethylene, which affords **2** in 70% yield,[15] and the oxidative addition of CO to $[Ru_6C(CO)_{16}]^{2-}$ **4**, which is nearly quantitative. The conversion of **4** to **2** involves oxidation with $FeCl_3$ or the ferrocenium cation, $[Fe(C_5H_5)_2][BF_4]$, in a dichloromethane solution saturated with carbon monoxide.[16,17] The dianion **4** was originally prepared in 60% yield from a one-step thermolysis of **3** with $Na[Mn(CO)_5]$ in diglyme[16]; however, **4** has since been isolated in 90% yield from a similar reaction in which sodium metal is used as the reducing agent.[17] The analogous thermal reaction employing $Na[Mn(CO)_5]$ as the reductant was originally used to prepare the hexairon cluster $[Fe_6C(CO)_{16}]^{2-}$.[18]

The development of these higher yielding routes has allowed the detail reactivity studies of $Ru_6C(CO)_{17}$ [and consequently $Ru_5C(CO)_{15}$, as it is derived from the hexamer] to be conducted. It is worth noting that in many thermal reactions commencing with **3,** especially those conducted in hydrocarbon solvents such as heptane or octane, with various organic ligands present, compound **2** is often produced in low yield together with other products usually in higher yield. Some of the products obtained from these types of reactions have given insights into the formation of the cluster and the origin of the carbide atom; these reactions are discussed in Section VI.

The pentaruthenium cluster **1** was first isolated in 1% yield together with **2** in higher yield from the reaction of $H_4Ru_4(CO)_{12}$ with ethylene under pressure.[19] It can now be prepared from **2** in near quantitative yield in a degradative carbonylation reaction in which **2** is heated in heptane to 70°C in an autoclave under 80 atmospheres of CO affording **1** and $Ru(CO)_5$.[20,21] The most convenient routes to **1** and **2** are illustrated in Scheme 1.

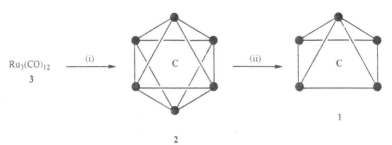

$Ru_3(CO)_{12}$    $\xrightarrow{\text{(i)}}$    C    $\xrightarrow{\text{(ii)}}$    C

3

2                                                                    1

SCHEME 1.   The synthesis of **1** and **2**. (i) ethylene (30 atm), heptane, 150°C, 3.5 h; (ii) CO (80 atm), heptane, 70°C, 3 h.

### B. *Spectroscopy*

The carbide atom in **1** is located in the center of the square face such that it is partially exposed whereas the carbide atom in **2** is completely encapsulated by the six ruthenium atoms. From a spectroscopic viewpoint, carbide atoms are very distinctive and the earlier reviews have dealt with these aspects in detail.[7,8] The IR spectrum of **1** contains peaks at 701(s) and 670(m) cm$^{-1}$, and **2** contains peaks at 717(sh), 703(s), 680(m), and 669(m) cm$^{-1}$.[22] $^{13}$C-NMR spectra of **1** and **2** do not appear to have been reported. This is probably due to the low yields in which these compounds were initially obtained at a time when $^{13}$C-NMR was still not in widespread use in cluster chemistry. In general, the $^{13}$C-NMR resonance of carbide atoms ranges from $\delta$ 250 to 500. The high frequency resonances exhibited in $^{13}$C-NMR spectra reflect the different diamagnetic and paramagnetic effects experienced by a nucleus in such an unusual chemical environment.[23]

The large number of derivatives of **1** and **2** in which the ruthenium core is preserved indicates that the presence of a carbide atom makes a significant contribution to the stability of the cluster.[24] Single crystal X-ray diffraction studies of some derivatives of **1** and **2** reveal that the carbide atom is located slightly closer to the new ligand. This has been attributed to an electronic compensatory effect of the carbide atom when the carbonyl ligands are replaced by poorer $\pi$-acceptor ligands.[25]

### C. *Solid-State Structure*

The structure of **2** was first determined in 1969 from the visual estimation of diffraction intensities obtained from Weissenburg rotational photographic data.[26] This study established that the octahedral cluster contained an interstitial C atom and also showed the distribution of ligands around the metal core to comprise 16 terminal carbonyl ligands and one bridging CO ligand. The structure of **1** has been established by single crystal X-ray diffraction and the crystal contains two independent molecules that are essentially the same.[20,21] The square pyramid of Ru atoms each have three terminal carbonyls and the carbide atom sits in the center of the square base, but displaced by 0.11(2) Å below the $Ru_4$ plane.

The structure of **2** has since been redetermined on two polymorphic crystalline modifications.[27] One crystal was grown from hot benzene and the other was obtained from a solution of dichloromethane–hexane by slow evaporation. The former crystal contains one molecule whereas the latter crystal contains two independent molecules. The structure of one of these molecules is depicted in Fig. 1 and key structural parameters for **1** and **2** are listed in Table I.

The differences between the cluster skeletons of the three molecules of **2** are very small with the mean values of the Ru–Ru distances being similar and the mean Ru–C(carbide) distances being identical. The most notable differences between the structures arise from the orientation of the tricarbonyl units attached to the apical Ru atoms above and below the molecular equator of the octahedral cluster (the molecular equator is defined as the $Ru_4$ plane in which the bridging carbonyl ligand is present). The two tricarbonyl units are almost exactly staggered in the crystal obtained from benzene, whereas they approach an eclipsed conformation in the other polymorph. Although the [13]C-NMR spectrum of **2** has not been recorded in solution (or in the solid state), it is not unreasonable to anticipate that

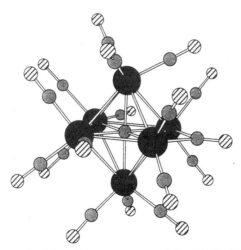

FIG. 1.    The solid-state structure of $Ru_6C(CO)_{17}$ **2**.

TABLE I
STRUCTURAL DATA FOR **1, 2A, 2B,** AND **2B'**

|      | Ru–Ru, range (Å)     | Ru–Ru, mean (Å) | Ru–carbide, mean (Å) |
|------|----------------------|-----------------|----------------------|
| **1**    | 2.800(2)–2.882(2)    | 2.84            | 2.04                 |
| **2A**   | 2.835(3)–2.967(3)    | 2.90            | 2.05                 |
| **2B**   | 2.826(1)–2.998(1)    | 2.89            | 2.05                 |
| **2B'**  | 2.803(1)–2.977(1)    | 2.89            | 2.05                 |

in solution the carbonyls undergo rapid rotation or perhaps more extensive carbonyl scrambling over the cluster framework.

### III

### ELECTROCHEMISTRY

The cyclic voltammogram of $Ru_6C(CO)_{17}$ **2** in $[Bu_4N][BF_4]$-$CH_2Cl_2$ at a platinum electrode displays an irreversible reduction at $E_p = -0.46$ V versus a Ag–AgCl reference electrode.[28,29] The process involves two electrons and results in the formation of the dianion $[Ru_6C(CO)_{16}]^{2-}$ **4.** An oxidation wave at $E_p = +0.45$ V (i.e., the oxidation wave of the reduced product **4**) is also observed. Under similar conditions the dianion, $[Ru_5C(CO)_{14}]^{2-}$ **5,** is oxidized at $E_p = +0.15$ V and the oxidized product is reduced at $E_p = -1.78$ V.[28] Under controlled electrochemical conditions, **4** and **5** undergo oxidative addition reactions with ligands such as CO, phosphines, and four electron donor ligands such as alkynes. With **4,** the same oxidative addition reactions can be carried out chemically on a preparative scale using $[Fe(\eta\text{-}C_5H_5)_2][BF_4]$ as the oxidant.[29]

A reinvestigation of the electrochemistry of **2** in an infrared spectroelectrochemical cell revealed some new insights into the redox processes that take place.[30] Under anaerobic conditions the reduction of **2** results in the formation of **4** and CO. However, the IR spectral changes accompanying the reduction of **2** in the presence of oxygen show that reduction to **4** is accompanied by the formation of $CO_2$. A mechanism has been proposed to account for these observations as well as others concerning the reoxidation of **4** to **2** (see Scheme 2). The first step in the reduction of **2** is believed to involve a one electron reduction to $[Ru_6C(CO)_{17}]^{\cdot-}$ whether $O_2$ is present or not. Although there is no spectral evidence for the formation of a radical anion in these experiments, the EPR spectrum of $[Ru_6C(CO)_{17}]^{\cdot-}$ has been reported.[31] From here the pathways must differ and it is conceivable that

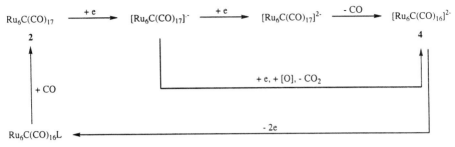

SCHEME 2. Mechanism for the reduction/oxidation of $Ru_6C(CO)_{17}$/$[Ru_6C(CO)_{16}]^{2-}$ under anaerobic and aerobic conditions.

in the absence of oxygen $[Ru_6C(CO)_{17}]^{2-}$ is produced followed by CO ejection from the electron-rich cluster to yield $[Ru_6C(CO)_{16}]^{2-}$. In the presence of $O_2$, localization of the odd electron in the radical anion onto a C atom of a carbonyl ligand would render the radical ·CO group reactive toward $O_2$, thereby generating $CO_2$. The IR spectrum recorded during the reoxidation of **4** to **2** in the presence of CO reveals the presence of an intermediate, believed to be $Ru_6C(CO)_{16}L$ (where L is a weakly stabilizing ligand). The tetraanion, $[Ru_6C(CO)_{15}]^{4-}$ **6,** may be generated electrochemically from **4** on further reduction and it has been found that this process only takes place when oxygen is present. The reason why $O_2$ is required in this step is not, as yet, understood.

On a preparative scale, **4** and **5** can be made from the appropriate neutral parent cluster by reduction with two equivalents of sodium metal in tetrahydrofuran or, more conveniently, by the addition of excess sodium carbonate or potassium hydroxide to **1** or **2** in methanol.[15]

## IV

## LIGAND REACTIONS OF $Ru_5C(CO)_{15}$ AND $Ru_6C(CO)_{17}$

### A. *Hydrides and Halides*

The dihydride $H_2Ru_5C(CO)_{14}$ **7** has been reported, although its characterization was based entirely on an analysis of its infrared spectrum.[20] In contrast, $H_2Ru_6C(CO)_{16}$ **8** has been prepared and fully characterized.[15] Compound **8** can be made by acidification of the dianion $[Ru_6C(CO)_{16}]^{2-}$ **4** with $H_2SO_4$ in acetonitrile at $-30°C$. Cluster **8** may also be prepared via stepwise protonation from **4**. Treatment of a dichloromethane solution of

**4** with *p*-toluenesulfonic acid in ethanol affords the monohydride $[HRu_6C(CO)_{16}]^-$ **9** in 54% yield.[32] Compound **9** readily decomposes in a solution of dichloromethane–water to yield both the dihydride **8** and $Ru_5C(CO)_{15}$ **1** (Scheme 3).

The nature of the product obtained from the direct reaction of **1** with the halogens $Cl_2$, $Br_2$, and $I_2$ depends on the halogen employed. These reactions are summarized in Scheme 4.[21] With $I_2$, an addition reaction takes place affording $Ru_5C(CO)_{15}I_2$ **10.** The structure of this compound remains uncertain. With $Cl_2$ and $Br_2$, cluster degradation takes place and the halide-bridged dimers $[(CO)_3RuX_2]_2$ (X = Cl or Br) are thought to be obtained.

Treatment of **1** with the halide ions $[Et_4N]X$ (X = F, Cl, Br, and I) in dichloromethane at room temperature affords the bridged butterfly anionic clusters $[Ru_5C(CO)_{15}X]^-$ (X = F **11**, Cl **12**, Br **13**, and I **14**).[21] These anions can be protonated with concentrated sulfuric acid to afford the neutral clusters $HRu_5C(CO)_{15}X$ (X = F **15**, Cl **16**, Br **17**, and I **18**). Deprotonation with the regeneration of the monoanion may be achieved using a proton sponge as shown in Scheme 4. Compounds **16** and **17** may also be prepared in a single step from **1** by reaction with HCl or HBr.

The hexaruthenium cluster $[Ru_6C(CO)_{16}]^{2-}$ **4** reacts with benzyl bromide at 60°C for 1 h to afford the monoanion $[Ru_6C(CO)_{16}(\mu\text{-}Br)]^-$ **19** in 45% yield (Scheme 5).[33] The structure of **19** has been established by single crystal X-ray diffraction, which shows that the bromine atom bridges an edge along the clusters molecular equator, causing the rupture of the underlying Ru–Ru bond. Further reaction of **19** with benzyl bromide under more forcing conditions, viz. at 130°C for 1 h, results in the formation of a pentaruthenium cluster $Ru_5C(CO)_{14}(\mu\text{-}Br)_2$ **20** in 17% yield. The metal-atom skeleton in **20** may be derived from a square pyramid by scission of two opposite base-to-apex edges; these edges are bridged by the two Br atoms.

## B. *Nitrogen and Phosphorus Donors*

Dissolving $Ru_5C(CO)_{15}$ **1** in acetonitrile results in an instant color change due to the formation of the addition product $Ru_5C(CO)_{15}(NCMe)$ **21,** which has a bridged butterfly ruthenium atom geometry.[34,21] Uptake of the aceto-nitrile ligand is completely reversible, and removal of the solvent under vacuum results in the regeneration of **1** (Scheme 6). Cluster **21** has also been reacted with two equivalents of trimethylamine *N*-oxide $(Me_3NO)$ in acetonitrile.[35] The $Me_3NO$ is used to oxidatively remove carbonyl ligands (as $CO_2$) in stoichiometric amounts. The compound obtained from this reaction has been formulated as $Ru_5C(CO)_{13}(NCMe)_2$ **22,** which would

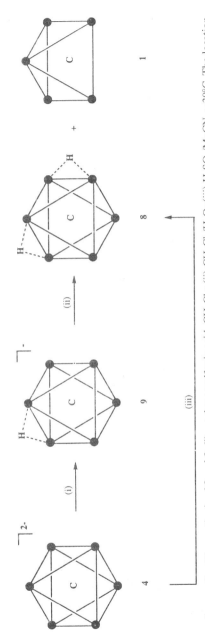

SCHEME 3. The synthesis of **8** and **9**. (i) *p*-toluenesulfonic acid, CH₂Cl₂; (ii) CH₂Cl₂/H₂O; (iii) H₂SO₄/MeCN, −30°C. The location of the hydride ligands in **8** are not known with certainty.

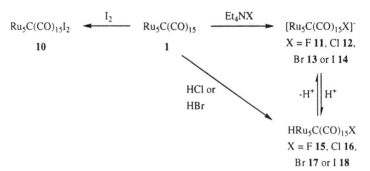

SCHEME 4.   The synthesis of $Ru_5C(CO)_{15}I_2$ **10**, $[Ru_5C(CO)_{15}X]^-$ (X = F **11**, Cl **12**, Br **13**, I **14**) and $HRu_5C(CO)_{15}X$ (X = F **15,** Cl **16,** Br **17,** I **18**).

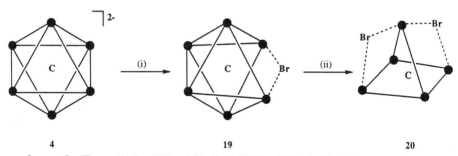

SCHEME 5.   The synthesis of **19** and **20.** (i) $PhCH_2Br$, 60°C, 1 h; (ii) $PhCH_2Br$, 130°C, 1 h.

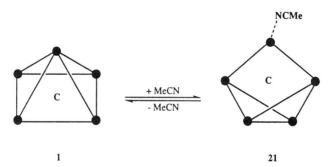

SCHEME 6.   Reversible uptake of MeCN by **1** to afford **21**.

have a square pyramidal structure. Given the stoichiometry of the reaction, the compound may well have an additional acetonitrile ligand yielding a cluster with bridged butterfly geometry. Because **22** has not been characterized, some doubt remains as to its precise nature. The hexanuclear cluster $Ru_6C(CO)_{17}$ **2** reacts with between one and three equivalents of $Me_3NO$ in the presence of acetonitrile, affording the clusters $Ru_6C(CO)_{16}(NCMe)$ **23**, $Ru_6C(CO)_{15}(NCMe)_2$ **24**, and $Ru_6C(CO)_{14}(NCMe)_3$ **25**.[36] These compounds are obtained as mixtures of products as the reactions do not proceed according to the precise stoichiometry of the $Me_3NO$ used. These "activated" derivatives (activated in the sense that the MeCN ligand is more labile than CO) are not very stable in solution or in the solid state and an *in situ* reaction is recommended.

A number of nitrosyl derivatives of $Ru_6C(CO)_{17}$ **2** have been prepared and the routes to these compounds are summarized in Scheme 7. The reaction of **2** with $[(Ph_3P)_2N][NO_2]$ in THF affords the anion $[Ru_6C(CO)_{15}(NO)]^-$ **26** in 85% yield.[37] Further reaction of **26** with $[NO]^+$ affords the neutral dinitrosyl derivative $Ru_6C(CO)_{14}(NO)_2$ **27**. Compound **26** also reacts with concentrated sulfuric acid in dichloromethane, affording $HRu_6C(CO)_{15}(NO)$ **28**.[37,38] The structures of **27** and **28** have been established by single crystal X-ray diffraction, and the NO ligands in **27** coordinate to the apical Ru atoms of the cluster whereas the NO ligand in **28** coordinates to a Ru atom on the cluster molecular equator.

The dianion $[Ru_6C(CO)_{16}]^{2-}$ **4** undergoes direct reaction with nitric oxide gas in dichloromethane, affording **26** in 85% yield.[39] Further reaction of **26** with NO gas results in the formation of the pentaruthenium cluster $Ru_5C(CO)_{14}(\mu\text{-}NO)(\mu\text{-}\eta^2\text{-}NO_2)$ **29** in 29% yield and contains both an NO group and an $NO_2$ ligand. Although the structure of **29** has not been determined in the solid state, a derivative in which one carbonyl has been replaced by a triphenylphosphine ligand, viz. $Ru_5C(CO)_{13}(\mu\text{-}NO)(\mu\text{-}\eta^2\text{-}NO_2)(PPh_3)$ **30**, has been analyzed by X-ray diffraction methods and two opposite basal-to-apex Ru–Ru edges have been cleaved and the NO and $NO_2$ ligands bridge these edges. The $NO_2$ ligand coordinates to the apical Ru atom via the N atom and to the basal Ru atom via an O atom as shown in Scheme 7. The simple, pentaruthenium nitrosyl complex $[Ru_5C(CO)_{13}(NO)]^-$ **31** is also known. It is obtained in 88% yield from the reaction of **1** with $[(Ph_3P)_2N][NO_2]$ in THF at room temperature.[40] Acidification of **31** with concentrated $H_2SO_4$ affords the neutral cluster $HRu_5C(CO)_{13}(NO)$ **32**.

The synthesis and structure of the triphenylphosphine species $Ru_5C(CO)_{14}(PPh_3)$ **33** were reported in the initial paper describing the high yield synthesis of **1**.[20] Direct reaction was noted to proceed under mild conditions (ca. at room temperature in dichloromethane) and the structure

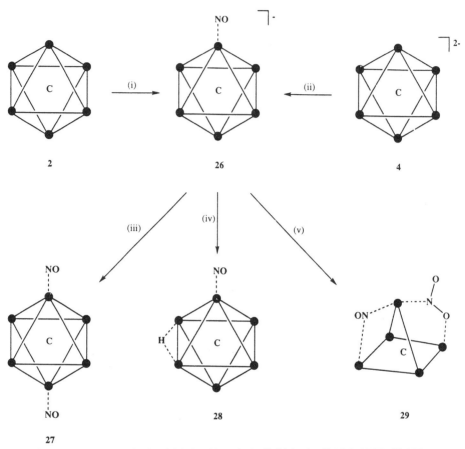

SCHEME 7.   The synthesis of **26, 27, 28,** and **29.** (i) [(Ph₃P)₂N][NO₂], THF; (ii) NO gas for 30 min, CH₂Cl₂, 4 h; (iii) [NO]⁺; (iv) conc. H₂SO₄, CH₂Cl₂; (v) NO gas for 30 min, CH₂Cl₂, 7 h.

of **33** is based on that of **1** in which one axial carbonyl group on a basal Ru atom has been replaced by the triphenylphosphine ligand. The kinetics of the reaction of Ru₅C(CO)₁₅ with 21 different phosphine and phosphite ligands has been studied.[41] It was found that two different mechanisms for substitution are in operation, depending on the cone angle of the ligand. With the smaller ligands ($\theta \leq 133°$) the reactions occur via two well-defined steps, initial adduct formation involving a bridged butterfly intermediate, followed by CO dissociation to form the product (Scheme 8). With larger ligands ($\theta \geq 136°$) the reaction involves a second order, one-step process with no spectral evidence for adduct formation and this mechanism is

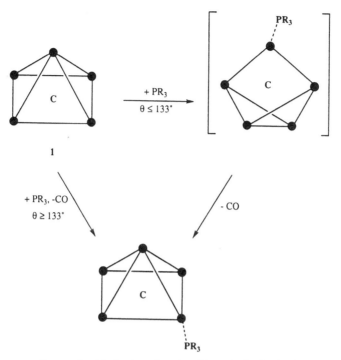

SCHEME 8.   Mechanisms for phosphine substitution in **1**.

comparable with phosphine substitution reactions that take place in many other cluster systems. It is also worth noting that substitution with the smaller nucleophiles is much more facile than comparable nucleophile-dependent reactions of other carbonyl clusters.

Cluster **1** may also form *bis*-phosphine derivatives[21] and its reactivity toward a series of chelating phosphines, $Ph_2P(CH_2)_nPPh_2$ (n = 1, dppm; 2, dppe; 3, dppp and 4, dppb), has been investigated.[42] Four different isomers have been identified with these chelating ligands and they are shown in Fig. 2. With $Ru_5C(CO)_{13}(dppm)$ **34**, $Ru_5C(CO)_{13}(dppe)$ **35,** and $Ru_5C(CO)_{13}(\mu\text{-dppp})$ **36** the predominant isomer in solution is type **A** in which the two phosphines coordinate to the same Ru atom in radial positions. This isomer tends to coexist in solution with lesser amounts of type **B**. The structure of **35** has been established by single crystal X-ray diffraction, and its structure is in fact that of type **B** with the P atoms replacing a radial CO and an axial CO from the same Ru atom.[29] In contrast, the phosphine ligand in $Ru_5C(CO)_{13}(\mu\text{-dppb})$ **37** spans the square cluster face

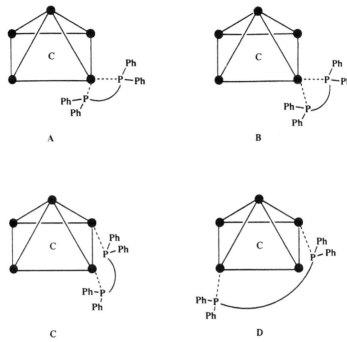

FIG. 2.  The four types of diphosphine complexes formed with $Ru_5C(CO)_{15}$ **1**: (**A**) equatorial–equatorial, (**B**) equatorial–axial, (**C**) $\mu$-axial–axial, and (**D**) trans-$\mu$-axial–axial.

with each P atom coordinating in axial positions to Ru atoms on opposite sides of the square base.

The reaction of **1** with diphenylphosphine, $PPh_2H$, in pentane at room temperature affords $Ru_5C(CO)_{14}(PPh_2H)$ **38** in 71% yield.[43] Heating **38** for an additional 2 h in hexane results in the transfer of the H atom from the phosphine to the cluster with the formation of $HRu_5C(CO)_{13}(\mu$-$PPh_2)$ **39** in 80% yield (Scheme 9). The X-ray structure of **39** reveals that the

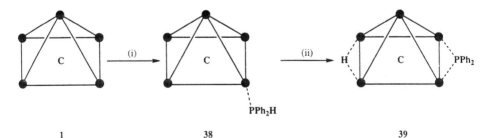

SCHEME 9.  The synthesis of **38** and **39**. (i) $PPh_2H$, pentane; (ii) hexane, $\Delta$, 2 h.

PPh2 group bridges a basal edge of the cluster. The related cluster $HRu_5C(CO)_{13}\{\mu\text{-}P(Ph)CH_2CH_2Si(OEt)_3\}$ **40** may be prepared in a similar two-step method to that of **39**. This phosphine was employed as the substituent group could potentially be used to attach the cluster to a silica or an alumina support.

The hexanuclear cluster **2** reacts with phosphines such as $PPh_3$ or $PPh_2Et$ in dichloromethane at room temperature to afford the mono- and *bis*-substituted derivatives $Ru_6C(CO)_{16}L$ (L = $PPh_3$ **41** and $PPh_2Et$ **42**) and $Ru_6C(CO)_{15}L_2$ (L = $PPh_3$ **43** and $PPh_2Et$ **44**).[44,45] The solid-state structure of $Ru_6C(CO)_{16}(PPh_2Et)$ **42** reveals that the phosphine ligand coordinates to an apical atom of the octahedral ruthenium skeleton. In solution, NMR studies indicate that **44** exists in two isomeric forms and these have been assigned to structures derived from **42** in which the second phosphine is located on a Ru atom *cis*- and to the original apically bound phosphine ligand in one isomer and *trans*- in the other isomer. This is illustrated in Fig. 3 for the $P(OMe)_3$ analogues. Mono- and *bis*-substituted derivatives of **2** have also been made with the phosphines $PPh_2CH_2CH_2Si(OEt)_x(OSiMe_3)_{3-x}$.[45] The location of these phosphines matches that observed for the simple phosphines adducts. Trimethylphosphite can replace up to four carbonyls in **2** by direct reaction in dichloromethane depending on the approximate stoichiometry of the ligand used. The reactions do not proceed according to the precise stoichiometry of the $P(OMe)_3$ employed, and mixtures of products are obtained with yields ranging from 32 to 53%.[44,45] Compounds $Ru_6C(CO)_{16}\{P(OMe)_3\}$ **45** and $Ru_6C(CO)_{15}\{P(OMe)_3\}_2$ **46** adopt structures analogous to clusters **41–44**. The *tris*- and *tetrakis*-substituted clusters $Ru_6C(CO)_{14}\{P(OMe)_3\}_3$ **47** and $Ru_6C(CO)_{13}\{P(OMe)_3\}_4$ **48** do not appear to exist as isomers in solution; in **47** one phosphite ligand coordinates to each apex of the cluster and one equatorial Ru atom and in **48** the additional $P(OMe)_3$ ligand bonds to a fourth Ru atom on the octahedral equator adjacent to the existing one as shown in Fig. 4. The mono-substituted clusters **41** and **45** have also been prepared by an oxidative addition reaction in dichloromethane from $[Ru_6C(CO)_{16}]^{2-}$ **4** using the ferrocenium cation as the oxidant.[29]

The chelating phosphines $Ph_2P(CH_2)_nPPh_2$ (n = 1, dppm; 2, dppe; 3, dppp and 4, dppb) react with **2** in hexane to afford products of formula $Ru_6C(CO)_{15}\{Ph_2P(CH_2)_nPPh_2\}$ in 74–94% yield.[46] Compounds $Ru_6C(CO)_{15}(\mu\text{-dppm})$ **49**, $Ru_6C(CO)_{15}(\mu\text{-dppe})$ **50**, and $Ru_6C(CO)_{15}(\mu\text{-dppp})$ **51** have similar structures with the ligand bridging an apex-to-equator edge of the cluster skeleton. With dppb, two isomers, $Ru_6C(CO)_{15}(dppb)$ **52** and $Ru_6C(CO)_{15}(\mu\text{-dppb})$ **53** are obtained. The minor isomer **53** is believed to have a structure analogous to that observed in clusters **49–51**, whereas

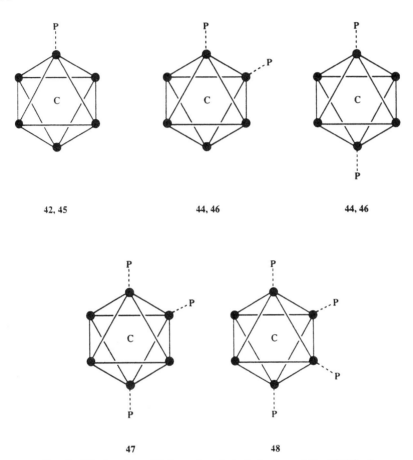

FIG. 3.   The location of P donor ligands in derivatives of $Ru_6C(CO)_{17}$ **2.**

in the major component the phosphine is thought to coordinate to the same Ru atom. The chelating phosphine 1,1′-bis(diphenylphosphino)-ferrocene (dppf) reacts with **2** to afford $Ru_6C(CO)_{16}(dppf)$ **54** and $Ru_6C(CO)_{15}(\mu\text{-}dppf)$ **55.**[47] Cluster **54** is the minor product, and the dppf ligand interacts with the cluster core via only one of the P atoms. The structure of **55** is related to that of **49–51** and **53,** although the cluster skeleton is considerably distorted. Two of the Ru–Ru edges have opened to distances beyond that usually considered to constitute a Ru–Ru bond, and experiments suggest that there is a strong interaction between the ferrocene and the cluster units.

## C. *Oxygen, Sulfur, and Selenium Donors*

The reaction of $Ru_5C(CO)_{15}$ **1** with two equivalents of oxalic acid $(H_2C_2O_4)$ in refluxing dichloromethane for 20 h affords two products, one of which has been characterized by single crystal X-ray diffraction as $\{HRu_5C(CO)_{14}\}_2(C_2O_4)$ **56.**[48] The two $Ru_5C$ cluster units are bridged by a tetradentate $C_2O_4$ group as shown in Fig. 4. The cluster skeletons adopt distorted bridged butterfly geometries in which the bridging Ru atom is chelated by two O atoms of the oxalato group.

Cluster **1** reacts rapidly with $H_2S$, $H_2Se$, and HSR (R = Me or Et) at room temperature to afford $HRu_5C(CO)_{14}(\mu\text{-SH})$ **57**, $HRu_5C(CO)_{14}(\mu\text{-SeH})$ 58, and $HRu_5C(CO)_{14}(\mu\text{-SR})$ (R = Me **59** and Et **60**).[49] Compounds **57–60** are structurally similar, and that of **60** has been established in the solid state by single crystal X-ray diffraction. The Ru atom skeleton adopts a bridged butterfly structure with the SEt group coordinating to a hinge Ru atom and the bridging Ru atom. The hydride ligand bridges the Ru–Ru hinge bond (Scheme 10). Heating **60** in cyclohexane for 30 min results in the expulsion of one CO ligand affording $HRu_5C(CO)_{13}(\mu\text{-SEt})$ **61.** The loss of the two electron donor ligand results in the closure of the cluster core in **61** reforming the square pyramid geometry. Cluster **61** undergoes an addition reaction with triphenylphosphine to afford the bridged butterfly cluster $HRu_5C(CO)_{13}(\mu\text{-SEt})(PPh_3)$ **62.** Heating **62** in cyclohexane for 8 h causes the loss of a carbonyl ligand and the formation of $HRu_5C(CO)_{12}(\mu\text{-SEt})(PPh_3)$ **63.** The X-ray structure of **63** reveals that the square pyramidal

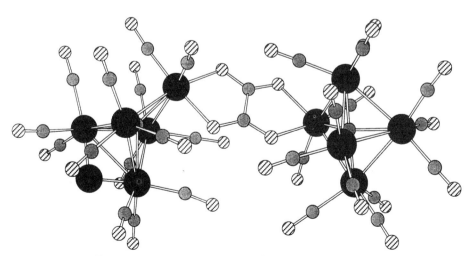

Fig. 4.   The solid-state structure of $\{HRu_5C(CO)_{14}\}_2(C_2O_4)$ **56.**

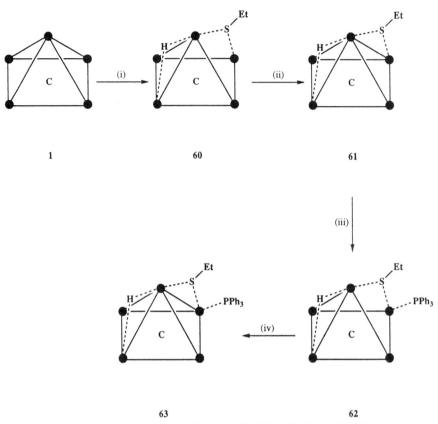

SCHEME 10.   The synthesis of **60, 61, 62**, and **63**. (i) HSEt, CH$_2$Cl$_2$, 5 min; (ii) cyclohexane, $\Delta$, 30 min; (iii) PPh$_3$, hexane, 6 h; (iv) cyclohexane, $\Delta$, 8 h.

Ru$_5$ core has reformed. The SEt group bridges a basal edge of the Ru$_5$C polyhedron and the triphenylphosphine ligand coordinates to one of the Ru atoms that interacts with the SEt group.

The reaction of Ru$_6$C(CO)$_{17}$ **2** with HSEt results in the formation of HRu$_6$C(CO)$_{15}$($\mu$-SEt)$_3$ **64**. This cluster has a very open metal polyhedron and is shown in Scheme 11.[44] The Ru atoms adopt a bridged butterfly geometry with one of the edges connecting the bridging atom with a wing-tip atom being bridged by the sixth Ru atom. The hydride ligand bridges the Ru–Ru hinge, and the three thiolato bridges coordinate along the edges of the triangle formed by the sixth Ru atom with the cluster bridge and wing-tip atoms.

The hexaruthenium dianion **4** reacts rapidly with phenylselenyl chloride (PhSeCl) in dichloromethane at room temperature to afford

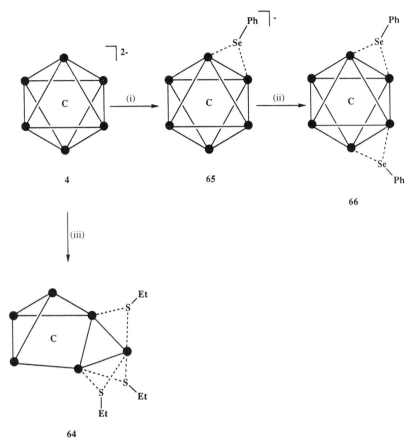

SCHEME 11.   The synthesis of **64, 65,** and **66.** (i) PhSeCl, CH$_2$Cl$_2$; (ii) *bis*(2-methoxyethyl)-ether, reflux, 1 h; (iii) HSEt.

[Ru$_6$C(CO)$_{15}$($\mu$-SePh)]$^-$ **65** in 65% yield (Scheme 11).[50] Heating **65** in *bis*(2-methoxyethyl)ether for 1 h produces the new cluster Ru$_6$C(CO)$_{14}$($\mu$-SePh)$_2$ **66** in low yield. Clearly, the expected reaction in which the SePh group is converted from a three electron donor to a five electron donor did not take place. The structure of **66** has been established by X-ray diffraction and reveals that the two SePh ligands bridge apex-to-equatorial edges as shown in the Scheme.

The macrocycle 1,5,9-trithiacyclododecane (12S3) reacts with Ru$_5$C(CO)$_{15}$ **1** in hexane under reflux to afford Ru$_5$C(CO)$_{13}$($\mu$-$\eta^1$-12S3) **67** in 86% yield.[51] Under more forcing conditions, viz. in refluxing octane,

$Ru_5C(CO)_{11}(\mu\text{-}\eta^3\text{-}12S3)$ **68** is produced in 75% yield (Scheme 12). The structures of clusters **67** and **68** have been determined by single crystal X-ray diffraction, and the 12S3 ligand in **67** bonds across a basal edge via one S atom, whereas in **68** a S atom again bridges a basal edge but the other two S atoms also coordinate to one of the Ru atoms involved with the bridging S atom. Compound **67** may be converted to **68** in high yield on further heating in octane.

The thermolysis of **2** with 12S3 in octane affords $Ru_6C(CO)_{13}(\mu\text{-}\eta^3\text{-}$12S3) **69** in 78% yield (Scheme 13).[52] Cluster **2** has also been reacted with 1,5,9,13-tetrathiacyclohexadecane (16S4) under similar conditions to afford $Ru_6C(CO)_{15}(\mu\text{-}\eta^2\text{-}16S4)$ **70** in 58% yield. The structures of **69** and **70** have been established in the solid state, the interaction of the ligand in **69** is closely related to that observed in **68,** and the 16S4 ligand in **70** coordinates to two consecutive Ru atoms of the molecular equator via two of the S atoms of the macrocycle. The reaction of **2** with 1,4,7-trithiacyclononane (9S3) in hexane under reflux affords $Ru_6C(CO)_{14}(\eta^3\text{-}9S3)$ **71** in 93% yield. Heating **71** in octane at reflux for 18 h affords the new cluster $Ru_6C(CO)_{14}(\mu_3\text{-}\eta^3\text{-}SCH_2CH_2SCH_2CH_2S)$ **72** in 68% yield. Cluster **72** may also be isolated in a single step from the reaction of **2** with 9S3 in octane in 36% yield. The solid-state structure of **71** reveals that all three S atoms of the 9S3 ligand coordinate to a single Ru atom on the molecular equator of the cluster. The cluster core in **72** has undergone a polyhedral rearrangement to a spiked square pyramid geometry (the Ru spike bonding to a basal Ru atom). The macrocycle has also cleaved and all three S atoms coordinate to the Ru spike as well as to three Ru atoms of a triangular face of the cluster as shown in Scheme 13.

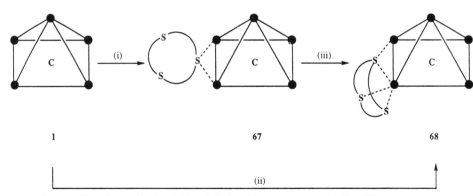

SCHEME 12. The synthesis of **67** and **68**. (i) 12S3, hexane, $\Delta$, 5 h; (ii) 12S3, octane, $\Delta$, 4 h; (iii) octane, $\Delta$, 3 h.

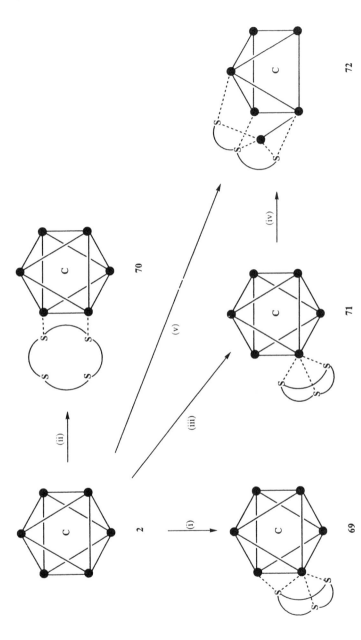

SCHEME 13. The synthesis of **69**, **70**, **71**, and **72**. (i) 12S3, octane, Δ, 4 h; (ii) 16S4, octane, Δ, 2 h; (iii) 9S3, hexane, Δ, 4 h; (iv) toluene, Δ, 18 h; (v) 9S3, octane, Δ, 2 h.

## D. *Alkyls*

Reaction of the dianion $[Ru_6C(CO)_{16}]^{2-}$ **4** with methyl iodide at 120°C affords the anionic cluster $[Ru_6C(CO)_{16}(Me)]^-$ **73** in 40% yield.[53] Cluster **73** undergoes a carbonyl insertion reaction under 50 atmospheres of CO to afford $[Ru_6C(CO)_{16}(COMe)]^-$ **74** in 56% yield (Scheme 14). The structures of **73** and **74** have been established by single crystal X-ray diffraction, which shows that the methyl group in **73** bonds to an apical Ru atom and does not change location on carbonylation.

## E. *Alkynes*

Oxidation of $[Ru_6C(CO)_{16}]^{2-}$ **4** with $FeCl_3$ or $[Fe(\eta\text{-}C_5H_5)_2]^+$ in the presence of a range of alkynes results in the formation of the neutral clusters $Ru_6C(CO)_{15}(\mu_3\text{-}C_2RR')$ (R = R' = H **75,** Me **76,** Et **77,** Ph **78,** R ≠ R' = H and Ph **79,** H and Et **80,** Me and Et **81,** Et and Ph **82**) with yields in excess of 80%.[29,54] The alkyne ligand in all these compounds bonds over a triruthenium face via one $\pi$ and two $\sigma$ bonds. The thermolysis of **79** in toluene for 3 days results in the formation of the new cluster $HRu_6C(CO)_{15}(CCPh)$ **83** in 82% yield. The precise nature of this compound is unknown.

Compound **78** may also be isolated in 46% yield from the photolysis of $Ru_6C(CO)_{17}$ **2** with diphenylacetylene in dichloromethane.[55] A second product, $Ru_6C(CO)_{13}(\mu_3\text{-}C_2Ph_2)_2$ **84,** is also obtained from this reaction and can be isolated in larger quantities relative to **78** as the time of the photolysis is extended (Scheme 15).

The reaction of **2** with alkynes using chemical activation methods is less straightforward than the methods outlined earlier (see Scheme 15). Cluster

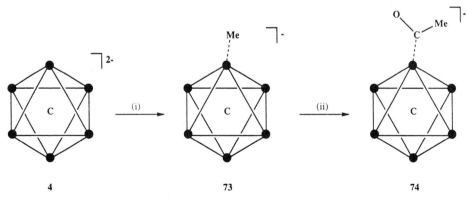

SCHEME 14.   The synthesis of **73** and **74.** (i) MeI, 120°C, 1 h; (ii) CO (50 atm), THF, 2 h.

**2** reacts with but-2-yne and two equivalents of $Me_3NO$ in dichloromethane to afford $Ru_6C(CO)_{15}(\mu_3\text{-}C_2Me_2)$ **76** in 37% yield together with the more highly substituted cluster $Ru_6C(CO)_{14}(\mu\text{-}C_2Me_2)(\mu_3\text{-}C_2Me_2)$ **85** in 12% yield.[56] The structure of **85** has been established by single crystal X-ray diffraction, which reveals that the cluster skeleton has undergone a transformation to a capped square pyramidal geometry. One alkyne ligand caps the triangular face formed by two basal Ru atoms and the Ru cap and the other alkyne bridges the Ru–Ru bond formed between the cluster apex and the capping Ru atom. From the X-ray structure it would appear that the edge bridging alkyne acts as a two electron donor and as such the cluster has 86 valence electrons, the required number for a cluster with a capped square pyramidal polyhedron. This is somewhat unusual in that alkynes in this bonding mode tend to provide four electrons to the underlying metal atoms. Heating **85** in heptane for 1 h results in the formation of $Ru_6C(CO)_{13}(\mu_3\text{-}C_2Me_2)_2$ **86,** which is isostructural to **84.** The same compound is obtained when **85** is treated with $Me_3NO$ in a noncoordinating solvent. The reaction of **85** with a further two equivalents of $Me_3NO$ in the presence of but-2-yne affords the tris-alkyne cluster $Ru_6C(CO)_{12}(\mu_3\text{-}C_2Me_2)_3$ **87** in 39% yield. The octahedral topology has been restored to **87** and the three alkyne ligands each bond over alternate trimetal faces. If each alkyne ligand formally acts as a four electron donor then the total electron count for this cluster would be 88, which is two more than that expected for an octahedron. Cluster **76** reacts with diphenylacetylene and $Me_3NO$ to afford $Ru_6C(CO)_{14}(\mu\text{-}C_2Ph_2)(\mu_3\text{-}C_2Me_2)$ **88** in 33% yield. This cluster is closely related to the capped square pyramidal cluster **85** with the $\mu\text{-}C_2Ph_2$ ligand bridging the apex-to-cap edge.

The main product obtained from the reaction of **2** with $Me_3NO$ and phenylacetylene in dichloromethane is **79,** although the yield is low compared with the oxidative addition route.[57,58] However, three other by-products are also isolated from this reaction, viz. $Ru_6C(CO)_{14}\{\mu_3\text{-}C(Ph)CHC(Ph)CH\}$ **89,** $Ru_6C(CO)_{14}\{\mu_3\text{-}C(Ph)CHCHC(Ph)\}$ **90,** and $Ru_6C(CO)_{13}(\eta\text{-}C_5H_3Ph_2\text{-}1,3)(\mu_3\text{-}CPh)$ **91** (Scheme 16). Clusters **89** and **90** are closely related and the solid-state structure of **90** reveals the presence of a face-bridging 1,3-diene ligand. This ligand is derived from the insertion of an alkyne into one already coordinated (as in cluster **79**). This has been shown to be the case, as the reaction of **79** with two equivalents of $Me_3NO$ in the presence of excess phenylacetylene affords **89** and **90** in about 10% yield. Reaction of **89** in a similar manner with HCCPh and $Me_3NO$ in dichloromethane affords **91** in 20% yield. This is quite an unusual reaction as one might envisage the formation of a $C_6$ ring, whereas the scission of a $C\equiv C$ triple bond takes place forming a 1,3-diphenylcyclopentadienyl ring and an alkylidyne ligand as shown in Scheme 16.

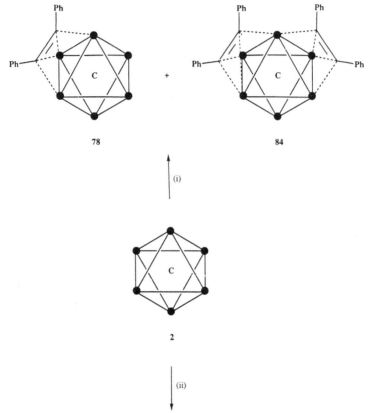

SCHEME 15.    The synthesis of **76, 78, 84, 85, 86,** and **87.** (i) $C_2Ph_2$, $CH_2Cl_2$, $h\nu$, 25°C, 14 h; (ii) $C_2Me_2$, $CH_2Cl_2$, 2 eq. $Me_3NO$, −78°C to RT, 35 min; (iii) heptane, Δ, 1 h or 1 eq. $Me_3NO$, $CH_2Cl_2$, RT; (iv) $C_2Me_2$, $CH_2Cl_2$, 2 eq. $Me_3NO$, −78°C to RT, 35 min.

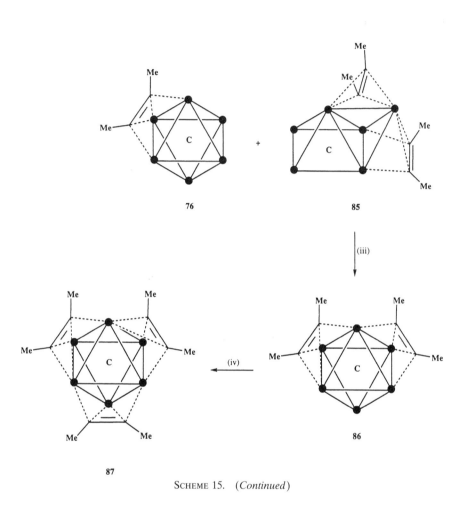

76 + 85

(iii)

86

(iv)

87

SCHEME 15. (*Continued*)

## F. *Allyls*

The reaction of $[Ru_6C(CO)_{16}]^{2-}$ **4** with allyl bromide at 85°C affords $[Ru_6C(CO)_{15}(\mu\text{-}C_3H_5)]^-$ **92** in 37% yield (Scheme 17).[53] The solid-state structure of **92** shows that the allyl ligand bridges an edge of the cluster

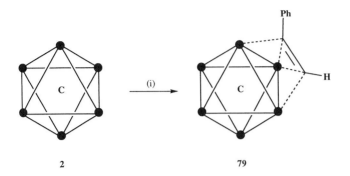

2

(i)

79

(ii)

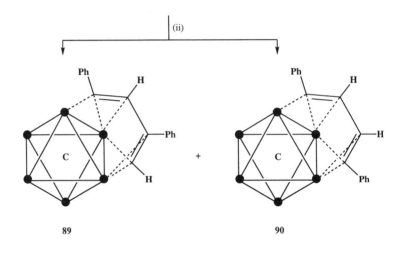

89

+

90

(iii)

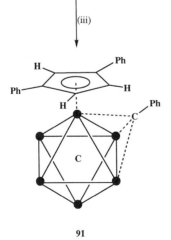

91

molecular equator. The analogous 2-methylallyl cluster $[Ru_6C(CO)_{15}(\mu\text{-}C_3H_4Me)]^-$ **93** can be prepared in similar yield using 2-methylallyl bromide.

Further reaction of **92** with allyl bromide under more forcing conditions, viz. heating at 115°C for 1 h, affords the bis-allyl compound $Ru_5C(CO)_{11}(\mu\text{-}Br)_2(\mu\text{-}C_3H_5)(\eta\text{-}C_3H_5)$ **94** in 37% yield.[33] The structure of **94** has been established by single crystal X-ray diffraction. The pentaruthenium core is derived from a square pyramid in which two adjacent basal-to-apex edges have been cleaved. These edges are bridged by Br atoms, and the two basal Ru atoms involved with these bridging atoms are bridged by the allyl ligand. The second allyl group coordinates to the Ru atom at the cluster apex. Cluster **92** also reacts with NO gas in dichloromethane to afford the neutral compounds $Ru_6C(CO)_{14}(NO)(\mu\text{-}C_3H_5)$ **95** and $Ru_5C(CO)_{11}(NO)(\mu\text{-}NO)(\mu\text{-}\eta^2\text{-}NO)_2(\eta\text{-}C_3H_5)$ **96** in 30 and 11% yield, respectively.[59,39] The NO ligand in **95** replaces a carbonyl ligand on one of the Ru atoms, which bonds to the allyl ligand. The solid-state structure of **96** reveals that the five Ru atoms are derived from a square pyramidal geometry in which two opposite apex-to-base edges have been removed. These are bridged by the NO and $NO_2$ ligands with the terminal NO ligand coordinating to the basal Ru atom, which interacts with the $NO_2$ bridge. The allyl ligand also coordinates to this Ru atom.

The reaction of **95** with further NO gas does not produce **96,** instead the new cluster $Ru_5C(CO)_{13}(\mu\text{-}\eta^2\text{-}NO_2)(\eta\text{-}C_3H_5)$ **97** is obtained in 37% yield.[39] The structure of **97** has been established in the solid state. The cluster skeleton adopts a bridged butterfly geometry with the $NO_2$ ligand connecting the bridging Ru atom to a hinge Ru atom. The allyl ligand coordinates to the bridging Ru atom. Analogues of compounds **92** and **95,** viz. $[Ru_6C(CO)_{15}(\mu\text{-}C_3H_4CO_2Me)]^-$ **98** and $Ru_6C(CO)_{14}(NO)(\mu\text{-}C_3H_4CO_2Me)$ **99,** have been prepared in 73 and 20% yield, respectively, using methyl-2-(bromomethyl)acrylate in place of allyl bromide. The substituted allyl ligand, $C_3H_4CO_2Me$, has a $CO_2Me$ group attached at the 2 position of the $C_3$-allyl backbone.[60] Another related cluster carrying an allyl ligand is $Ru_6C(CO)_{14}(\mu\text{-}SePh)(\eta\text{-}C_3H_5)$ **100**, which is isolated from the reaction of $[Ru_6C(CO)_{15}(\mu\text{-}SePh)]^-$ **65** with allyl bromide in 32% yield.[50]

## G. Cyclobutadienes

The dianions $[Ru_5C(CO)_{14}]^{2-}$ **5** and $[Ru_6C(CO)_{16}]^{2-}$ **4** undergo reaction with the dicationic complex $[Pd(\eta\text{-}C_4Ph_4)(OCMe_2)_2]^{2+}$ in refluxing dichlo-

---

SCHEME 16.   The synthesis of **79, 89, 90,** and **91.** (i) and (ii) $C_2HPh$, $CH_2Cl_2$, 2 eq. $Me_3NO$, −78°C to RT, 30 min; (iii) $C_2HPh$, $CH_2Cl_2$, 1 eq. $Me_3NO$, −78°C to RT, 30 min.

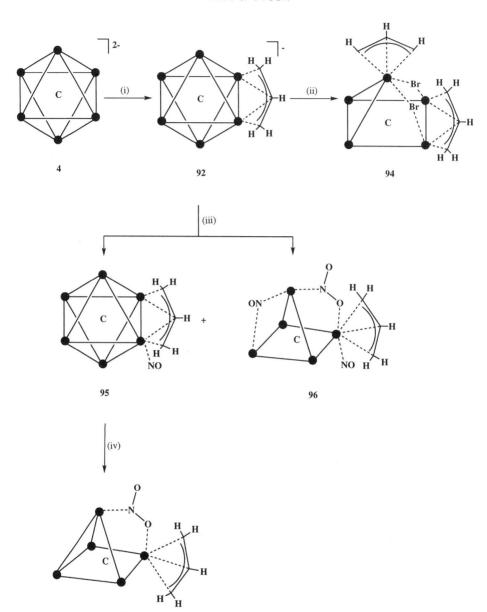

SCHEME 17.   The synthesis of **92, 94, 95, 96,** and **97.** (i) $C_3H_5Br$, $CH_2Cl_2$, 85°C, 1 h; (ii) $C_3H_5Br$, $CH_2Cl_2$, 115°C, 1 h; (iii) NO gas, $CH_2Cl_2$, 2 h; (iv) NO gas (2.5 ml), $CH_2Cl_2$, 0°C, 30 min.

romethane to afford the neutral clusters $Ru_5C(CO)_{13}(\eta\text{-}C_4Ph_4)$ **101** and $Ru_6C(CO)_{15}(\eta\text{-}C_4Ph_4)$ **102** in high yield (Scheme 18).[61,62] In these reactions the tetraphenylcyclobutadiene ring migrates from the palladium complex to the ruthenium cluster. With the square pyramidal cluster an intermediate is observed by IR spectroscopy, which is believed to consist of a hetero-nuclear cluster in which the palladium caps the $Ru_5C$ skeleton prior to transfer of the $C_4Ph_4$ ring.

The solid-state structure of **102** reveals that the $C_4Ph_4$ ring is coordinated to a Ru atom on the apex of the cluster octahedron, and one of the Ru–Ru edges between this Ru atom and a Ru atom on the cluster molecular equator is longer than that usually considered to constitute a Ru–Ru bond,

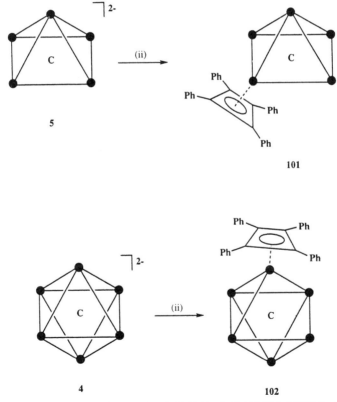

SCHEME 18.    The synthesis of **101** and **102**. (i) $[Pd(\eta\text{-}C_4Ph_4)(OCMe_2)_2][BF_4]_2$, $CH_2Cl_2$, $\Delta$, 30 min, then RT for 24 h; (ii) $[Pd(\eta\text{-}C_4Ph_4)(OCMe_2)_2][BF_4]_2$; $CH_2Cl_2$, $\Delta$, 30 min.

viz. 3.213 Å. This feature has been investigated by extended Hückel molecular orbital theory, which shows that replacing two COs with the bulky $C_4Ph_4$ ligand modifies the carbonyl shell and consequently weakens the Ru–Ru bond. The $C_4Ph_4$ ring in **101** coordinates to a basal Ru atom of the square pyramidal metal skeleton.

### H. *Cyclopentadienyls and Fulvenes*

The reaction of $Ru_5C(CO)_{15}$ **1** with an excess of $NaC_5H_5$ in THF affords an anionic cluster formulated at $[Ru_5C(CO)_{13}(C_5H_5)]^-$. Protonation of this species with $HBF_4$ affords $HRu_5C(CO)_{13}(\eta\text{-}C_5H_5)$ **103** in high yield (Scheme 19).[63] The structure of **103** has been established by single crystal X-ray diffraction,[64] which reveals that the five Ru atoms adopt a bridged butterfly geometry with the cyclopentadienyl ligand coordinated to the bridging Ru atom. The hydride ligand spans the Ru–Ru hinge. Cluster **1** also reacts with cyclopentadiene ($C_5H_6$) and $Me_3NO$ in dichloromethane to afford the *bis*-cyclopentadienyl cluster $Ru_5C(CO)_{10}(\eta\text{-}C_5H_5)_2$ **104** in 14% yield.[65] The solid-state structure of **104** shows that the two cyclopentadienyl groups coordinate to basal Ru atoms at opposite corners of the square base.

The oxidative addition reaction of $[Ru_6C(CO)_{16}]^{2-}$ **4** and pentamethylcyclopentadiene in the presence of $[Fe(\eta\text{-}C_5H_5)]^+$ in dichloromethane at room temperature affords $HRu_6C(CO)_{14}(\eta\text{-}C_5Me_5)$ **105** in 10% yield together with larger quantities of the homoleptic cluster $Ru_6C(CO)_{17}$ **2**.[29]

Cluster **2** reacts with three equivalents of $Me_3NO$ and triphenylphospho-

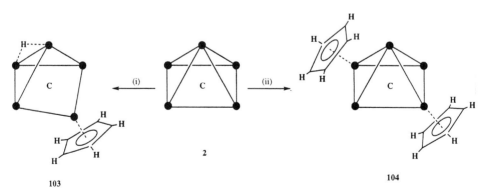

SCHEME 19.   The synthesis of **103** and **104**. (i) $NaC_5H_5$, THF, $\Delta$ followed by addition of $HBF_4$ in ether; (ii) $C_5H_6$, $CH_2Cl_2$, 2 eq. $Me_3NO$, $-78°C$ to RT, 30 min.

niocyclopentadienide ($C_5H_4PPh_3$) in dichloromethane to afford the zwit-terionic cluster $Ru_6C(CO)_{14}(\eta\text{-}C_5H_4PPh_3)$ **106** in 64% yield (Scheme 20).[66] The structure of **106** shows that the $C_5$ ring coordinates to an apical Ru atom. Cluster **2** also reacts with cyclopentadiene and two equivalents of $Me_3NO$ to afford $Ru_6C(CO)_{12}(\eta\text{-}C_5H_5)_2$ **107** in 17% yield together with some by-products, including **104** and $Ru_6C(CO)_{14}(\mu\text{-}C_5H_4CH_2)$ **108**.[65] The structure of **107** has been determined by single crystal X-ray diffraction. One cyclopentadienyl ligand coordinates to an apical Ru atom and the other to a Ru atom on the cluster molecular equator. Compound **108** contains an edge bridging $C_5H_4CH_2$ fulvene ligand.[67] The fulvene bridges an apex-to-equator edge of the cluster with the ring forming a typical $\eta^5$ interaction with an apical Ru atom and the $CH_2$ portion of the ligand bonding to an adjacent Ru atom. It is not certain how the fulvene ligand was formed in this reaction, although it is suggested that the $-CH_2$ unit is derived from the reduction of a coordinated CO ligand by the $C_5H_6$ moiety.

Other fulvene derivatives of **2** have been prepared from more ap-propriate precursors. For example, the reaction of **2** with two equivalents of $Me_3NO$ in the presence of diphenylfulvene in dichloromethane affords $Ru_6C(CO)_{15}(\mu\text{-}C_5H_4CPh_2)$ **109** in 42% yield together with smaller amounts of $Ru_6C(CO)_{14}(\mu_3\text{-}C_5H_4CPh_2)$ **110** (Scheme 21, page 76).[67] The structures of **109** and **110** have been established by single crystal X-ray diffraction. The $C_5H_4CPh_2$ ligand in **109** bonds as a 1,3-diene across a Ru–Ru edge, whereas in **110** the five C atoms of the ring and the exocyclic C atom bond over a triruthenium face. If three equivalents of $Me_3NO$ are used in this reaction (rather than just two equivalents), then cluster **110** is isolated in 71% yield. In a related reaction employing dimethylfulvene, an analogue of **110**, viz. $Ru_6C(CO)_{14}(\mu_3\text{-}C_5H_4CMe_2)$ **111,** has been prepared in 24% yield. Compound **111** undergoes near quantitative conversion to $Ru_6C(CO)_{13}(\mu\text{-}C_5H_4C\{CH_2\}_2)$ **112** on standing in chloroform for about 1 week. The solid-state structure of **112** shows that the fulvene ligand coordi-nates to the cluster via the $C_5$ ring and the $C\{CH_2\}_2$ part of the ligand forms an allylic-type interaction as the methyl groups in the precursor have lost one H atom each. Clearly, this latter interaction is formed in order to compensate for the loss of the CO ligand from the ligand shell. Compound **111** reacts with further dimethylfulvene together with $Me_3NO$ and water in a solution of dimethylformamide and dichloromethane to afford $Ru_6C(CO)_{12}(\eta\text{-}C_5H_4CMe_2H)(\eta\text{-}C_5H_4CMe_2OH)$ **113** in 30% yield. The solid-state structure of **113** shows that two $C_5$ rings coordinate to the cluster in analogous positions to those found in **107**. In addition, water has inserted into the fulvene ligands to generate the substituted cyclopentadienyl li-gands.

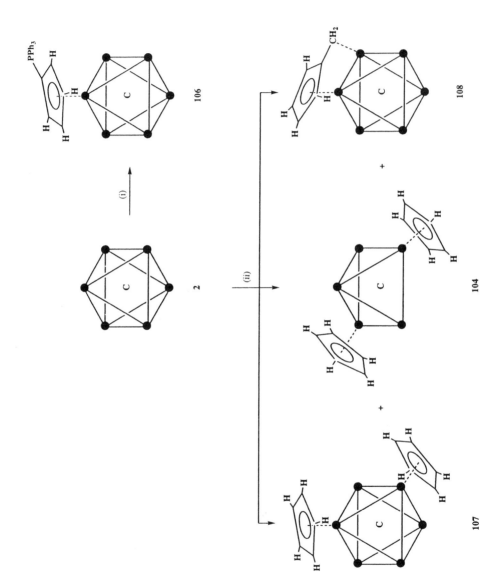

I. *Arenes*

Section II,A has already described how **2** was first isolated from the thermolysis of $Ru_3(CO)_{12}$ **3** in an arene solvent; a reaction that also produced the related arene-coordinated cluster $Ru_6C(CO)_{14}(\eta\text{-arene})$.[9,10] A number of arene derivatives of **1** and **2** can also be made by alternative routes. For example, the ionic coupling reactions between the pentanuclear dianion $[Ru_5C(CO)_{14}]^{2-}$ **5** and the dicationic ruthenium–arene fragments $[Ru(\eta\text{-arene})(NCMe)_3]^{2+}$ (arene $=$ $C_6H_6$, $C_6H_5Me$, $C_6H_4Me_2\text{-}1,3$ and $C_6H_3Me_3\text{-}1,3,5$) in dichloromethane under reflux affords the hexaruthenium clusters $Ru_6C(CO)_{14}(\eta\text{-arene})$ (arene $=$ $C_6H_6$ **114**, $C_6H_5Me$ **115**, $C_6H_4Me_2\text{-}$1,3 **116**, and $C_6H_3Me_3\text{-}1,3,5$ **117**) in yields of ca. 80%.[68,69] The arene ligands in **114–117** are coordinated to an apical Ru atom (see Scheme 22). Reduction of the benzene derivative **114** with sodium carbonate in methanol affords the dianion $[Ru_6C(CO)_{13}(\eta\text{-}C_6H_6)]^{2-}$ **118**, which reacts further with $[Ru(\eta\text{-}C_6H_6)(NCMe)_3]^{2+}$ in dichloromethane to afford the bis-benzene cluster $Ru_6C(CO)_{11}(\eta\text{-}C_6H_6)(\mu_3\text{-}C_6H_6)$ **119**.[68] The solid-state structure of **119** shows that one benzene coordinates to an apical Ru atom and the other coordinates over a trimetal face of the octahedron. This reaction is related to the redox-mediated ligand exchange reaction that takes place with the palladium dication $[Pd(\eta\text{-}C_4Ph_4)(OCMe_2)_2]^{2+}$ (see Section IV,G) as the nuclearity of the cluster remains unchanged. In general, ionic reactions of this nature usually result in an increase in the nuclearity due to condensation of the cluster with the metal fragment.

Chemical activation using $Me_3NO$ may also be used to prepare the bis-benzene cluster **119** as well as benzene derivatives of **1**. The synthesis of benzene derivatives of **1** using this method will first be described, followed by those of **2**. A solution of **1** in dichloromethane containing an excess of cyclohexa-1,3-diene reacts with two equivalents of $Me_3NO$ to afford $Ru_5C(CO)_{13}(\mu\text{-}C_6H_8)$ **120** in 35% yield.[70,71] The structure of **120** has been established by single crystal X-ray diffraction, which reveals that the diene ligand has replaced two radial carbonyls on two consecutive Ru atoms of the cluster square base (Scheme 23).[71] Under similar conditions, **1** reacts with cyclohexa-1,4-diene to afford a related product, viz. $Ru_5C(CO)_{13}(\mu\text{-}C_6H_8\text{-}1,4)$ **121,** in which the diene has retained the 1,4 stereochemistry of the unsaturated bonds.[72,73] A second cyclohexa-1,4-diene ligand may be introduced into the ligand shell of the cluster by further reaction with $Me_3NO$ in the presence of the ligand. An X-ray diffraction analysis of the

SCHEME 20.    The synthesis of **104, 106, 107,** and **108.** (i) $C_5H_4PPh_3$, 3 eq. $Me_3NO$; (ii) $C_5H_6$, $CH_2Cl_2$, 2 eq. $Me_3NO$, $-78°C$ to RT, 30 min.

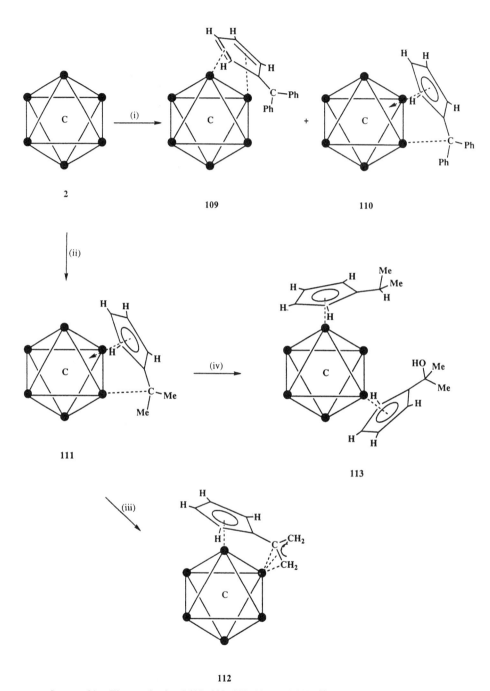

SCHEME 21. The synthesis of **109, 110, 111, 112,** and **113.** (i) dpf, CH$_2$Cl$_2$, 2 eq. Me$_3$NO, −78°C to RT; (ii) dmf, CH$_2$Cl$_2$, 3 eq. Me$_3$NO, −78°C to RT; (iii) CH$_2$Cl$_2$/MeCN, Δ, 20 min; (iv) dmf, H$_2$O (1 μl), CH$_2$Cl$_2$, 2 eq. Me$_3$NO; −78°C to RT, 40 min.

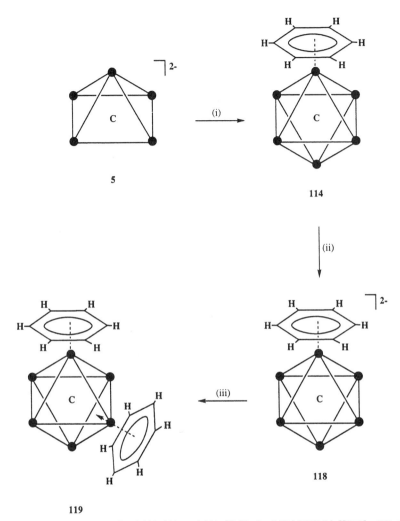

SCHEME 22. The synthesis of **114, 118,** and **119.** (i) [Ru($\eta$-C$_6$H$_6$)(NCMe)$_3$][BF$_4$]$_2$, CH$_2$Cl$_2$, $\Delta$, 20 min; (ii) Na$_2$CO$_3$, MeOH; (iii) [Ru($\eta$-C$_6$H$_6$)(NCMe)$_3$][BF$_4$]$_2$.

new bis-diene cluster Ru$_5$C(CO)$_{11}$($\mu$-C$_6$H$_8$-1,4)$_2$ **122** reveals that the second diene moiety coordinates across the opposite edge of the cluster base to the first one, but instead of replacing two radial carbonyls, it has been replaced by two axial ones as shown in Scheme 23.

Cluster Ru$_5$C(CO)$_{13}$($\mu$-C$_6$H$_8$) **120** reacts with Me$_3$NO in dichloromethane (in the absence of any potential ligands) to afford two benzene-

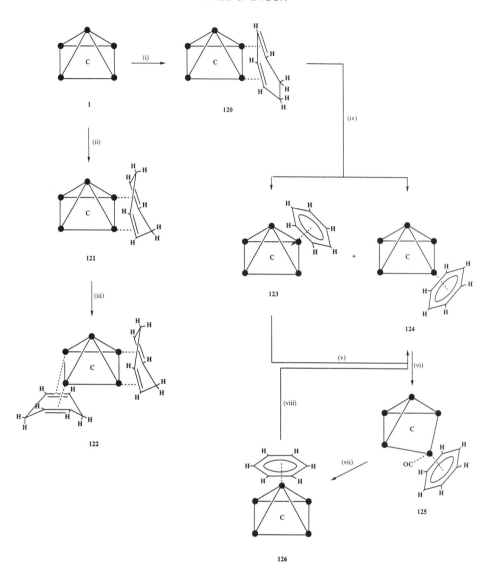

SCHEME 23.   The synthesis of **120, 121, 122, 123, 124, 125,** and **126.** (i) $C_6H_8$-1,3, $CH_2Cl_2$, 2 eq. $Me_3NO$, $-78°C$ to RT, 35 min; (ii) $C_6H_8$-1,4, $CH_2Cl_2$, 2 eq. $Me_3NO$, $-78°C$ to RT, 35 min; (iii) $C_6H_8$-1,4, $CH_2Cl_2$, 2 eq. $Me_3NO$, $-78°C$ to RT, 35 min; (iv) $CH_2Cl_2$, 1 eq. $Me_3NO$, RT, 15 min or hexane, $\Delta$, 18 h; (v) hexane, $\Delta$, 4 h; (vi) $CH_2Cl_2$, CO, 5 min; (vii) $CH_2Cl_2$ (viii) hexane, $\Delta$, 40 h.

coordinated cluster isomers $Ru_5C(CO)_{12}(\mu_3\text{-}C_6H_6)$ **123** (in 45% yield) and $Ru_5C(CO)_{12}(\eta\text{-}C_6H_6)$ **124**.[70,71] The solid-state structures of these clusters differ with respect to the bonding mode of the benzene ligand. The benzene straddles a triruthenium face in **123** and in **124** the benzene bonds to a basal Ru atom.

Heating **123** in hexane under reflux for 4 h results in the migration of the benzene ring to a terminal position affording **124**. This isomerization process is quantitative and has been monitored using variable temperature $^1$H-NMR spectroscopy, which indicates that the benzene ligand remains associated to the cluster during migration.[74]

Cluster **124** undergoes an addition reaction with CO in dichloromethane to give $Ru_5C(CO)_{13}(\eta\text{-}C_6H_6)$ **125** in quantitative yield. The solid-state structure of this cluster reveals a bridged butterfly array of Ru atoms with the benzene and one CO ligand coordinated to the bridging Ru atom.[71] Freshly prepared samples of **125** readily evolve carbon monoxide regenerating the starting material **124**. However, if **125** is prepared and crystallized from solution under a CO atmosphere and then redissolved in dichloromethane, the new compound $Ru_5C(CO)_{12}(\eta\text{-}C_6H_6)$ **126** is produced. It is believed that the benzene ligand in **126** coordinates to the apical ruthenium atom of the $Ru_5C$ square pyramid. The isomer in which the benzene bonds to the basal ruthenium atom, **123,** can be regenerated from this new isomer by thermolysis in hexane for 40 h. The reverse isomerization from **123** to **126** takes place under photolytic conditions when the compounds are embedded within a polymer film.[75] The apparent conflict between thermal and photolytic processes is probably due to heterolytic and homolytic Ru–Ru bond fission brought about by the respective initiation techniques.

Clusters **123** and **124** can be produced from **1** in a single step together with the new cluster $Ru_5C(CO)_{11}(\eta^4\text{-}C_6H_8)_2$ **127** in 15% yield. This is achieved by adding three equivalents of $Me_3NO$ to a dichloromethane solution of **1** and cyclohexa-1,3-diene. The structure of **127** has been established by single crystal X-ray diffraction and is illustrated in Fig. 5.[76] The two cyclohexa-1,3-diene ligands coordinate to basal Ru atoms on opposite edges of the square base. Cluster **127** reacts with CO under ambient conditions to afford $Ru_5C(CO)_{12}(\mu_3\text{-}C_6H_6)$ **123** in 70% yield. The proposed mechanism for this reaction is believed to involve the initial addition of CO bringing about a change in the geometry of the cluster core from square pyramidal to a bridged butterfly followed by the subsequent ejection of one of the $C_6H_8$ ligands and aromatization of the remaining ring.

Clusters **123** and **124** react further with cyclohexa-1,3-diene and two equivalents of $Me_3NO$ to afford the benzene-diene derivatives $Ru_5C(CO)_{10}(\mu_3\text{-}C_6H_6)(\mu\text{-}C_6H_8)$ **128** and $Ru_5C(CO)_{10}(\eta\text{-}C_6H_6)(\mu\text{-}C_6H_8)$ **129** in 40 and 34% yield, respectively.[76] Using chemical and thermal methods

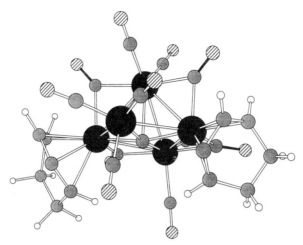

FIG. 5.   The solid-state structure of $Ru_5C(CO)_{11}(\eta^4\text{-}C_6H_8)_2$ **127**.

it has not been possible to generate *bis*-benzene adducts from either of these clusters.

The reaction of $Ru_6C(CO)_{17}$ **2** toward cyclohexadienes using a stepwise chemical activation pathway proceeds in a similar manner to the reactions of **1** (Scheme 24).[77] Cluster **2** reacts with either cyclohexa-1,3-diene or cyclohexa-1,4-diene in dichloromethane when activated with two equivalents of $Me_3NO$ to yield $Ru_6C(CO)_{15}(\mu_2\text{-}C_6H_8)$ **130** and $Ru_6C(CO)_{14}(\eta^6\text{-}C_6H_6)$ **114** in 20 and 25% yield, respectively. The solid-state structure of **130** reveals that the coordinated cyclohexadiene ligand bridges an apex-to-equator edge of the cluster. Compound **130** may be converted into **114** by the addition of one further equivalent of $Me_3NO$ in a noncoordinating solvent such as dichloromethane.

In turn, the treatment of **114** with two equivalents of $Me_3NO$ in dichloromethane containing excess cyclohexa-1,3-diene or cyclohexa-1,4-diene affords $Ru_6C(CO)_{12}(\eta\text{-}C_6H_6)(\mu\text{-}C_6H_8)$ **131** (30%), $Ru_6C(CO)_{12}(\mu_3\text{-}C_6H_6)(\mu\text{-}C_6H_8)$ **132** (5%), and $Ru_6C(CO)_{11}(\eta\text{-}C_6H_6)(\mu_3\text{-}C_6H_6)$ **119** (5%). The structures of **131** and **132** have been established by single crystal X-ray diffraction and in both clusters the diene spans an apex-to-equator edge. The benzene ring in **131** coordinates to an apical Ru atom, whereas in **132** it coordinates over a triruthenium face. The bis-benzene cluster **119** may be generated directly from **131** in 50% yield by treatment with a further equivalent of $Me_3NO$. Alternatively, $[Ph_3C][BF_4]$ may be used to abstract a hydride from the coordinated cyclohexa-1,3-diene ring in **131** to afford the benzene-cyclohexadienyl species $[Ru_6C(CO)_{12}(\eta\text{-}$

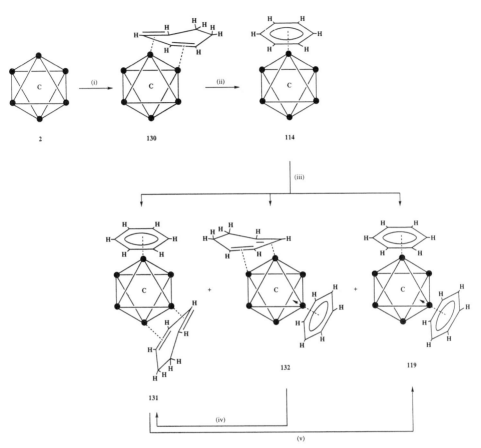

Scheme 24. The synthesis of **114, 119, 130, 131,** and **132.** (i) $C_6H_8$, $CH_2Cl_2$, 2 eq. $Me_3NO$, −78°C to RT, 30 min; (ii) $CH_2Cl_2$, 1 eq. $Me_3NO$, RT, 25 min; (iii) $C_6H_8$, $CH_2Cl_2$, 2 eq. $Me_3NO$, −78°C to RT, 30 min; (iv) hexane, Δ, 18 h; (v) $CH_2Cl_2$, 1 eq. $Me_3NO$, RT, 15 min or $[Ph_3C][BF_4]$ followed by DBU, $CH_2Cl_2$.

$C_6H_6)(\eta_3\text{-}C_6H_7)][BF_4]$ **133** in 30% yield. Reaction of **133** with the nonnucleophilic base 1,8-diazabicyclo[5.4.0]undeca-7-ene (DBU) removes a proton from the ring affording **119** in quantitative yield.

A number of related arene-cyclohexadiene and arene-benzene derivatives of **2** have been prepared.[69] For example, the arene-clusters $Ru_6C(CO)_{14}(\eta\text{-arene})$ (arene = $C_6H_5Me$ **115,** $C_6H_4Me_2\text{-}1,3$ **116,** and $C_6H_3Me_3\text{-}1,3,5$ **117**) react with cyclohexa-1,3-diene and two equivalents of $Me_3NO$ in dichloromethane to afford analogues of **131,** viz. $Ru_6C(CO)_{12}(\eta\text{-arene})(\mu\text{-}C_6H_8)$ (arene = $C_6H_5Me$ **134,** $C_6H_4Me_2\text{-}1,3$ **135,** and $C_6H_3Me_3$-

1,3,5 **136**), in yields of 20–30%. Further reaction of these clusters with
Me$_3$NO in dichloromethane affords the arene-benzene species Ru$_6$C(CO)$_{11}$($\eta$-
arene)($\mu_3$-C$_6$H$_6$) (arene = C$_6$H$_5$Me **137**, C$_6$H$_4$Me$_2$-1,3 **138** and C$_6$H$_3$Me$_3$-
1,3,5 **139**) in 5–10% yield. Spectroscopic and structural studies show that
the methyl-substituted arenes always adopt the terminal bonding mode
whereas the benzene straddles a trimetal face.

Clusters **119, 137–139** slowly isomerize in dichloromethane to form the
new clusters Ru$_6$C(CO)$_{11}$($\eta$-arene)($\eta$-C$_6$H$_6$) (arene = C$_6$H$_6$ **140**, C$_6$H$_5$Me
**141**, C$_6$H$_4$Me$_2$-1,3 **142**, and C$_6$H$_3$Me$_3$-1,3,5 **143**). The solid-state structure of
Ru$_6$C(CO)$_{11}$($\eta$-C$_6$H$_3$Me$_3$-1,3,5)($\eta$-C$_6$H$_6$) **143** has been established by single
crystal X-ray diffraction and is shown in Fig. 6.[78] The two C$_6$ rings bond
to consecutive Ru atoms on the cluster. In **143** the original mesitylene
ligand remains coordinated to an apical Ru atom and the benzene ligand
bonds to a Ru atom on the cluster molecular equator.

A number of other *bis*-arene clusters have also been prepared with the
introduction of the second arene ring achieved using Me$_3$NO activation.
For example, the *bis*-mestiylene complex Ru$_6$C(CO)$_{11}$($\eta$-C$_6$H$_3$Me$_3$-1,3,5)$_2$
**144** is prepared from Ru$_6$C(CO)$_{14}$($\eta$-C$_6$H$_3$Me$_3$-1,3,5) **117** and C$_6$H$_5$Me$_3$ in
dichloromethane by the addition of three equivalents of Me$_3$NO.[78,79] Cluster

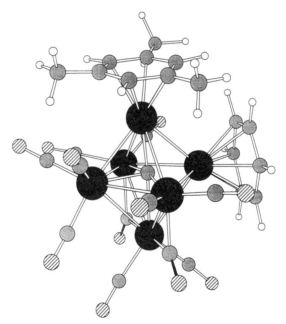

FIG. 6.   The solid-state structure of Ru$_6$C(CO)$_{11}$($\eta$-C$_6$H$_3$Me$_3$-1,3,5)($\eta$-C$_6$H$_6$) **143**.

**144** differs from the other *bis*-arene structures described thus far in that the two rings coordinate to the apical Ru atoms of the cluster octahedron forming a type of sandwich structure as shown in Fig. 7. The rings are not exactly planar and form angles of 5.6° and 4.4° with respect to the molecular equatorial plane.

The *bis*-toluene cluster $Ru_6C(CO)_{11}(C_6H_5Me)_2$ **145** is made in a similar way from $Ru_6C(CO)_{14}(\eta\text{-}C_6H_5Me)$ **115** by reaction with $C_6H_7Me$ and $Me_3NO$.[80] In solution **145** exists in two isomeric forms, viz. one like cluster **119** and one like **144**. Spectroscopic evidence suggests that the toluene ligands migrate over the cluster surface; this phenomenon is described in Section IX. The solid-state structure of **144** is similar to that of **119** with one toluene ring adopting a face-capping coordination mode and the other bonding to an apical Ru atom. Other related *bis*-arene clusters include the *m*-xylene derivative $Ru_6C(CO)_{11}(\eta\text{-}C_6H_4Me_2\text{-}1,3)_2$ **146,** which has a sandwich structure similar to that of the mesitylene derivative,[81] and the two *m*-xylene-toluene isomers, $Ru_6C(CO)_{11}(\eta\text{-}C_6H_4Me_2\text{-}1,3)(\eta\text{-}C_6H_5Me)$ **147** and **148,** in which the rings are located on *cis* and *trans* Ru atoms, respectively, with respect to their location on the $Ru_6C$ octahedral core.[82] The sandwich clusters **146** and **148** were not prepared in the same way as

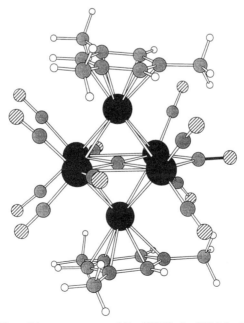

FIG. 7.   The solid-state structure of $Ru_6C(CO)_{11}(\eta\text{-}C_6H_3Me_3\text{-}1,3,5)_2$ **144.**

the *bis*-mesitylene cluster **144**, i.e., using the route shown in Scheme 24. The second ring is introduced directly as the arene (as opposed to initial coordination as a diene and subsequent conversion to the arene) by reaction of the appropriate cluster with three equivalents of $Me_3NO$ in a dichloromethane–propanone solution. The propanone is believed to stabilize any coordinatively unsaturated intermediates generated on route to the *bis*-arene derivative.

The benzene cluster **114** can also be prepared from hexa-1,3,5-triene.[83] Reaction of **2** with hexa-1,3,5-triene (a mixture of *cis* and *trans* isomers) in a sealed system in heptane at 155°C for 21 h results in the formation of the two isomers $Ru_6C(CO)_{14}\{\mu\text{-}CH_2(CH)_4CH_2\}$ **149** and **150** in 3 and 19% yield, respectively, together with the benzene derivative **114** in 22% yield (Scheme 25).

Although one can envisage "dehydrogenation" and rearrangement of the coordinated trienes in **149** and **150** to yield **114,** no such process has actually been observed to take place. Because both triene complexes are derived from the *trans* ligand, and not the *cis* isomer, it is possible that it is the *cis*-hexa-1,3,5-triene ligand that produces the benzene ring in cluster **114,** probably via initial conversion to cyclohexadiene. This step is proposed as cyclohexa-1,4-diene reacts with the parent cluster **2** under almost identical conditions to afford **114** together with the *bis*-benzene clusters **119** and $Ru_6C(CO)_{11}(\eta\text{-}C_6H_6)_2$ **151.** Cluster **151** has a sandwich-type structure analogous to that of **144.**[84] In a related thermal reaction, heating **2** with dimethyl cyclohexa-1,3-diene-1,4-dicarboxylate, $C_6H_6(CO_2Me)_2\text{-}1,4$, in dibutylether gives two isomeric products, $Ru_6C(CO)_{14}\{\mu_3\text{-}C_6H_4(CO_2Me)_2\text{-}1,4\}$ **152** and $Ru_6C(CO)_{14}\{\eta\text{-}C_6H_4(CO_2Me)_2\text{-}1,4\}$ **153**, in 15–25% yield.[85] The structures of these clusters have been established by single crystal X-ray diffraction, and the most significant difference between them involves the mode of coordination of the arene ligand to the cluster unit. In **152** the ring bonds over a trimetal face, whereas in **153** it bonds to an apical Ru atom. The synthesis of related isomers has already been described, including the pentaruthenium isomers $Ru_5C(CO)_{12}(\mu_3\text{-}C_6H_6)$ **123** and $Ru_5C(CO)_{12}(\eta\text{-}C_6H_6)$ **124** and the hexaruthenium isomers $Ru_6C(CO)_{11}(\eta\text{-}C_6H_6)(\mu_3\text{-}C_6H_6)$ **119** and $Ru_6C(CO)_{11}(\eta\text{-}C_6H_6)_2$ **140** and **151.** The isomers with terminally bonded benzene ligands are the most stable in solution. Extended Hückel theory has been used to rationalize this phenomenon and traces it to an increased destruction of C–C bonding character in facially coordinated rings compared with rings that are terminally bonded.[86]

The arene clusters $Ru_6C(CO)_{14}(\eta\text{-arene})$ (arene = $C_6H_6$ **114**, $C_6H_5Me$ **115**, $C_6H_4Me_2\text{-}1,3$ **116**, and $C_6H_3Me_3\text{-}1,3,5$ **117**) react with two equivalents of $Me_3NO$ in the presence of but-2-yne to afford $Ru_6C(CO)_{12}(\eta\text{-arene})(\mu_3\text{-}C_2Me_2)$ (arene = $C_6H_6$ **154**, $C_6H_5Me$ **155**, $C_6H_4Me_2\text{-}1,3$ **156**, and $C_6H_3Me_3$-

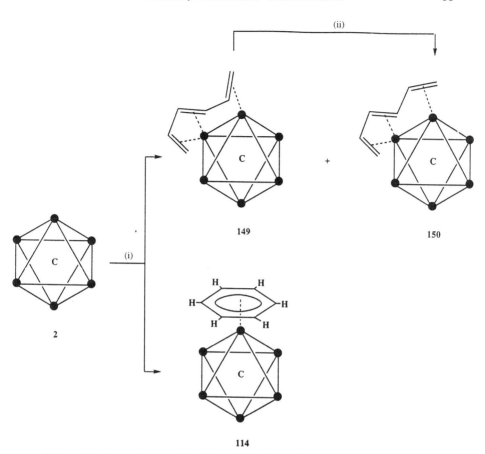

**SCHEME 25.** The synthesis of **114, 149,** and **150.** (i) 1,3,5-hexatriene, heptane, 155°C, 21 h; (ii) heptane, 155°C, 14 h.

1,3,5 **157**) in yields ranging from 32 to 43%.[87] The bonding mode of the arene is not affected by the substitution of two carbonyl ligands for the alkyne that adopts a face-capping bonding mode.

Reactivity studies involving the benzene ring in clusters **114, 119,** and **131** have been undertaken and the results are summarized in Scheme 26.[88] The reaction of these clusters with 1.2–1.5 equivalents of phenyl lithium results in *exo* attack at the coordinated ring with the formation of $[Ru_6C(CO)_{14}(\eta^4\text{-}C_6H_6Ph_2\text{-}1,4)]^{2-}$ **158,** $[Ru_6C(CO)_{11}(\eta^5\text{-}C_6H_6Ph)(\mu_3\text{-}C_6H_6)]^-$ **159,** and $[Ru_6C(CO)_{12}(\eta^5\text{-}C_6H_6Ph)(\mu\text{-}C_6H_8)]^-$ **160.** Given the stoichiometry of the reagent added, the expected dienyl clusters **159** and **160** are formed from **119** and **131**; however, **114** gives a diene derivative due to a double nucleophilic addition reaction. Further reaction of these

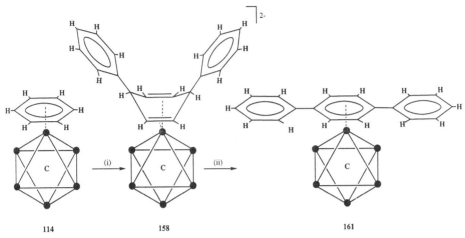

SCHEME 26. The synthesis of **158** and **161**. (i) LiPh, −78°C followed by addition of [N(PPh$_3$)$_2$]Cl; (ii) [Ph$_3$C][BF$_4$], −78°C.

anions with [Ph$_3$C][BF$_4$] affords the neutral substituted derivatives Ru$_6$C(CO)$_{14}$($\eta$-C$_6$H$_4$Ph$_2$-1,4) **161,** Ru$_6$C(CO)$_{11}$($\eta$-C$_6$H$_5$Ph)($\mu_3$-C$_6$H$_6$) **162,** and Ru$_6$C(CO)$_{12}$($\eta$-C$_6$H$_5$Ph)($\mu$-C$_6$H$_8$) **163**. This study indicates that the reactivity of benzene coordinated to the hexaruthenium cluster unit is not dissimilar to that when benzene is bonded to a Cr(CO)$_3$ group.

### J. Cycloheptatrienyls

A redox reaction takes place between [Ru$_6$C(CO)$_{16}$]$^{2-}$ **4** and two equivalents of tropylium bromide, [C$_7$H$_7$]Br, to afford the neutral bitropyl cluster Ru$_6$C(CO)$_{14}$($\mu_3$-C$_{14}$H$_{14}$) **164** in high yield.[16] The structure of **164** has been established by single crystal X-ray diffraction and is shown in Fig. 8.[89] Two C$_7$ rings have fused with the formation of a C–C single bond. The coordinated cycloheptatrienyl ring bonds over a triruthenium face of the cluster via six of the seven C atoms of the ring.

### K. Other Ligands

The reaction of Ru$_5$C(CO)$_{15}$ **1** with an excess of pyridine produced an equimolar mixture composed of two isomers of HRu$_5$C(CO)$_{14}$($\mu$-C$_5$H$_4$N) **165** and **166** together with a minor *bis*-pyridine product HRu$_5$C(CO)$_{13}$($\mu$-C$_5$H$_4$N)(C$_5$H$_5$N) **167**.[90] The solid-state structures of **165** and **166** show that the cluster skeleton consists of a bridged butterfly geometry. The pyridine

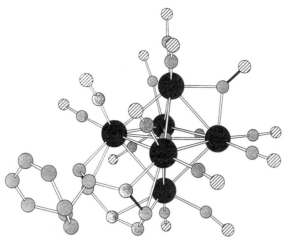

FIG. 8. The solid-state structure of $Ru_6C(CO)_{14}(\mu_3\text{-}C_{14}H_{14})$ **164**. The H atoms are not included.

ligand bonds via the lone pair on the N atom and via a $\sigma$ bond from an adjacent C atom to the bridging Ru atom and a hinge Ru atom. In **165** the N atom coordinates to the bridging atom and the C atom to the cluster hinge, whereas in **166** the reverse interactions are observed. The solid-state structure of **165** is shown in Fig. 9.

The activation of **1** with two equivalents of $Me_3NO$ in acetonitrile followed by the addition of dimethyl maleate, $(CHCO_2Me)_2$, affords

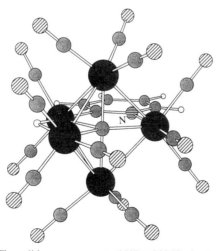

FIG. 9. The solid-state structure of $HRu_5C(CO)_{14}(\mu\text{-}C_5H_5N)$ **165**.

$Ru_5C(CO)_{13}\{\mu\text{-}(CHCO_2Me)_2\}$ **168** in 51% yield (Scheme 27).[35] The solid-state structure of **168** reveals that the five Ru atoms adopt a bridged butterfly geometry. The dimethyl maleate ligand coordinates to the cluster core via two C atoms that bond to a wing-tip Ru atom and via the O atoms of the ligand that coordinate to the bridging Ru atom. In an analogous reaction employing dimethylacetylene dicarboxylate, $(C_2O_2Me)_2$, in place of the dimethyl maleate ligand, a closely related bridged butterfly cluster $Ru_5C(CO)_{15}\{\mu\text{-}(C_2O_2Me)_2\}$ **169** is produced in 8% yield. The solid-state structure of **169** shows that the alkene ligand spans a hinge-bridge vector of the bridged butterfly cluster. The ligand in **168** is displaced readily by CO regenerating the starting material **1** in 49% yield.

The reaction of the anion $[Ru_6C(CO)_{16}(Me)]^-$ **73** with MeI at 150°C in a sealed system affords the pentanuclear cluster $[Ru_5C(CO)_{13}I(\mu\text{-}I)(\mu\text{-}COMe)]^-$ **170** in 32% yield.[33] The structure of **170** has been determined by single crystal X-ray diffraction and is shown in Fig. 10. The metal skeleton is derived from a square pyramid by removal of two *trans* apex-to-base edges. These edges are bridged by one I atom and the COMe group. The other I atom coordinates to a single Ru atom on the base of the cluster.

## V

### DERIVATIVES OF $Ru_5C(CO)_{15}$ AND $Ru_6C(CO)_{17}$ PREPARED FROM OTHER CLUSTERS

#### A. *From Triruthenium Clusters*

The reaction of $Ru_3(CO)_{12}$ **3** with ethylene affords $Ru_6C(CO)_{17}$ **2** as the major product. Other by-products have been isolated from this reaction,

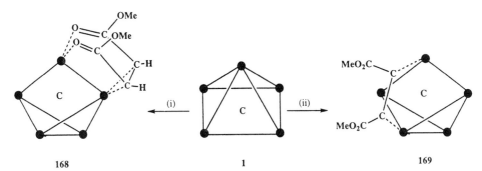

| 168 | 1 | 169 |

SCHEME 27.    The synthesis of **168** and **169**. (i) $MeCN/CH_2Cl_2$, 2 eq. $Me_3NO$, RT, 30 min followed by addition of 1 eq. $(CHCO_2Me)_2$, 10 min; (ii) $MeCN/CH_2Cl_2$, 1 eq. $Me_3NO$, RT, 30 min followed by addition of 1 eq. $(C_2O_2Me_2)_2$, 20 min.

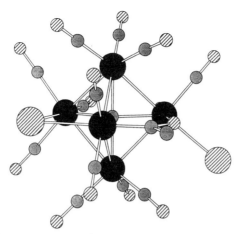

FIG. 10. The solid-state structure of $[Ru_5C(CO)_{13}I(\mu\text{-}I)(\mu\text{-}COMe)]^-$ **170**. The H atoms are not included.

including the tetranuclear butterfly cluster $Ru_4(CO)_{12}(\mu_4\text{-}C_2Me_2)$ and $Ru_6C(CO)_{15}(\mu\text{-}MeCH=CHCH=CHMe)$ **171**.[91] This reaction demonstrates that oligomerization of ethylene may take place (Scheme 28). The structure of **171** has been established by single crystal X-ray diffraction, which reveals that the hexa-2,4-diene ligand coordinates across an apex-to-equator edge of the cluster relative to the bridging CO ligand in an analogous position to that of the cyclohexa-1,3-diene ligand in **130**.

The thermolysis of **3** with cyclooctene in refluxing octane for 5 h affords a number of products, including $Ru_6C(CO)_{15}(\mu_3\text{-}C_8H_{12})$ **172** in 27% yield.[92]

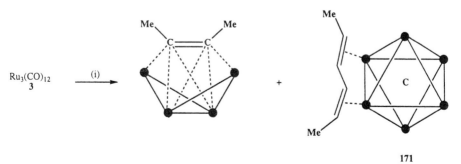

SCHEME 28. The synthesis of **171**. (i) $C_2H_4$.

The $C_8$ ring in **172** has undergone "dehydrogenation" to form a cyclooctyne ligand that coordinates to a cluster face in the analogous bonding mode to that observed in the alkyne clusters **75–82**.

The clusters $Ru_6C(CO)_{14}(\eta\text{-arene})$ (arene $= C_6H_6$ **114**, $C_6H_5Me$ **115**, $C_6H_4Me_2$-1,3 **116**, and $C_6H_3Me_3$-1,3,5 **117**) may be prepared in high yield from the ionic coupling reaction described in Section IV,I. However, they were originally prepared from the thermolysis of **3** with the appropriate arene used as the solvent in yields ranging from 3 to 16%.[9,10] The solid-state structures of **115** and **117** were determined on crystals obtained using this method.[93,11] More recently, the reaction between **3** and mesitylene, $C_6H_3Me_3$-1,3,5, has been carried out in several different hydrocarbon solvents, and it has been found that the yield of **117** is strongly dependent on the reflux temperature, with higher yields obtained in nonane and octane and low yields obtained from hexane.[94,95] When the reaction is carried out in heptane, two other products are obtained; these are described in Section VI as they provide information regarding the mechanism by which the carbide atom is produced. The direct reaction of arenes with **3** in octane has been used to prepare the compounds $Ru_6C(CO)_{14}(\eta\text{-arene})$ (arene $= Ph_2$ **173**,[96,97] $Ph_2O$ **174**,[96] $Ph_2CH_2$ **175**,[97] $Ph_2(CH_2)_2$ **176**,[97] PhOMe **177**,[98] PhCOOEt **178**,[98] and $C_6H_3Et_3$-1,3,5 **179**[99]). The hexamethylbenzene derivative has also been proposed,[98] although a more recent study indicates that the compound is in fact $Ru_6(\mu_4\text{-}\eta^2\text{-CO})(CO)_{13}(\eta\text{-}C_6Me_6)$ **180**.[100] The solid-state structure of **180** shows that the six Ru atoms adopt a bi-edged bridged tetrahedral arrangement as shown in Scheme 33 for the mesitylene analogue **202**. Two *pseudo* butterfly cavities are formed that are bridged by a carbonyl that bonds via both C and O atoms. Heating this cluster further does not result in a transformation to the octahedral carbide cluster, although the analogous mesitylene derivative undergoes a polyhedral rearrangement of this type (see Section VI).

Other $Ru_6C(CO)_{14}(\eta\text{-arene})$ clusters have been made from arenes that originally had unsaturated substituents attached to the ring. In these reactions a number of other clusters with nuclearities varying from three to seven are also obtained. Reactions with this type of ligand include the thermolysis of $Ru_3(CO)_{12}$ **3** with isopropenylbenzene or 4-phenyl-1-butene in octane affording $Ru_6C(CO)_{14}(\eta\text{-}C_6H_5CHMe_2)$ **181** and $Ru_6C(CO)_{14}(\eta\text{-}C_6H_5C_4H_9)$ **182**, in 2 and 3% yield, respectively.[101] The unsaturated substituent groups attached to the rings in compounds **181** and **182** were hydrogenated during the course of the reaction. With allylbenzene, both hydrogenation and isomerization take place to afford $Ru_6C(CO)_{14}(\eta\text{-}C_6H_5C_3H_7)$ **183** and $Ru_6C(CO)_{14}(\eta\text{-}C_6H_5C_3H_5)$ **184** in 3 and 5% yield, respectively.[101] The structure of **182** has been established by single crystal X-ray diffraction, which reveals an unusual, distorted octahedral $Ru_6C$ core in which one of the Ru–Ru edges is significantly longer than that distance considered to constitute a bond.

The thermolysis of **3** with the diethylferrocenylphosphine in octane under reflux for 18 h affords several compounds, including $Ru_6C(CO)_{14}(\mu\text{-}PEt_2)_2$ **185** and $Ru_6C(CO)_{13}(\mu\text{-}PEt_2)(\eta\text{-}C_5H_5)$ **186** in 15 and 25% yield, respectively (Scheme 29).[102] Their structures have been established by single crystal X-ray diffraction and both are based on the familiar octahedral metal skeleton. The phosphine has lost the ferrocene moiety in these clusters and coordinates as a phosphido ligand. Two such $PEt_2$ ligands are present in **185**: one bridges an edge of the molecular equator and the other bridges an apex-to-equator edge of the cluster. In **186,** only one phosphido ligand bridges two Ru atoms on the molecular equator. The cyclopentadienyl ligand in this cluster coordinates to an apical Ru atom.

The vacuum pyrolysis of the trinuclear cluster $HRu_3(CO)_{10}(\mu\text{-}CNMe_2)$ **187** at 185°C for 90 min affords the hexanuclear cluster $Ru_6C(CO)_{14}(\mu\text{-}CNMe_2)_2$ **188** in 20% yield.[103] The solid-state structure of **188** is closely related to that of **185** with one of the carbyne ligands bridging the molecular equator of the cluster and the other bridging an apex-to-equator edge as shown in Fig. 11.

## B. *From Pentaruthenium Clusters*

The thermolysis of the pentaruthenium cluster $Ru_5(\mu_5\text{-}C_2PPh_2)(CO)_{13}(\mu\text{-}PPh_2)$ **189** under varying conditions affords a number of products, including some containing a $Ru_5C$ cluster skeleton. The carbide atom is derived from

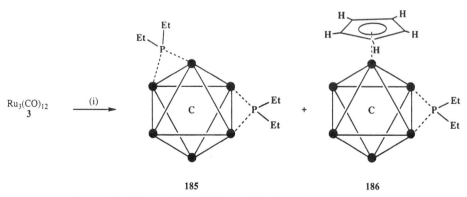

**185**          **186**

SCHEME 29.   The synthesis of **185** and **186**. (i) $PEt_2Fc$, octane, $\Delta$, 18 h.

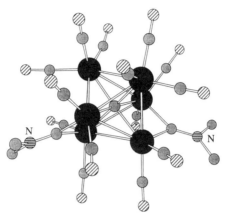

FIG. 11.    The solid-state structure of $Ru_6C(CO)_{14}(\mu\text{-}CNMe_2)_2$ **188.** The H atoms are not included.

the $\mu_5\text{-}C_2PPh_2$ ligand by cleavage of the C–C bond. This is brought about by the sequential hydrogenation of the $\beta$–C atom according to the equation:

$$C \equiv CPPh_2 + 3\ H \rightarrow C + H_3CPPh_2$$

Compound **189** reacts with hydrogen gas in cyclohexane under reflux for 5 h to afford three main products, all of which have been analyzed by single crystal X-ray diffraction (Scheme 30).[104,105] One of these products is the square pyramidal cluster $H_3Ru_5C(CO)_{11}(PPh_2Me)(\mu\text{-}PPh_2)$ **190** obtained in 54% yield. The other two products obtained from this reaction, viz. $HRu_5(\mu_5\text{-}C_2HPPh_2)(CO)_{13}(\mu\text{-}PPh_2)$ **191** and $H_2Ru_5(\mu_4\text{-}C_2H_2PPh_2)(CO)_{12}(\mu\text{-}PPh_2)$ **192,** have metal polyhedra, which consist of bi-edge bridged triangular arrangements of Ru atoms similar to that observed in the starting material **189.** The thermolysis of **191** affords **192** and **190** and thermolysis of **192** gives **190,** demonstrating them to be intermediates on route to the square pyramidal cluster **190** as shown in Scheme 30.

The reaction of **189** with terminal alkenes results in the addition of the alkene to the $\alpha$ C atom of the $\mu_5\text{-}C_2PPh_2$ moiety, which causes the C–C bond to cleave (Scheme 31).[106] The reaction of **189** with propene in toluene at 90°C for 4 days affords four products, including the carbide cluster $Ru_5C(CO)_{11}(\mu\text{-}PPh_2)(\mu\text{-}PPh_2C_3H_3Me)$ **193** in 8% yield. The structure of **193** has been established by single crystal X-ray diffraction, which shows that the five Ru atoms adopt a square pyramidal geometry. The phosphido ligand bridges a basal edge of the cluster and the phosphine ligand, $PPh_2C_3H_3Me$, coordinates to the apical Ru atom. The allyl portion of the phosphine ligand also forms a $\eta^3$ interaction with one of the basal Ru

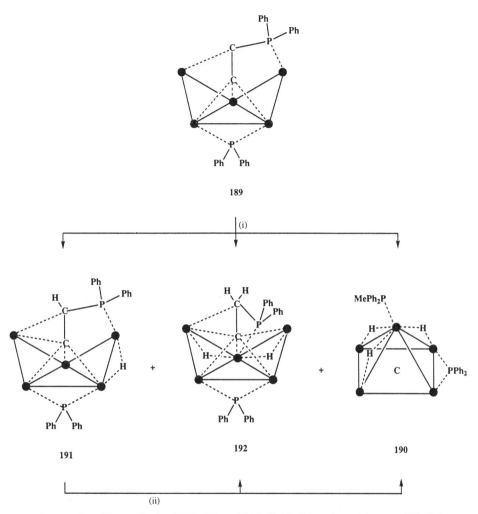

SCHEME 30. The synthesis of **190, 191,** and **192.** (i) $H_2$ (10 atm), cyclohexane, RT, 18 h;
(ii) $H_2$ (10 atm), cyclohexane, $\Delta$, 45 min.

atoms. An analogue of **193** in which the methyl group of the $PPh_2C_3H_3Me$
ligand is replaced by a phenyl ring may be isolated in 32% yield from the
reaction of **189** with styrene. This compound, $Ru_5C(CO)_{11}(\mu\text{-}PPh_2)(\mu\text{-}$
$PPh_2C_3H_3Ph)$ **194,** has been reacted with trimethylphosphite, which af-
fords $Ru_5C(CO)_{10}(\mu\text{-}PPh_2)(\mu\text{-}PPh_2)(\mu\text{-}PPh_2C_3H_3Ph)$ $\{P(OMe)_3\}$ **195** and
$Ru_5C(CO)_{11}(\mu\text{-}PPh_2)(\mu\text{-}PPh_2C_3H_3Ph)$ $\{P(OMe)_3\}$ **196** in 24 and 64% yield,
respectively. The solid-state structures of **195** and **196** are closely related;

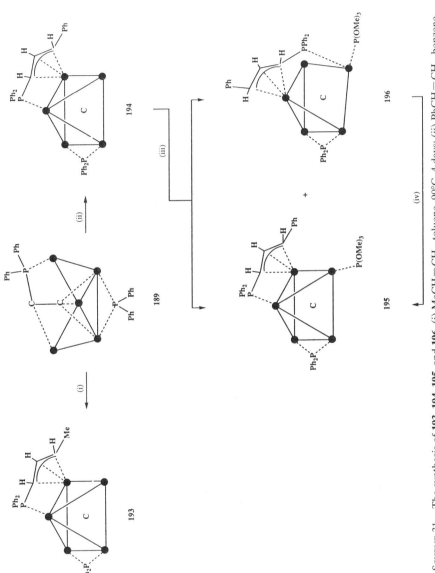

SCHEME 31. The synthesis of **193**, **194**, **195**, and **196**. (i) MeCH=CH₂, toluene, 90°C, 4 days; (ii) PhCH=CH₂, benzene, 90°C, 3 days; (iii) P(OMe)₃, THF, 40°C, 3.5 h; (iv) toluene, 90°C, 2.5 h.

in **195** one axial carbonyl attached to a basal Ru atom of the square pyramid has been substituted by a P(OMe)$_3$ ligand, whereas cluster **196** is formed as a result of an addition of the phosphite ligand (causing the square pyramidal cluster to open to a bridged butterfly). The phosphite coordinates to the bridging Ru atom. Cluster **196** may be converted to **195** in 75% yield by thermal decarbonylation in toluene at 90°C for 2.5 h.

The reaction of **189** with C$_5$H$_6$, C$_5$H$_5$Me, or C$_5$HMe$_5$ affords the hexaruthenium compounds Ru$_6$C(CO)$_{10}$($\mu$-PPH$_2$)$_2$($\mu_3$-CH)($\eta$-C$_5$H$_{5-x}$Me$_x$) (X = 0 **197**, 1 **198**, and 5 **199**) with yields ranging from 15 to 32% (Scheme 32).[107]

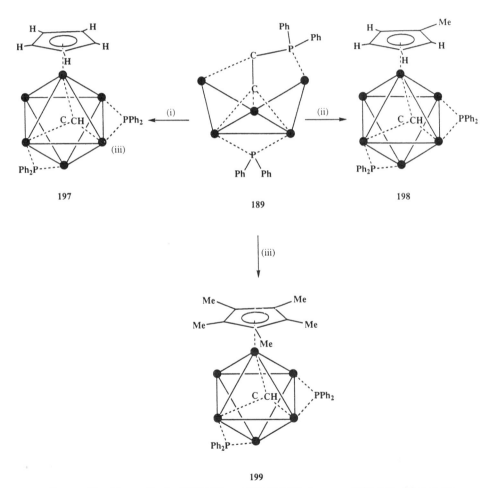

197          189          198

199

SCHEME 32.   The synthesis of **197, 198,** and **199.** (i) C$_5$H$_6$, benzene, 95°C, 58 h; (ii) C$_5$H$_5$Me, toluene, 97°C, 44 h; (iii) C$_5$HMe$_5$, benzene, 90°C, 44 h.

The structure of the cyclopentadienyl derivative **197** has been established by single crystal X-ray diffraction, which reveals that the cluster has an octahedral arrangement of Ru atoms. An equatorial edge is bridged by one of the phosphinidene ligands whereas the other phosphinidene bridges an apex-to-equator edge. A trimetal face is capped by a C–H ligand, and the cyclopentadienyl ring coordinates to an apical Ru atom. An increase in cluster nuclearity is also observed when the pentaruthenium cluster $Ru_5(CO)_{14}(CNBu^t)(\mu_5\text{-}CNBu^t)$ **200** is heated in nonane for 2 h, affording the hexanuclear cluster $Ru_6C(CO)_{16}(CNBu^t)$ **201** in 13% yield.[108] The isonitrile ligand in this cluster coordinates to an apical Ru atom of the octahedral skeleton. It has been established from $^{13}C$-labeling experiments that the carbide atom in this cluster is derived from the isonitrile ligand.

It is also possible to generate $Ru_6C$-based compounds from higher nuclearity clusters. For example, $[Ru_{10}C(CO)_{24}]^{2-}$ reacts with carbon monoxide to form $[Ru_6C(CO)_{16}]^{2-}$ and $H_4Ru_8(CO)_{18}(\eta\text{-}C_{16}H_{16})$ undergoes a similar reaction with CO to form $Ru_6C(CO)_{14}(\mu_3\text{-}C_{16}H_{16})$.[109,110] This latter reaction is described in more detail in the following section, as it is unusual in that not only does the nuclearity of the cluster decrease, but the carbide atom is generated simultaneously.

## VI

### ORIGIN OF THE CARBIDE ATOM IN $Ru_6C(CO)_{17}$ AND SOME OF ITS DERIVATIVES

Shortly after the initial synthesis of $Ru_6C(CO)_{17}$ **2**, $^{13}C$-labeling experiments established that the carbide atom originated from the disproportionation of a coordinated CO ligand.[44,111] After much speculation concerning the mechanism by which the C≡O bond is cleaved, it is now considered to take place via an intermediate in which both the C and O atoms are coordinated to metals, as this has the effect of significantly weakening the triple bond.[112,113] Several reactions have provided further insight into this process and are described in this section.

The thermolysis of $Ru_3(CO)_{12}$ **3** in heptane containing mesitylene $(C_6H_3Me_3\text{-}1,3,5)$ affords $Ru_6(\eta\text{-}\mu_4\text{-}CO)_2(CO)_{13}(\eta\text{-}C_6H_3Me_3\text{-}1,3,5)$ **202** in 15% yield together with $HRu_6(\eta\text{-}\mu_4\text{-}CO)(CO)_{13}(\mu_2\text{-}C_6H_3Me_2CH_2\text{-}1,3,5)$ **203** and $Ru_6C(CO)_{14}(\eta\text{-}C_6H_3Me_3\text{-}1,3,5)$ **117** in low yield (Scheme 33).[94,95] The structure of **202** is closely related to that of **180** with two $\eta^2$ CO ligands bonded between the wings of the *pseudo* butterfly cavities. Experiments have shown that **202** is an intermediate of the carbide cluster **117** as it undergoes conversion to **117** together with the formation of **203** (in equal

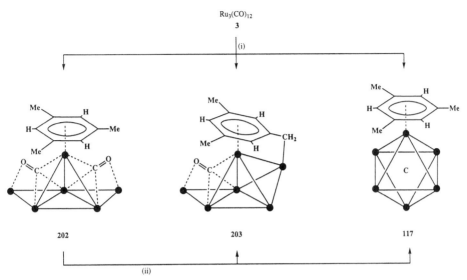

SCHEME 33.    The synthesis of **117, 202,** and **203.** (i) $C_6H_3Me_3$-1,3,5, heptane, $\Delta$; (ii) $C_6H_3Me_3$-1,3,5, $\Delta$.

amounts) on thermolysis in mesitylene. Carbon dioxide is also evolved during the reaction. From these observations it is suggested that the formation of the carbide atom involves the cleavage of a $\eta$ CO ligand in **202** via nucleophilic attack of its oxygen on a terminal CO carbon of a second cluster molecule. Elimination of $CO_2$ from this intermediate would generate the carbide atom and subsequent rearrangement of the metal polyhedron would give **117.**

The reaction between **3** and [2.2]paracyclophane ($C_{16}H_{16}$) in refluxing octane for about 3 h affords several compounds, including $Ru_6C(CO)_{15}(\mu_3\text{-}C_{16}H_{16}\text{-}\mu\text{-}O)$ **204,** $Ru_6C(CO)_{14}(\mu_3\text{-}C_{16}H_{16})$ **205,** $Ru_6C(CO)_{11}(\mu_3\text{-}C_{16}H_{16})(\eta\text{-}C_{16}H_{16})$ **206,** and $H_4Ru_8(CO)_{18}(\eta\text{-}C_{16}H_{16})$ **207** (Scheme 34).[99,110,114–116] The yield of these compounds ranges from 3 to 25% and their distribution can be varied by careful control of the reaction time.

The structures of **204–207** have been established in the solid state by single crystal X-ray diffraction. Clusters **205** and **206** are based on the $Ru_6C$ octahedral skeleton. The [2.2]paracyclophane ligand in **205** coordinates over a triangular metal face, and in **206** one [2.2]paracyclophane adopts a similar face-capping bonding mode and the other coordinates in a terminal mode to an apical Ru atom. The metal polyhedron in **204** is relatively open in comparison to the octahedron having only nine Ru–Ru bonds. A carbide atom occupies the central cavity and interacts with five of the six Ru atoms.

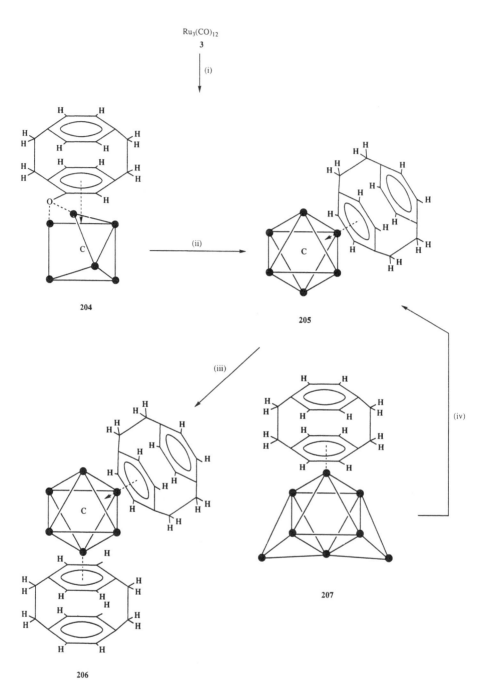

SCHEME 34. The synthesis of **204, 205, 206,** and **207.** (i) $C_{16}H_{16}$, octane, $\Delta$; (ii) octane, $\Delta$, 3 h or pyrolyse at 200°C for 2 min; (iii) $C_{16}H_{16}$, octane, $\Delta$; (iv) CO, $CH_2Cl_2$, RT, 5 min.

An oxo atom spans two Ru atoms and bonds to the facially coordinated [2.2]paracyclophane ring. It has been proposed that the carbide and oxo atoms are both derived from the same CO ligand. This seems reasonable given that pyrolysis of **204** affords **205** and $CO_2$ in quantitative yield. In contrast to the generation of **117** from **202** the transformation here appears to involve an *intra*molecular process.[115]

The cluster skeleton in **207** consists of a *cis*-bicapped octahedron array of Ru atoms. The [2.2]paracyclophane ligand is coordinated to the apical Ru atom of the octahedron subunit. It does not contain a carbide atom in the octahedral cavity, but on reaction with CO in dichloromethane at room temperature the clusters **205** and $Ru_3(CO)_{12}$ **3** are obtained in quantitative yield. Cluster **3** is probably derived from a recombination of ruthenium carbonyl fragments generated from the carbonylation of the capping groups on the starting material. The formation of the carbide atom in **205** is less easy to envisage as the octahedron is already present. However, it has been noted that heating $[Ru_6(CO)_{18}]^{2-}$ **208** in refluxing diglyme for 1 h affords $[Ru_6C(CO)_{16}]^{2-}$ **4** in 96% yield.[17] Cluster **208** also undergoes conversion to the neutral carbide cluster **2** on reaction with trifluoromethanesulfonic or trifluoroethanoic anhydrides.[117]

The reactivity of $Ru_6C(CO)_{14}(\mu_3\text{-}C_{16}H_{16})$ **205** has been studied in some detail. Using either chemical or thermal activation, one carbonyl may be replaced by triphenylphosphine or tricyclohexylphos phine to afford $Ru_6C(CO)_{13}(\mu_3\text{-}C_{16}H_{16})(PPh_3)$ **209** and $Ru_6C(CO)_{13}(\mu_3 C_{16}H_{16})$ (PCy3) **210,** respectively. The method involving $Me_3NO$ affords the product in ca. 30% yield whereas thermolysis gives a lower yield of ca. 10%.[118] Two carbonyl ligands may be displaced using $Me_3NO$ in the presence of but-2-yne[87] or cyclohexa-1,3-diene[116] to afford $Ru_6C(CO)_{12}(\mu_3\text{-}C_{16}H_{16})$ $(\mu_3\text{-}C_2Me_2)$ **211** and $Ru_6C(CO)_{12}(\mu_3\text{-}C_{16}H_{16})(\mu\text{-}C_6H_8)$ **212,** in 38 and 25% yield, respectively. The face-capping bonding mode of the [2.2]paracyclophane ligand does not change in any of these derivatives, although the precise orientation of the $C_6$ ring over the $Ru_3$ face does vary.

VII

## REACTIONS OF $Ru_5C(CO)_{15}$ AND $Ru_6C(CO)_{17}$ THAT RESULT IN HIGHER NUCLEARITY CLUSTERS

The thermolysis of $Ru_3(CO)_{12}$ **3** with isopropenylbenzene ($C_9H_{10}$) or 1,3-diisopropenylbenzene ($C_{12}H_{14}$) in octane results in the formation of several products, including one compound from each reaction with a nuclearity of seven characterized as $Ru_7C(CO)_{16}(\mu_4\text{-}C_9H_8)$ **213** and $Ru_7C(CO)_{16}(\mu_4\text{-}$

$C_{12}H_{12}$) **214**, respectively.[119] The structures of **213** and **214** have been established by single crystal X-ray diffraction and that of **213** is shown in Fig. 12. The metal core in these compounds comprises a $Ru_6C$ octahedron with a seventh Ru atom protruding from a vertex of the octahedron. It is conceivable that the $Ru_6C$ octahedron is initially produced in the reaction and the seventh Ru atom attached to the cluster at a later stage; however, experiments have not been carried out to test this hypothesis. The Ru spike participates in the coordination of the ligand via the formation of two $\sigma$ bonds. In addition, the aromatic ring and the $\pi$ bond from the unsaturated fragment of the side arm coordinate over a face of the cluster. In **214** the second isopropenyl unit is not involved in bonding to the cluster.

The octanuclear dicarbide cluster $Ru_8C_2(CO)_{17}(\mu\text{-}PPh_2)_2$ **215** has been isolated from the thermolysis of $Ru_5(\mu_5\text{-}C_2PPh_2)(CO)_{13}(\mu\text{-}PPh_2)$ **189** with **3** in cyclohexane under an atmosphere of CO (45 atm) in 21% yield.[120] The structure of **215** has been established by single crystal X-ray diffraction and is shown in Fig. 13. The metal core can be regarded as an octahedron and square pyramid fused through a triangular face. Each of these cluster units contains a carbide atom. One of the phosphido ligands bridges an edge on the octahedron and the other bridges a basal edge on the square pyramidal unit.

The thermolysis of $[Ru_6C(CO)_{16}]^{2-}$ **4** in tetraglyme at 210–230°C for 80 h affords the decanuclear dicarbide cluster $[Ru_{10}C_2(CO)_{24}]^{2-}$ **216** in 35% yield (Scheme 35).[121,122] The structure has been established by single crystal X-ray diffraction, which reveals that the metal core comprises two

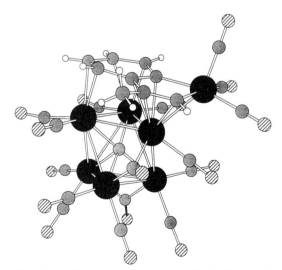

FIG. 12.    The solid-state structure of $Ru_7C(CO)_{16}(\mu_4\text{-}C_9H_8)$ **213.**

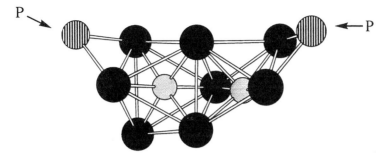

FIG. 13. The solid-state structure of $Ru_8C_2(CO)_{17}(\mu\text{-}PPh_2)_2$ **215.** The phenyl rings and COs have been omitted for clarity.

octahedral carbide clusters fused through an edge. The metal core in **216** undergoes a reversible skeletal transformation on reaction with diphenylace-tylene.[123] Heating **216** with an excess of $C_2Ph_2$ in diglyme for 5 days affords $[Ru_{10}C_2(CO)_{22}(\mu\text{-}C_2Ph_2)]^{2-}$ **217** in 70% yield. The solid-state structure of **217** is based on that of **216** with an additional bond formed between two apical Ru atoms of the adjacent octahedra. The alkyne ligand bridges this new edge. The homoleptic cluster **216** can be regenerated from **217** by heating it at 125°C in diglyme under an atmosphere of carbon monoxide for 3 days.

The thermolysis of **4** with $Ru_3(CO)_{12}$ **3** in *bis*(2-methoxyethyl)ether under reflux for 3 h affords the decanuclear cluster $[Ru_{10}C(CO)_{24}]^{2-}$ **218** in 81% yield (Scheme 36).[124] The same cluster has also been isolated from the thermolysis of **3** with mesitylene.[95] The structure of **218** has been established by single crystal X-ray diffraction, which shows that the metal skeleton consists of a tetra-capped octahedron decorated with terminal carbonyl ligands. Cluster **218** reacts with CO in dichloromethane under ambient conditions to regenerate **4** and **3** in quantitative yield.[109] The decanuclear cluster **218** also undergoes a reversible reaction with two equivalents of iodine to afford $[Ru_{10}C(CO)_{24}I]^-$ **219.**[109] At higher temperatures further reaction occurs with iodine to produce a species tentatively characterized as the hexamer $[Ru_6C(CO)_{16}I_2]^-$ **220.**

Large aggregates of $Ru_6C(CO)_{17}$ **2** have been detected by time-of-flight mass spectrometry when the cluster is desorbed/ionized with a 337-nm $N_2$

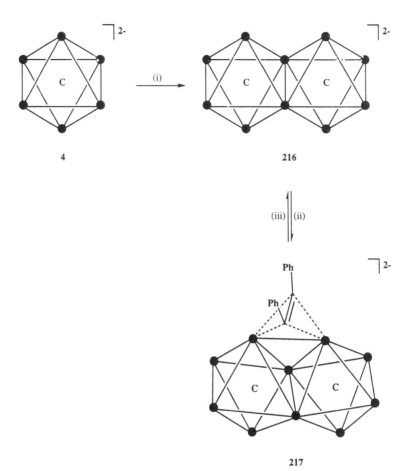

SCHEME 35. The synthesis of **216** and **217**. (i) diglyme, $\Delta$, 80 h; (ii) $C_2Ph_2$, diglyme, $\Delta$, 5 days; (iii) CO, diglyme, $\Delta$, 3 days.

laser.[125] In a typical spectrum the molecular parent ion is observed together with cluster aggregates at higher masses that extend to ca. $m/z$ 35,000. The precise nature of these aggregates remains uncertain and this behavior is not exclusive to **2**. Some derivatives of **2**, including $Ru_6C(CO)_{14}(\eta\text{-}C_6H_5Me)$ **115** and $Ru_6C(CO)_{14}(\mu_3\text{-}C_{16}H_{16})$ **205**, also show similar aggregation, although not to such high masses.[125–127] It is proposed that the desorption/ionization process strips carbonyl ligands from the cluster, thereby generating highly reactive fragments in the gas phase that may combine to form fused octahedral cluster aggregates.

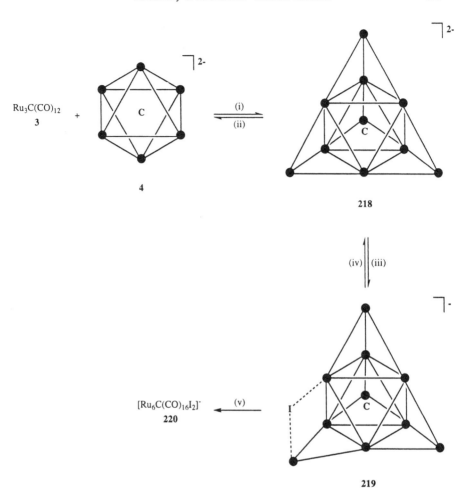

$Ru_3C(CO)_{12}$
3

$[Ru_6C(CO)_{16}I_2]^-$
220

SCHEME 36.   The synthesis of **218, 219,** and **220.** (i) *bis*(2-methoxyethyl)ether, $\Delta$, 3 h; (ii) CO, $CH_2Cl_2$; (iii) $I_2$; (iv) $CH_2Cl_2$, RT; (v) $I_2$, elevated temperature.

## VIII

## HETERONUCLEAR DERIVATIVES

### A. *Condensation Reactions of* $[Ru_5C(CO)_{14}]^{2-}$

The square face of cluster **5** is readily capped by a number of different metal fragments affording octahedral heteronuclear clusters in high yield. The dianion **5** reacts with the Group 6 complexes $M(CO)_3L_3$ (M = Cr,

L = C₅H₅N; M = Mo or W, L = MeCN) in THF under reflux to afford
the dianionic species $[Ru_5CrC(CO)_{17}]^{2-}$ **222,** $[Ru_5MoC(CO)_{17}]^{2-}$ **223,** and
$[Ru_5WC(CO)_{17}]^{2-}$ **224.**[128] Although the structures of **222–224** have not been
established in the solid state, a gold derivative, viz. **267** (Fig. 16), has been
analyzed by single crystal X-ray diffraction, and the central cluster unit
consists of a Ru₅WC octahedron. A similar structure would be expected
for **222–224.**

Cluster **5** reacts with $Pt(\eta^4\text{-}C_8H_{12}\text{-}1,5)Cl_2$ in methanol at room temperature to afford the heteronuclear cluster $Ru_5PtC(CO)_{14}(\eta^4\text{-}C_8H_{12}\text{-}1,5)$ **225**
in 41% yield (Scheme 37). Two by-products are also isolated from this

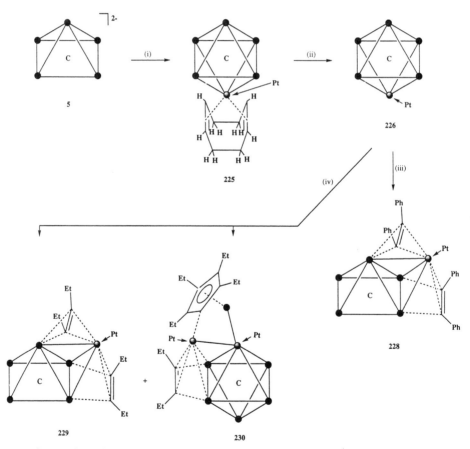

SCHEME 37. The synthesis of **225, 226, 228, 229,** and **230.** (i) $Pt(\eta^4\text{-}C_8H_{12}\text{-}1,5)_2$, CH₂Cl₂,
RT, 1 h; (ii) CO, CH₂Cl₂, RT, 15 min; (iii) C₂Ph₂, CH₂Cl₂, $h\nu$, RT, 10 h; (iv) C₂Et₂, CH₂Cl₂,
$h\nu$, RT, 10 h.

reaction, viz. $Ru_5PtC(CO)_{16}$ **226** and $Ru_5Pt_2C(CO)_{13}(\eta^4\text{-}C_8H_{12}\text{-}1,5)_2$ **227**, in 1 and 2% yield, respectively.[129] The structures of **226** and **227** have been established by single crystal X-ray diffraction. The metal atom topology in **226** comprises an octahedron with the Pt atom occupying an apical position, whereas the cluster skeleton in **227** adopts a monocapped octahedral geometry.

Cluster **225** may be converted to the homoleptic species **226** in 97% yield by passing CO through a dichloromethane solution of **225** at 25°C for 15 min. Cluster $Ru_5PtC(CO)_{16}$ **226** undergoes a polyhedral rearrangement to a monocapped square pyramidal on reaction with alkynes. The irradiation of **226** and diphenylacetylene in dichloromethane with UV light affords $Ru_5PtC(CO)_{13}(\mu\text{-}C_2Ph_2)(\mu_3\text{-}C_2Ph_2)$ **228** in 37% yield. This cluster is isostructural with **85** and **88** described earlier. Treatment of **228** with carbon monoxide displaces the alkyne ligands regenerating the starting material **226.** The cluster $Ru_5PtC(CO)_{13}(\mu\text{-}C_2Et_2)(\mu_3\text{-}C_2Et_2)$ **229** (an analogue of **228**) may be isolated in 20% yield from the reaction of **226** with 3-hexyne under photochemical conditions.[130] This cluster is isolated together with $Ru_6Pt_2C(CO)_{17}(\mu\text{-}\eta^5\text{-}C_5Et_4)(\mu_3\text{-}C_2Et_2)$ **230** in 7% yield. The solid-state structure of **230** reveals an unusual metal atom geometry based on a central $Ru_5PtC$ octahedron with additional Ru and Pt atoms bonding to the Pt atom of the central cluster unit, but not to each other. The face-capping alkyne ligand, $C_2Et_2$, forms a $\pi$ bond with the Pt atom and two $\sigma$ bonds with the Ru atoms, whereas the $\mu\text{-}\eta^5\text{-}C_5Et_4$ ligand adopts a $\eta^5$ bonding mode to the Ru spike and a $\sigma$ bond with the Pt cap.

The reaction of $Ru_5PtC(CO)_{16}$ **226** with two equivalents of $Me_3NO$ in dichloromethane followed by the addition of $Pt(\eta^4\text{-}C_8H_{12}\text{-}1,5)_2$ affords **227** and $Ru_5Pt_3C(CO)_{14}(\eta^4\text{-}C_8H_{12}\text{-}1,5)_2$ **231** in 8 and 5% yield, respectively.[129] Cluster **231** can also be made directly from **1** in 4% yield by reaction with $Pt(\eta^4\text{-}C_8H_{12}\text{-}1,5)_2$ in dichloromethane at 25°C for 30 min. The solid-state structure of **231** reveals a *cis*-bicapped octahedral metal framework that may be constructed from **225** (and subsequently **227**) by sequential capping of the $Ru_5PtC$ octahedral core by the $Pt(\eta^4\text{-}C_8H_{12}\text{-}1,5)$ fragment.

The pentaruthenium–rhodium cluster $Ru_5RhC(CO)_{14}(\eta\text{-}C_5Me_5)$ **232** may be isolated in 60% yield from the ionic coupling reaction between the dianion **5** and the monometallic dication, $[Rh(\eta\text{-}C_5Me_5)(NCMe)_3]^{2+}$ (Scheme 38).[131] The structure of **232** has been established by single crystal X-ray diffraction. The metal core comprises an octahedron with the pentamethylcyclopentadienyl ring remaining coordinated to the Rh atom.[131,132] This cluster reacts with excess cyclopentadiene and three equivalents of $Me_3NO$ to afford $Ru_5RhC(CO)_9(\eta\text{-}C_5Me_5)(\eta\text{-}C_5H_5)_2$ **233** in 20% yield.[131] Alternatively, the reaction of **232** with $Me_3NO$ in the presence of cyclohexa-1,3-diene affords $Ru_5RhC(CO)_{11}(\eta\text{-}C_5Me_5)(\mu_3\text{-}C_6H_6)$ **234** in 12% yield. The

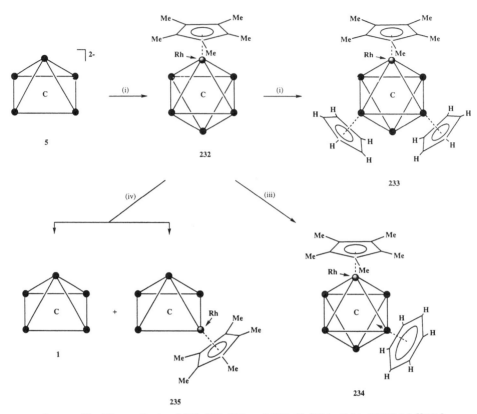

SCHEME 38.   The synthesis of **232, 233, 234,** and **235.** (i) [Rh($\eta$-C$_5$Me$_5$)(NCMe)$_3$][BF$_4$]$_2$, CH$_2$Cl$_2$, $\Delta$, 30 min; (ii) C$_5$H$_6$, CH$_2$Cl$_2$, 3 eq. Me$_3$NO, $-78°$C to RT; (iii) C$_6$H$_8$-1,3, CH$_2$Cl$_2$, 3 eq. Me$_3$NO, $-78°$C to RT; (iv) CO (80 atm), heptane, 80°C, 4 h.

solid-state structure of **233** shows that the two cyclopentadienyl rings bond to two consecutive Ru atoms on the cluster molecular equator. Cluster **232** undergoes fragmentation when heated in heptane at 80°C under 80 atmospheres of CO to afford **1** in 51% yield together with the pentanuclear cluster Ru$_4$RhC(CO)$_{12}$($\eta$-C$_5$Me$_5$) **235** in 36% yield.[133] The metal core in **235** adopts a square pyramidal geometry with the Rh atom occupying a position on the square base. As in **1** the carbide atom is located near the center of the square base but is displaced slightly below the Ru$_3$Rh plane.

The dianion [Ru$_5$C(CO)$_{14}$]$^{2-}$ **5** also undergoes an ionic coupling reaction with the monocation [Rh($\eta^4$-C$_8$H$_{12}$-1,5)$_2$]$^+$ in dichloromethane to afford the monoanionic cluster [Ru$_5$RhC(CO)$_{14}$($\eta^4$-C$_8$H$_{12}$-1,5)]$^-$ **236** in 64% yield.[134]

Cluster **236** may be protonated with $HBF_4 \cdot Et_2O$ in dichloromethane to afford the neutral cluster $HRu_5RhC(CO)_{14}(\eta^4\text{-}C_8H_{12}\text{-}1,5)$ **237** in 72% yield.

### B. *Reactions with Gold-Phosphine Fragments*

The reaction of $Ru_5C(CO)_{15}$ **1** with $XAuPPh_3$ (X = Cl or Br) in dichloromethane affords $Ru_5C(CO)_{15}(AuPPh_3)X$ (X = Cl **238** or Br **239**) (Scheme 39).[135] The solid-state structure of **238** is based on a bridged butterfly arrangement of Ru atoms with the $AuPPh_3$ fragment bridging the Ru–Ru hinge bond. The Cl atom coordinates to the bridging Ru atom. Clusters **238** and **239** readily lose one carbonyl ligand when stirred in a solution of dichloromethane for 28 h through which a stream of $N_2$ is passed forming $Ru_5C(CO)_{14}(AuPPh_3)(\mu\text{-}X)$ (X = Cl **240** or Br **241**). The structure of **241** has been established by single crystal X-ray diffraction, which shows that the loss of CO has been accompanied by a change in bonding of the Br atom from a terminal position to a bridging site. In this transformation the bromine changes from a one electron donor to a three electron donor and as such it is not necessary for the metal core to undergo a polyhedral rearrangement (i.e., closure to a square pyramidal structure). The reaction of **1** with $IAuPPh_3$ in dichloromethane affords $Ru_5C(CO)_{14}(AuPPh_3)(\mu\text{-}I)$ **242**, which is isostructural to clusters **240** and **241**.[49] Heating **242** in heptane under reflux for 2 h results in decarbonylation to yield $Ru_5C(CO)_{13}(AuPPh_3)(\mu\text{-}I)$ **243**. Cluster **243** may undergo an addition reaction with $PPh_3$ to afford $Ru_5C(CO)_{13}(PPh_3)(AuPPh_3)(\mu\text{-}I)$ **244**. The solid-state structure of **244** is based on a bridged butterfly arrangement

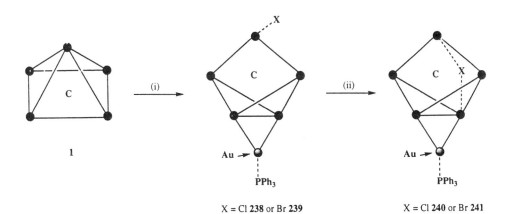

SCHEME 39.    The synthesis of **238/239** and **240/241**. (i) $XAuPPh_3$ (X = Cl or Br), $CH_2Cl_2$; (ii) $XAuPPh_3$ (X = Cl or Br), $CH_2Cl_2$ followed by $N_2$, 28 h.

of Ru atoms with the I atom bridging a hinge-to-bridge bond. The Au atom bridges the opposite hinge-hinge edge and the triphenylphosphine ligand coordinates to the bridging Ru atom. Alternatively, $[Ru_5C(CO)_{14}]^{2-}$ **5** reacts with two equivalents of $ClAuPR_3$ ($PR_3$ = $PEt_3$, $PPh_3$ or $PMe_2Ph$) in dichloromethane to afford $Ru_5C(CO)_{14}(AuPR_3)_2$ ($PR_3$ = $PEt_3$ **245**, $PPh_3$ **246**, or $PMe_2Ph$ **247**).[136] The structures of **245–247** are not known with certainty. However, these clusters react with CO (80 atm) in toluene at 60°C to afford the tetraruthenium-digold clusters $Ru_4C(CO)_{12}(AuPR_3)_2$ ($PR_3$ = $PEt_3$ **248**, $PPh_3$ **249**, or $PMe_2Ph$ **250**) in ca. 80% yield. The single crystal X-ray structure of **250** reveals that the four Ru atoms adopt a butterfly geometry with one $AuPMe_2Ph$ fragment bridging the wing tips of the butterfly and the other bridging the hinge.

The dianion **5** reacts with the chelating digold complex $[Au_2(Ph_2PCH_2)_2]^{2+}$ (generated from the dichloride by the addition of excess $TiPF_6$) in dichloromethane to afford $Ru_5C(CO)_{14}\{Au_2(Ph_2PCH_2)_2\}$ **251** in 80% yield together with the linked cluster compound $[\{Ru_5C(CO)_{14}\}_2\{Au_2(Ph_2PCH_2)_2\}]^{2-}$ **252** obtained in 10% yield.[137] The structure of **251** has been determined by single crystal X-ray diffraction and is shown in Fig. 14. The Ru₅C unit

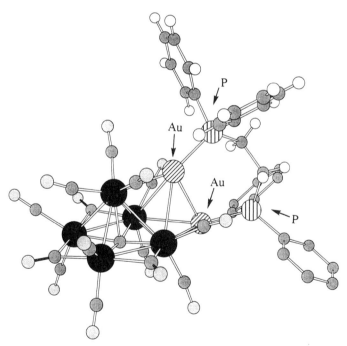

FIG. 14.   The solid-state structure of $Ru_5C(CO)_{14}\{Au_2(Ph_2PCH_2)_2\}$ **251**.

adopts a square pyramidal geometry. The two Au atoms cap one of the triangular faces of the cluster and also bond to each other.

The reaction of $[Ru_6C(CO)_{16}]^{2-}$ **4** with two equivalents of ClAuPR$_3$ (PR$_3$ = PMe$_3$, PEt$_3$, PPh$_3$, or PPh$_2$Me) affords the neutral clusters Ru$_6$C(CO)$_{16}$(AuPR$_3$)$_2$ (PR$_3$ = PMe$_3$ **253**, PEt$_3$ **254**, PPh$_3$ **255**, PPh$_2$Me **256**) (Scheme 40).[128] The solid-state structure of **256** has been established by single crystal X-ray diffraction and shows that two opposite apex-to-equator edges of the ruthenium octahedron are bridged by the AuPPh$_2$Me fragments. The reaction of **4** with the digold complex Cl$_2$Au$_2$(Ph$_2$PCH$_2$)$_2$ in dichloromethane in the presence of an excess of TiPF$_6$ affords Ru$_6$C(CO)$_{16}$ {Au$_2$(Ph$_2$PCH$_2$)$_2$} **257** in 80% yield.[138] The solid-state structure of **257** reveals that the Au$_2$(Ph$_2$PCH$_2$)$_2$ fragment bonds over a trimetal face of the Ru$_6$C cluster but not in the same way that it bonds to the Ru$_5$C cluster face in **251**.

The monoanion $[Ru_5C(CO)_{13}(NO)]^-$ **32** reacts with [AuPR$_3$][ClO$_4$] (R = Et or Ph) to afford Ru$_5$C(CO)$_{13}$(NO)(AuPR$_3$) (R = Et **258** or Ph **259**) in ca. 70% yield.[40] The structure of the triethylphosphine derivative has been established by single crystal X-ray diffraction, which shows that the square pyramidal core found in the parent compound remains intact. The NO ligand coordinates to a basal Ru atom in a radial position and the AuPEt$_3$ fragment straddles one of the triangular faces of the cluster not involved in bonding to the NO ligand. Similarly, $[Ru_6C(CO)_{15}(NO)]^-$ **26** reacts with ClAuPPh$_3$ in the presence of TiPF$_6$ to afford Ru$_6$C(CO)$_{15}$(NO)(AuPPh$_3$) **260**.[37,38] The structure of **260** is closely related to that of **258** with the AuPPh$_3$ unit capping a triangular face of the cluster. The anionic cluster $[Ru_6C(CO)_{15}(\mu\text{-SePh})]^-$ **65** also reacts with ClAuPPh$_3$ in dichloromethane in the presence of AgBF$_4$ to afford

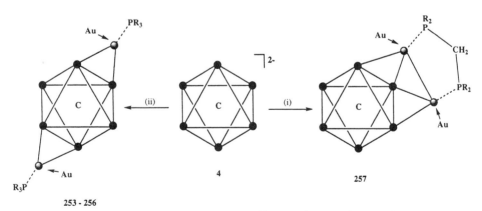

SCHEME 40.  The synthesis of **253–256** and **257**. (i) Cl$_2$Au$_2$(Ph$_2$PCH$_2$)$_2$, 2 eq. TlPF$_6$, CH$_2$Cl$_2$, overnight; (ii) ClAuPR$_3$ (PR$_3$ = PMe$_3$, PEt$_3$, PPh$_3$ or PPh$_2$Me).

$Ru_6C(CO)_{15}(\mu\text{-SePh})(AuPPh_3)$ **261** in 65% yield.[50] The single crystal X-ray structure of **261** reveals that the $AuPPh_3$ fragment bonds across an edge of the molecular equator. This differs to the bonding observed in clusters **258–260** where the gold fragment bonds over a trimetal face.

The reaction of **1** with an excess of $NaC_5H_5$ to afford $[Ru_5C(CO)_{13}(C_5H_5)]^-$ followed by protonation to give the neutral cluster $HRu_5C(CO)_{13}(\eta\text{-}C_5H_5)$ **103** was described in Section IV,H. If, however, $[AuPR_3][ClO_4]$ (R = Et or Ph) is added to the anion $[Ru_5C(CO)_{13}(C_5H_5)]^-$ instead of $H^+$ the clusters $Ru_5C(CO)_{13}(\eta\text{-}C_5H_5)(AuPR_3)$ (R = Et **262** or Ph **263**) are obtained in high yield.[63] The structure of the triphenylphosphine derivative **263** has been established by single crystal X-ray diffraction and can be derived from that of **103** by replacing the bridging hydride ligand by the isolobal $AuPPh_3$ fragment. In a related reaction, treatment of **1** with methyl lithium in diethyl ether followed by the addition of one equivalent of $ClAuPPh_3$ affords $Ru_5C(CO)_{14}(\mu\text{-}\eta^2\text{-COMe})(AuPPh_3)$ **264**.[63] Cluster **264** may also be prepared directly from the reaction of **1** with $MeAuPPh_3$. The solid-state structure of **264** reveals that the cluster skeleton adopts a distorted bridged butterfly geometry with the COMe ligand bridging one of the hinge-to-bridge edges: the C atom bonding to the hinge and the O atom coordinating to the bridging Ru atom. The Ru–Ru hinge is bridged by the gold unit.

Some clusters containing three different metals have also been prepared from some of the heteronuclear species described in Section VIII,A. The reaction of $[Ru_5MC(CO)_{17}]^{2-}$ (M = Cr **222**, Mo **223**, or W **224**) with $ClAuPEt_3$ affords the neutral clusters $Ru_5MC(CO)_{17}(AuPEt_3)_2$ (M = Cr **265**, Mo **266**, or W **267**) in 60–90% yield.[128] The structure of the tungsten derivative **267** has been established in the solid state and is shown in Fig. 15. The octahedral $Ru_5WC$ core is bridged asymmetrically by the two $AuPEt_3$ fragments interacting with one face of the cluster and each other. From IR data of **265–267** recorded in solution it is thought that different isomers are present.

The reaction of the monoanionic cluster $[Ru_5RhC(CO)_{14}(\eta^4\text{-}C_8H_{12})]^-$ **236** with $ClAuPR_3$ (R = Et or Ph) in dichloromethane affords $Ru_5RhC(CO)_{14}(\eta^4\text{-}C_8H_{12})(AuPR_3)$ (R = Et **268** or Ph **269**) in ca. 65% yield (Scheme 41).[134] The structure of the triphenylphosphine derivative **269** has been established by single crystal X-ray diffraction. The central cluster unit comprises an octahedron with the cycloocta-1,5-diene ligand coordinating to the Rh atom and the $AuPPh_3$ fragment bonds over a triruthenium face. The cycloocta-1,5-diene ligand in **268** and **269** may be displaced by two carbonyl ligands when heated to 80°C for 4 h under 25 atmospheres of CO. Several products are obtained from this reaction, including the simple substitution products $Ru_5RhC(CO)_{16}(AuPR_3)$ (R = Et **270** or Ph **271**)

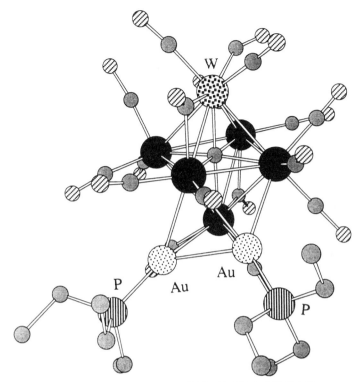

FIG. 15.   The solid-state structure of $Ru_5WC(CO)_{17}(AuPEt_3)_2$ **267.** The H atoms are not included.

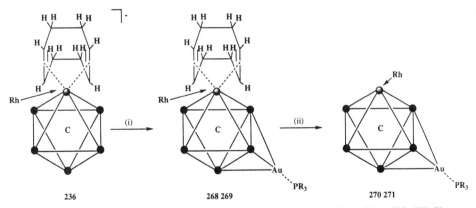

SCHEME 41.   The synthesis of **268/269** and **270/271.** (i) $ClAuPR_3$ (R = Et or Ph), $CH_2Cl_2$,
10 min; (ii) CO (80 atm), heptane, 80°C, 4 h.

isolated in ca. 30% yield and the breakdown products $Ru_4RhC(CO)_{14}$ $(AuPR_3)$ ($R$ = Et **272** or Ph **273**) obtained in ca. 30% yield.

### C. Reactions with Mercury and Copper Reagents

The reaction of $Ru_5C(CO)_{15}$ **1** with an equimolar equivalent of $Hg(CF_3)(CF_3CO_2)$ in dichloromethane at room temperature for 1 h affords $Ru_5C(CO)_{15}(HgCF_3)(OCOCF_3)$ **274** in quantitative yield (Scheme 42).[139] The solid-state structure of **274** may be viewed as consisting of a bridged butterfly arrangement of Ru atoms with the hinge edge bridged by the $HgCF_3$ unit. The $OCOCF_3$ moiety bonds to the bridging Ru atoms via one of the O atoms. Cluster **1** also reacts with $HgCl_2$ in dichloromethane at room temperature for 20 h affording the mercury-linked cluster $\{Ru_5C(CO)_{14}(\eta\text{-}Cl)\}_2Hg_2Cl_2$ **275** in ca. 45% yield.[140] The solid-state structure of **275** shows that two bridged butterfly $Ru_5C$ units are connected via $Hg_2Cl_2$ unit in which the Hg atoms bond across the hinges of the clusters. In the gas phase the Hg–Cl bonds cleave and the cluster dissociates into two monomeric units formulated as $Ru_5C(CO)_{14}(\mu\text{-}Cl)(HgCl)$ **276**.

The hexanuclear dianion $[Ru_6C(CO)_{16}]^{2-}$ **4** reacts with $Hg(CF_3CO_2)_2$ in equimolar quantities to afford a polymeric material that has not, as yet, been fully characterized.[140] However, if two equivalents of **4** are reacted with $Hg(CF_3CO_2)_2$, the mercury-linked cluster $[\{Ru_6C(CO)_{16}\}_2Hg]^{2-}$ **277** is isolated in ca. 80% yield (Scheme 43). Cluster **277** may also be isolated in 82% yield from the reaction of **4** and $HgCl_2$. The structure of **277** has been established by single crystal X-ray diffraction. Two octahedral $Ru_6C$ units are linked via a Hg atom that bridges an edge on the molecular equator of both the clusters. The dianion **4** reacts with $Hg(CF_3)(CF_3CO_2)$ in dichloromethane at room temperature for 30 min to afford the monoanion $[Ru_6C\text{-}(CO)_{16}(HgCF_3)]^-$ **278** in quantitative yield.[140] The location of the $HgCF_3$ unit in **278** is not known with certainty, but by analogy to the structure of **277**, it is possible that it bridges two consecutive Ru atoms on the molecular equator of the $Ru_6C$ octahedron.

Treatment of **4** with $[Cu(NCMe)_4][BF_4]$ in acetone affords $Ru_6Cu_2C\text{-}(CO)_{16}(NCMe)_2$ **279** in 77% yield.[141] The structure of **279** has been established by single crystal X-ray diffraction. The metal core consists of a central $Ru_6C$ octahedron with the two Cu atoms bonding over a trimetal face of the cluster as well as to each other as shown in Scheme 43. Cluster **4** also reacts with five equivalents of CuCl in refluxing THF for 4 h to afford $[\{Ru_6Cu_2C(CO)_{16}\}_2Cl_2]^{2-}$ **280** in quantitative yield.[142] The solid-state structure of **280** comprises two octahedral $Ru_6C$ units linked by a rectangular planar $Cu_4$ array.

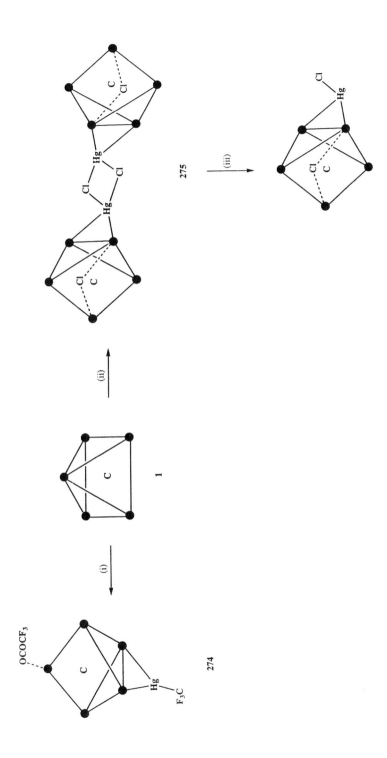

SCHEME 42. The synthesis of **274**, **275**, and **276**. (i) Hg(CF$_3$)(CF$_3$CO$_2$), CH$_2$Cl$_2$, RT, 1 h; (ii) HgCl$_2$, CH$_2$Cl$_2$, RT, 20 h; (iii) FAB mass spectrometry.

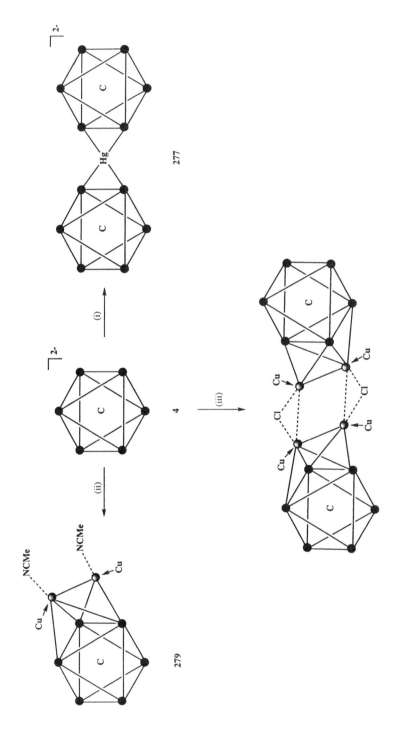

SCHEME 43. The synthesis of **277**, **279**, and **280**. (i) 2 eq. Hg(CF₃CO₂)₂, CH₂Cl₂, RT, 30 min or HgCl₂, CH₂Cl₂, RT, 15 min; (ii) [Cu(NCMe)₄][BF₄], propanone; (iii) CuCl, THF, Δ, 4 h.

## D. *Other Methods*

The thermolysis of $Ru_3(CO)_{12}$ **3** with the cluster-acetylide complex $Ru_2W$-$(CO)_8(\eta\text{-}C_5H_5)(\mu_3\text{-}C_2Ph)$ in heptane under reflux for 17 h affords the penta-nuclear cluster $Ru_4WC(CO)_{12}(\eta\text{-}C_5H_5)(\mu\text{-}CPh)$ **281** in 68% yield and the hexanuclear cluster $Ru_5WC(CO)_{14}(\eta\text{-}C_5H_5)(\mu\text{-}CPh)$ **282** in 19% yield (Scheme 44).[143] The structures of these clusters have been established by single crystal X-ray diffraction and are closely related. The metal skeleton of cluster **281** consists of a square pyramid with the W atom occupying a position in the base of the cluster. The CPh ligand forms a double bond to the W atom and a single bond to the apical Ru atom. These same features are also observed in **282** but in addition the square face of the cluster is capped by a further Ru atom, thereby forming an octahedron. Cluster **282** may be generated from **281** in 60% yield when it is heated with **3** in toluene under reflux.

The dimer $Mo_2(CO)_4(\eta\text{-}C_5H_5)_2$ contains a reactive $Mo\equiv Mo$ triple bond and when heated with **3** in toluene under reflux for 5 h affords the octahedral

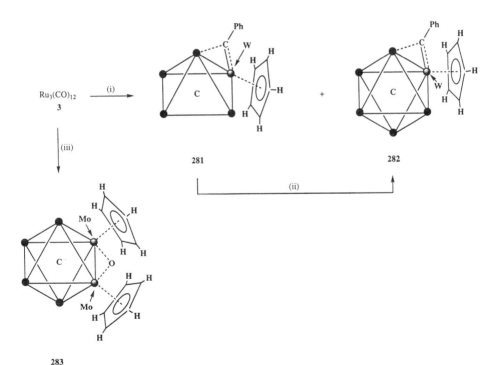

SCHEME 44.   The synthesis of **281, 282,** and **283.** (i) $Ru_2W(CO)_8(\eta\text{-}C_5H_5)(\mu_3\text{-}C_2Ph)$, heptane, $\Delta$, 17 h; (ii) $Ru_3(CO)_{12}$, toluene, $\Delta$; (iii) $Mo_2(CO)_4(\eta\text{-}C_5H_5)_2$, toluene, $\Delta$, 24 h.

cluster $Ru_4Mo_2C(\mu\text{-}O)(CO)_{12}(\eta\text{-}C_5H_5)_2$ **283** in 41% yield (see Scheme 45).[144] The solid-state structure of **283** reveals that the cluster skeleton is based on an octahedron comprising four Ru atoms and two Mo atoms, which accommodates a carbide atom. A cyclopentadienyl ligand coordinates to each of the Mo atoms (as in the precursor compound) and the Mo–Mo bond is bridged by the oxo atom.

The thermolysis of $Ru_5(\mu_5\text{-}C_2PPh_2)(CO)_{13}(\mu\text{-}PPh_2)$ **189** with $Fe_2(CO)_9$ in toluene at 100°C for 40 h affords several compounds, including the heteronuclear species $Ru_5FeC(CO)_{14}(\mu\text{-}PPh_2)_2$ **284** and $Ru_5Fe_3C_2(CO)_{17}(\mu\text{-}PPh_2)_2$ **285** both in ca. 35% yield. The by-products $Ru_4FeC(CO)_{13}(\mu\text{-}dppm)$ **286** and $Ru_5C(CO)_{13}(\mu\text{-}dppm)$ **287** are also obtained in 1 and 4% yield, respectively.[145,146] These clusters have been characterized in the solid state by single crystal X-ray diffraction. The dicarbide cluster **285** is isostructural to the homonuclear cluster $Ru_8C_2(CO)_{17}(\mu\text{-}PPh_2)_2$ **215** (see Fig. 13) consisting of face-sharing octahedral and square pyramidal cluster units, each incorporating a carbide atom. The cluster core of **284** adopts an octahedral geometry with the Fe atom located in an apical position. The phosphido ligands adopt bridging coordination modes, one across an edge of the molecular equator and the other bridging an apex-to-equator edge of the cluster. Cluster **286** has a square pyramidal metal polyhedron with the Fe atom occupying the apical site. The dppm ligand bridges two *cis*-basal Ru atoms in axial positions below the $Ru_4$ plane. The homonuclear cluster **287** is isostructural to the mixed-metal cluster **286.**

## IX

## STEREOCHEMICAL NONRIGID BEHAVIOR

The preceding sections described the synthesis and structures of a large number of ruthenium–carbide clusters derived from $Ru_5C(CO)_{15}$ **1** and $Ru_6C(CO)_{17}$ **2**. Due to the diverse ways in which ligands can bond to these metal cores (sometimes altering the structure of the metal skeleton itself), most effort has been directed toward establishing their structures in the solid state using single crystal X-ray diffraction. Some of these clusters have also been found to exhibit stereochemical nonrigid behavior in solution and these processes are described in this section.

The single crystal X-ray structure of the *bis*-cyclohexa-1,4-diene cluster $Ru_5C(CO)_{11}(\mu\text{-}C_6H_8\text{-}1,4)_2$ **122** shows that the two diene ligands occupy different positions on the square pyramidal $Ru_5C$ core (see Scheme 23). The ligands coordinate across opposite edges of the square base with one having replaced two radial carbonyl ligands and the other two axial carbonyl ligands. It was found that the cyclohexa-1,4-diene ligands undergo site exchange on the NMR time scale, and at room temperature the rate of exchange was estimated using EXSY spectroscopy to be approximately $31\ s^{-1}$.[73] As the temperature is decreased to ca. 233 K, the $^1$H-NMR spectrum indicates that the two $C_6H_8$ ligands are in such slow exchange that the molecule is essentially the same as that of the crystallographically determined structure. At higher temperatures ($>$330 K) the situation is less clear due to sample decomposition; however, it would appear that in addition to the site exchange outlined earlier, the ligands undergo rapid rotation and flexing.

The fulvene ligand in the related clusters $Ru_6C(CO)_{14}(\mu_3\text{-}C_5H_4CPh_2)$ **110** and $Ru_6C(CO)_{14}(\mu_3\text{-}C_5H_4CMe_2)$ **111** undergo fluxional behavior.[67,147] $^1$H-NMR spectra of **110** and **111** at several different temperatures between 208 and 298 K show changes in the number of signals present and their broadness. This has been attributed to a swiveling movement of the ring relative to the underlying triruthenium face. The type of swiveling motion proposed is illustrated in Scheme 45. The transition state free energy for the swivel process has been estimated from the coalescence temperature of equivalent signals and for **110** and **111** is 52 and 47 $kJmol^{-1}$, respectively.

Migration of ligands has been observed in some of the reactions outlined in Section IV,I. For example, the benzene ligand in $Ru_5C(CO)_{12}(\mu_3\text{-}C_6H_6)$ **123** migrates from the cluster face to a basal Ru atom affording $Ru_5C(CO)_{12}(\eta\text{-}C_6H_6)$ **123**.[70,71] This process is not reversible under thermal conditions. However, variable temperature $^1$H-NMR spectroscopy recorded between 295 and 385 K reveals that the toluene ligands in the hexaruthenium cluster $Ru_6C(CO)_{11}(C_6H_5Me)_2$ **145** migrate over the cluster surface in solution.[80] Two structures are believed to exist in equilibrium in solution. One is like the crystallographically determined structure (**145a**) with one toluene ring bonded in terminal mode and the other over a trimetal face, whereas in the other molecule (**145b**) the two toluene rings are thought to coordinate to the apices of the cluster as shown in Scheme 46. At

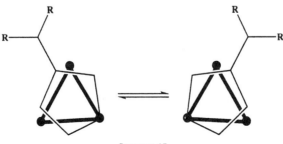

SCHEME 45

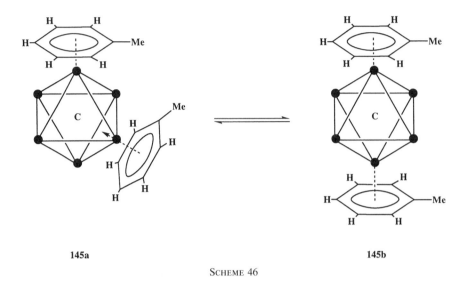

145a                                                   145b

SCHEME 46

295 K the mixture of **145a : 145b** is estimated to be 9 : 1 and this ratio changes to approximately 1 : 1 at 385 K. From these data, $\Delta H^{\dagger}$ was estimated as $16\,kJmol^{-1}$, a value that suggests the process is fluxional, although the actual process appears to be somewhat slower. Saturation transfer experiments showed no exchange between the $\mu_3$ and $\eta^6$ toluene groups, even with a 30s mixing time. However, the partly deuterated cluster $Ru_6C(CO)_{11}(\eta^6\text{-}C_6D_5CD_3)(\mu_3\text{-}C_6H_5Me)$ was prepared, and the initial $^1H$-NMR spectrum only contained peaks that corresponded to the face-capping ring. After a few hours, peaks corresponding to both rings were observed. This implies that scrambling of the toluene ligand between the two potential sites takes place and these observations are consistent with a unimolecular process of the type shown in Scheme 46.

# X

## CONCLUDING REMARKS

Clusters $Ru_5C(CO)_{15}$ **1** and $Ru_6C(CO)_{17}$ **2** undergo a wide range of reactions, including many which are also common to mononuclear species, such as substitution, addition, and insertion reactions. Some of these reactions are accompanied by rearrangements of the cluster polyhedron. Other reactions that take place are typically encountered within cluster chemistry, including reactions involving a change of nuclearity and also the formation of heteronuclear species. The complexity of reactions in cluster chemistry often means that several products are produced in a reaction and yields are relatively low. Although certain reactions of **1** and **2** are of this nature, others have been shown to give a single product in near quantitative yield. It is perhaps reactions of this type that will be given most attention in the future. Another aspect that may also attract attention is the use of these clusters, or their derivatives, as catalysts. Clearly, the ability of **1** and some of its derivatives to undergo facile nucleophilic addition reactions could prove to be a valuable step in catalytic processes as it may be possible for the nucleophile to undergo a transformation once coordinated to the cluster prior to removal. A number of organic transformations also take place while the organic molecule is coordinated to the $Ru_5C$ or $Ru_6C$ unit and this could potentially be used to prepare new compounds or alternative routes to known ones. For example, preliminary studies indicate that the $Ru_6C(CO)_{14}$ unit is not dissimilar to the $Cr(CO)_3$ fragment when an arene ring is bonded to it. It may even be possible to introduce potential reactants via initial coordination to an adjacent metal on the cluster, which is not possible using the $Cr(CO)_3$ group.

ACKNOWLEDGEMENTS

I thank the Royal Society for a University Research Fellowship. I also thank Dr. Philip Dyer and Anna Williamson for their assistance in preparing some of the figures.

REFERENCES

(1) Chini, P.; Longoni, G.; Albano, V. G. *Adv. Organomet. Chem.* **1976**, *14*, 285.
(2) Johnson, B. F. G.; Lewis, J. *Adv. Inorg. Chem. Radiochem.* **1980**, *24*, 225.
(3) *Transition Metal Clusters;* Johnson, B. F. G., Ed.; Wiley: Chichester, 1980.
(4) Vargas, M. D.; Nicholls, J. N. *Adv. Inorg. Chem. Radiochem.* **1986**, *30*, 123.
(5) *Introduction to Cluster Chemistry,* Mingos, D. M. P.; Wales, D. J. Prentice-Hall, New Jersey, 1990.
(6) *The Chemistry of Metal Cluster Complexes;* Shriver, D. F.; Kaesz, H. D.; Adams, R. D., Eds.; VCH: Weinheim, 1990.
(7) Tachikawa, M.; Muetterties, E. L. *Prog. Inorg. Chem.* **1981**, *28*, 203.
(8) Bradley, J. S. *Adv. Organomet. Chem.* **1983**, *22*, 1.
(9) Johnson, B. F. G.; Johnston, R. D.; Lewis, J. *J. Chem. Soc., Chem. Commun.* **1967**, 1057.
(10) Johnson, B. F. G.; Johnston, R. D.; Lewis, J. *J. Chem. Soc. (A)* **1968**, 2865.
(11) Mason, R.; Robinson, W. R. *J. Chem. Soc., Chem. Commun.* **1968**, 468.
(12) Braye, E. H.; Hubel, W.; Dahl, L. F.; Wampler, D. L. *J. Am. Chem. Soc.* **1962**, *84*, 4633.
(13) Piacenti, F.; Bianchi, M.; Benedetti, E. *J. Chem. Soc., Chem. Commun.* **1967**, 775.
(14) Eady, C. R.; Johnson, B. F. G.; Lewis, J. *J. Chem. Soc., Dalton Trans.* **1975**, 2606.
(15) Johnson, B. F. G.; Lewis, J.; Sankey, S. W.; Wong, K.; McPartlin, M.; Nelson, W. J. H. *J. Organomet. Chem.* **1980**, *191*, C3.

(16) Bradley, J. S.; Ansell, G. B.; Hill, E. W. *J. Organomet. Chem.* **1980**, *184*, C33.

(17) Hayward, C.-M. T.; Shapley, J. R. *Inorg. Chem.* **1982**, *21*, 3816.

(18) Churchill, M. R.; Wormald, J. *J. Chem. Soc., Dalton Trans.* **1974**, 2410.

(19) Eady, C. R.; Johnson, B. F. G.; Lewis, J.; Mather, T. *J. Organomet. Chem.* **1973**, *57*, C82.

(20) Farrar, D. H.; Jackson, P. F.; Johnson, B. F. G.; Lewis, J.; Nicholls, J. N.; McPartlin, M. *J. Chem. Soc., Chem. Commun.* **1981**, 415.

(21) Johnson, B. F. G.; Lewis, J.; Nicholls, J. N.; Puga, J.; Raithby, P. R.; McPartlin, M.; Clegg, W. *J. Chem. Soc., Dalton Trans.* **1983**, 277.

(22) Humphrey, D. G. Monash University. Unpublished results.

(23) Heaton, B. T.; Iggo, J. A.; Longoni, G.; Mulley, S. *J. Chem. Soc., Dalton Trans.* **1995**, 1985.

(24) Halet, J. F.; Evans, D. G.; Mingos, D. M. P. *J. Am. Chem. Soc.* **1988**, *110*, 87.

(25) Elian, M.; Chen, M. M. L.; Mingos, D. M. P.; Hoffmann, R. *Inorg. Chem.* **1976**, *15*, 1148.

(26) Sirigu, A.; Bianchi, M.; Benedetti, E. *J. Chem. Soc., Chem. Commun.* **1969**, 596.

(27) Braga, D.; Grepioni, F.; Dyson, P. J.; Johnson, B. F. G.; Frediani, P.; Bianchi, M.; Piacenti, F. *J. Chem. Soc., Dalton Trans.* **1992**, 2565.

(28) Drake, S. R.; Johnson, B. F. G.; Lewis, J. *J. Chem. Soc., Chem. Commun.* **1988**, 1033.

(29) Drake, S. R.; Johnson, B. F. G.; Lewis, J. *J. Chem. Soc., Dalton Trans.* **1989**, 243.

(30) Clark, R. J. H.; Dyson, P. J.; Humphrey, D. G.; Johnson, B. F. G. *Polyhedron* (in press).

(31) Peake, B. M.; Robinson, B. H.; Simpson, J.; Watson, D. J. *J. Chem. Soc., Chem. Commun.* **1974**, 945.

(32) Chihara, T.; Yamazaki, H. *J. Cluster Sci.* **1992**, *3*, 489.

(33) Chihara, T.; Yamazaki, H. *J. Chem. Soc., Dalton Trans.* **1995**, 1369.

(34) Johnson, B. F. G.; Lewis, J.; Nicholls, J. N.; Oxton, I. A.; Raithby, P. R.; Rosales, M. J. *J. Chem. Soc., Chem. Commun.* **1982**, 289.

(35) Way, C.-J.; Chi, Y.; Mavunkal, I. J.; Wang, S.-L.; Liao, F.-L.; Peng, S.-M.; Lee, G.-J. *J. Cluster Sci.* **1997**, *8*, 61.

(36) Dyson, P. J. Imperial College of Science, Technology and Medicine. Unpublished results.

(37) Johnson, B. F. G.; Lewis, J.; Nelson, W. J. H.; Puga, J.; Raithby, P. R.; Braga, D.; McPartlin, M.; Clegg, W. *J. Organomet. Chem.* **1983**, *243*, C13.

(38) Johnson, B. F. G.; Lewis, J.; Nelson, W. J. H.; Puga, J.; McPartlin, M.; Sironi, A. *J. Organomet. Chem.* **1983**, *253*, C5.

(39) Chihara, T.; Sawamura, K.; Ikezawa, H.; Ogawa, H.; Wakatsuki, Y. *Organometallics* **1996**, *15*, 415.

(40) Hendrick, K.; Johnson, B. F. G.; Lewis, J.; Mace, J.; McPartlin, M.; Morris, M. *J. Chem. Soc., Chem. Commun.* **1985**, 1617.

(41) Farrar, D. H.; Poë, A. J.; Zheng, Y. *J. Am. Chem. Soc.* **1994**, *116*, 6252.

(42) Evans, J.; Gracey, B. P.; Gray, L. R.; Webster, M. *J. Organomet. Chem.* **1982**, *240*, C61.

(43) Cook, S. L.; Evans, J.; Gray, L. R.; Webster, M. *J. Chem. Soc., Dalton Trans.* **1986**, 2149.

(44) Johnson, B. F. G.; Lewis, J.; Wong, K.; McPartlin, M. *J. Organomet. Chem.* **1980**, *185*, C17.

(45) Brown, S. C.; Evans, J.; Webster, M. *J. Chem. Soc., Dalton Trans.* **1981**, 2263.

(46) Adatia, T.; Conole, G.; Drake, S. R.; Johnson, B. F. G.; Kessler, M.; Lewis, J.; McPartlin, M. *J. Chem. Soc., Dalton Trans.* **1997**, 669.

(47) Blake, A. J.; Harrison, A.; Johnson, B. F. G.; McInnes, E. J. L.; Parsons, S.; Shephard, D. S.; Yellowlees, L. J. *Organometallics* **1995**, *14*, 3160.

(48) Johnson, B. F. G.; Lewis, J.; Saharan, V. P.; Wong, W. T. *J. Chem. Soc., Chem. Commun.* **1991**, 365.

(49) Cowie, A. G.; Johnson, B. F. G.; Lewis, J.; Nicholls, J. N.; Raithby, P. R.; Rosales, M. J. *J. Chem. Soc., Dalton Trans.* **1983**, 2311.

(50) Chihara, T.; Yamazaki, H. *J. Organomet. Chem.* **1992**, *428*, 169.

(51) Adams, R. D.; Falloon, S. B.; McBride, K. T. *Organometallics* **1994**, *13*, 4870.
(52) Adams, R. D.; Falloon, S. B.; McBride, K. T.; Yamamoto, J. H. *Organometallics* **1995**, *14*, 1739.
(53) Chihara, T.; Aoki, K.; Yamazaki, H. *J. Organomet. Chem.* **1990**, *383*, 367.
(54) Drake, S. R.; Johnson, B. F. G.; Lewis, J.; Conole, G.; McPartlin, M. *J. Chem. Soc., Dalton Trans.* **1990**, 995.
(55) Adams, R. D.; Wu, W. *Organometallics* **1993**, *12*, 1238.
(56) Mallors, R. L.; Blake, A. J.; Dyson, P. J.; Johnson, B. F. G.; Parsons, S. *Organometallics* **1997**, *16*, 1668.
(57) Haggitt, J. L.; Johnson, B. F. G.; Blake, A. J.; Parsons, S. *J. Chem. Soc., Chem. Commun.* **1995**, 1263.
(58) Blake, A. J.; Haggitt, J. L.: Johnson, B. F. G.; Parsons, S. *J. Chem. Soc., Dalton Trans.* **1997**, 991.
(59) Chihara, T.; Sawamura, K.; Ogawa, H.; Wakatsuki, Y. *J. Chem. Soc., Chem. Commun.* **1994**, 1179.
(60) Chihara, T.; Jesorka, A.; Ikezawa, H.; Wakatsuki, Y. *J. Chem. Soc., Dalton Trans.* **1997**, 443.
(61) Bailey, P. J.; Blake, A. J.; Dyson, P. J.; Ingham, S. L.; Johnson, B. F. G. *J. Chem. Soc., Chem. Commun.* **1994**, 2233.
(62) Dyson, P. J.; Ingham, S. L.; Johnson, B. F. G.; McGrady, J. E.; Mingos, D. M. P.; Blake, A. J. *J. Chem. Soc., Dalton Trans.* **1995**, 2749.
(63) Cowie, A. G.; Johnson, B. F. G.; Lewis, J.; Nicholls, J. N.; Raithby, P. R.; Swanson, A. G. *J. Chem. Soc., Chem. Commun.* **1984**, 637.
(64) Dyson, P. J.; Johnson, B. F. G.; Blake, A. J.; Braga, D.; Byrne, J. J.; Grepioni, F. *Polyhedron* **1995**, *14*, 2697.
(65) Blake, A. J.; Dyson, P. J.; Gash, R. C.; Johnson, B. F. G.; Trickey, P. *J. Chem. Soc., Dalton Trans.* **1994**, 1105.
(66) Blake, A. J.; Johnson, B. F. G.; Parsons, S.; Shephard, D. S. *J. Chem. Soc., Dalton Trans.* **1995**, 495.
(67) Blake, A. J.; Dyson, P. J.; Johnson, B. F. G.; Parsons, S.; Reed, D.; Shephard, D. S. *Organometallics* **1995**, *14*, 4199.
(68) Gomez-Sal, M. P.; Johnson, B. F. G.; Lewis, J.; Raithby, P. R.; Wright, A. H. *J. Chem. Soc., Chem. Commun.* **1985**, 1682.
(69) Dyson, P. J.; Johnson, B. F. G.; Reed, D.; Braga, D.; Grepioni, F.; Parisini, E. *J. Chem. Soc., Dalton Trans.* **1993**, 2817.
(70) Bailey, P. J.; Braga, D.; Dyson, P. J.; Grepioni, G.; Johnson, B. F. G.; Lewis, J.; Sabatino, P. *J. Chem. Soc., Chem. Commun.* **1992**, 177.
(71) Braga, D.; Grepioni, F.; Sabatino, P.; Dyson, P. J.; Johnson, B. F. G.; Lewis, J.; Bailey, P. J.; Raithby, P. R.; Stalke, D. *J. Chem. Soc., Dalton Trans.* **1993**, 985.
(72) Dyson, P. J.; Johnson, B. F. G.; Lewis, J.; Braga, D.; Sabatino, P. *J. Chem. Soc., Chem. Commun.* **1993**, 301.
(73) Dyson, P. J.; Johnson, B. F. G.; Reed, D.; Braga, D.; Sabatino, P. *Inorg. Chim. Acta* **1993**, *213*, 191.
(74) Dyson, P. J.; Johnson, B. F. G.; Braga, D. *Inorg. Chim. Acta* **1994**, *222*, 299.
(75) Brown, D. B.; Dyson, P. J.; Johnson, B. F. G.; Parker, D. *J. Organomet. Chem.* **1995**, *491*, 189.
(76) Braga, D.; Sabatino, P.; Dyson, P. J.; Blake, A. J.; Johnson, B. F. G.; *J. Chem. Soc., Dalton Trans.* **1994**, 393.
(77) Dyson, P. J.; Johnson, B. F. G.; Lewis, J.; Martinelli, M.; Braga, D.; Grepioni, F. *J. Am. Chem. Soc.* **1993**, *115*, 9062.

(78) Braga, D.; Grepioni, F.; Righi, S.; Dyson, P. J.; Johnson, B. F. G.; Bailey, P. J.; Lewis, J. *Organometallics* **1992**, *11*, 4042.

(79) Braga, D.; Grepioni, F.; Righi, S.; Johnson, B. F. G.; Bailey, P. J.; Dyson, P. J.; Lewis, J.; Martinelli, M. *J. Chem. Soc., Dalton Trans.* **1992**, 2121.

(80) Braga, D.; Grepioni, F.; Parisini, E.; Dyson, P. J.; Johnson, B. F. G.; Reed, D.; Shephard, D. S.; Bailey, P. J.; Lewis, J. *J. Organomet. Chem.* **1993**, *462*, 301.

(81) Braga, D.; Grepioni, F.; Martin, C. M.; Parisini, E.; Dyson, P. J.; Johnson, B. F. G. *Organometallics* **1994**, *13*, 2170.

(82) Dyson, P. J.; Johnson, B. F. G.; Braga, D.; Grepioni, F.; Martin, C. M.; Parisini, E. *Inorg. Chim. Acta* **1995**, *235*, 413.

(83) Adams, R. D.; Wu, W. *Organometallics* **1993**, *12*, 1243.

(84) Adams, R. D.; Wu, W. *Polyhedron* **1992**, *2*, 2123.

(85) Edwards, A. J.; Johnson, B. F. G.; Parsons, S.; Shephard, D. S. *J. Chem. Soc., Dalton Trans.* **1996**, 3837.

(86) Braga, D.; Dyson, P. J.; Grepioni, F.; Johnson, B. F. G.; Calhorda, M. *Inorg. Chem.* **1994**, *33*, 3218.

(87) Mallors, R. L.; Blake, A. J.; Parsons, S.; Johnson, B. F. G.; Dyson, P. J.; Braga, D.; Grepioni, F.; Parisini, E. *J. Organomet. Chem.* **1997**, *532*, 133.

(88) Borchert, T.; Lewis, J.; Pritzkow, H.; Raithby, P. R.; Wadepohl, H. *J. Chem. Soc., Dalton Trans.* **1995**, 1061.

(89) Ansell, G. B.; Bradley, J. S. *Acta Cryst.* **1980**, *B36*, 1930.

(90) Conole, G.; McPartlin, M.; Powell, H. R.; Dutton, T.; Johnson, B. F. G.; Lewis, J. *J. Organomet. Chem.* **1989**, *379*, C1.

(91) Jackson, P. F.; Johnson, B. F. G.; Lewis, J.; Raithby, P. R.; Will, G. J. *J. Chem. Soc., Chem. Commun.* **1980**, 1190.

(92) Brown, D. B.; Johnson, B. F. G.; Martin, C. M.; Parsons, S. *J. Organomet. Chem.* **1997**, *536–537*, 285.

(93) Farrugia, L. J. *Acta Cryst.* **1988**, *C44*, 997.

(94) Anson, C. E.; Bailey, P. J.; Conole, G.; Johnson, B. F. G.; Lewis, J.; McPartlin, M.; Powell, H. R. *J. Chem. Soc., Chem. Commun.* **1989**, 442.

(95) Bailey, P. J.; Duer, M. J.; Johnson, B. F. G.; Lewis, J.; Conole, G.; McPartlin, M.; Powell, H. R.; Anson, C. E. *J. Organomet. Chem.* **1990**, *383*, 441.

(96) Kolehmainen, E.; Laihia, K.; Korvola, J.; Kaganovich, V. S.; Rybinsaya, M. I.; Kerzina, Z. A. *J. Organomet. Chem.* **1995**, *487*, 215.

(97) Blake, A. J.; Johnson, B. F. G.; Reed, D.; Shephard, D. S. *J. Chem. Soc., Dalton Trans.* **1995**, 843.

(98) Kagaonvich, V. S.; Kerzina, Z. A.; Asunta, T.; Weikström, K.; Rybinsaya, M. I. *J. Organomet. Chem.* **1991**, *421*, 117.

(99) Braga, D.; Grepioni, F.; Parisini, E.; Dyson, P. J.; Blake, A. J.; Johnson, B. F. G.; *J. Chem. Soc., Dalton Trans.* **1993**, 2951.

(100) Blake, A. J.; Dyson, P. J.; Ingham, S. L.; Johnson, B. F. G.; Martin, C. M. *Inorg. Chim. Acta* **1995**, *240*, 29.

(101) Dyson, P. J.; Geade, P. E.; Johnson, B. F. G.; McGrady, J. E.; Parsons, S. *J. Cluster Sci.* **1997**, *18*, 533.

(102) Zheng, T. C.; Cullen, W. R.; Rettig, S. J. *Organometallics* **1994**, *13*, 3594.

(103) Adams, R. D.; Babin, J. E.; Tanner, J. *Organometallics* **1988**, *7*, 765.

(104) Bruce, M. I.; Skelton, B. W.; White, A. H.; Williams, J. L. *J. Chem. Soc., Chem. Commun.* **1985**, 744.

(105) Bruce, M. I.; Williams, M. L.; Skelton, B. W.; White, A. H. *J. Organomet. Chem.* **1989**, *369*, 393.

(106) Adams, C. J.; Bruce, M. I.; Skelton, B. W.; White, A. H. *J. Organomet. Chem.* **1993**, *445*, 199.
(107) Adams, C. J.; Bruce, M. I.; Skelton, B. W.; White, A. H. *J. Cluster Sci.* **1994**, *5*, 419.
(108) Adams, R. D.; Mathur, P.; Segmüller, B. E. *Organometallics* **1983**, *2*, 1258.
(109) Coston, T.; Lewis, J.; Wilkinson, D.; Johnson, B. F. G. *J. Organomet. Chem.* **1991**, *407*, C13.
(110) Braga, D.; Grepioni, F.; Parisini, E.; Dyson, P. J.; Johnson, B. F. G.; Martin, C. M. *J. Chem. Soc., Dalton Trans.* **1995**, 909.
(111) Eady, C. R.; Johnson, B. F. G.; Lewis, J. *J. Organomet. Chem.* **1972**, *39*, 329.
(112) Muetterties, E. L. *Bull. Soc. Chim. Belg.* **1975**, *84*, 959.
(113) Muetterties, E. L. *Bull. Soc. Chim. Belg.* **1976**, *85*, 451.
(114) Dyson, P. J.; Johnson, B. F. G.; Martin, C. M.; Braga, D.; Grepioni, F. *J. Chem. Soc., Chem. Commun.* **1995**, 771.
(115) Dyson, P. J.; Johnson, B. F. G.; Martin, C. M.; Reed, D.; Braga, D.; Grepioni, F. *J. Chem. Soc., Dalton Trans.* **1995**, 4113.
(116) Dyson, P. J.; Johnson, B. F. G.; Martin, C. M.; Blake, A. J.; Braga, D.; Grepioni, F. *Organometallics* **1994**, *13*, 2113.
(117) Bailey, P. J.; Johnson, B. F. G.; Lewis, J. *Inorg. Chim. Acta* **1994**, *227*, 197.
(118) Blake, A. J.; Dyson, P. J.; Ingham, S. L.; Johnson, B. F. G.; Martin, C. M. *Transition Met. Chem.* **1995**, *20*, 577.
(119) Blake, A. J.; Dyson, P. J.; Gaede, P. E.; Johnson, B. F. G.; Braga, D.; Parasini, E. *J. Chem. Soc., Dalton Trans.* **1995**, 3431.
(120) Adams, C. J.; Bruce, M. I.; Skelton, B. W.; White, A. H. *Aust. J. Chem.* **1993**, *46*, 1811.
(121) Hayward, C.-M. T.; Shapley, J. R.; Churchill, M. R.; Bueno, C.; Rheingold, A. L. *J. Am. Chem. Soc.* **1982**, *104*, 7347.
(122) Churchill, M. R.; Bueno, C.; Rheingold, A. L. *J. Organomet. Chem.* **1990**, *395*, 85.
(123) Ma, L.; Rodgers, D. P. S.; Wilson, S. R.; Shapley, J. R. *Inorg. Chem.* **1991**, *30*, 3591.
(124) Chihara, T.; Komoto, R.; Kobayashi, K.; Yamazaki, H.; Matsuura, Y. *Inorg. Chem.* **1989**, *28*, 964.
(125) Dale, M. J.; Dyson, P. J.; Johnson, B. F. G.; Langridge-Smith, P. R. R.; Yates, H. T. *J. Chem. Soc., Dalton Trans.* **1996**, 771.
(126) Dale, M. J.; Dyson, P. J.; Johnson, B. F. G.; Martin, C. M. Langridge-Smith, P. R. R.; Zenobi, R. *J. Chem. Soc., Chem. Commun.* **1995**, 1689.
(127) Dyson, P. J.; Johnson, B. F. G.; Langridge-Smith, P. R. R. Unpublished results.
(128) Bunkhall, S. R.; Holden, H. D.; Johnson, B. F. G.; Lewis, J.; Pain, G. N.; Raithby, P. R.; Taylor, M. J. *J. Chem. Soc., Chem. Commun.* **1984**, 25.
(129) Adams, R. D.; Wu, W. *J. Cluster Sci.* **1991**, *2*, 271.
(130) Adams, R. D.; Wu, W. *J. Cluster Sci.* **1993**, *4*, 245.
(131) Bailey, P. J.; Blake, A. J.; Dyson, P. J.; Johnson, B. F. G.; Lewis, J.; Parisini, E. *J. Organomet. Chem.* **1993**, *452*, 175.
(132) Adatia, T.; McPartlin, M.; Morris, J. *Acta Cryst.* **1994**, *C50*, 1233.
(133) Adatia, T.; Curtis, H.; Johnson, B. F. G.; Lewis, J.; McPartlin, M.; Morris, J. *J. Chem. Soc., Dalton Trans.* **1994**, 243.
(134) Adatia, T.; Curtis, H.; Johnson, B. F. G.; Lewis, J.; McPartlin, M.; Morris, J. *J. Chem. Soc., Dalton Trans.* **1994**, 3069.
(135) Johnson, B. F. G.; Lewis, J.; Nicholls, J. N.; Puga, J.; Whitmire, K. H. *J. Chem. Soc., Dalton Trans.* **1983**, 787.
(136) Cowie, A. G.; Johnson, B. F. G.; Lewis, J.; Raithby, P. R. *J. Chem. Soc., Chem. Commun.* **1984**, 1710.

(137) Amoroso, A. J.; Edwards, A. J.; Johnson, B. F. G.; Lewis, J.; Al-Mandhary, M. R.; Raithby, P. R.; Saharan, V. P.; Wong, W. T. *J. Organomet. Chem.* **1993**, *443*, C11.

(138) Bailey, P. J.; Beswick, M. A.; Lewis, J.; Raithby, P. R.: Ramirez de Arellano, M. C. *J. Organomet. Chem.* **1993**, *459*, 293.

(139) Johnson, B. F. G.; Lewis, J.; Raithby, P. R.; Saharan, V. P.; Wong, W. T. *J. Organomet. Chem.* **1992**, *434*, C10.

(140) Johnson, B. F. G.; Kwik, W.-L.; Lewis, J.; Raithby, P. R.; Saharan, V. P. *J. Chem. Soc., Dalton Trans.* **1991**, 1037.

(141) Bradley, J. S.; Pruett, R. L.; Hill, E.; Ansell, G. B.; Leonowicz, M. E.; Modrick, M. A. *Organometallics* **1982**, *1*, 748.

(142) Beswick, M. A.; Lewis, J.; Raithby, P. R.; Ramirez de Arellano, M. C. *J. Chem. Soc., Dalton Trans.* **1996**, 4033.

(143) Chiang, S.-J.; Chi, Y.; Su, P.-C.; Peng, S.-M.; Lee, G.-H. *J. Am. Chem. Soc.* **1994**, *116*, 11181.

(144) Adams, H.; Gill, L. J.; Morris, M. J. *Organometallics* **1996**, *15*, 464.

(145) Adams, C. J.; Bruce, M. I.; Skelton, B. W.; White, A. H. *Inorg. Chem.* **1992**, *31*, 3336.

(146) Adams, C. J.; Bruce, M. I.; Skelton, B. W.; White, A. H. *J. Organomet. Chem.* **1992**, *423*, 105.

(147) Blake, A. J.; Dyson, P. J.; Johnson, B. F. G.; Reed, D.; Shephard, D. S. *J. Chem. Soc., Chem. Commun.* **1994**, 1347.

ADVANCES IN ORGANOMETALLIC CHEMISTRY, VOL. 43

# Transition Metal Heteroaldehyde and Heteroketone Complexes

## HELMUT FISCHER, RÜDIGER STUMPF, and GERHARD ROTH

*Fakultät für Chemie*
*Universität Konstanz*
*Fach M727*
*D-78457 Konstanz, Germany*

## I

## INTRODUCTION

Thiocarbonyl, selenocarbonyl, and tellurocarbonyl compounds, $R^1(R^2)$-$C=E$ ($E=S$, Se, Te), are formal analogues of carbonyl compounds. Therefore, thioaldehydes, selenoaldehydes, and telluroaldehydes and the corresponding ketones should constitute useful sources of sulfur, selenium, and tellurium heteroatoms in the synthesis of heterocycles and other compounds containing these heteroatoms.[1-9] However, the usually very high reactivity of these heterocarbonyl compounds considerably restricts their use. When generated, most heteroaldehydes as well as selenoketones quickly oligomerize. Telluroaldehydes and -ketones often rapidly decompose by elimination of elemental tellurium. For example, thiobenzaldehyde was reported to be stable only at $-206°C$, polymerizing readily above $-163°C$.[10] Therefore,

125

only few thio- and selenoaldehydes as well as selenoketones have been isolated until now.

Three strategies for the stabilization of these highly reactive heterocarbonyl species have proven to be successful: (a) electronic stabilization by $\pi$ interaction with heteroatoms such as nitrogen, oxygen, and sulfur via unsaturated linking groups, (b) kinetic stabilization by bulky substituents, and (c) coordination to transition metals.

Thioaldehydes stabilized by mesomeric effects include **1**,[11] **2**,[12] and **3**.[13] Corresponding selenoaldehydes, e.g., **4**,[11] have also been described.

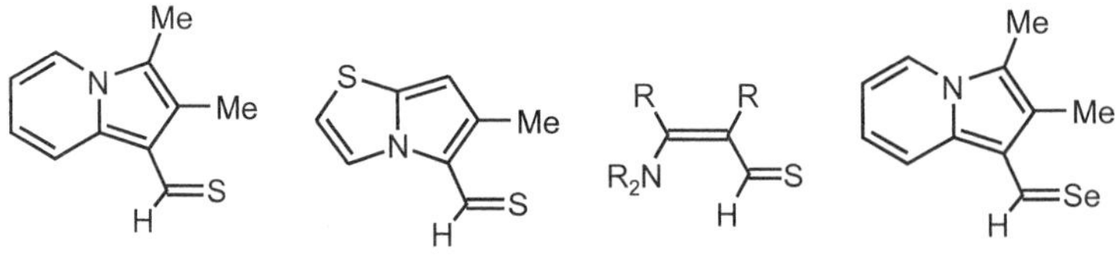

only very few electronically unperturbed isolable thio- and selenoaldehydes are known, e.g., *o,o',p*-tris[bis(trimethylsilyl)methyl]thiobenzaldehyde[14] and its selenium analogue,[15,16] *o,o',p-tert*-butylthiobenzaldehyde[17] and its selenium analogue[18] as well as tris(trimethylsilyl)ethanethial[19] and bis(2-*tert*-butyl)ethanethial.[20] Thiopivaldehyde was generated and could be handled in solution but was not isolated.[21,22] Until now, there is no report on the isolation of a telluroaldehyde, although the generation and trapping of telluroaldehydes has been described.[23–25]

Selenobenzophenone could be isolated as a dimer that dissociates in solution.[26] Only recently has the isolation of two telluroketones, 1,1,3,3-tetramethylindantellone[27,28] and bis(ferrocenyl)telluroketone,[29] been reported.

Usually these heteroaldehydes and heteroketones are generated in solution in the presence of suitable substrates and immediately trapped. The synthesis, properties, and reactivity of thioaldehydes and thioketones have been the subject of several reviews.[1–5] The chemistry of selenocarbonyl compounds has likewise been reviewed in recent years.[6–9]

The stabilization of these heteroaldehydes and -ketones by coordination to transition metals is the subject of this review. Many problems connected with the high reactivity of these heterocarbonyl compounds can be circumvented by using their transition metal complexes. The chemistry of organosulfur and organoselenium transition metal complexes in more general terms[30,31] and some aspects of thio- and selenoaldehydes and -ketones as ligands[32] have been reviewed. This review focuses on the syntheses and properties of chalcogenoaldehyde and -ketone complexes. The coverage is essentially limited to mono- and binuclear complexes with $R^1(R^2)C=E$ ligands ($R^1$, $R^2$ = H, alkyl, aryl, alkenyl). Complexes with $\pi$-donor hetero-atom functionalities bonded to the $C=E$ carbon atom ($R^1$, $R^2$ = $NR_2$, OR, SR . . .) have been omitted.

## II

## BONDING IN HETEROALDEHYDES AND HETEROKETONES

The highest occupied molecular orbital (HOMO) in formaldehyde and heteroaldehydes, $H_2C=E$, is the lone pair at E ($n_E$), and the second highest MO (SOMO) is the $C=E$ $\pi$-bonding orbital. The LUMO is the $\pi^*_{CE}$ orbital composed of the antibonding combination of $p_z(C)$ and $p_z(E)$. The ionization energy of the HOMO in formaldehyde is 10.88 eV and of the SOMO 14.5 eV, as determined by photoelectron spectroscopy.[33] The ionization energy of the HOMO and the SOMO both decrease considerably when the oxygen atom in formaldehyde is replaced by sulfur or selenium (see Fig. 1, data are compiled from Refs. 33–37).

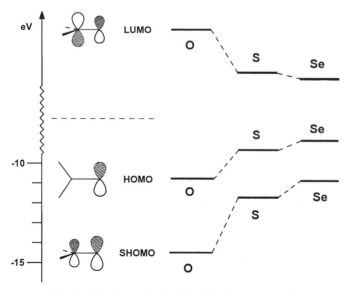

FIG. 1.   Frontier orbitals of $E=CH_2$ (E = O, S, Se).

In the series $H_2C=O > H_2C=S > H_2C=Se$ the difference in energy between the HOMO and the SOMO decreases (3.63, 2.44, and 2.35 eV, respectively).[35] From these data it follows that the donor properties of the lone pair and the $\pi$ orbital increase in the series $H_2C=O > H_2C=S > H_2C=Se$, those of the $\pi$ orbital more so than those of the lone pair. Simultaneously, the LUMO significantly decreases in energy (Fig. 1). The vacant $\pi^*_{CS}$ orbital calculated for thioformaldehyde is 2.8 eV lower in energy than the vacant $\pi^*_{CO}$ orbital of formaldehyde.[37] Therefore, thiocarbonyl compounds are both more electrophilic (lower LUMO) and more nucleophilic (higher HOMO) than carbonyl compounds. This is probably even more pronounced with selenocarbonyl and tellurocarbonyl compounds when compared to carbonyl compounds. The tendency of $R^1(R^2)C=E$ compounds to convert $C=E$ double bonds into more stable $C-E$ single-bonded species (by, e.g., oligomerization or polymerization) or to extrude the heteroatom (especially when $E = Te$) increases in the series O, S, Se, Te. This trend is readily understood when considering the increasing nucleophilicity and the simultaneously decreasing electrophilicity in the same series.

The substituent effects on the HOMO and LUMO energies of thiocarbonyl compounds were also studied and found to be in general accord with the substituent effects on carbonyl compounds. Electron-donating groups such as methyl raise the energy of the $\pi_{CS}$ orbital the most and the $\pi^*_{CS}$ orbital to a lesser extent. Electron-withdrawing groups (CHO, CN) lower the energies of all orbitals in the order $\pi^*_{CS} > n_S > \pi_{CS}$. Conjugating substituents (Ph, vinyl) lower the $\pi^*_{CS}$ orbital energy and dramatically raise the $\pi_{CS}$ orbital energy.[37]

## III

### BONDING AND STRUCTURE IN HETEROALDEHYDE AND HETEROKETONE COMPLEXES

#### A. *Bonding*

In mononuclear complexes the heterocarbonyl ligand can coordinate to the metal in either a $\eta^1$ mode (**A**) or a $\eta^2$ mode (**B**).

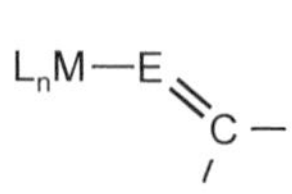

(A) 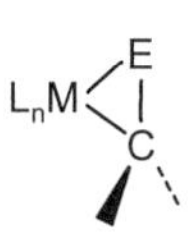                    (B)

In addition, $\eta^1$ aldehyde complexes can adopt a $\eta^1$-$E$ (**A**-$E$) or a $\eta^1$-$Z$ (**A**-$Z$) configuration.

$$(\textbf{A}\text{-}E) \qquad\qquad (\textbf{A}\text{-}Z)$$

In the $\eta^1$ mode ($E$ and $Z$) the coordinate bond between the heterocarbonyl ligand and the metal may essentially be described in terms of a $\sigma$ donation from the ligand (lone electron pair at the heteroatom) to an empty d orbital of the metal (d $\leftarrow$ $n_E$) (Fig. 2). $\pi$-Back-donation from an occupied metal d orbital to the $\pi^*_{CE}$ orbital of the ligand (d $\rightarrow$ $\pi^*_{CE}$) in the $\eta^1$ mode is rather unimportant.

In the $\eta^2$ mode the bonding takes place through interaction of the occupied $\pi_{CE}$ orbital with an empty d orbital of the metal (d $\leftarrow$ $\pi_{CE}$) and $\pi$-back-donation from an occupied metal d orbital to the $\pi^*_{CE}$ orbital of the ligand (d $\rightarrow$ $\pi^*_{CE}$) (Fig. 3).

In the $\eta^2$ mode $\pi$-back-donation is the main bonding interaction between the $L_nM$ fragment and the heterocarbonyl ligand.

Both coordination modes, $\eta^1$ and $\eta^2$, are combined in binuclear $\mu$-$\eta^1$:$\eta^2$ complexes (**C** and **C'**) in which the heterocarbonyl coordinates to one metal in the $\eta^1$ and to the other in the $\eta^2$ fashion. Complexes with (**C'**) and complexes without a metal–metal bond (**C**) are known. Complexes of type **D** ($\mu$-$\eta^1$:$\eta^1$) have not been structurally characterized presumably due to the poorer donor properties of the "second lone pair" at the heteroatom E ($\sigma_{CE}$) as compared to the $\pi_{CE}$ orbital. The energy of the $\sigma_{CE}$ orbital is well below that of the $\pi_{CE}$ orbital. In heteroformaldehydes, e.g., the energy separation between the $\sigma_{CE}$ and the $\pi_{CE}$ orbital is 1.25 eV (E = S) and 1.85 eV (E = Se).[33] However, complexes of type **D** might be intermediates in a possible isomerization process of complexes **C** (site exchange of $L_nM^1$ and $L_nM^2$).

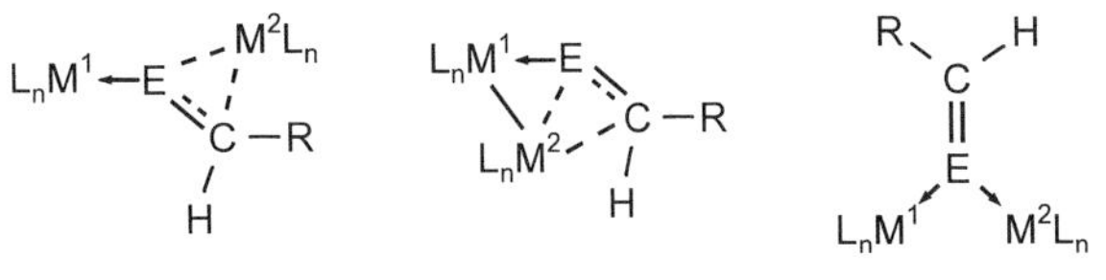

$$(\textbf{C}) \qquad\qquad (\textbf{C'}) \qquad\qquad (\textbf{D})$$

Likewise, trinuclear and polynuclear complexes of type **E** have not been

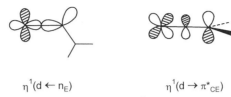

$$\eta^1(d \leftarrow n_E) \qquad\qquad \eta^1(d \rightarrow \pi^*_{CE})$$

FIG. 2.  $\sigma$-Donation and $\pi$-back donation in $\eta^1$-heteroaldehyde and heteroketone complexes.

reported, although closely related $\mu$ alkylthiolato triosmium complexes have been described in which the bridging $H_2C{=}S$ ligand coordinates via sulfur to $Os^1$ and $Os^2$ and via $CH_2$ to $Os^3$ (see **F:** E = S, there is no bonding interaction between S and $Os^3$).[38,39]

$$
\begin{array}{cc}
\text{(E)} & \text{(F)}
\end{array}
$$

The major bonding interaction between the heterocarbonyl ligand and the metal in $\eta^2$ complexes is the back-bonding interaction $\eta^2(d \rightarrow \pi^*_{CE})$, whereas in $\eta^1$ complexes the $\sigma$-donation $\eta^1(d \leftarrow n_E)$ dominates. Therefore, from the trends in orbital energies of the heterocarbonyl compounds (see Section II), one would expect that in mononuclear complexes $\eta^2$ coordination is increasingly favored within the series O, S, Se, Te. This expectation is conformed by the experimental results (see Section III,B).

The interaction of formaldehyde and thioformaldehyde with some group 8 metal-ligand fragments $[Ru(CO)_4,^{40} Fe(CO)_2(PH_3)_2^{41}]$ in the $\eta^2$ conformation has been analyzed by the Hartree–Fock–Slater transition-state method

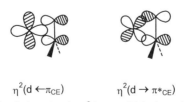

$$\eta^2(d \leftarrow \pi_{CE}) \qquad \eta^2(d \rightarrow \pi^*_{CE})$$

FIG. 3.   Donation and $\pi$-back donation in $\eta^2$-heteroaldehyde and heteroketone complexes.

and by ab initio calculations, respectively. In addition, the preferences for $\eta^1$ and $\eta^2$ coordination of aldehydes and ketones on several types of organometallic fragments have been studied by extended Hückel MO calculations[42] as well as by the ab initio method (for $[Ni(PH_3)_2(H_2C=O)]$).[43]

From the analysis of $[Ru(CO)_4(\eta^2-E=CH_2)]$ (E = O, S) it was concluded that the hydrogens of $H_2C=E$ are strongly bent away from the metal-ligand fragment and that the back-donation in terms of energy is more important for the stability of $[Ru(CO)_4(S=CH_2)]$ than the donation. In terms of charge, 0.30e (E = O) and 0.46e (E = S) were donated from $\pi_{CE}$ to the LUMO of $Ru(CO)_4$, and 0.74e (E = O) and 0.76e (E = S) were back-donated from the HOMO of $Ru(CO)_4$ to $\pi^*_{CE}$. Thioaldehyde was calculated to be more strongly bound to $Ru(CO)_4$ than formaldehyde as a result of more favorable contributions from $\eta^2(d \rightarrow \pi^*_{CE})$, particularly from $\eta^2(d \leftarrow \pi_{CE})$. The contribution for $\eta^2(d \leftarrow \pi_{CE})$ is larger for E = S than for E = O because $\pi_{CS}$ is higher in energy than $\pi_{CO}$ and thus is better able to interact with the LUMO of $Ru(CO)_4$. The overlaps between the HOMO of $Ru(CO)_4$ and $\pi^*_{CE}$ are more favorable for E = S than for E = O as $\pi^*_{CS}$ is lower in energy than $\pi^*_{CO}$.[40]

Similar results were obtained with $[(CO)_2(PH_3)_2Fe(\eta^2-E=CH_2)]$. For E = S a substantially larger binding energy and a more pronounced deformation of the $H_2C=E$ ligand was calculated than for E = O. It was concluded that the $\pi$-back-donation is the main bonding interaction between the metal and the $H_2C=E$ molecule in the complexes. In the case of the thioformaldehyde complex, a nonnegligible ligand-to-metal $\sigma$-dative contribution to the interaction between thioformaldehyde and iron exists.[41]

From the extended Hückel MO calculation of $\eta^1$ and $\eta^2$ coordination of aldehydes and ketones on several types of organometallic fragments it follows (a) that increasing energy of the metal d orbitals (more electropositive metal) favors the $\eta^2$ form compared to the $\eta^1$ form and (b) that increasing energy of the C=E ligand orbitals on the organic part (donor substituents $R^1$, $R^2$) favors the $\eta^1$ form with respect to the $\eta^2$ form.[42]

## B. *Ligand Dynamics*

### 1. $\eta^1/\eta^2$ *Isomerization*

Mononuclear heterocarbonyl complexes usually adopt either a $\eta^1$ or a $\eta^2$ configuration. Complexes of $d^6$ metal-ligand fragments such as $M(CO)_5$ or $ML^1L^2(\eta^5-C_5H_5)$ may be regarded as borderline cases between $\eta^1$ and $\eta^2$ preferences. The coexistence of the two forms was reported for the first time in 1984.[44] An equilibrium between $\eta^1$ and $\eta^2$ isomers of $[M(CO)_5\{Se=$

C(C$_6$H$_4$R-$p$)H}] (M = Cr, W) in solution was observed.[44,45] Later, $\eta^1$ and $\eta^2$ isomers of some thiobenzaldehyde complexes of tungsten [W(CO)$_5$ {S=C(C$_6$H$_4$R-$p$)H}][46] and of ruthenium [Ru(PR$_3$)$_2$\{S=C(C$_6$H$_4$R-$p$)H\} ($\eta^5$-C$_5$H$_5$)]PF$_6$ [(PR$_3$)$_2$ = (PMe$_3$)$_2$, dmpe][47] were also found to coexist and rapidly interconvert.

The $\eta^1/\eta^2$ equilibrium in [M(CO)$_5$\{E=C(C$_6$H$_4$R'-$p$)H\}] complexes (M = Cr, W; E = S, Se; R' = OMe, Me, H, CF$_3$) (Scheme 1) was studied in detail.[44–46]

At room temperature the interconversion of the isomers was found to be rapid on the $^1$H-nuclear magnetic resonance (NMR) time scale. However, infrared (IR) spectra of the $\eta^1$ forms and the $\eta^2$ form differ significantly. Thus, it was possible to determine the equilibrium constants $K = \{[\eta^1\text{-}E] + [\eta^1\text{-}Z]\}/[\eta^2]$ (Scheme 1) and their dependence on the solvent, the temperature, the metal, the heteroatom, and the *para* substituent. The $\eta^1/\eta^2$ equilibrium constant $K$ critically depends on the electron density at the metal and the electron-accepting properties of the heteroaldehyde ligand. Several trends have been identified. The $\eta^1/\eta^2$ equilibrium shifts toward the $\eta^2$ isomer (a) with increasing electron-accepting ability of the *para* substituent R', (b) with increasing back-donating potential of the metal, (c) with decreasing temperature, and (d) with increasing polarity of the solvent.

A good linear correlation between the equilibrium constant log($K$) and the Hammett constant $\sigma_p^+$ was observed (Fig. 4). Replacement of sulfur in the heteroaldehyde ligand by selenium favored the $\eta^2$ isomer by a factor of at least 30 (Fig. 4). When the stronger back-bonding fragment W(CO)$_5$ was substituted for Cr(CO)$_5$ in [M(CO)$_5$\{Se=C(Ph)H\}] the equilibrium

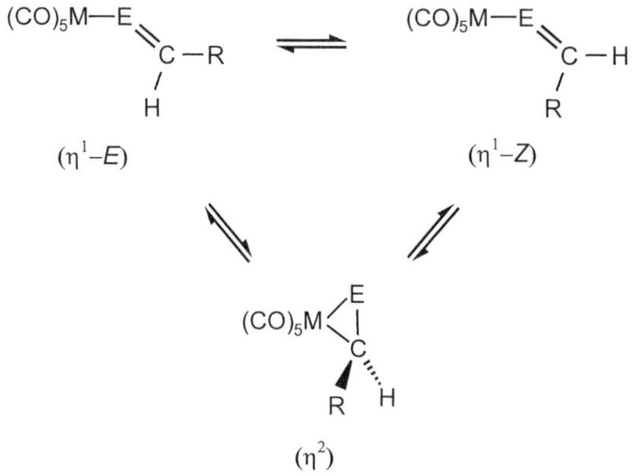

SCHEME 1

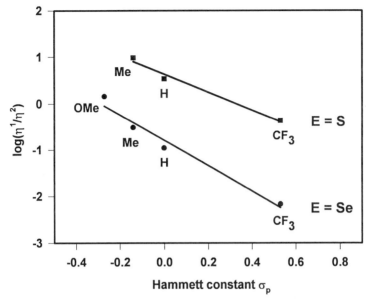

FIG. 4.   Plot of $\log(\eta^1/\eta^2)$ versus the Hammett constant $\sigma^+$ of various *para* substituents in $[W(CO)_5\{E=C(C_6H_4R\text{-}p)H\}]$ (E = S, Se) in *n*-hexane at 25°C.

constant decreased by a factor of approximately 500. The solvent dependence indicated that the $\eta^1$ isomers are slightly more polar than the $\eta^2$ isomer. These observations confirm the conclusions from the theoretical analyses described in Section III,A.

It was later found that some aldehyde complexes of rhenium, $[Re(NO)(PPh_3)\{O=C(Aryl)H\}(\eta\text{-}C_5H_5)]BF_4$, are also present in solution as a rapidly interconverting mixture of $\eta^1$ isomers and the $\eta^2$ isomer.[48,49] With the exception of solvent dependence, similar dependencies of the equilibrium constants on the aryl substituents and on the temperature were observed. The solvent dependence was found to be opposite, which, however, is easily understood when taking into account that the rhenium aldehyde complexes are cationic whereas the chromium and tungsten thio- and selenoaldehyde complexes are neutral.

The activation barrier for the $\eta^1/\eta^2$ interconversion is usually very low. With $[W(CO)_5\{Se=C(C_6H_4Me\text{-}p)H\}]$ and $[Ru(PR_3)_2\{S=C(C_6H_4Cl\text{-}p)H\}(\eta\text{-}C_5H_5)]^+$ $[(PR_3)_2 = (PMe_3)_2,$ dmpe] it was possible to determine the free energy of activation $\Delta G^{\neq}$. For the tungsten selenoaldehyde complex, $\Delta G^{\neq}$ *was calculated to be 44 ± 2 kJ/mol for the $\eta^1 \rightarrow \eta^2$ isomerization and 42 ± 2 kJ/mol for the $\eta^2 \rightarrow \eta^1$ isomerization, both at* −30°C in $(CD_3)_2O$.[44]

For the ruthenium thioaldehyde complexes, $\Delta G^{\neq}$ values of 48 kJ/mol [(PR$_3$)$_2$ = (PMe$_3$)$_2$] and 48 kJ/mol [(PR$_3$)$_2$ = dmpe] were reported.[47]

### 2. $\eta^1$-E/$\eta^1$-Z Isomerization

The barrier for the $\eta^1$-E/$\eta^1$-Z isomerization is presumably even lower than that for the $\eta^1$/$\eta^2$ interconversion. With both types of complexes it was not possible to detect both $\eta^1$ isomers, $\eta^1$-E and $\eta^1$-Z, in solution even at $-100°$C. Therefore, either the $\eta^1$-E/$\eta^1$-Z isomerization is still very fast even at that temperature or the second isomer is present in too small a concentration to be detected. The presence of two isomers at temperatures below $-80°$C was detected for the unsymmetrically substituted thioketone complex [W(CO)$_5$\{S=C(C$_6$H$_4$Me-$p$)Ph\}][50]; however, the barrier to interconversion was not determined. The site exchange of the two methyl groups in the thioacetone complexes [M(CO)$_5$(S=CMe$_2$)\}] was also reported to be very fast.[51] For M = W only one singlet in the $^1$H-NMR spectrum was observed for the methyl groups down to $-100°$C. It was not possible to decide whether the interconversion proceeds by rapid rotation around the S=C bond or by inversion at the sulfur atom. A mechanism that involves a rapid dissociation and readdition of the thioacetone ligand could be discarded as replacement of coordinated thioacetone by acetonitrile in acetonitrile is slow even at room temperature.[51]

In contrast, both isomers of [Mn(CO)$_2$\{S=C(C$_6$H$_4$R-$p$)(C$_5$H$_4$Mn(CO)$_3$)\} ($\eta^5$-C$_5$H$_5$)] (R = H, Me), E-**5** and Z-**5,** are detectable at low temperature [Eq. (1)].

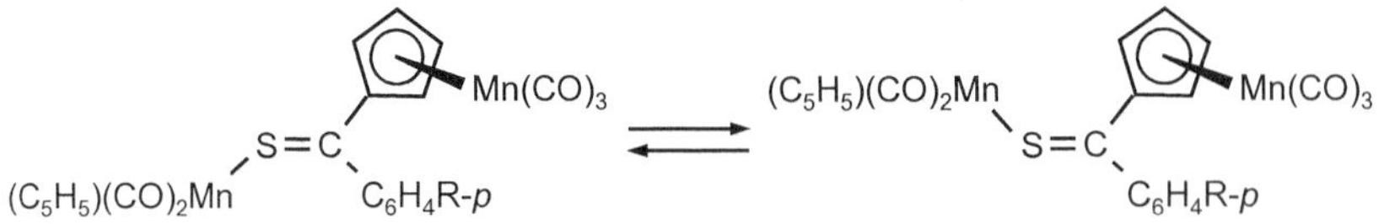

$$E\text{-}(\mathbf{5}) \qquad\qquad\qquad Z\text{-}(\mathbf{5}) \tag{1}$$

The E isomer is more stable than the Z isomer by 2.9–4.6 kJ/mol (at 25°C) depending on the solvent and on R. The activation enthalpy $\Delta H^{\neq}$ is in the range of 44.8–58.6 kJ/mol, whereas the activation entropy $\Delta S^{\neq}$ varies between $-11.7$ and $-36.0$ J/(mol K).[52] The $\eta^1$-E/$\eta^1$-Z isomerization was proposed to proceed by inversion at the sulfur atom.

### C. Spectroscopic Properties

The spectroscopic properties of these heterocarbonyl compounds vary considerably depending on the coordination mode of the heterocarbonyl ligand.

From $\nu$(CO) spectra of [M(CO)$_5${E=C(Aryl)H}] (M = Cr, W; E = S, Se) it follows that the heteroaldehyde ligand acts essentially as a $\sigma$ donor in the $\eta^1$ coordination mode. The $\sigma$-donor/$\pi$-acceptor properties and thus $\nu$(CO) spectra are similar to those of [M(CO)$_5$(thioether)] and [M(CO)$_5$ (phosphine)] complexes. In the $\eta^2$-bonding mode the heteroaldehyde ligands are strong $\pi$ acceptors and their IR spectra resemble those of, e.g., alkyne and olefin complexes. Because [W(CO)$_5${Se=C(Aryl)H}] complexes are present in solution as rapidly interconverting mixtures of the $\eta^1$ isomers and the $\eta^2$ isomer their IR spectra are composed of both types of $\nu$(CO) spectra. The intensity ratio of the $\nu$(CO) absorptions depends on the temperature due to the temperature dependence of the $\eta^1/\eta^2$ equilibrium.

In $\eta^1$ complexes the $^{13}$C-NMR resonance signal of the heterocarbonyl carbon atom C=E is at low field and appears for neutral complexes in the range $\delta$ = 210–245 and for cationic complexes in between $\delta$ = 195 and 215. The C=E resonance is almost independent of the heteroatom. In cases where a direct comparison between $\eta^1$ heteroaldehyde and $\eta^1$ heteroketone complexes is possible (analogous ML$_n$ fragment) the $^{13}$C(C=E)-NMR signal of heteroketone complexes is usually found at a slightly higher field ($\Delta\delta$ = 15–20 ppm). Complexation of free heteroketones also causes a slight upfield shift.

In contrast to $\eta^1$ complexes the $^{13}$C-NMR heterocarbonyl resonance signal in $\eta^2$ complexes is observed at a much higher field in the range between $\delta$ = 27 and 75. In [W(CO)$_5${Se=C(Aryl)H}] complexes the resonance of the C=Se carbon is temperature dependent due to the temperature dependence of the $\eta^1/\eta^2$ equilibrium and rapid interconversion of the isomers. The position of the C=Se resonance of binuclear $\mu$-$\eta^1$:$\eta^2$ complexes usually resembles that of the corresponding $\eta^2$ complexes. Two sets of signals for the ML$_n$ fragments in binuclear $\mu$-$\eta^1$:$\eta^2$ complexes indicate that the isomerization **G/H** is slow on the $^{13}$C-NMR time scale [Eq. (2)].

$$\text{(G)} \qquad\qquad\qquad \text{(H)} \qquad\qquad\qquad (2)$$

Analogously to the $^{13}$C-NMR resonances the position of the $^1$H-NMR signals of heteroaldehyde complexes also depends on the coordination mode. In $\eta^1$ complexes the signal is observed in the range $\delta$ = 10 to 12. In $\eta^2$ complexes it is at $\delta$ < 10 (usually between $\delta$ = 3 and 7) and strongly dependent on the metal and the coligands. In analogous aldehyde complexes the $^1$H-NMR aldehyde resonance signal in the series S, Se, Te usually shifts increasingly upfield (see, e.g., Refs. 53–56). Again, the position of the $^1$H-NMR resonance of the bridging aldehyde ligand in $\mu$-$\eta^1$:$\eta^2$ complexes is similar to that in $\eta^2$ complexes.

In solution [M(CO)$_5${Se=C(Aryl)H}] complexes show a thermochromic behavior. The color of the compounds is caused by a MLCT of d electrons into the LUMO localized mainly in the Se=C(Aryl)H ligand. In $\eta^1$ complexes this transition is at considerably lower energy than in $\eta^2$ complexes. The observed color of solutions of [M(CO)$_5${Se=C(Aryl)H}] therefore depends on the $\eta^1$/$\eta^2$ equilibrium, which is temprature dependent. Thus, solutions of, e.g., [W(CO)$_5${Se=C(Ph)H}] are blue at room temperature and green at $-78°$C.[45]

### D. *Structure*

Table I summarizes the structural studies of mononuclear $\eta^1$ and $\eta^2$ thio-, seleno-, and telluroaldehyde complexes, Table II those of simple binuclear $\mu$-$\eta^1$:$\eta^2$ heteroaldehyde complexes without a metal–metal bond, and finally Table III those of heteroketone complexes. Figure 5 shows the structure of a representative $\mu$-$\eta^1$:$\eta^2$ complex, [{(CO)$_5$W}$_2${Te=C(Ph)H}].

As expected, the C=E distance and the M-E-C angle vary considerably depending on the $\eta^1$- or $\eta^2$-bonding mode. In $\eta^1$ complexes the E-C distance is short and the M-E-C angle is large, whereas in $\eta^2$ complexes the E-C distance is long and the M-E-C angle is small.

Upon coordination in the $\eta^1$ mode in mononuclear complexes, the C=E bond only slightly elongates. The elongation in heteroaldehydes is usually 0.01–0.03 Å and that in heteroketones is 0.02–0.05 Å. An additional elongation is observed in case of $\pi$ conjugation between the heterocarbonyl group C=E and a strong $\pi$-donor substituent D as, e.g., in E=C-C=C-D (D = NR$_2$, SR). Unless specific steric requirements are to be met, the M-E-C angle is close to 120° (in the range 111.2–123.1°), depending on the substituents and the coligands.

In contrast, the alteration of the C=E bond length on coordination in the $\eta^2$ mode is considerably more pronounced. In $\eta^2$ complexes the C=E bond is 0.09–0.15 Å (E = S) and 0.10–0.20 Å (E = Se) longer than in the free heterocarbonyl compounds and is nearly comparable to a C(sp$^3$)-E single bond.[57] Because of the metallaheterocyclopropane structure the M-E-C angle is close to 60° (59–66.1° for E = S and 55.8–62.0° for E = Se).

TABLE I

E=C Distance and M-E-C Angle in Selected $\eta^1$ and $\eta^2$ Mononuclear Thio-, Seleno-, and Telluroaldehyde Complexes (Distances in Å, Angles in Degrees)

| $L_nM$ | R | d(E=C) | M-E-C | Ref. |
|---|---|---|---|---|
| **a. Thioaldehydes** | | | | |
| | S=CH$_2$ | 1.611[a] | | 58 |
| | | 1.6108(9)[a] | | 59 |
| | S=C(Me)H | 1.610[a] | | 60 |
| | S=C(CH=CH$_2$)H | 1.62(2)[a] | | 61 |
| **$\eta^1$-Thioaldehyde complexes** | | | | |
| Cr(CO)$_5$ | S=C[CH=C(SEt)$_2$]H | 1.642(4) | 118.7(2) | 62 |
| W(CO)$_5$ | S=C(Ph)H | 1.62(1) | 116.7(4) | 63 |
| W(CO)$_5$ | S=C[CPh=C(Ph)OEt]H | 1.63(2) | 112(1) | 64 |
| W(CO)$_5$ | S=C(R$^1$)H[b] | 1.68(1) | 114.3(3) | 65 |
| [Ru(dppe)(C$_5$H$_5$)]$^+$ | S=C(C$_6$H$_4$OMe-$p$)H | 1.632(5) | 112.0(2) | 47,66 |
| **$\eta^2$-Thioaldehyde complexes** | | | | |
| Ti(PMe$_3$)(C$_5$H$_5$)$_2$ | S=CH$_2$ | 1.744(3) | 62.0(1) | 67 |
| Ti(C$_5$H$_5$) | S=C(CH$_2$CH$_2$S-)H[c] | 1.741(6) | 62.7(2) | 68,69 |
| Ti(MeC$_5$H$_4$) | S=C(CH$_2$CH$_2$S-)H[c] | 1.80(1) | 62.6(4) | 69 |
| Zr(PMe$_3$)(C$_5$H$_5$)$_2$ | S=C(Me)H | 1.785(11) | 63.1(3) | 70 |
| | | 1.739(13) | 63.4(3) | |
| Ta(H)(C$_5$Me$_5$)$_2$ | S=C(CH$_2$Ph)H | 1.86(2) | 62.8(7) | 71 |
| W(CO)(Et$_2$NCS)(Et$_2$NCS$_2$) | S=C(Ph)H | 1.739(7) | 66.1(2) | 72 |
| Fe(CO)$_2$(PPh$_3$) | S=C(CH=CH$_2$)H | 1.794(10) | 58.5(6) | 73,74 |
| [Re(NO)(PPh$_3$)(C$_5$H$_5$)]$^+$ | S=CH$_2$ | 1.742(9) | 62.2(3) | 75,76 |
| [Re(NO)(PPh$_3$)(C$_5$H$_5$)]$^+$ | S=C(Ph)H | 1.70(1) | 63.5(5) | 77 |
| Rh[(PPh$_2$CH$_2$)$_3$CMe] | S=C(CH=CHCPh$_3$)H | 1.72(3) | 59(1) | 78 |
| Rh(C$_5$Me$_5$) | S=C(R$^2$)H[d] | 1.65(8) | 59.9 | 79 |
| **b. Selenoaldehydes** | | | | |
| | Se=CH$_2$ | 1.753[a] | | 80 |
| | | 1.759[a] | | 81 |
| | Se=C(Me)H | 1.758(10)[a] | | 82 |
| **$\eta^1$-Selenoaldehyde complexes** | | | | |
| W(CO)$_5$ | Se=C{C$_6$H$_2$[CH(SiMe$_3$)$_2$]$_3$}H[e] | 1.783(15) | 107.8(4) | 15,16 |
| **$\eta^2$-Selenoaldehyde complexes** | | | | |
| W(CO)$_5$ | Se=C(Ph)H | 1.864(13) | 62.0(4) | 45 |
| | | 1.876(3) | 61.1(5) | |
| W(CO)(HC≡CBu$^t$)$_2$ | Se=C(Ph)H | 1.961(6) | 58.0(2) | 56 |
| [Re(NO)(PPh$_3$)Cp]$^+$ | Se=CH$_2$ | 1.879(6) | 57.0(2) | 83 |
| Rh(PPr$^i_3$)(C$_5$H$_5$) | Se=C(Me)H | 1.917(5) | 55.8(1) | 54,55 |
| **c. $\eta^1$-Telluroaldehyde complexes** | | | | |
| Ta(H)(C$_5$H$_5$)$_2$ | Te=CH$_2$ | 2.210(15) | 52.4(3) | 84 |

[a] Determined by microwave spectroscopy.

[b] R$^1$ =

[c] Dimer

[d] R$^2$ =

[e] Two isomers.

## TABLE II

### $E{=}C$ Distance and M-E-C Angle in Selected Binuclear $\eta^1{:}\eta^2$ Thio-, Seleno-, and Telluroaldehyde Complexes $(L_nM)_2[\eta^1{:}\eta^2\text{-}E{=}C(R)R']$ without Metal–Metal Bond (Distances in Å, Angles in Degrees)

| $L_nM(\eta^1)$ | $L_nM(\eta^2)$ | $E{=}C(R)R'$ | $E{=}C$ | $M(\eta^1)$-E-C | $M(\eta^2)$-E-C | Ref. |
|---|---|---|---|---|---|---|
| $Ti(OC_6H_3Pr^i_2)(C_5H_5)$ | $Ti(OC_6H_3Pr^i_2)(C_5H_5)$ | $[S{=}C(Me)H]_2$ | 1.78(1) | 111.2(4) | 58.5(4) | 85 |
| | | | 1.76(1) | 111.7(4) | 58.6(4) | |
| $Mo(\mu\text{-}SMe)(CO)(C_5H_5)$ | $Mo(CO)(\mu\text{-}S_2CH_2)(C_5H_5)$ | $S{=}CH_2$ | 1.826(8) | 105.6(3) | 64.0(2) | 86 |
| $Mn(CO)_2(C_5H_5)$ | $Mn(CO)_2(C_5H_5)$ | $S{=}C(Ph)H$ | 1.725(2) | 115.3(1) | 61.6(1) | 87 |
| $Fe(CO)_3$ | $Fe(CO)_3$ | $[S{=}C(Me)H]_2$ | 1.753(4) | 102.62(7) | 59.22(6) | 88 |
| $Fe(CO)_2(PPh_3)$ | $Fe(CO)_3$ | $S{=}C(CMe_2CH_2)_2CH_2$ | 1.761(4) | 103.0(1) | 62.8(1) | 89 |
| $Ir[(PPh_2CH_2)_3CMe]$ | $Ir[(PPh_2CH_2)_3CMe]$ | $S{=}C(CHCHCH\text{-})H$ | 1.76(4) | 89(1) | 58(1) | 90 |
| $Mn(CO)_2(C_5H_5)$ | $Mn(CO)_2(C_5H_5)$ | $Se{=}CH_2$ | 1.900(11) | 112.2(3) | 55.4(4) | 91 |
| $Fe(CO)_3$ | $Fe(CO)_3$ | $[Se{=}CH_2]_2$ | 1.902(6) | 100.4(2) | 55.3(2) | 92 |
| | | | 1.901(6) | 99.9(2) | 54.2(2) | |
| $Fe(CO)_3$ | $Fe(CO)_3$ | $[Se{=}C(Me)H]_2$ | 1.914(4) | 100.7(1) | 55.5(1) | 88 |
| $W(CO)_5$ | $W(CO)_5$ | $Te{=}C(Ph)H$ | 2.112(7) | 112.7(2) | 54.6(2) | 93 |
| $Fe(CO)_3$ | $Fe(CO)_3$ | $[Te{=}CH_2]_2$ | 1.979(14) | 97.4(4) | 52.8(4) | 94 |
| | | | 1.989(16) | 96.3(5) | 52.2(4) | |

TABLE III

E=C Distance and M-E-C Angle in Selected Mononuclear $\eta^1$ and $\eta^2$-Thio- and Telluroketone Complexes (Distances in Å, Angles in Degrees)

| $L_nM$ | R | $d(E=C)$ | M-E-C | Ref. |
|---|---|---|---|---|
| **a. Thioketone** | | | | |
| | $S=CPh_2$ | 1.636(9) | | 94 |
| $\eta^1$-**thioketone complexes** | | | | |
| $Cr(CO)_5$ | $S=CMe_2$ | 1.618(8) | 120.8(4) | 95 |
| $Cr(CO)_5$ | $S=C[CH=C(Me)NHCHMe_2](C_6H_4Ph\text{-}o)$ | 1.712(3) | 119.0(1) | 96 |
| $Cr(CO)_5$ | $S=C(Ph)C_5H_4Fe(C_5H_5)$ | 1.667(2) | 121.9(1) | 97 |
| $W(CO)_5$ | $S=C[CH=C(Ph)OEt]Ph$ | 1.667(6) | 119.5(2) | 64 |
| $W(CO)_5$ | $S=C[CMe=C(Ph)H]Fe(CO)_2indenyl$ | 1.658(7) | 119.8(3) | 98 |
| $Mn(CO)_2(C_5H_5)$ | $S=C(Ph)C_5H_4Mn(CO)_3$ | 1.656(6) | 123.1(2) | 99 |
| $Fe(CO)_4$ | $S=C(CPh)_2$ | 1.652(7) | 111.2(2) | 100 |
| $Fe[C(S)OMe](PPh_3)$ | $S=C(Bu^t)CH_2PPh_2\text{-}$ | 1.647(14) | 107.7 | 101 |
| $ZnCl_2$ | $S=C(CH_2CHMe)_2O$ | 1.692(8) | 112.5(3) | 102 |
| | | 1.689(7) | 111.6(3) | |
| $\eta^2$-**thioketone complexes** | | | | |
| $V(C_5H_5)_2$ | $S=CPh_2$ | 1.762(4) | 65.9(1) | 103 |
| $Co(CO)(C_5Me_5)$ | $S=C(CN)CH_2Bu^t$ | 1.742(11) | 60.5 | 104 |
| $Rh(C_5Me_5)$ | $S=C(Me)CMe=C(Me)[C(O)Me]^a$ | 1.76(1) | 60.0 | 105 |
| | | 1.72(1) | 62.7 | |
| $Pd(PPh_3)_2$ | $S=C(CF_3)_2$ | 1.730(4) | 61.5(1) | 106 |
| $Pt(PPh_3)_2$ | $S=C(CF_3)_2$ | 1.762(7) | 60.5(2) | 107 |
| **b. $\eta^1$-Telluroketone complexes** | | | | |
| $W(CO)_5$ | $Te=C(CMe_2)_2(o\text{-}C_6H_4)$ | 1.987(5) | 124.0(1) | 108 |

$^a$ Three independent molecules per unit cell, one of which is severely disordered.

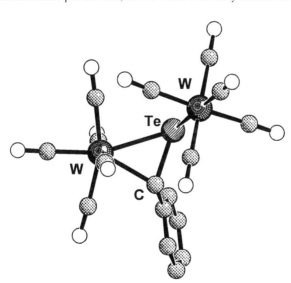

Fig. 5. Structure of the binuclear complex $\mu\text{-}\eta^1\text{:}\eta^2\text{-}[\{W(CO)_5\}_2\{Te=C(Ph)H\}]$ in the crystal.

The smaller median value for the angle in the selenium compounds can be attributed to the larger radius of selenium as compared to that of sulfur.

In binuclear $\mu$-$\eta^1$:$\eta^2$ complexes the C=E distance (Table II) resembles that in mononuclear $\eta^2$ complexes. The M($\eta^1$)-E-C and M($\eta^2$)-E-C angles are close to the corresponding angle in mononuclear $\eta^1$ and $\eta^2$ complexes, respectively. The M($\eta^1$)-E=C-C torsion angle in binuclear complexes with a single bridging heteroaldehyde is in the range 103.8 (in [{W(CO)$_5$}$_2$ {$\mu$ - Te=C(Ph)H}]) to 118.2° (in [{Mn(CO)$_2$($\eta^5$-C$_5$H$_5$)}$_2${$\mu$-S=C(Ph)H}]). Consequently, the heteroatom is sp$^3$-hybridized. In contrast, in $\eta^1$ heteroaldehyde complexes the M-E=C-C fragment is almost planar (torsion angle: 173.7–180°).

In addition to those complexes summarized in Tables I–III, X-ray structural analyses of some bi- and trinuclear complexes with a supporting metal–metal bond,[109–115] a trinuclear complex without a metal–metal bond supporting the $\eta^1$,$\eta^2$ bridge,[38,39,116] and two trinuclear osmium complexes with a terminal diphenylthioketone ligand[117] have been reported.

## IV

## SYNTHESIS OF HETEROALDEHYDE AND HETEROKETONE COMPLEXES

### A. *Complexes with $\eta^1$- and $\eta^2$-Bound Heterocarbonyl Ligands*

The major synthetic routes may be divided into three categories: replacement of a coordinated ligand by a heterocarbonyl compound E=C(R$^1$)R$^2$, synthesis from heterocarbonyl precursors and metal complexes, and transformation of a coordinated ligand into a heterocarbonyl ligand.

### 1. *Replacement of a Coordinated Ligand by a Heterocarbonyl Compound E=C(R$^1$)R$^2$*

The most straightforward route to heteroaldehyde and heteroketone complexes is the substitution of a heterocarbonyl compound for another coordinated ligand. This method is naturally restricted to heteroaldehydes and heteroketones stable in the uncoordinated form, i.e., is usually restricted to thioketones and a few stable seleno- and tellurocarbonyl compounds.[14,17,18,27,118] In most cases, metal carbonyls or solvent complexes of metal carbonyls were used as the complex precursors. The photochemically or thermally induced loss of the ligand to be replaced is followed by coordination of the heterocarbonyl compound [Eq. (3)].

$$L_nM\text{-}L^* \xrightarrow[-L^*]{E=C(R^1)R^2} L_nM[E=C(R^1)R^2] \qquad (3)$$

Thus, pentacarbonyl(thioketone) complexes $[M(CO)_5L]$ were prepared by irradiation of solutions of $[M(CO)_6]$ (M = Cr, Mo, W) in the presence of thioketones L (L = thiobenzophenone, adamantanethione)[119] [Eq. (4)] and by refluxing solutions of $[Mo(CO)_6]$ in tetrahydrofuran (THF) in the presence of L = $p,p'$-disubstituted thiobenzophenones, thiofenchnone, and thiocamphor, respectively.[120] The formation of two diastereomers of the thiocamphor complex was observed, probably due to a high inversion barrier at sulfur. Only one isomer of the thiofenchnone complex could be detected. However, fluxional behavior could not be ruled out.[120]

$$M(CO)_6 \xrightarrow[-CO]{h\nu \;/\; S=CR_2} \begin{array}{c} (CO)_5M-S \\ \diagdown \\ C-R \\ / \\ R \end{array} \qquad (4)$$

Pentacarbonyl complexes of chromium, molybdenum, and tungsten with the organometallic thioketones thiobenzoyl ferrocene, thiobenzoyl cymantrene, thioacetyl ferrocene, and thioacetyl cymantrene[121] were formed on treatment of $[M(CO)_5(thf)]$ (M = Cr, Mo, W) with the thioketones.[97,122,123] Analogously, bis(ferrocenyl)thioketone complexes and a bis(ferrocenyl)-selenoketone complex $[M(CO)_5\{E=C(Fc)_2\}]$ (M = Cr, W: E = S; M = W: E = Se; Fc = ferrocenyl) were obtained from $[M(CO)_5(thf)]$ and the corresponding heteroketone.[29] However, the reaction of $Te=C(Fc)_2$ with $[W(CO)_5(thf)]$ gave only tetraferrocenylethene. In contrast, the reaction of 1,1,3,3-tetramethylindanetellone[27] with $[W(CO)_5(thf)]$ at elevated temperatures afforded at $\eta^1$ telluroketone complex. The telluroketone is weakly bound and easily displaced by stronger coordinating ligands like MeCN.[108]

Binuclear thioketone complexes containing an S=C-C=C fragment were obtained when $[\{(\eta^5\text{-}C_5H_5)Mo(CO)_3\}_2]$ was treated in refluxing benzene with the thiones thiosantonin and androsta-1,4-diene-17-one-3-thione. The heterodiene coordinates via the S=C moiety in a $\eta^2$ mode to one metal and via the C=C bond to the other metal.[124]

In addition to group 6 metal heterocarbonyl complexes, some thioketone complexes of other metals have also been prepared by the substitution route. The synthesis of the binuclear complex $[Mn(CO)_2\{S=C(Ph)[(\eta^5\text{-}C_5H_4)Mn(CO)_3]\}(\eta^5\text{-}C_5H_5)]$ (5) was achieved by reaction of $[Mn(CO)_2(thf)(\eta^5\text{-}C_5H_5)]$ with cymantrenyl(phenyl)thioketone.[99] Similarly, the tetracar-

bonyl iron complex **6** was synthesized from $[Fe_2(CO)_9]$ and the cyclic thioketone diphenylcyclopropenethione at room temperature.[100]

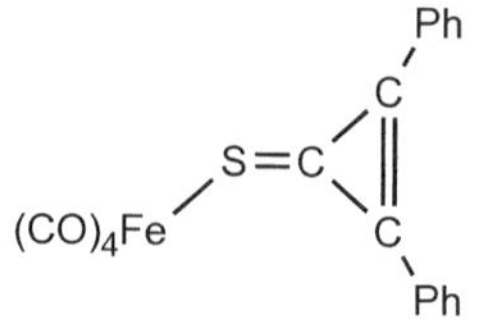

6

With some complexes, thiobenzophenone coordination is succeeded by
an orthometallation. Photolysis of $[Re_2(CO)_{10}]$ in the presence of various
thiobenzophenones yielded, as expected, the binuclear complexes $[Re_2$
$(CO)_9\{\eta^1\text{-}S{=}C(C_6H_4R\text{-}p)_2\}]$ (**7**). On subsequent thermolysis the monomeric orthometallated thioketone complexes **8** were formed (Scheme 2).[125]
In the thermal reaction at elevated temperatures the orthometallated com-

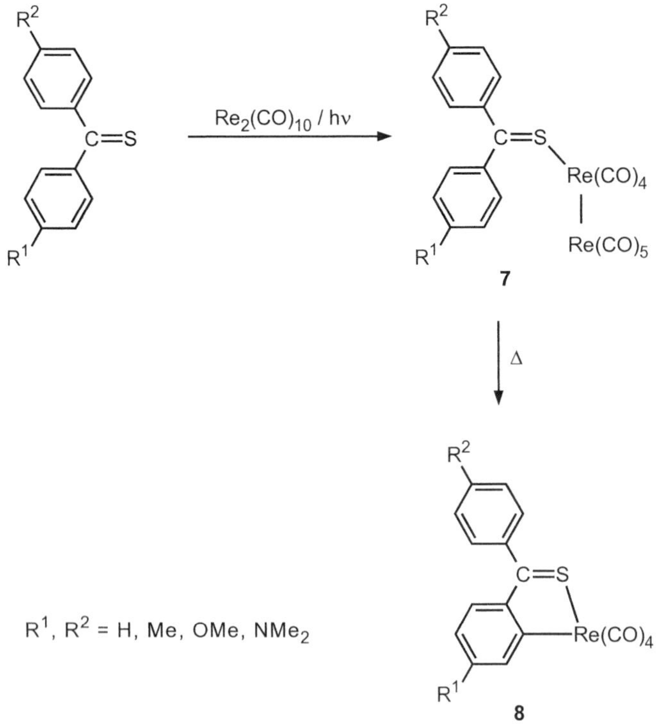

SCHEME 2

plexes were obtained directly from $[Re_2(CO)_{10}]$ and thioketones. The homologous manganese complex $[Mn_2(CO)_{10}]$ thermally reacted with *para* amino-substituted thioketones, $S=C(C_6H_4R\text{-}p)_2$ (R = NEt$_2$, NMe$_2$), to give the analogous mononuclear orthometallated compounds. In contrast, with R = H, Me, OMe, or F only substituted tetraphenylethenes could be isolated.[126]

The unusual binuclear orthometallated complexes **9** were obtained on treatment of $[Ru_3(CO)_{12}]$ with $S=C(C_6H_4R\text{-}p)_2$ (R = Me, OMe) at 70°C.[127]

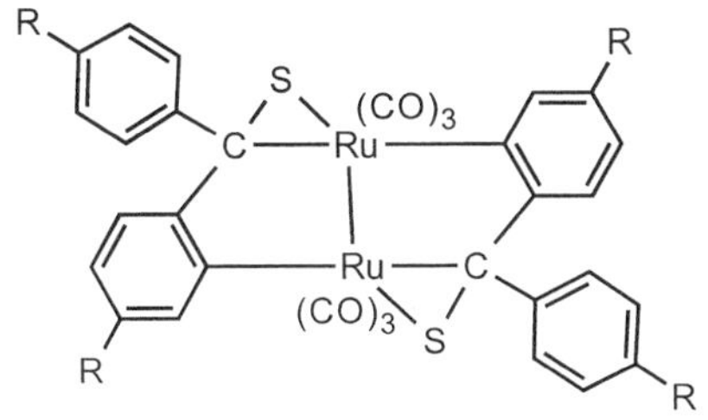

**9a,b**

R = OMe (**a**),  Me (**b**)

Modifications of the route described earlier are (a) the conversion of $[M(CO)_6]$ (M = Mo, W) into the chloro-bridged binuclear trianions **10,** which subsequently react with $S=C(C_6H_4OMe\text{-}p)_2$ to give the thioketone complexes, and (b) reduction of $[W(CO)_6]$ by sodium amalgam to give the dianion $[W_2(CO)_{10}]^{2-}$ and treatment of the latter with thiobenzophenones, adamantanethione, or thiocamphor (Scheme 3).[128]

Compared with the reactions of $[Re_2(CO)_{10}]$, those of the isoelectronic anion $[W_2(CO)_{10}]^{2-}$ are much faster,[128] presumably due to the stronger metal–metal bond in $[Re_2(CO)_{10}]$.[129]

Instead of metal carbonyls, a variety of other complexes can also be used as complex precursors. $[MCl_4]^{2-}$ anions (M = Pd, Pt) were found to react readily with $p,p'$-disubstituted thiobenzophenones to form a mixture of the bis(thiobenzophenone) complexes $[MCl_2\{\eta^1\text{-}S=C(C_6H_4R\text{-}p)_2\}_2]$ (R = H, Me, OMe, NMe$_2$) and the orthometallated chloride-bridged dimers. Cleavage of the dimers by triphenylphosphine afforded the monomeric orthometallated complexes **11.**[130]

The moderate yield of cyclopalladated products could be increased using ligand exchange reactions starting from another cyclopalladated complex.[131] Other thioketones such as thioxanthone, dithioxanthone, and thiochromone derivatives react in a similar way, affording cyclopalladated compounds.[132]

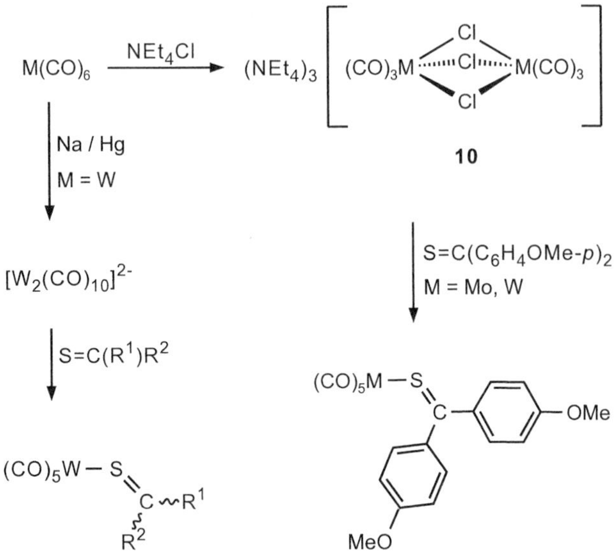

SCHEME 3

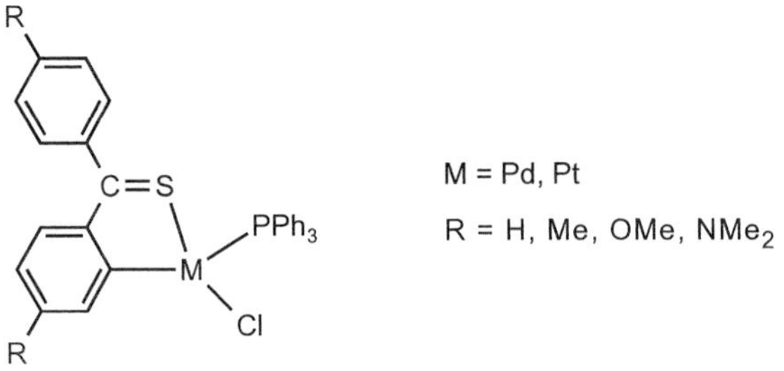

11

An intermediate in the reaction of bis(hexafluoroacetylacetonato)-palladium with $S=C(C_6H_4OMe-p)_2$ to give the cyclopalladated complex **12** could be observed at low temperatures. In the intermediate the thiobenzophenone is $\eta^1$ coordinated to the metal and one acetylacetonato ligand via the central carbon atom. Subsequent loss of the $\sigma$-coordinated acetylacetonate afforded **12** (Scheme 4).[133]

Complexes of the type $[MX_2(dmpt)_2]$ (dmpt = 2,6-dimethyl-4$H$-pyrane-4-thione; X = Cl, Br, I) were obtained by treatment of the anhydrous halides of palladium and platinum,[134] zinc and cadmium,[102] and mercury[135]

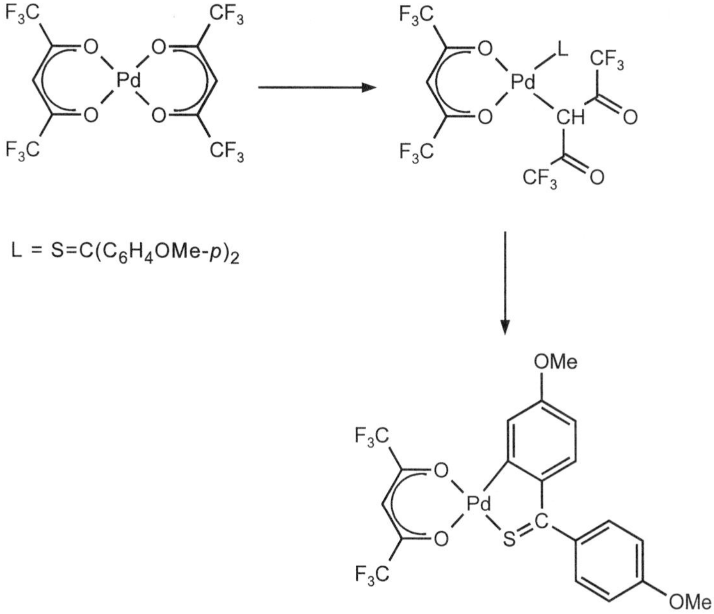

L = S=C(C$_6$H$_4$OMe-$p$)$_2$

**12**

SCHEME 4

with DMPT. In contrast to the diene thione molybdenum complexes mentioned earlier, DMPT is $\eta^1$ coordinated to the metal in all cases.

When [PdCl$_2$(NCPh)$_2$] was mixed with an excess of cycloheptatrienethione the red dichlorobis(cycloheptatrienethione) complex [PdCl$_2$
($\eta^1$-SC$_7$H$_6$)$_2$] precipitated. A *cis* configuration in the solid state was proposed. A cyclopalladated product was not observed.[136]

The dimeric thioketone complex **13** was obtained from [PdCl$_4$]$^{2-}$ and
thiopivaloylferrocene [Eq. (5)].[137]

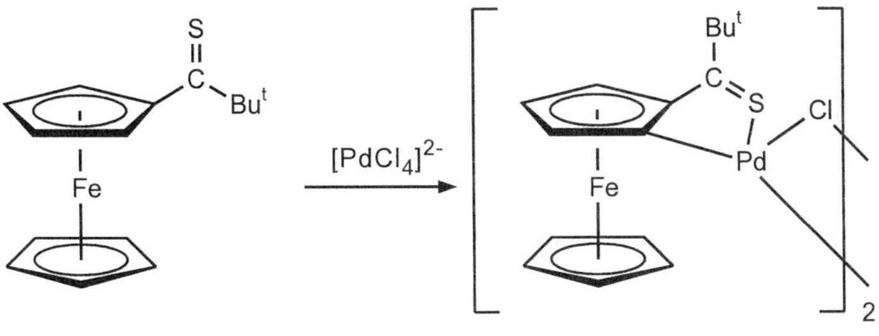

$$(5)$$

**13**

Coordinated olefins are also readily displaced by thioketones, e.g., stilbene, in the molybdocene complex **14** by thiobenzophenone to form the
complex **15** [Eq. (6)].[138] Similarly, thiobenzophenone is substituted for
ethene in $[Rh(C_2H_4)(PMe_3)(\eta^5-C_5H_5)]$ to yield $[Rh(PMe_3)(\eta^1-S=CPh_2)(\eta^5-C_5H_5)]$.[139]

$$\text{(6)}$$

**14**                                        **15**

Even one phosphine ligand in $[Co(PR_3)_2(\eta^5-C_5H_5)]$ $(PR_3 = PMe_3,$
$PMe_2Ph)$ could be replaced by $S=CPh_2$ to afford $[Co(PR_3)(\eta^1-S=CPh_2)(\eta^5-C_5H_5)]$ complexes in which the thioketone ligand is bound via a
sulfur lone pair.[139]

The vanadocene complex $[(\eta^5-C_5H_5)_2V(\eta^2-S=CPh_2)]$ is related to the
molybdenum complex **15,** although it is obtained by a different route.
The vanadium complex is formed by the addition of thiobenzophenone to
coordinatively unsaturated vanadocene.[103] In contrast, permethylvanadocene does not react with thiobenzophenone even under more forcing conditions. Steric rather than electronic effects may be responsible because the
less crowded thiocarbonyl group of PhNCS was shown to react even with
permethylvanadocene to give the $\eta^2$-$C,S$-isothiocyanate complex.[140]

Cluster complexes can also be used as starting compounds for the synthesis of heterocarbonyl compounds via ligand replacement. Treatment of
$[Os_3(H)_2(CO)_9(NMe_3)]$ with $S=CPh_2$ afforded two different osmium clusters depending on the reaction conditions. Each contained a $\eta^1$-bound
thiobenzophenone ligand.[117]

The trapping of thioketones generated *in situ* by suitable complexes has
also been reported. Ketones, e.g., react at $-80°C$ to $-40°C$ with $H_2S$ in acidic
solution to form thioketones.[141,142] These thioketones could be trapped by
$(CO)_5M$ fragments. Treatment of $[M(CO)_5I]^-$ (M = Cr, Mo, W) with two
equivalents of $AgBF_4$ in acetone at $-72°C$ and then with $H_2S$ afforded the
thioacetone complexes $[(CO)_5M\{\eta^1-S=CMe_2\}]$ (Scheme 5). When only one
equivalent of $AgBF_4$ was employed, the reaction stopped at the stage
of the acetone complex. Ethyl methyl thioketone and thiobenzophenone
tungsten complexes were obtained analogously.[51,143]

The direct approach to heterocarbonyl complexes by ligand exchange
reactions is usually restricted to heteroketone complexes, as suitable and

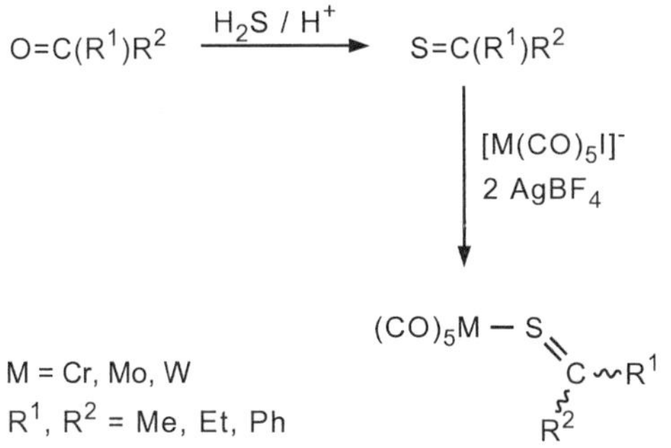

$$O=C(R^1)R^2 \xrightarrow{\text{H}_2\text{S / H}^+} S=C(R^1)R^2$$

$$\downarrow \begin{array}{c} [\text{M(CO)}_5\text{I}]^- \\ 2\ \text{AgBF}_4 \end{array}$$

M = Cr, Mo, W
$R^1$, $R^2$ = Me, Et, Ph

(CO)$_5$M — S
$$\overset{\Vert}{\underset{R^2}{C}} \rightsquigarrow R^1$$

SCHEME 5

stable-free heteroaldehydes are very scarce. However, it was possible to isolate the sterically highly congested selenobenzaldehyde Se=C(H)-$C_6H_2R'_3$-$o,o',p$ [R' = CH(SiMe$_3$)$_2$] as two separate rotational isomers. When treated with [W(CO)$_5$(thf)] at room temperature, both isomers formed selenobenzaldehyde complexes (**16-A** and **16-B**). In solution, the isomer **16-B** slowly converted into **16-A** (Scheme 6). Isomer **16-A** was the first $\eta^1$-selenoaldehyde complex characterized by a crystal structure determination.[15,16]

Similar to thioketones, the trapping of several selenoaldehydes generated *in situ* has been achieved. When treated with aqueous or methanolic sodium hydrogen sulfide, the Vilsmaier salts **17** were solvolysed to form the enamino thioaldehydes **18** in moderate to good yield. The selenoaldehyde analogues, however, were rather unstable. When exposed to light or air they rapidly decomposed. However, when their diluted solutions were treated immediately with [W(CO)$_5$I]$^-$/AgNO$_3$ they could be trapped as pentacarbonyl tungsten complexes **19** (Scheme 7).[13]

### 2. Synthesis from Heterocarbonyl Precursors and Metal Complexes

Reactions of metal complexes with suitable heterocarbonyl precursors constitute another route to heteroaldehyde and heteroketone complexes. Very often, these reactions involve the cleavage of a heterocyclic compound.

[Pt(PPh$_3$)$_3$] or [Pt(C$_2$H$_4$)(PPh$_3$)$_2$], [Pd(PPh$_3$)$_3$], [Ni(cod)$_2$] (cod = 1,5-cyclooctadiene), and [IrH(CO)(PPh$_3$)$_3$] already react at or even below room temperature with 2,2,4,4-tetrakis(trifluoromethyl)-1,3-dithietane by ring cleavage to yield the thioacetone complexes [Pt(PPh$_3$)$_2$\{$\eta^2$-S=C(CF$_3$)$_2$\}],[107,144] [Pd(PPh$_3$)$_2$\{$\eta^2$-S=C(CF$_3$)$_2$\}],[106] [Ni(cod)\{$\eta^2$-S=C(CF$_3$)$_2$\}],[145] and *cis*-[IrH(CO)(PPh$_3$)$_2$\{$\eta^2$-S=C(CF$_3$)$_2$\}],[146] respectively. In contrast, the thermal cleavage of the 1,3-dithietane requires temperatures of about

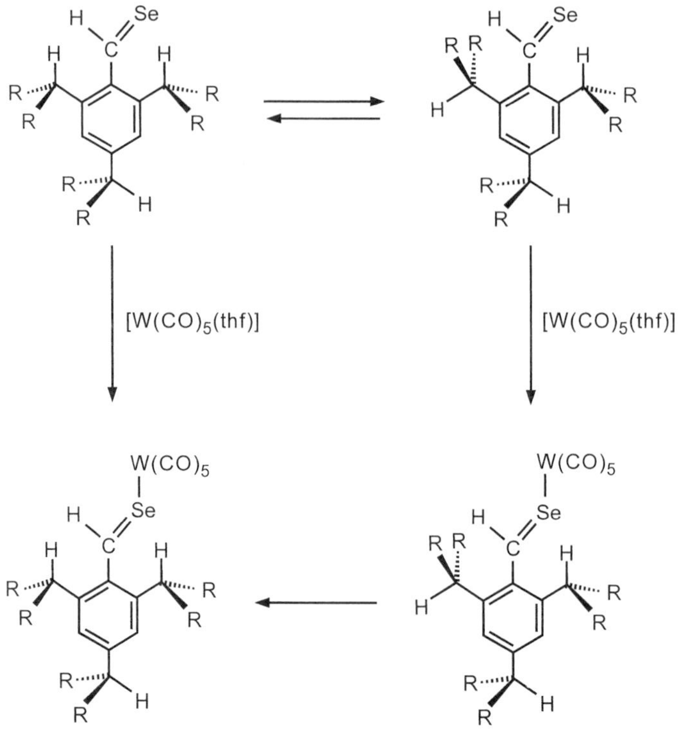

R = SiMe₃

SCHEME 6

600°C. Other hexafluorothioacetone nickel complexes are accessible by substitution of $PPh_3$, $P(OPh)_3$, 1,2-bis(diphenylphosphino)ethane, bipyridyls, or $EtC(CH_2O)_3P$ for the 1,5-cyclooctadiene ligand in $[Ni(cod)\{\eta^2-S=C(CF_3)_2\}]$.[145]

$\eta^4$-Thioacroleine complexes (**21a–c, 22a,b,d**) were obtained by photolysis of thietes (**20a–d**) in the presence of $[Fe(CO)_5]$ or $[Co(CO)_2(\eta^5-C_5H_5)]$ or by thermolytic ring cleavage in the presence of $[Fe_2(CO)_9]$.[73,147,148] Replacement of a CO ligand in **21a** by $PPh_3$ gave **23a** (Scheme 8).[73,147]

Related to these routes is the synthesis of the complexes **25** by reaction of the 1,6,6a-$\lambda^4$-trithiapentalenes **24** with $[NEt_4][W(CO)_5I]/AgNO_3$. In solution the complexes are fluxional due to a rapid site exchange (Scheme 9).[149]

The reactivity of the complexes **25** is similar to that of a dimeric-orthometallated thioketone complex obtained by the reaction of 2,5-

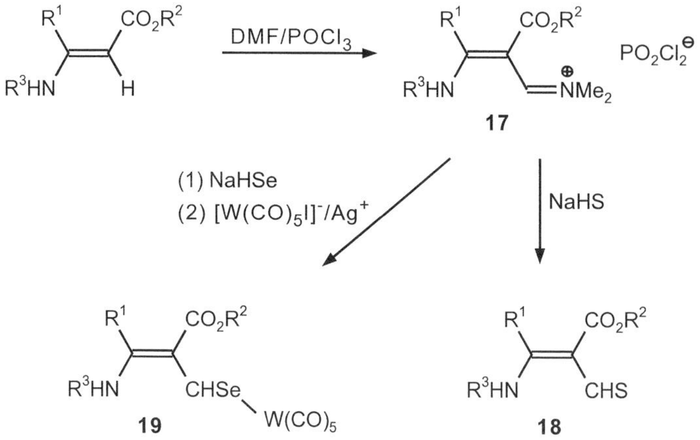

R¹, R², R³ = H, Me, Ph

SCHEME 7

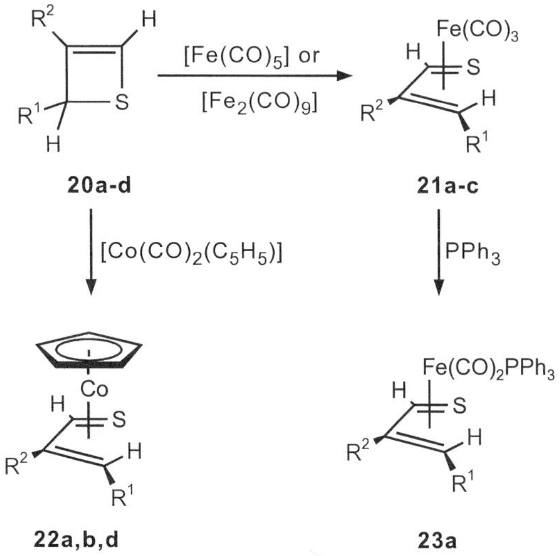

$R^1 = R^2 = H$ (**a**)     $R^1, R^2 = (CH_2)_5$ (**c**)
$R^1 = Me, R^2 = Et$ (**b**)     $R^1 = H, R^2 = Ph$ (**d**)

SCHEME 8

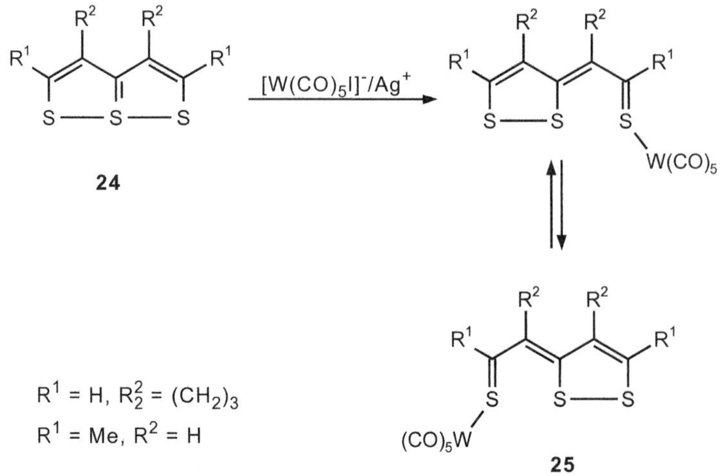

$R^1$ = H, $R^2_2$ = $(CH_2)_3$
$R^1$ = Me, $R^2$ = H

SCHEME 9

diphenyl-1,6,6a-$\lambda^4$-trithiapentalene with $PdCl_2$ and subsequent treatment with $PPh_3$.[150]

The cleavage of thiophene by coordination and the hydrolysis of coordinated thiophene represent models for the commercially important hydrodesulfurization process. Because the initially formed thiophene complexes are not isolable generally, such reactions are discussed in this section for reasons of convenience. A rhodium complex with a $\eta^4$-bound $\alpha,\beta$-unsaturated thioketone ligand was formed by an unusual ring cleavage of a thiophene ligand. When $[Rh(C_4Me_4S)(\eta^5\text{-}C_5Me_5)][OTf]_2$ (OTf = triflate) (**26**) was dissolved in 0.03 $M$ aqueous KOH, red-orange crystals of the complex **27** precipitated [Eq. (7)].[105,151]

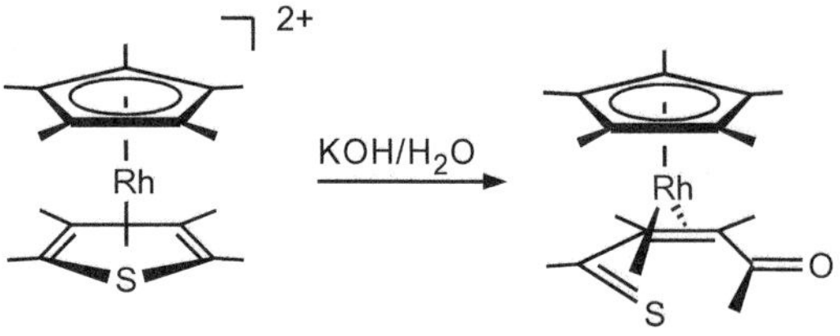

The reaction of $[Rh(C_2H_4)_2(\eta^5\text{-}C_5Me_5)]$ with an excess of thiophene in benzene solution led to dimerization of thiophene and to formation of the unusual binuclear complex **28**. The bridging ligand can be considered as an $\omega$-mercapto $\alpha,\beta$-unsaturated thioaldehyde.[79,152] When $[Co(C_2H_4)_2$

$(\eta^5\text{-}C_5Me_5)]$ was substituted for $[Rh(C_2H_4)_2(\eta^5\text{-}C_5Me_5)]$ instead of the dimerization of the thiophene, only the cleavage of the thiophene and the formation of a binuclear complex were observed, which, on treatment with $H_2S$, afforded **29**. Complex **29** contains a coordinated bisthioaldehyde (Scheme 10).[153]

Several other reactions of thiophene complexes lead to thioaldehyde or thioketone complexes.[78,113,114,154–162] Some examples are summarized in Scheme 11.

Two different thioacetaldehyde complexes (**31** and **32**) were obtained from **30** by a deprotonation/reprotonation sequence depending on the base used (KOH/CH$_2$Cl$_2$ or KO$^t$Bu/THF) (Scheme 12).[163]

### 3. Transformation of a Coordinated Ligand into a Heterocarbonyl Ligand

a. *Transformation of Alkyl Ligands.* Many chalcogenoaldehyde and chalcogenoketone complexes, including heteroformaldehyde complexes,

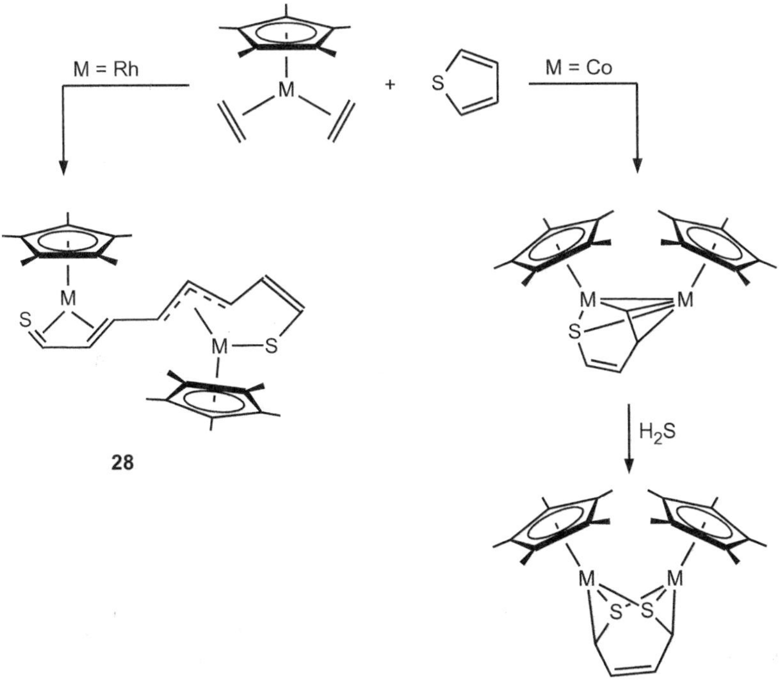

SCHEME 10

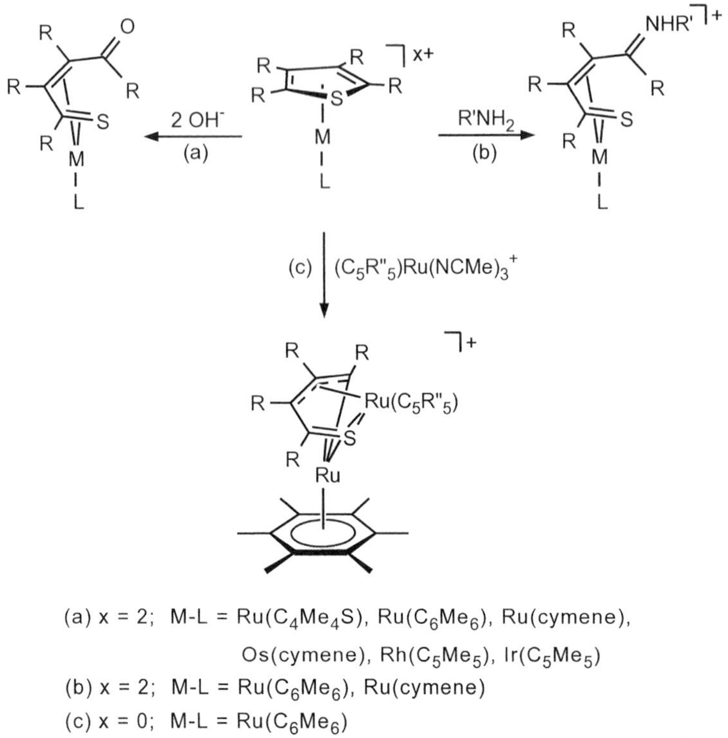

(a) x = 2;  M-L = Ru(C$_4$Me$_4$S), Ru(C$_6$Me$_6$), Ru(cymene),

Os(cymene), Rh(C$_5$Me$_5$), Ir(C$_5$Me$_5$)

(b) x = 2;  M-L = Ru(C$_6$Me$_6$), Ru(cymene)

(c) x = 0;  M-L = Ru(C$_6$Me$_6$)

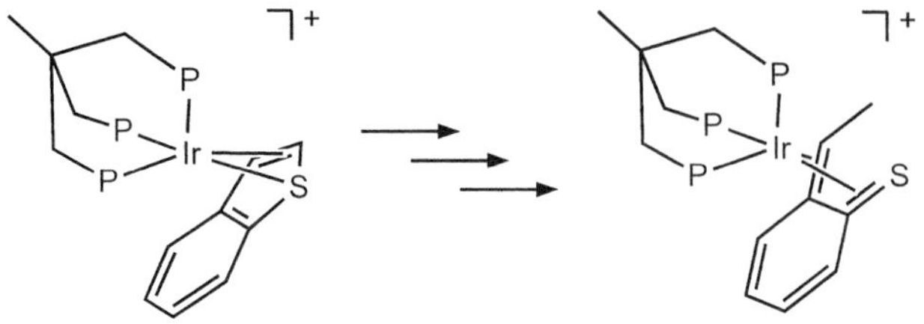

SCHEME 11

were synthesized from halogeno(halogenoalkyl) complexes of the type
[L$_n$M(CH$_2$X)X] (M = Rh, Os; L = $\eta^5$-C$_5$H$_5$, $\eta^5$-C$_5$Me$_5$, CO, and phosphines)
and sulfur, selenium, or tellurium sources. With a 10-fold excess of NaEH
(E = S, Se, Te) the rhodium complex **33**[164] afforded the thio-, seleno-, or
telluroformaldehyde complexes **35** in low to moderate yield. The reaction
is probably initiated by a nucleophilic substitution of the I$^-$ ligand by EH$^-$

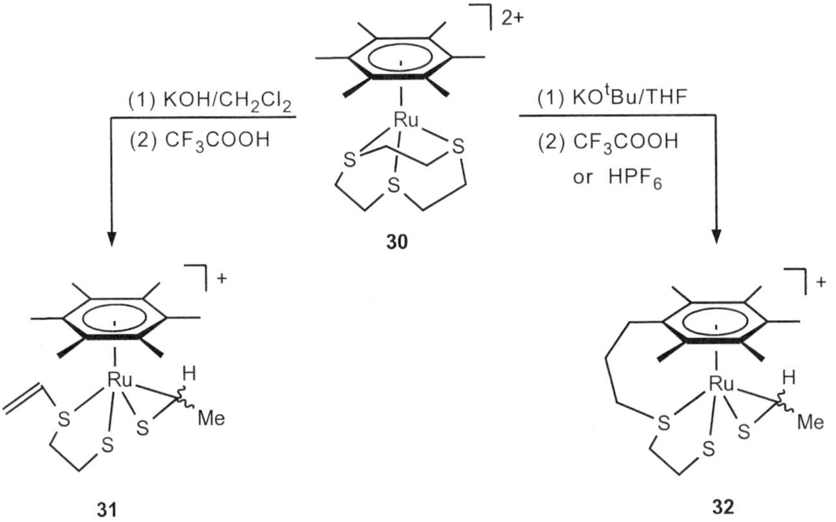

SCHEME 12

followed by deprotonation of the intermediates **34** by excess EH⁻. Final ring closure produces the complexes **35** (Scheme 13).[165]

The complexes $[Rh(PR_3)\{\eta^1\text{-}S\!=\!CH_2\}(\eta^5\text{-}C_5H_5)]$ bearing more bulky phosphines such as $PMe_2Ph$ or $PPr^i_3$ were prepared analogously. As an alternative starting material, $[Rh(CH_2I)Cl(PMe_3)(\eta^5\text{-}C_5H_5)]$ was also used and gave satisfactory yields. Obviously, the reactions are not influenced by the steric requirements of the $PR_3$ ligand and the type of the metal-bound halide. Similarly, the pentamethylcyclopentadienyl analogues $[Rh(CO)\{\eta^1\text{-}S\!=\!CH_2\}(\eta^5\text{-}C_5Me_5)]$ (E = S, Se, Te) and $[Rh\{P(OMe)_3\}\{\eta^1\text{-}Se\!=\!CH_2\}$

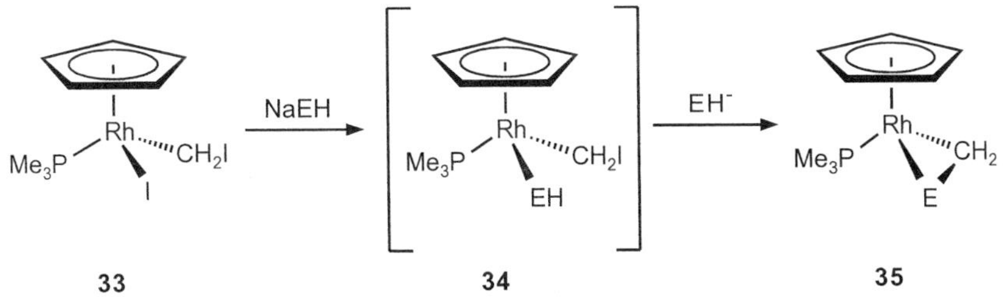

E = S (**a**), Se (**b**), Te (**c**)

SCHEME 13

$(\eta^5\text{-}C_5Me_5)]$ are also accessible by this method. The stability of these complexes decreases significantly in the series S, Se, Te. The telluroformaldehyde complex $[Rh(CO)\{\eta^1\text{-}Te\!=\!CH_2\}(\eta^5\text{-}C_5Me_5)]$ has been detected only spectroscopically.[54,166] By an analogous procedure the osmium complexes $[Os(CO)_2(PPh_3)\{\eta^2\text{-}E\!=\!CH_2\}]$ (E = Se, Te)[167] have been synthesized from $[Os(CH_2I)I(CO)_2(PPh_3)_2]$[168] and $EH^-$.

Cobalt complexes of the type $[L_nCo(CH_2X)X]$ (X = Br, I) (**36**), which are unknown until now, are likely intermediates in the formation of the cobalt chalcogenoaldehyde complexes **37** from $[Co(CO)(PMe_3)(\eta^5\text{-}C_5H_5)]$, $EH^-$ (E = S, Se), and $CH_2I_2$ or $CH_2Br_2$ (Scheme 14).[169] Surprisingly, no aldehyde complex was obtained when starting from the isolable compound $[Co(CH_2Cl)I(PMe_3)(\eta^5\text{-}C_5H_5)]$.[170] When $CH_2Br_2$ was replaced by $MeCHBr_2$, the thio- and selenoacetaldehyde complexes **38** were obtained. The reaction was highly stereoselective, the methyl group being adjacent to the cyclopentadienyl ring. However, due to unavoidable side reactions, the yields were rather poor (Scheme 14).[55,170]

Although this route also seemed to be suitable for rhodium compounds, the readily accessible complex $[Rh(CHClMe)Cl(PPr^i_3)(\eta^5\text{-}C_5H_5)]$ did not afford the thioacetaldehyde complex when treated with NaHS but rather the vinyl complex $[Rh(CH\!=\!CH_2)Cl(PPr^i_3)(\eta^5\text{-}C_5H_5)]$.[55]

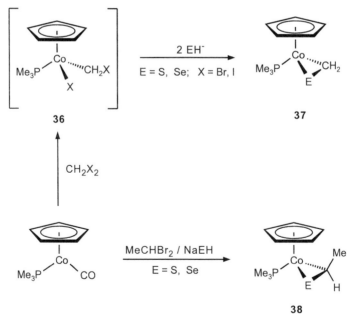

SCHEME 14

b. *Transformation of Carbene and Vinylidene Ligands.* The insertion of sulfur, selenium, and also tellurium into the metal–carbon bond of carbene complexes is one of the most versatile routes to chalcogenoaldehyde and -ketone complexes. As sources of sulfur, selenium, and tellurium the elemental chalcogens as well as chalcogen compounds such as cyclohexene sulfide, $Ph_3P=E$, or $[ECN]^-$ can be employed. However, chalcogen compounds are usually preferred due to their better solubility in organic solvents.

The first examples of a chalcogen insertion into a transition metal–carbene bond were reported in 1981. Stirring solutions of the pentacarbonyl-(diarylcarbene)tungsten complexes **39** in pentane in the presence of solid sulfur afforded the thioketone complexes **40** [Eq. (8)].[171] The reaction rate increased considerably when pentane was replaced by $CS_2$ as the solvent due to the higher solubility of elemental sulfur in $CS_2$.

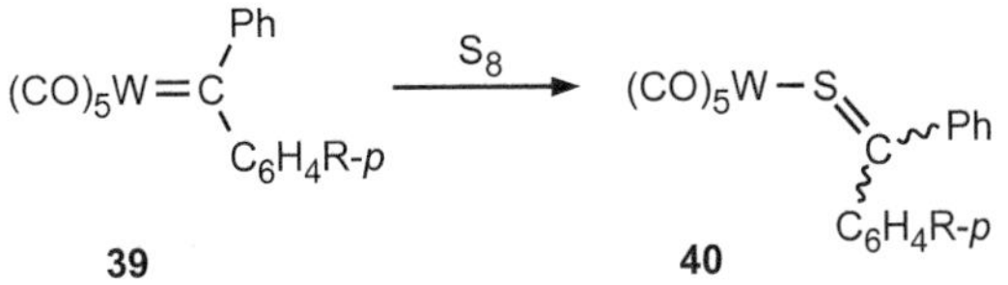

$$(CO)_5W=C\begin{subarray}{l} Ph \\ \\ C_6H_4R\text{-}p \end{subarray} \xrightarrow{S_8} (CO)_5W-S\begin{subarray}{l} \\ C\sim Ph \\ \\ C_6H_4R\text{-}p \end{subarray} \quad (8)$$

**39**  **40**

R = H, OMe, $CF_3$

The reaction of the benzylidene manganese complex **41** with elemental sulfur or selenium afforded the $\eta^2$-thio- and selenobenzaldehyde cymantrene derivatives **42a,b** [Eq. (9)].[172]

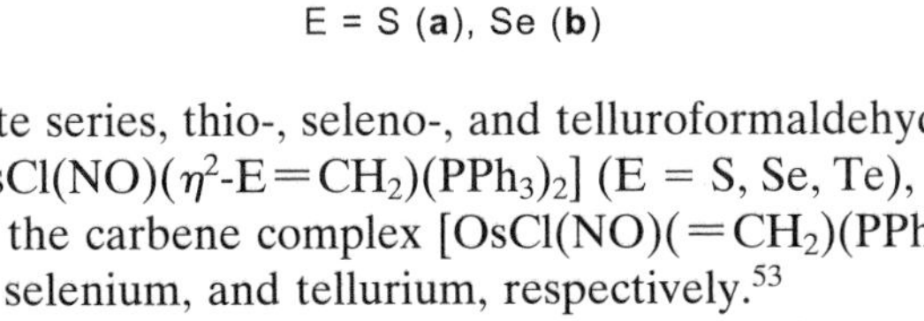

$$\text{41} \xrightarrow{E} \text{42a,b} \quad (9)$$

**41**  **42a,b**

E = S (a), Se (b)

The complete series, thio-, seleno-, and telluroformaldehyde complexes of osmium $[OsCl(NO)(\eta^2\text{-}E=CH_2)(PPh_3)_2]$ (E = S, Se, Te), was obtained by reaction of the carbene complex $[OsCl(NO)(=CH_2)(PPh_3)_2]$ with elemental sulfur, selenium, and tellurium, respectively.[53]

The diarylcarbene complexes **39** (R = H, Me, OMe, Br, $CF_3$) also reacted with various organyl isothiocyanates R′NCS (R′ = Ph, Et, Me) to give the corresponding diarylthioketone complexes **40**.[50,173] On the basis of a kinetic study of these reactions and the dependence of the reaction rate on the substitutents R and R', a rate-determining nucleophilic attack of the isothio-cyanate at the carbene carbon was proposed. The addition step is followed by elimination of R'NC and formation of a $\eta^2$-coordinated thioketone that rapidly rearranges to give the $\eta^1$ isomer.[174]

The route could be extended to the syntheses of selenoketone complexes of chromium and tungsten (**43**) from the corresponding diarylcarbene complexes (e.g., **39**: R = H, Me, OMe, Br, $CF_3$) and PhNCSe. Later on, the use of tetraethylammonium or potassium salts of [ECN]⁻ as the heteroatom sources turned out to be superior to R'NCE.[175] Thus it was even possible to synthesize a pentacarbonyl tellurobenzophenone complex of tungsten (**44**) from K⁺[TeCN]⁻ and **39** (R = H).[176]

R = H, Me, OMe, Br, $CF_3$

The reaction of [ECN]⁻ (E = S, Se) with the strongly electrophilic benzyli-dene(pentacarbonyl) complexes of chromium, molybdenum, and tungsten $(CO)_5M=C(C_6H_4R$-$p)H$ (M = Cr, Mo, W)[177–179] provided the correspond-ing thio- and selenoaldehyde complexes $[(CO)_5M\{E=C(C_6H_4R$-p$)H\}]$.[44–46,179] Pentacarbonyl(tellurobenzaldehyde)tungsten, $[(CO)_5W\{Te=C(Ph)H\}]$, could also be generated by reaction of $[(CO)_5W=C(Ph)H]$ with [TeCN]⁻ but could not be isolated because of its instability even at −80°C. However, the complex could be detected spectroscopically and be trapped by cycload-dition reactions with dienes.[180] In solution the thio- and selenobenzaldehyde complexes are present as rapidly interconverting mixtures of the two $\eta^1$ isomers and the $\eta^2$ isomer (see Scheme 1, Section III,B,1).

The carbene complex **45** added the nucleophile PhC≡CS⁻ to the carbene carbon to yield the adduct **46** that readily rearranged on protonation to give the thioaldehyde complex **47**.[181,182] When the adduct was first treated with carbon disulfide and then protonated, **47** was isolated in much higher yield, along with the unexpected thioketone complex **48**. The chromium carbene complex $[(CO)_5Cr=C(OEt)Ph]$ reacted similarly with PhC≡CS⁻/H⁺. $[(CO)_5W\{\eta^1$-Se$=CHC(Ph)=C(Ph)OEt\}]$ was prepared by the same method. When the corresponding precursor adduct $[(CO)_5W$-C(OEt)(Ph)-C(Ph)=C=Se]⁻ was treated with $CS_2$/H⁺, **48** was formed along with the seleno analogue of **47**. A selenoketone complex analogous to **48** was not detected, indicating that the sulfur in **48** originated from $CS_2$ (Scheme 15).[64]

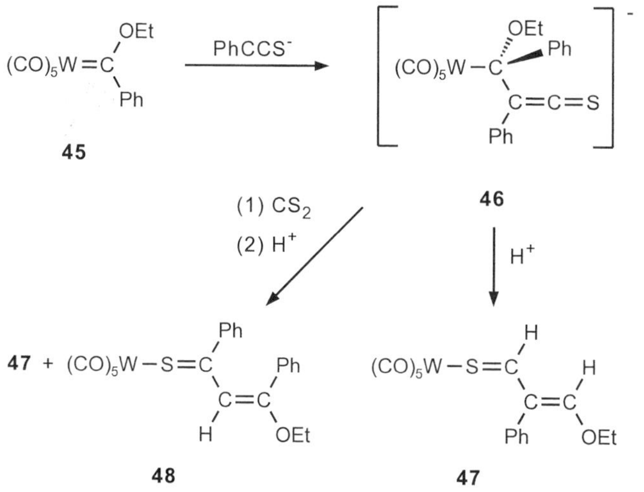

SCHEME 15

Complexes with $\alpha,\beta$-unsaturated thioketones as ligands (49) were obtained in addition to other compounds when $[(CO)_5M=C(OEt)Aryl]$ (M = Cr, Aryl = Ph, $C_4H_6Ph$-o; M = W, Aryl = Ph) were treated at $-78°C$ with $LiC(S)NMe_2$ [prepared from $HC(S)NMe_2$ and $LiNEt_2$ at $-100°C$] and then protonated at ambient temperature with acetic acid [Eq. (10)]. However, the yield of 49 was rather low (3–8%).[96]

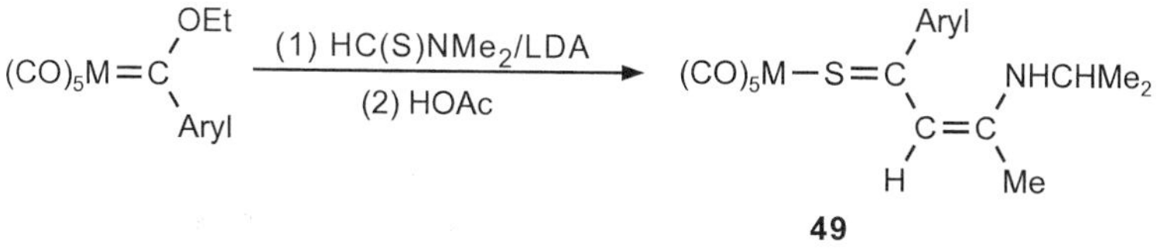

M = Cr, Aryl = Ph, $C_6H_4Ph$-o;   M = W, Aryl = Ph

Thio- and selenoformaldehyde complexes of rhenium were synthesized from 50 generated *in situ*. The reaction of 50 with cyclohexene sulfide as the sulfur source afforded the thioformaldehyde complex 51a in excellent yield. Complex 51a could also be prepared from $Ph_3P=S$ and 50. However, the yield was much lower due to the reaction of the coproduct $PPh_3$ with 50 to form the ylide complex 52.[75,76] The selenoformaldehyde complex 51b was obtained from 50 and $Ph_3P=Se$ in 39% yield together with 41% of 52. When $Ph_3P=Se$ was replaced by $K^+[SeCN]^-$ as the selenium source in

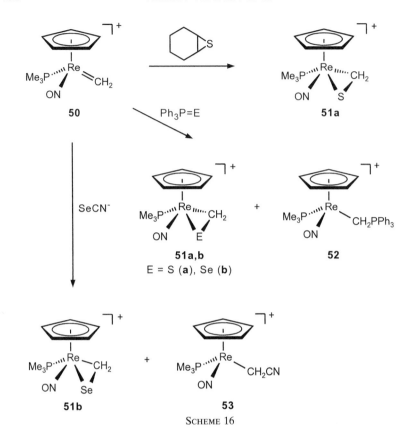

SCHEME 16

addition to 45% of **51b,** 13% of the cyanomethyl complex **53** was isolated (Scheme 16).[83]

A thioformaldehyde complex of titanium (**55**) was similarly obtained from the titanocene methylene complex **54** and either elemental sulfur or sulfur sources such as propene sulfide, cyclohexene sulfide, styrene sulfide, or $Ph_3P=S$. Best yields were achieved with propene sulfide [Eq. (11)].[67]

$$(11)$$

The telluroformaldehyde complex **57** was obtained by a $PMe_3$-catalyzed insertion of Te into the methylidene unit of **56** [Eq. (12)],[84] and the seleno-formaldehyde complex **59** was formed when the selenide **58** was treated with MeLi [Eq. (13)]. The different synthetic paths led to a different coordination mode of the heteroformaldehyde ligands.[183]

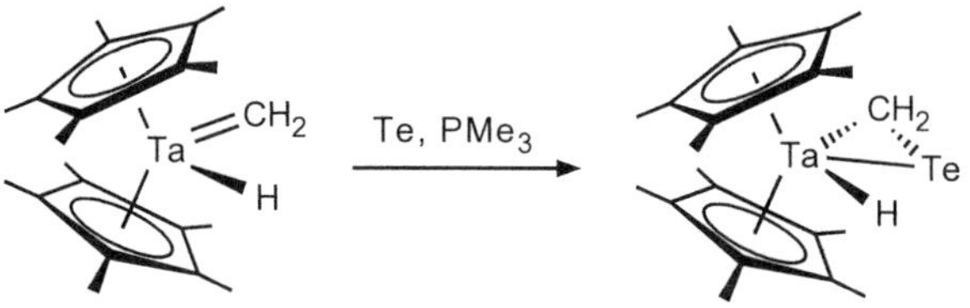

$$(12)$$

**56**          **57**

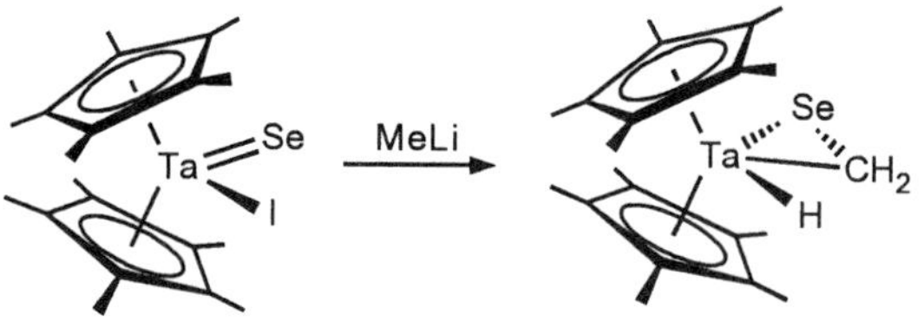

$$(13)$$

**58**          **59**

The reaction of an osmium carbene cluster containing the cyclic carbene ligand $=C(CH=CH)_2C(Ph)H$ which is "side on" bound to a nonacarbo-nyltriosmium unit with either elemental sulfur/$NEt_3$ or cyclohexene sulfide also afforded a $\eta^2$-thioketone complex.[184]

The synthesis of heterocarbonyl complexes from carbene complexes and heteroatoms or heteroatom sources is not confined to terminal carbene ligands. Bridging carbene ligands can also be transformed into heteroalde-hyde ligands. This was demonstrated by the reaction of the cluster complex $[Os_3(\mu\text{-}CH_2)(CO)_{11}]$ with ethene sulfide or cylohexene sulfide to give a complex with a bridging thioformaldehyde ligand (see Section IV,B).

Insertion of chalcogens into an $M=C$ bond is not restricted to carbene complexes. The synthesis of rhodium heteroacetaldehyde complexes could be accomplished by starting from the vinylidene complex **60**.[185] Treatment with sulfur, selenium, or tellurium gave the heteroketene complexes **61**.[186] The chalcogenoacetaldehyde complexes **62** were finally obtained by subse-quent hydrogenation catalyzed by Wilkinsons catalyst $[RhCl(PPh_3)_3]$. The telluroaldehyde complex was too unstable for isolation but it could be identified spectroscopically. Analogously to the corresponding cobalt com-plexes, the rhodium compounds were obtained each as a single diastereo-

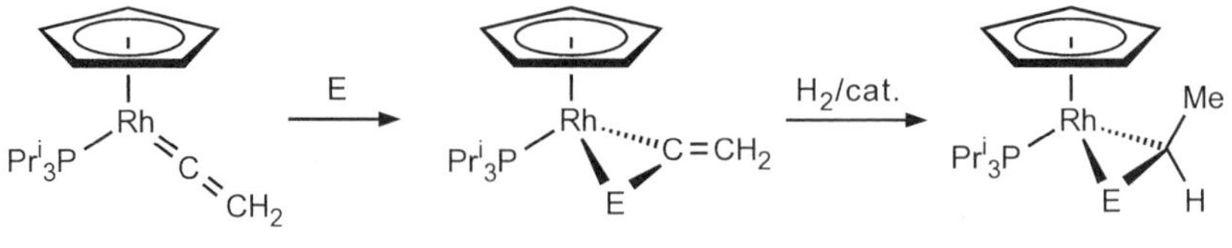

60                          61                          62

E = S, Se, Te

SCHEME 17

mer, its methyl group adjacent to the cyclopentadienyl ligand (Scheme 17).[54,55]

The synthesis of the thioaldehyde complex **63** somewhat paralleled the synthesis outlined in Scheme 17. In this case, sulfur and hydrogens were transferred to the vinylidene ligand in a single step [Eq. (14)].[187]

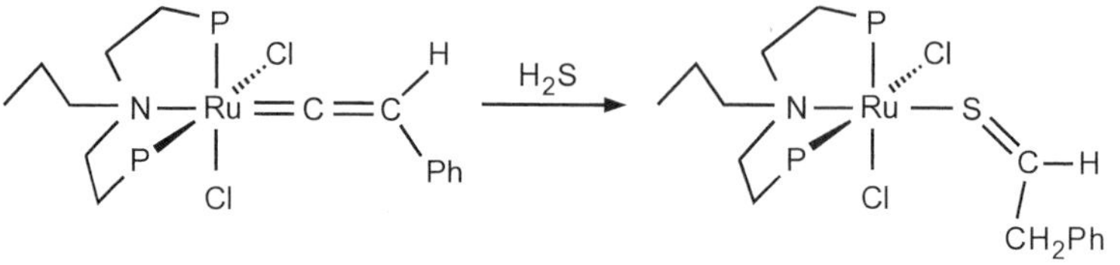

(14)

63

Similarly, the thioaldehyde complexes **66a,b** were formed via addition of thiols to the vinylidene complex **64** and elimination of ethene. The vinylidene complex **64** reacted with benzyl mercaptan to yield the isolable intermediate **65**. On heating **65** lost ethene in a first-order reaction to give **66b**.[188] The complexes **66a–d** could also be prepared from the benzyne-(hydrido) complex **67** and thiols $RCH_2SH$ (Scheme 18).[188]

In solution there is a rapid equilibrium between the thioaldehyde-(hydrido) complexes **66a–d** and their thiolato tautomers. At room temperature the thioaldehyde(hydrido) form dominates. However, the equilibrium strongly influences the reactivity.[188,189]

c. *Transformation of Carbyne Ligands.*   Carbyne complexes were also employed as precursors of thioaldehyde complexes. The reaction of the carbyne complexes **68** with $[H_2NEt_2][S_2CNEt_2]$ afforded the thioaldehyde complexes **69**.[72] Mechanistic conclusions concerning this reaction could be drawn from the reaction of the ynol ester complex **70**, which was deprotonated by $HNMe_2$ to the intermediate **71**. Complex **71** slowly transformed into the thioaldehyde complex **69** ($R = C_6H_4Bu^t$-$p$) (Scheme 19).[190]

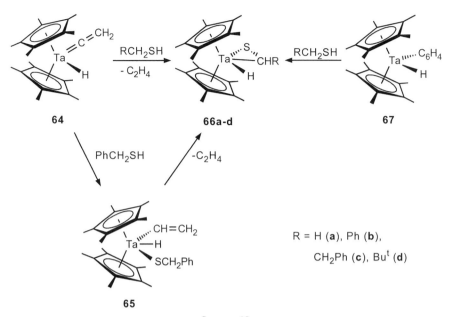

**64**

RCH₂SH, −C₂H₄

**66a-d**

RCH₂SH

**67**

PhCH₂SH

−C₂H₄

**65**

R = H (**a**), Ph (**b**),
CH₂Ph (**c**), Buᵗ (**d**)

Scheme 18

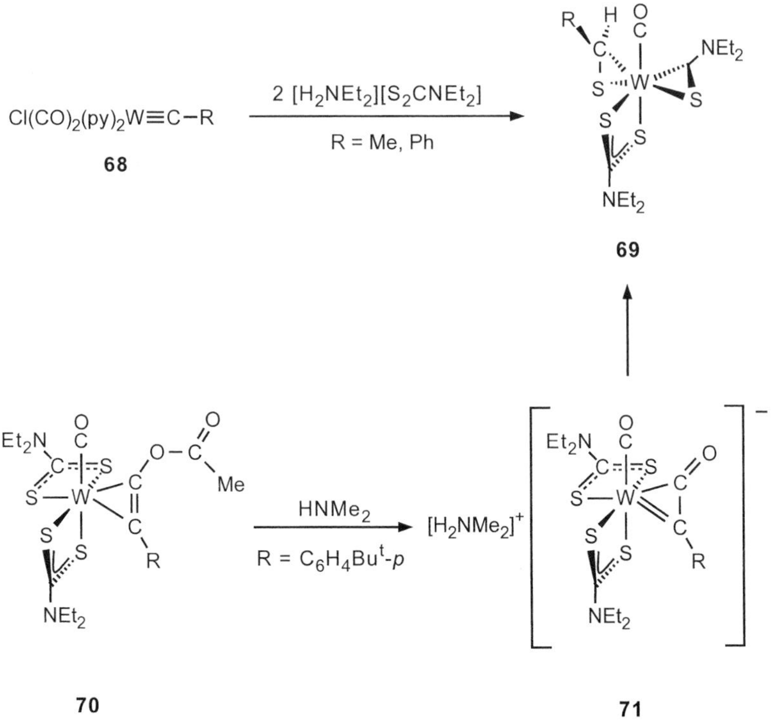

$Cl(CO)_2(py)_2W \equiv C-R$

**68**

$2\ [H_2NEt_2][S_2CNEt_2]$

R = Me, Ph

**69**

**70**

$HNMe_2$

R = C₆H₄Buᵗ-p

$[H_2NMe_2]^+$

**71**

Scheme 19

Reaction of the carbyne complex **72** analogously led to the complex **73** with the unusual thioaldehyde ligand $(\eta^5\text{-}C_5H_4CHS)Mn(CO)_3$ [Eq. (15)].[191]

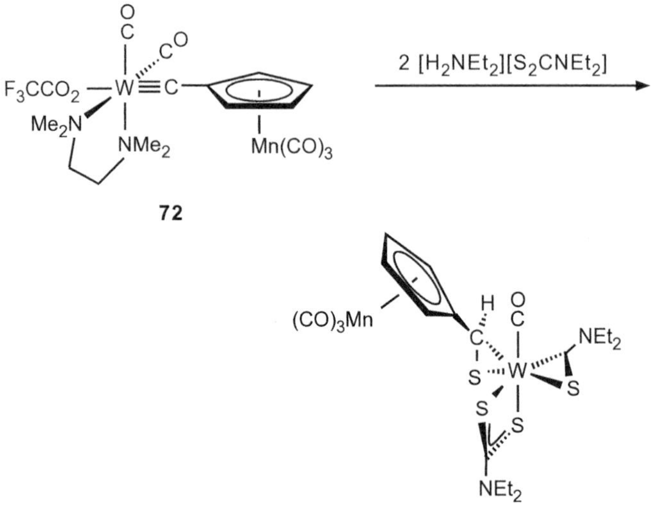

(15)

d. *Transformation of Thiocarbonyl Ligands.* The first thioformaldehyde complex $[Os(CO)_2(PPh_3)_2(\eta^2\text{-}S\text{=}CH_2)]$ (**75**) was prepared via reduction of the thioformyl ligand of the osmium complex $[Os(CHS)Cl(CO)_2(PPh_3)_2]$ (**74**) by $BH_4^-$.[192-194] Complex **75** was also formed in the reaction of $[OsH_2(CS)(PPh_3)_3]$ (**76**)[195] with CO under vigorous conditions (Scheme 20).[191,193]

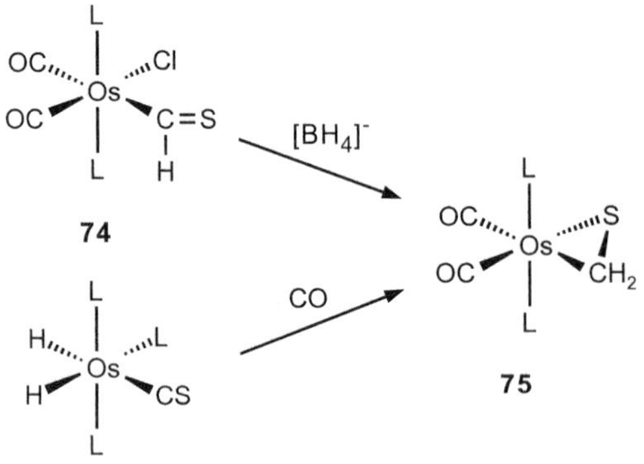

L = PPh$_3$

SCHEME 20

A thioformaldehyde complex was proposed as an intermediate in the formation of the thiomethyl complex $[IrH_2(SMe)(PPh_3)_3]$ from $[IrH(CS)(PPh_3)_3]$ and $H_2$. The reaction was thought to proceed by a stepwise transfer of coordinated hydride to the CS group to form the $S=CH_2$ ligand via the intermediates mer-$[IrH_2(CHS)(PPh_3)_3]$ and mer-$[IrH_2(PPh_3)_2(\eta^2-S=CH_2)]$. Because these intermediates could not be detected, the hydride transfer was assumed to be fairly rapid.[196]

e. *Transformation of Thiolato Ligands.*   Thiolato ligands were transformed into thioaldehyde ligands by $\beta$-hydride abstraction either by use of a hydride acceptor or by other ligands bound to the metal. For example, this same method was applied to the synthesis of the chiral rhenium thioaldehyde complexes $[Re(NO)(PPh_3)\{S=C(R)H\}(\eta-C_5H_5)]PF_6$ (R = Ph, $C_6H_4Cl$-p, $C_6H_4OMe$-p, H, Me).[77] Similarly, the cationic thiobenzaldehyde ruthenium complexes **78** were obtained from the thiolato complexes **77** by $\beta$-hydride abstraction with trityl hexafluorophosphate [Eq. (16)].[47,66]

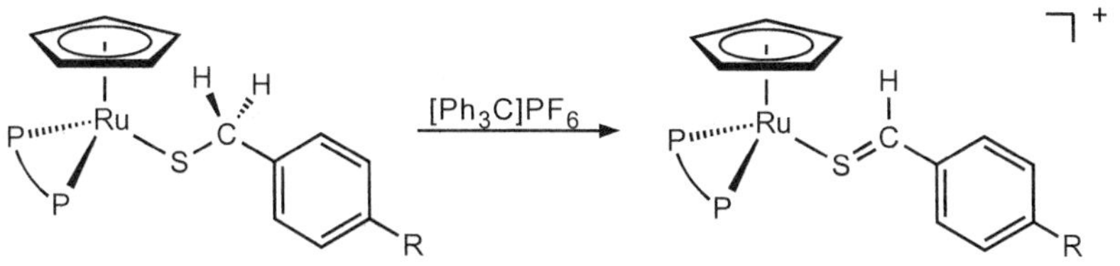

$$(16)$$

**77**                **78**

P-P = $Ph_2PCH_2PPh_2$, $Ph_2PCH_2CH_2PPh_2$;   R = H, Cl, OMe

The thioaldehyde complexes **81a,b** were formed from the thiolato complexes **80a,b** on treatment with $PMe_3$ (Scheme 21). The thiolato compounds were obtained from **79** and $MeCH_2SH$ or $PhCH_2SH$. Although the complexes **81a,b** were stable even in refluxing benzene for several hours, they

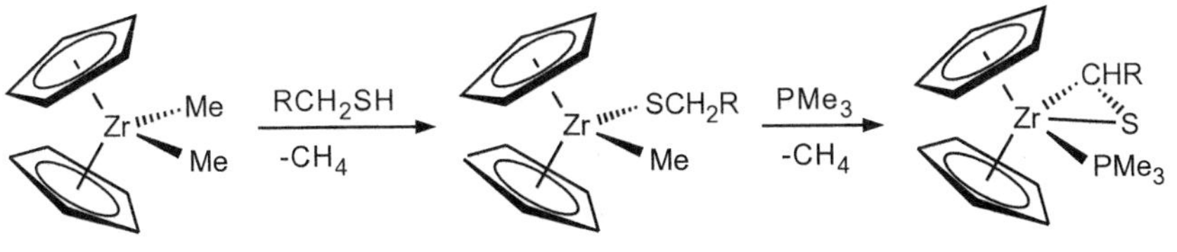

**79**                **80**                **81**

R = Me (**a**), Ph (**b**)

SCHEME 21

reacted readily with many substrates.[70] A variety of thioaldehyde zircono-cene complexes were prepared by this method. Kinetic measurements indi-cated that the formation of the thioaldehyde complexes proceeds via a concerted four-center cyclometallation process in which the transition state is polarized so that hydrogen moves as a proton. A stepwise $\beta$-hydride elimination/reductive elimination sequence was regarded as less likely.[197]

The protonation of the enethiolato complex **82** with $HBF_4$ led to **83** with an $\alpha,\beta$-unsaturated thioaldehyde ligand. The initially formed *syn* product could be partially converted to the less hindered *anti* isomer by heating in refluxing acetone (Scheme 22).[198]

Another route to thioaldehyde and thioketone complexes uses the tung-state $[W(CO)_5SH]^-$ as the precursor. When $[W(CO)_5SH]^-$ was treated with ketones in the presence of trifluoroacetic acid, thioketone complexes $[W(CO)_5\{S{=}C(R^1)R^2\}]$ were formed. Analogously, thiobenzaldehyde complexes were obtained from $[W(CO)_5SH]^-$ and benzaldehydes bearing electron-releasing *para* substituents. Benzaldehyde, benzaldehydes with electron-withdrawing substituents, and aliphatic aldehydes did not re-act. Presumably, the binuclear dianion $[\mu\text{-}S\{W(CO)_5\}_2]^{2-}$ is an intermedi-ate in the reaction, as on acidification of a mixture of $[\mu\text{-}S\{W(CO)_5\}_2]^{2-}$ and *p-N,N'*-dimethylaminobenzaldehyde the thioaldehyde complex $[W(CO)_5\{S{=}C(C_6H_4NMe_2\text{-}p)H\}]$ is formed.[143]

f. *Transformation of Miscellaneous Ligands.*    Cleavage of the Se–Se bridge in **84** by diazomethane afforded the paramagnetic selenoformalde-hyde complex **85** [Eq. (17)].[199] It was not possible to extend this route to the synthesis of the corresponding $\eta^5\text{-}C_5H_5$ complex as the binuclear precursor containing an unsymmetrically bridging $Se_2$ unit was inaccessible.

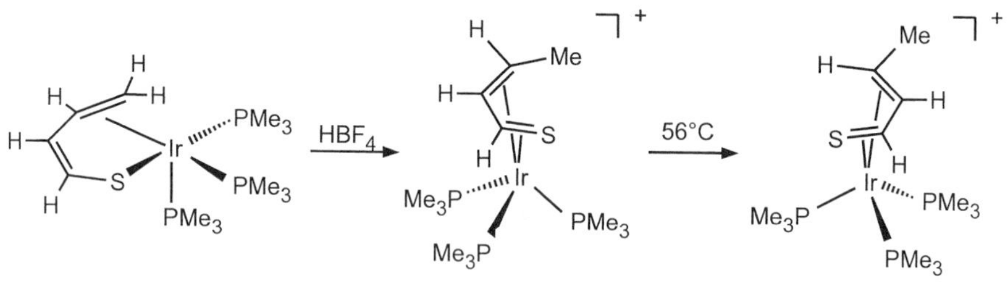

82                    syn-**83**                    anti-**83**

SCHEME 22

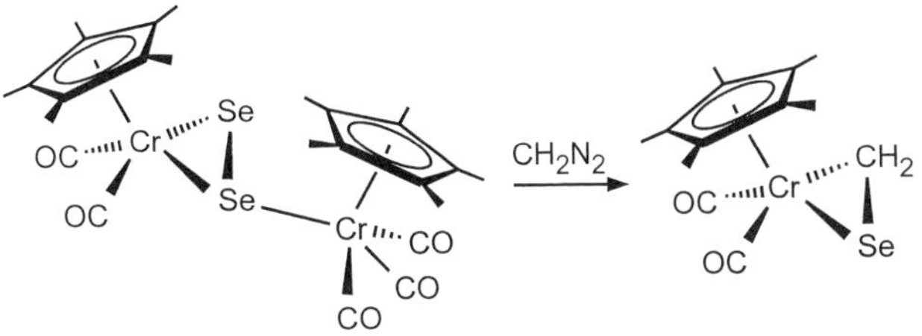

$$(17)$$

**84**                                              **85**

The reduction of **86** by sodium hydrido borate afforded the dithioformate complex **87** in addition to small amounts of the thioformaldehyde complex **88** (5%).[200,201] By modification of the procedure the low yield of **88** could be raised to 25%. Neutral complexes related to **86** form isolable adducts with $BH_3$ (see **90**).[202] Based on this observation a mechanism was proposed that involves initial formation of the $BH_3$ adducts (**89a** and **89b**) (Scheme 23).[201]

The structurally characterized compound **90**[201] reacted with phenylacetylene to give **91** with a chelating thioketone ligand [Eq. (18)].[203]

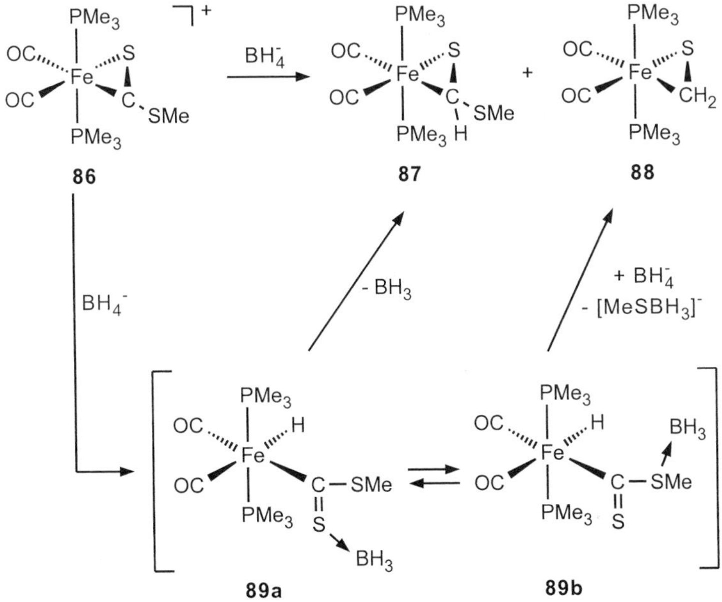

SCHEME 23

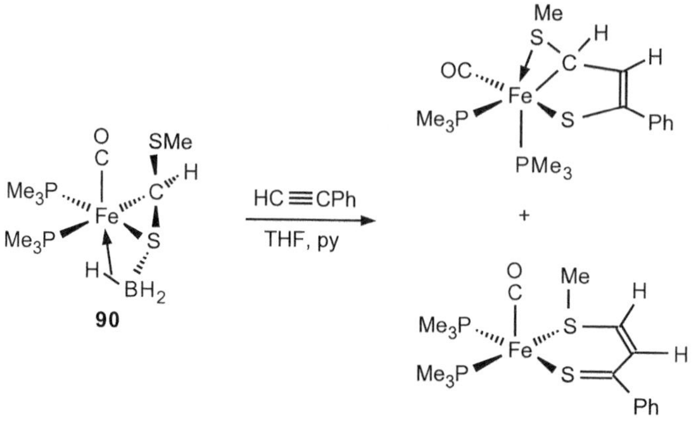

$$(18)$$

Complexes with related chelating thioketone ligands are reversibly formed on protonation of enethiolato complexes of iron.[101]

A thioaldehyde complex (**93**) was formed, although in very poor yield (<1%), when the 1,3-dithiolane-2-thione complex **92** was deprotonated with LDA at low temperature, treated with $CS_2$, and finally alkylated with $[Et_3O]BF_4$ [Eq. (19)]. No reaction took place in the absence of $CS_2$. Studies using $^{14}C$-labeled $CS_2$ confirmed that $CS_2$ was incorporated into the complex.[62]

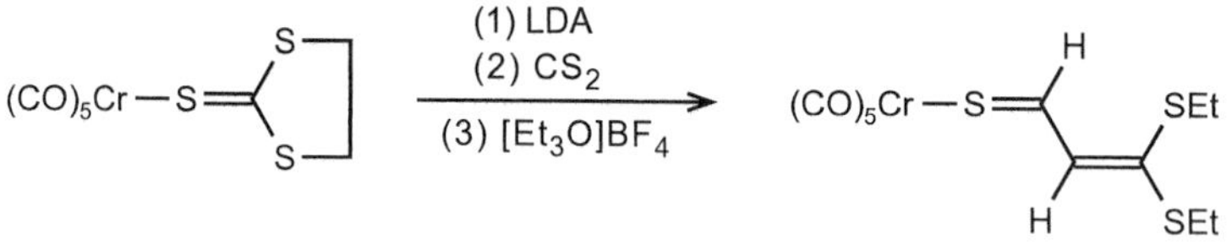

$$(19)$$

### B. Complexes with a μ Heterocarbonyl E=C Bond Bridging Two Metal Centers

As for mononuclear complexes, the reaction of a stable thioketone with a suitable complex precursor is the most simple route to binuclear complexes with bridging heterocarbonyl ligands. For example, heating or irradiating solutions of **94** in benzene in the presence of various thioketones (diaryl, arylalkyl, or cycloalkyl thioketones) gave rise to the formation of usually a mixture of the complexes **95** and **96** with a bridging thioketone ligand and, in the case of **96,** additionally a terminally coordinated thioketone ligand. In some cases one single product is formed. The same products, although in different ratios, were obtained from the binuclear complexes **97** containing a metal–metal triple bond (Scheme 24).[115]

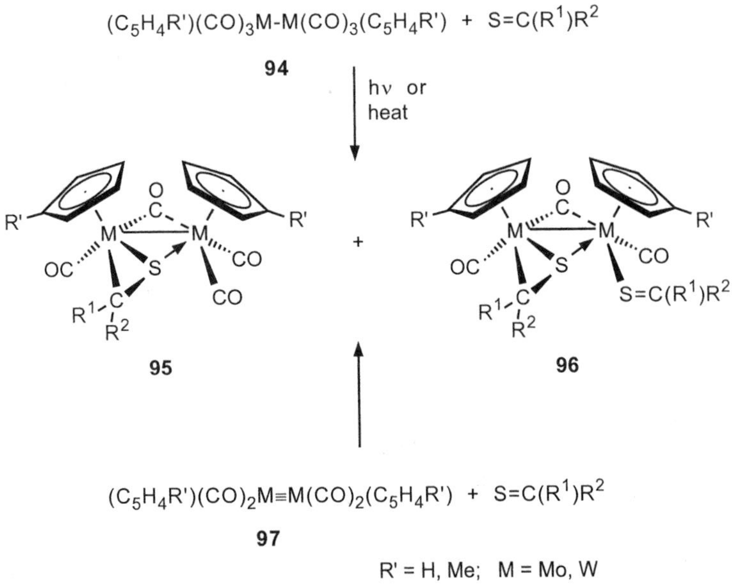

$$(C_5H_4R')(CO)_3M\text{-}M(CO)_3(C_5H_4R') \ + \ S=C(R^1)R^2$$

**94**

hν or
heat

**95**                    +                    **96**

$$(C_5H_4R')(CO)_2M\equiv M(CO)_2(C_5H_4R') \ + \ S=C(R^1)R^2$$

**97**

R' = H, Me;   M = Mo, W

SCHEME 24

The most straightforward route to heterobinuclear complexes with bridging heteroaldehyde or -ketone ligands is the addition of a metal-ligand fragment to the heteroatom of a heteroaldehyde or heteroketone complex. The reaction of the complexes **35** with the THF complexes [Cr(CO)$_5$(thf)], [W(CO)$_5$(thf)], [Mn(CO)$_2$(thf)($\eta^5$-C$_5$H$_5$)], or [Mn(CO)$_2$(thf)($\eta^5$-C$_5$H$_4$Me)] yielded the binuclear $\mu$-$\eta^1$:$\eta^2$-thio- and selenoformaldehyde complexes **98** [Eq. (20)]. The corresponding pentamethylcyclopentadienyl(selenoformaldehyde) rhodium complexes were obtained from [Rh(CO){Se=CH$_2$} ($\eta^5$-C$_5$Me$_5$)] and [M(CO)$_5$(thf)] (M = Cr, Mo, W; generated *in situ* by irradiation of [M(CO)$_6$] in THF).[54,166]

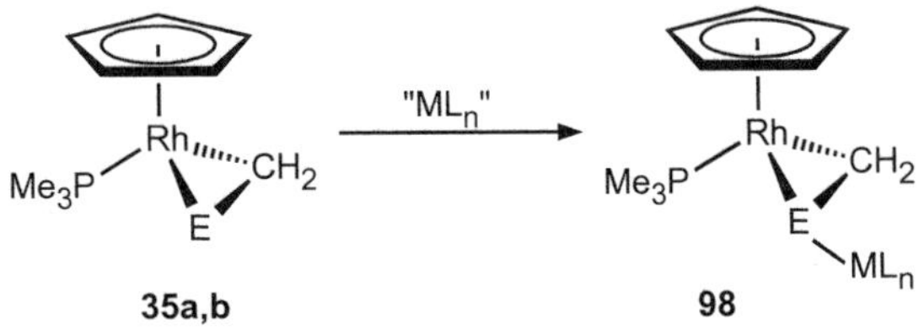

**35a,b**                    **98**                    (20)

E= S:    ML$_n$ = Cr(CO)$_5$

E = Se:   ML$_n$ = Cr(CO)$_5$, W(CO)$_5$, Mn(CO)$_2$(C$_5$H$_5$)

Mn(CO)$_2$(C$_5$H$_4$Me)

Related to these reactions is the formation of the homobinuclear complex **100** as the major thermolysis product of the mononuclear heterobenzalde-hyde complexes $[W(CO)_5\{X{=}C(Ph)H\}]$ [**99:** X = S (**a**) Se (**b**), Te (**c**)] [Eq. (21)].[204,205]

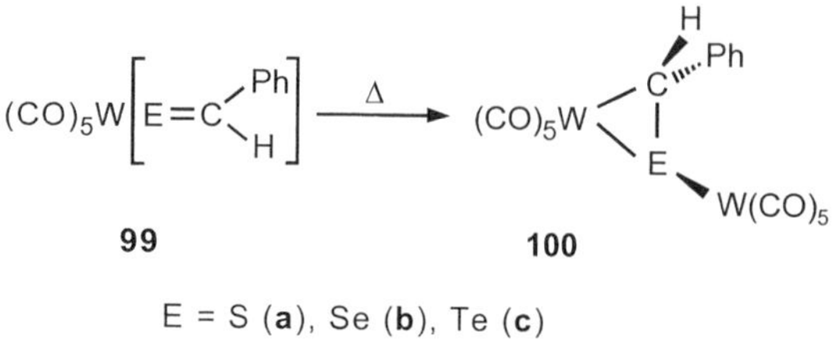

$$(CO)_5W\left[E{=}C{<}^{Ph}_{H}\right] \xrightarrow{\Delta} (CO)_5W{<}^{\overset{H}{\underset{E}{\overset{|}{C}}{\cdots}^{Ph}}}_{W(CO)_5} \tag{21}$$

**99**　　　　　　　　　　　　**100**

E = S (**a**), Se (**b**), Te (**c**)

A homobinuclear complex with a bridging thiobenzaldehyde ligand (**101**) was obtained when $[W(CO)_5\{S{=}C(Ph)H\}]$ (**99a**) was treated with $[Mn(CO)_2(thf)(\eta^5\text{-}C_5H_5)]$ in excess [Eq. (22)].[206] The reaction presumably proceeds by an initial addition of $Mn(CO)_2(\eta^5\text{-}C_5H_5)$ to $[W(CO)_5\{S{=}C(Ph)H\}]$ to give a heterobinuclear complex $[(CO)_5W\{S{=}C(Ph)H\}Mn(CO)_2(\eta^5\text{-}C_5H_5)]$ and subsequent substitution of $Mn(CO)_2(\eta^5\text{-}C_5H_5)$ for $W(CO)_5$.

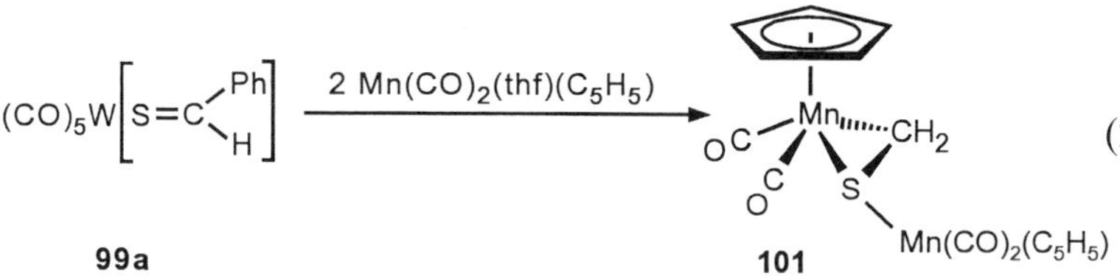

$$(CO)_5W\left[S{=}C{<}^{Ph}_{H}\right] \xrightarrow{2\ Mn(CO)_2(thf)(C_5H_5)} \underset{\overset{\displaystyle C}{O}}{\overset{\displaystyle OC}{}}{>}Mn{\cdots}CH_2 \tag{22}$$

**99a**　　　　　　　　　　　　　　　　　**101**　　$Mn(CO)_2(C_5H_5)$

The insertion of sulfur, selenium, or tellurium into metal–carbon multiple bonds, especially of alkylidene complexes, is a versatile direct route to mononuclear complexes of the heteroaldehydes and -ketones (see Section IV,A). In contrast, there are only very few examples for the trans-formation of a bridging alkylidene group into a heteroaldehyde or -ketone ligand. The cluster complex $[Os_3(\mu\text{-}SCH_2)(CO)_{11}]$ was formed along with $[Os_3(\mu_3\text{-}SCH_2)(CO)_{10}]$ on treatment of $[Os_3(\mu\text{-}CH_2)(CO)_{11}]$ with ethene sulfide. $[Os_3(\mu_3\text{-}SCH_2)(CO)_{10}]$ was obtained by decarbonylation of $[Os_3(\mu\text{-}SCH_2)(CO)_{11}]$. The structure of both cluster compounds has been con-firmed by X-ray structural analyses.[207,208]

The formation of the iron cluster complexes **103** with bridging thioalde-hyde ligands is thought to proceed by a similar insertion mechanism. Heat-ing of PhNSO led to fragmentation and reduction of PhNSO to a nitrene and a sulfido fragment. These fragments then inserted into the metal carbon bond of **102** [Eq. (23)].[110]

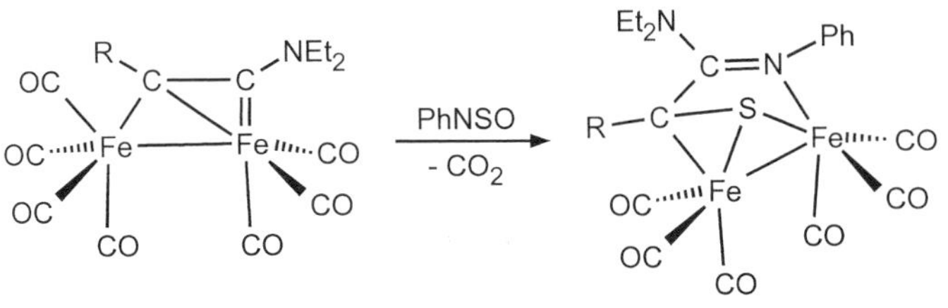

$$(23)$$

**102**                 **103**

Complexes with a bridging vinylidene ligand were also used as the starting compound for the synthesis of thioaldehyde complexes. Sequential treatment of **104** with cyclohexene sulfide, $HBF_4$, and $Li[BHEt_3]$ gave the thioaldehyde complexes **105** [Eq. (24)].[209]

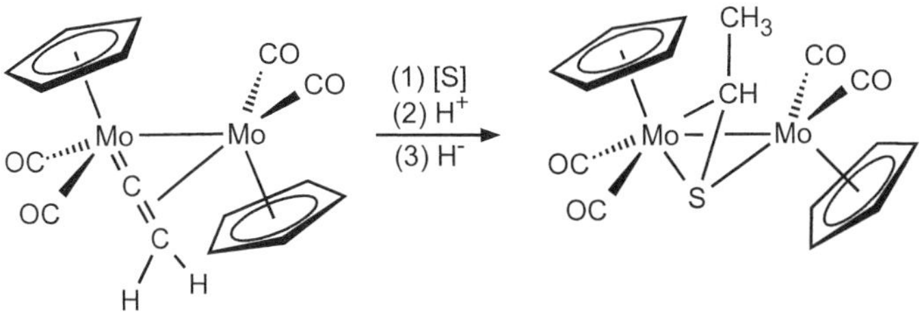

$$(24)$$

**104**                 **105**

This route is similar to those syntheses starting from mononuclear complexes with a terminal vinylidene ligand (see Scheme 17).[54,55] In both cases a thioketene complex is formed as an intermediate, which is then reduced.

In a closely related reaction the thioketene complex **106**[210] was reduced by sequential addition of $H^-$ and $H^+$ to give the thioketone complex **107**. However, compared to the reaction of **104** the sequence of $H^+$ and $H^-$ addition is inverse. Complex **107** is coordinatively unsaturated and therefore added $PPh_3$ rapidly (Scheme 25).[109]

The reduction of coordinated $CS_2$ also led to thioaldehyde complexes. In the reactions of the cluster complex $[Os_3(\mu\text{-}H)_2(CO)_9(PMe_2Ph)]$ with one molecule of $CS_2$ the hydrido ligands were transferred to the thiocarbonyl carbon to give a thioformaldehyde ligand bound to two osmium centers. One sulfur–carbon bond was cleaved and the second sulfur atom appeared as a $\mu_3$-sulfido ligand in the product $[Os_3(\mu_3\text{-}S{=}CH_2)(\mu_3\text{-}S)(CO)_9(PMe_2Ph)]$. In addition, a product was obtained containing a $\mu_3\text{-}S{=}CH_2$ ligand formed by subsequent decarbonylation of $[Os_3(\mu_3\text{-}S{=}CH_2)(\mu_3\text{-}S)(CO)_9(PMe_2Ph)]$.[38,39] When the cluster complex $[Os_3(\mu\text{-}H)_2(CO)_{10}]$ was treated with $CS_2$ only a methanedithiolato-bridged dicluster complex was isolated.[211]

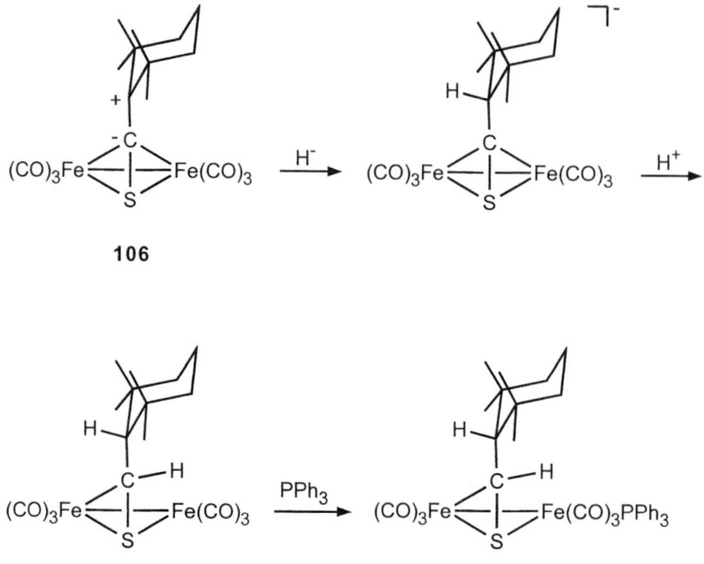

SCHEME 25

The thioformaldehyde complex **108** was detected by $^1$H-NMR spectroscopy as a minor by-product when the dimer $[\{Mo(\mu\text{-}\eta^2:\eta^2\text{-}S_2CH_2)(\eta^5\text{-}C_5H_5)\}_2]$ was protonated at -70°C and the solution then warmed up to -50°C.[86] Several thioaldehyde and thioketone complexes (**109:** $R^1, R^2 = (CH_2)_4, (CH_2)_6$; $R^1 = H$: $R^2 = Me, Bu^t$; $R^1 = Ph$: $R^2 = C_6H_4Bu^t\text{-}p$) along with other binuclear products were prepared by reductive cleavage of thiadiazoles in the presence of $[Fe_2(CO)_9]$.[111] Another binuclear thioketone iron complex (**110**) was obtained when a solution of the substituted ethene $H_2C=C(SBu^t)CN$ and $[\{Fe(CO)_2(\eta^5\text{-}C_5H_5)\}_2]$ in toluene was heated to 110°C for 12 hr. The structure of complex **110** was determined by an X-ray structural analysis.[104]

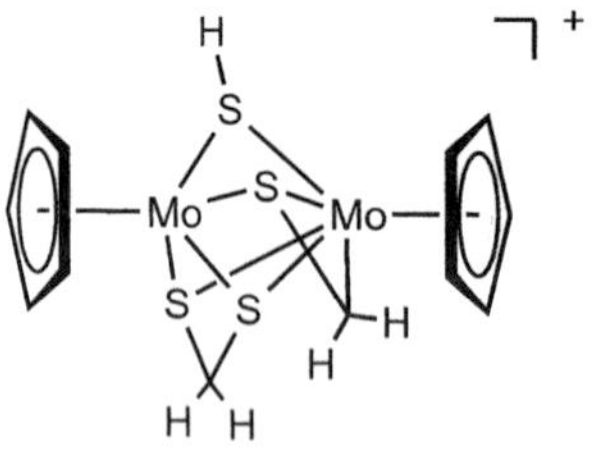

**108** 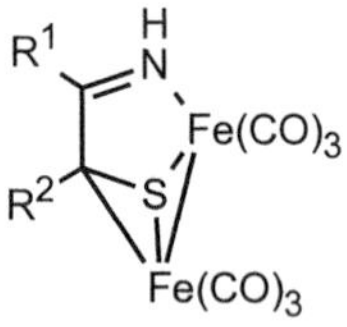                        **109** 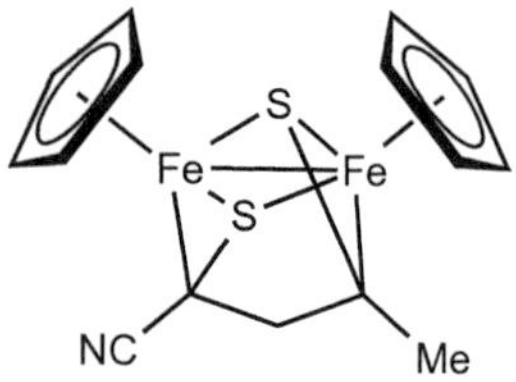                        **110**

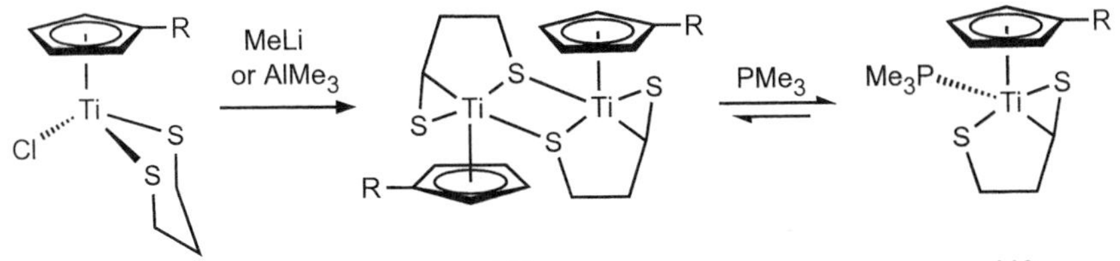

**111**                              **112**                              **113**

R = H, Me

SCHEME 26

Treatment of the dithiolato complexes **111** with MeLi or AlMe$_3$ afforded the unusual thioaldehyde(thiolato) complexes **113,** which in solution and in the crystal form dimers (**112**). The dimers could reversibly be cleaved by addition of PMe$_3$ to the metal center (Scheme 26).[68,69]

Methane elimination from the thermally unstable alkoxy(thiolato) complex [Ti(Me)(OC$_6$H$_3$Pr$_2^i$-$o,o'$)(SEt)($\eta^5$-C$_5$H$_5$)] yielded a similar thioacetaldehyde complex.[85] In the reaction of hexafluorobut-2-yne with the thiolato complex [Co$_2$(CO)$_6$(SC$_6$F$_5$)$_4$], a dinuclear complex with a bridging thioketone ligand was also formed.[212]

A route that has turned out to be generally applicable to the synthesis of both mono- and binuclear complexes is the reaction of complexes containing sulfur, selenium, or tellurium ligands with diazoalkanes [see also Eq. (17)]. Binuclear complexes of manganese, chromium, tungsten, and iron have been prepared by this method. However, different starting compounds had to be used in the reactions with CH$_2$N$_2$ to form the homobinuclear heteroformaldehyde complexes **114** (Scheme 27).[213–215]

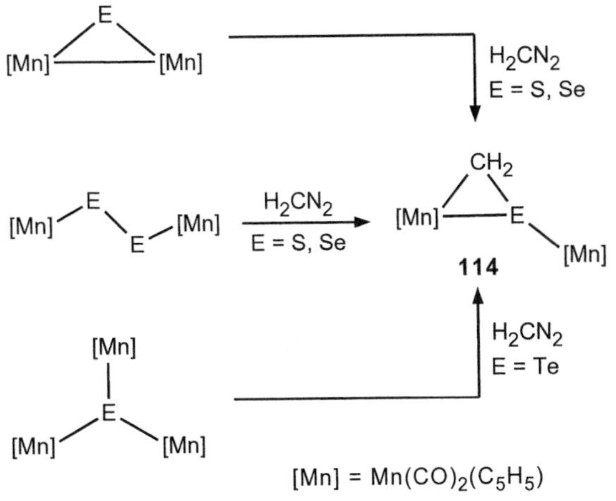

[Mn] = Mn(CO)$_2$(C$_5$H$_5$)

SCHEME 27

By a similar reaction of $CMe_2N_2$ with $[(\mu\text{-}E)\{Mn(CO)_2(\eta^5\text{-}C_5Me_5)\}_2]$ (E = Se, Te) the corresponding seleno- and telluroacetone complexes $[(\mu\text{-}\eta^1:\eta^2\text{-}E=CMe_2)\{Mn(CO)_2(\eta^5\text{-}C_5Me_5)\}_2]$ were prepared.[213,216] The related selenoformaldehyde complex $[(\mu\text{-}\eta^1:\eta^2\text{-}Se=CH_2)\{Mn(CO)_2(\eta^5\text{-}C_5H_4Me)\}_2]$ was formed as a by-product in the reaction of $[Se_2\{Mn(CO)_2(\eta^5\text{-}C_5H_4Me)\}_2]$ with $H_2AsC_6H_{11}$.[217]

Chromium complexes of type **115a–c** have also been prepared from $R_2CN_2$ and either $[(\mu\text{-}Se)\{Cr(CO)_2(\eta^5\text{-}C_5H_5)\}_2]$ or $[(\mu\text{-}S)\{W(CO)_3(\eta^5\text{-}C_5H_5)\}_2]$.[218–220] The synthesis of the pentamethylcyclopentadienyl(thio-formaldehyde) complex related to **115a** required the unsymmetrically $S_2$-bridged complex $[(\eta^5\text{-}C_5Me_5)(CO)_2Cr(\mu\text{-}S_2)Cr(CO)_3(\eta^5\text{-}C_5Me_5)]$ as the starting compound, as the sulfido-bridged compound $[(\mu\text{-}S)\{Cr(CO)_2(\eta^5\text{-}C_5Me_5)\}_2]$ did not react with $CH_2N_2$.[218,219]

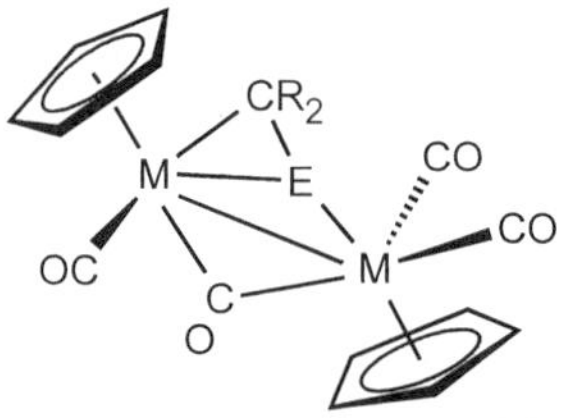

a: M = Cr, E = Se, R = H
b: M = Cr, E = Se, R = Me
c: M = W, E = S, R = H

**115**

Diiron complexes containing two equal or two different $\mu\text{-}\eta^1:\eta^2$-bridging heteroaldehyde ligands also have been prepared by the diazoalkane route. Reaction of $[Fe_2(CO)_6(\mu\text{-}EE')]$ (**116**) with diazoethane gave $[Fe_2(CO)_6\{\mu\text{-}\eta^1:\eta^2\text{-}E=C(Me)H\}\{\mu\text{-}\eta^1:\eta^2\text{-}E'=C(Me)H\}]$ (**117**). With the exception of the bis(telluroacetaldehyde) compound the complete series of complexes with E = E' and E ≠ E' was accessible [Eq. (25)].[88] Surprisingly, the reaction of $[Fe_2(CO)_6(\mu\text{-}STe)]$ with diazomethane afforded a small amount (4.1%) of the bis(telluroformaldehyde) complex $[Fe_2(CO)_6\{\mu\text{-}\eta^1:\eta^2\text{-}Te=CH_2\}_2]$. The major product was $[Fe_2(CO)_6\{\mu\text{-}S=CH_2Te\}]$.[93] The bis(selenoformaldehyde) complex $[Fe_2(CO)_6\{\mu\text{-}\eta^1:\eta^2\text{-}SeCH_2\}_2]$ was obtained from $[Fe_3(CO)_9(\mu_3\text{-}Se)_2]$ and diazomethane.[91]

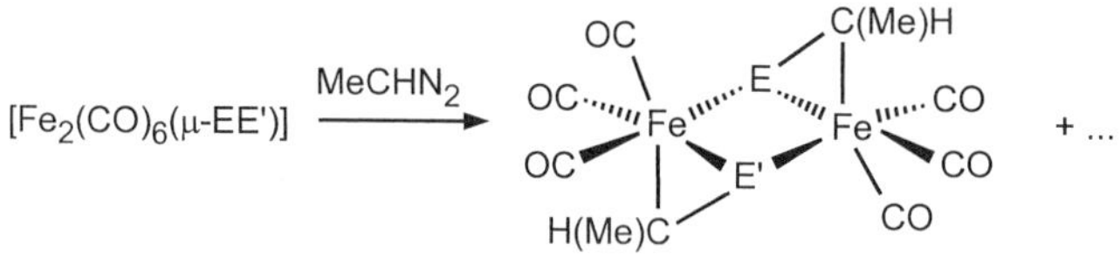

$$[Fe_2(CO)_6(\mu\text{-}EE')] \xrightarrow{MeCHN_2} \quad + \ldots \quad (25)$$

**116**          **117**

E = E' = S, Se          E ≠ E': E,E' = S,Se; S,Te; Se,Te

## V

## Reactions of Heteroaldehyde and Heteroketone Complexes

### A. Reactions with Electrophiles or Nucleophiles

1. *Reactions with Electrophiles*

When coordinated to the metal in the $\eta^2$ form heteroaldehyde ligands react readily with electrophiles. As expected, carbon electrophiles add to the chalcogen atom. Very likely, the site of the initial attack of protons is also the chalcogen atom. However, rapid succeeding rearrangement and addition of an anion usually lead to chalcogenolato complexes.

For instance, the chalcogenoformaldehyde complexes **118** were found to be alkylated readily by MeI or CF$_3$SO$_3$Me at the chalcogen atom. However, the isolated products of the reaction with HCl were the chalcogenolato complexes **119** presumably formed by protonation at the chalcogen atom, rearrangement of the protonated species, and addition of Cl$^-$ to the metal (Scheme 28).[167,192,193] Chalcogenolato(methyl) complexes [OsMe(EMe)(CO)$_2$(PPh$_3$)$_2$], related to **119**, were obtained when **118** was treated with MeI and NaBH$_4$ then added to the resulting adduct.[167]

On protonation with HBF$_4$ or CF$_3$SO$_3$H, the cobalt and rhodium complexes **120** behaved similarly to **118**. However, because of the lack of a well-coordinating anion the initially formed 16-electron chalcogenolato intermediate dimerized to form the chalcogenolato-bridged complexes **121** (Scheme 29).[54,170]

As observed with the osmium complexes **118,** the alkylation of [M(PMe$_3$)($\eta^2$-E=CH$_2$)($\eta^5$-C$_5$H$_5$)] (M = Co, Rh; E = S, Se) as well as of [Rh(CO){E=CH$_2$}($\eta^5$-C$_5$Me$_5$)] (E = Se) with MeI or methyl triflate afforded E-methylated cationic adducts.[54,169,170] Subsequent addition of CF$_3$COOH/NaI to the adduct (M = Co, E = S) gave rise to a cleavage of the Co-CH$_2$ bond and formation of the dimethyl thioether complex [Co(SMe$_2$)(PMe$_3$)(I)($\eta^5$-C$_5$H$_5$)].[169,170]

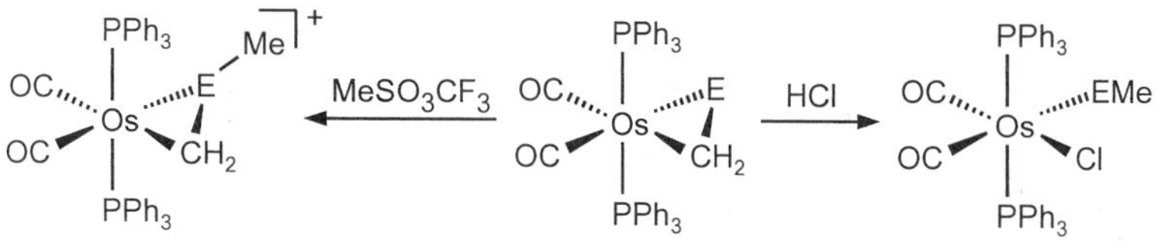

E = S, Se, Te

SCHEME 28

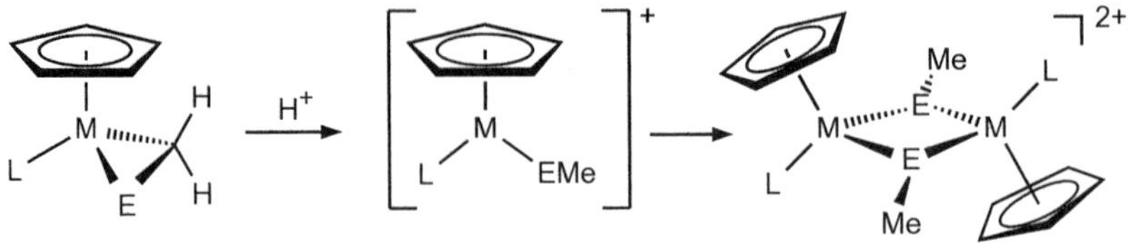

**120**                                              **121**

E = S;        M = Co,   L = PMe$_3$

E = S, Se;   M = Rh,   L = PMe$_3$, PPr$^i_3$

SCHEME 29

The alkylation of the zirconocene thioacetaldehyde complex **81a** with excess MeI also yielded an *S*-methylated product (**122**) [Eq. (26)]. In contrast, the protonation by methanol in excess gave ethanethiol, trimethylphosphine, and dimethoxyzirconocene, presumably via a [Zr(OMe)(SEt) ($\eta^5$-C$_5$H$_5$)$_2$] intermediate.[70]

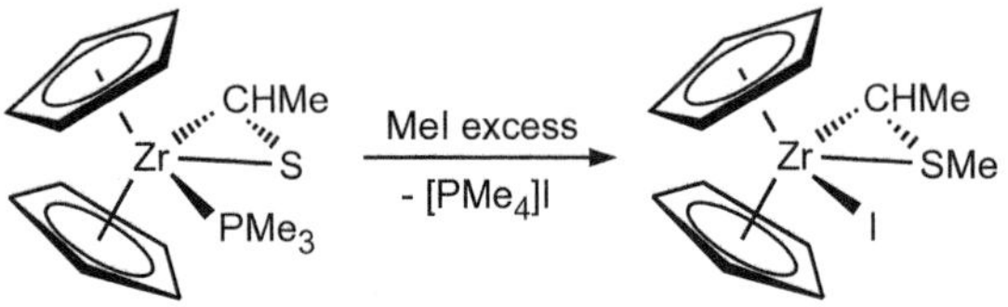

**81a**                              **122**

(26)

The protonation of the analogous titanocene complex **55** with trifluoroacetic acid led to similar products: methanethiol and bis(trifluoroacetato) titanocene. When **55** was treated with MeI, rather than the corresponding iodide complex, the cationic Me$^+$ adduct **123** was obtained with iodide as a counterion. However, the acylation of **55** with acetyl chloride afforded the chloride **124** related to **122** (Scheme 30).[67]

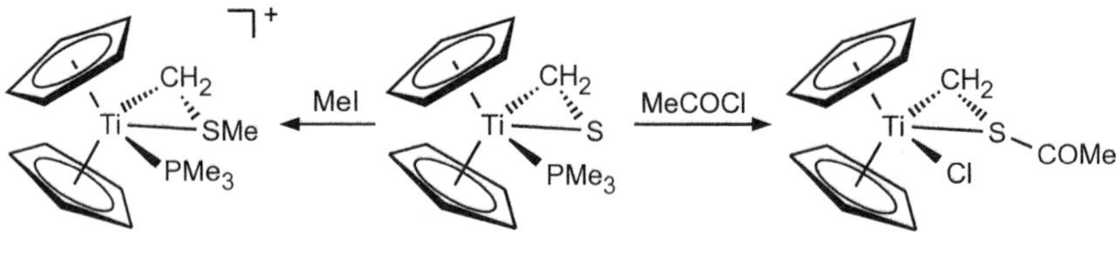

**123**                         **55**                         **124**

SCHEME 30

Thioacrolein derivatives $\eta^4$ coordinated to cobalt were similarly $S$-methylated by [Me$_3$O]BF$_4$ or MeI to give cationic compounds.[221] The alkylated ligands were displaced readily by cyanide ions and allyl and vinyl thioethers were formed.[222] Weak Lewis acids such as HgCl$_2$ also added to sulfur. HgCl$_2$ could subsequently be removed by PPh$_3$.[147,221] Moreover, the sulfur atom in thioacrolein complexes is probably the site of attack of lanthanide shift reagents in NMR experiments. The proton on the neighboring carbon exhibits the most pronounced shift on treatment with the shift reagents. In contrast, the related iron complexes did not show any effect and alkylation of them proved to be unsuccessful. However, the interpretation of these results remains difficult.[223]

In addition to carbon electrophiles and protons, organometallic electrophiles have also been employed. Thus, the addition of M(CO)$_5$ (M = Cr, Mo, W) or Mn(CO)$_2$($\eta^5$-C$_5$H$_4$R) (R = H, Me) to thio- and selenoaldehyde complexes of rhodium provides a convenient route to heterobinuclear $\mu$-$\eta^1$ : $\eta^2$-thio- and selenoaldehyde complexes [see Section IV,B, Eq. (20)].[54,166]

Due to the reduced availability of the lone pair at the chalcogen atom, the sulfur or selenium atom in $\eta^1$-chalcogenoaldehyde or -ketone complexes cannot be alkylated by the usual alkylating agents.

### 2. Reactions with Nucleophiles

In contrast to electrophiles, nucleophiles are expected to react preferentially with $\eta^1$-bonded heteroaldehyde and -ketone ligands. The site of attack is expected to be the carbon atom of the E=C group. $\eta^2$-coordinated ligands should be less reactive.

Pyridine was found to add reversibly to the thiocarbonyl carbon atom of [W(CO)$_5${S=C(C$_6$H$_4$R-$p$)H}]. The equilibrium constant increased in the series R = OMe, Me, H.[224] Isolable adducts, phosphoniothiolato complexes, were formed when [W(CO)$_5${$\eta^1$-S=C(Ph)H}] (**99a**) was treated with tertiary phosphines. The corresponding adducts obtained from **99a** and primary or secondary phosphines rearranged rapidly to form phosphine complexes (Scheme 31).[63] The formation of adducts from PPh$_3$ and [M(CO)$_5$ ($\eta^1$-S=CPh$_2$)] (M = Cr, Mo, W) might be the reason why earlier attempts failed to synthesize carbene complexes by PPh$_3$-induced extrusion of sulfur.[119]

When two equivalents of PPh$_3$ were added to the cationic $\eta^2$-S=CH$_2$ complex **51a,** the ylide complex **52** together with triphenylphosphine sulfide was formed. The reaction very likely proceeds via the carbene complex intermediate [Re(=CH$_2$)(NO)(PMe$_3$)($\eta^5$-C$_5$H$_5$)]$^+$ (**50**), which is rapidly trapped by PPh$_3$. The trapping reaction is significantly faster than the for-

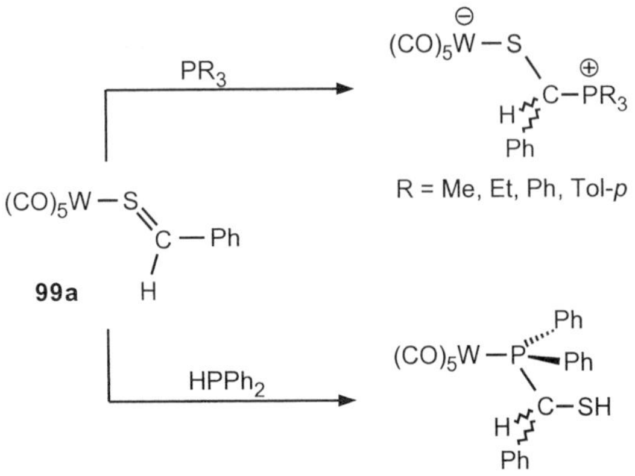

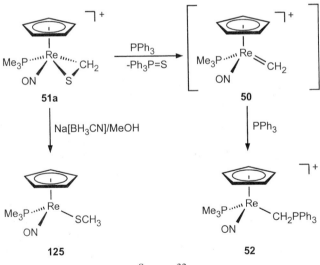

SCHEME 31

mation of the carbene complex as only half of the complex **51a** was consumed when one equivalent of PPh$_3$ was employed.[76] Treatment of **51a** with Na[BH$_3$CN]/MeOH yielded the methylthiolato complex **125** (Scheme 32).[75,76]

In contrast to **51a** but analogous to [W(CO)$_5${S=C(Ph)H}] (**99a**), the cationic thiobenzaldehyde ruthenium complexes **78** [see Eq. (16)] added PMe$_3$ to give phosphoniothiolato complexes.[47] Complexes **78** also reacted

SCHEME 32

readily with a variety of anionic nucleophiles such as $OC_6H_4Me^-$, $SMe^-$, $SCH_2C_6H_4OMe$-$p^-$, $SCH_2Ph^-$, $Bu^-$, $CH=CH_2^-$, and $CH(OR)_2^-$ ($R = Me$, OMe) by addition of the nucleophile to the thiocarbonyl carbon atom. The reaction with hydride regenerated the thiolato complexes **77**.[47]

The reactions of dinuclear thioaldehyde and -ketone complexes with $PR_3$ take a different course. When the dimeric complexes **112** are treated with $PMe_3$ a Ti–S bond in **112** is cleaved and the monomeric complexes **113** are reversibly formed (see Scheme 26).[69] The reaction of **112** with benzene-thiol led to a Ti–C cleavage and formation of 1,3-propanedithiolato-(phenylthiolato) complex.[69] The reaction of binuclear $\mu$-$\eta^1$:$\eta^2$-thioketone molybdenum complexes related to **115** ($M = Mo$, $E = S$, $R = Aryl$) with $P(OEt)_3$ also gave mononuclear complexes $[Mo(CO)_2(Et)\{\eta^2\text{-}S=C(Aryl)_2\}$ $(\eta^5\text{-}C_5H_5)]$ by a rather complicated reaction sequence. When other phosphites were employed additionally binuclear complexes derived from addition of the phosphite to one molybdenum atom were isolated.[225]

## B. *Oxidation and Reduction*

Several modes of reactions were observed when heterocarbonyl complexes were exposed to oxidizing or reducing agents. The oxidation of the thioformaldehyde ligand in the osmium complex $[OsCl(NO)(\eta^2\text{-}S=CH_2)$ $(PPh_3)_2]$[53] with $m$-chloroperbenzoic acid afforded the $\eta^2$-$S,C$ complex of the unsubstituted sulfine $H_2C=S=O$, a molecule that is unknown in the uncoordinated form.[226,227]

A similar oxidation of the thioacrolein cobalt (**22a,b,d**) and iron complexes (**23a–c,** see Scheme 8) led to complexes of the corresponding $S$-oxides. $H_2O_2$/acetic acid or $m$-chloroperbenzoic acid was used as the oxidant. The oxidation of **22b** was achieved by heating the compound in air.[147]

Oxidation of the thioacetaldehyde zirconocene complex **81** (see Scheme 21: $R = Me$) by $I_2$ yielded $[Zr(I)_2(\eta^5\text{-}C_5H_5)_2]$ and thioformaldehyde, which trimerized to a mixture of the $\alpha$ and $\beta$ isomers of 2,4,6-trimethyl-1,3,5-trithiane ("thioparaldehyde").[70]

The cleavage of the $S=C$ thioaldehyde bond was observed when the manganese thioketone complex related to **8** (Scheme 2: $R^1 = R^2 = NMe_2$) was treated with $LiAlD_4$. The $C(C_6H_4OMe$-$p)_2$-moiety of the complex was recovered as tetra(*para*-methoxyphenyl)ethane bearing a deuterium in the *ortho* position of two vicinal phenyl groups each and additionally two deuterium atoms on the ethane skeleton.[127]

## C. *Rearrangement and Displacement Reactions*

In solution, the selenobenzaldehyde complexes $[M(CO)_5\{Se=C$ $(C_6H_4$-$R$-$p)H\}]$ ($M = Cr$, $W$), some thiobenzaldehyde complexes of tungsten, and the ruthenium complexes **78** are dynamic. The $\eta^1$ and $\eta^2$ isomers are in equilibrium and the isomers interconvert rapidly (see Section III,B,1).

Another type of isomerization was observed with the hydrido(thioalde-hyde) permethyltantalocene complexes $[TaH\{\eta^2\text{-}S=C(R)H\}(\eta^5\text{-}C_5Me_5)_2]$ (R = H, Ph, CH$_2$Ph, CH$_2$Bu$^t$) (**66a–d,** see Scheme 18). These complexes were shown to be in rapid equilibrium with the corresponding 16-electron thiolato species $[Ta(SCH_2R)(\eta^5\text{-}C_5Me_5)_2]$ through a $\beta$-H migratory inser-tion/elimination process. When the thioaldehyde complexes were heated to ca. 100°C, they rearranged to form the thermodynamically favored alkyl(chalcogenido) tautomer $[Ta(CH_2R)(=S)(\eta^5\text{-}C_5Me_5)_2]$ (**126a–d**) (Scheme 33).[188,189]

The activation-free energy for the conversion of **66a** to **126a** was deter-mined to be $\Delta G^{\neq} = 31.8(1)$ kcal/mol at 100°C.[189] The related seleno- and telluroformaldehyde complexes $[TaH\{\eta^2\text{-}E=CH_2\}(\eta^5\text{-}C_5Me_5)_2]$ [E = Se (**59**), Te (**57**)] also rearranged to form chalcogenido(methyl) complexes $[Ta(CH_3)(=E)(\eta^5\text{-}C_5Me_5)_2]$. However, slightly higher temperatures were required: 110°C for E=Se[183] and 130°C for E=Te.[84]

There are only few reports on the substitution of nucleophiles for coordi-nated heteroaldehydes or -ketones.

At slightly elevated temperature in acetonitrile the 1,1,3,3-tetramethyl-indanetellone complex gradually underwent decomplexation via ligand exchange and quantitative regeneration of the telluroketone along with $[W(CO)_5(NCMe)]$.[108] The thioacetone ligand in $[W(CO)_5(\eta^1\text{-}S=CMe_2)]$ was also displaced readily by a number of nucleophiles such as MeCN, I$^-$, N$_3^-$, and H$_2$NC$_6$H$_{11}$ in CCl$_4$. However, when treated with the amine in pentane, a yellow precipitate formed that reacted with MeI or $[Et_3O]BF_4$ to give $[W(CO)_5(SMe_2)]$ and $[W(CO)_5(SEt_2)]$, respectively. Presumably, the precipitate is an adduct. When it was stirred in diethyl ether under an atmosphere of gaseous HBr, the original thioketone complex was regen-erated.[51]

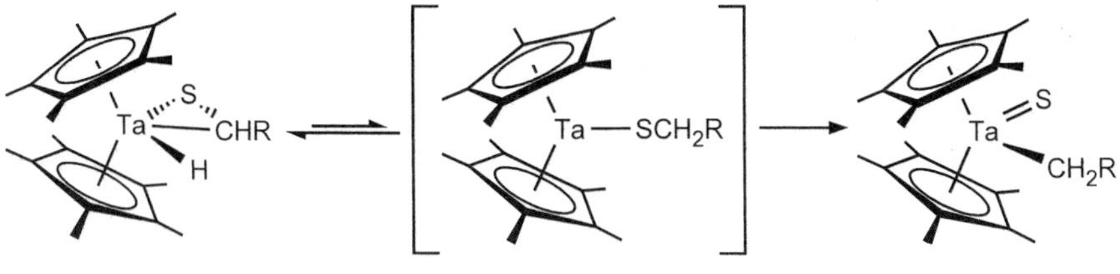

**66a–d**                                           **126a–d**

R = H (**a**), Ph (**b**), CH$_2$Ph (**c**), Bu$^t$ (**d**)

SCHEME 33

Displacement of the thioaldehyde ligand was similarly observed when acetic acid was added to the dinuclear complex **112** (R = H, see Scheme 26) and the tris(acetato) complex $[Ti(O_2CCH_3)_3(\eta^5\text{-}C_5H_5)_2]$ was formed.[69]

## D. Cycloaddition Reactions of $\eta^1$-Heteroaldehyde and Heteroketone Complexes

Thioaldehydes and thioketones that cannot be isolated because of rapid oligomerization are usually trapped by [4+2] cycloadditions with conjugated dienes.[1-9] Similarly, chalcogenoaldehydes and -ketones coordinated to pentacarbonyl chromium or tungsten also react with a variety of 1,3-dienes to form chalcogenacycles. The coordination to the transition metal, however, modifies the reactivity of the E=C functionality and the stereoselectivity of the reactions. In addition to 1,3-dienes, electron-rich olefins and alkynes can be employed as substrates. Compared to the reactivity of the uncoordinated molecules, that of the metal-coordinated heteroaldehydes and -ketones is reduced. However, their reactivity is still high, thus rendering them easily accessible and easy to handle starting materials for the synthesis of three-, four-, five-, and six-membered thia- and selenacycles and other sulfur- and selenium-containing compounds.

### 1. Reactions with Conjugated Dienes

A large number of metal-coordinated chalcogenacycles were prepared from the heteroaldehyde and -ketone complexes $[M(CO)_5\{E=C(Ph)R\}]$ (M = Cr, Mo, W; E = S, Se, Te; R = H, Aryl) and 1,3-dienes. The cycloadditions proceeded rapidly, even at low temperatures.[179,180,228-233] As dienes 2,3-dimethyl-1,3-butadiene [for an example see Eq. (27)], isoprene, *trans*-1,3-pentadiene, cyclopentadiene, pentamethylcyclopentadiene, and 1,3-cyclohexadiene were used. Decomplexation with pyridine/THF yielded the free heterocycles. Although the tellurobenzaldehyde complex $[W(CO)_5\{Te=C(Ph)H\}]$ (**99c**) proved too unstable for isolation, it could be generated and employed in subsequent cycloadditions with dienes.[180]

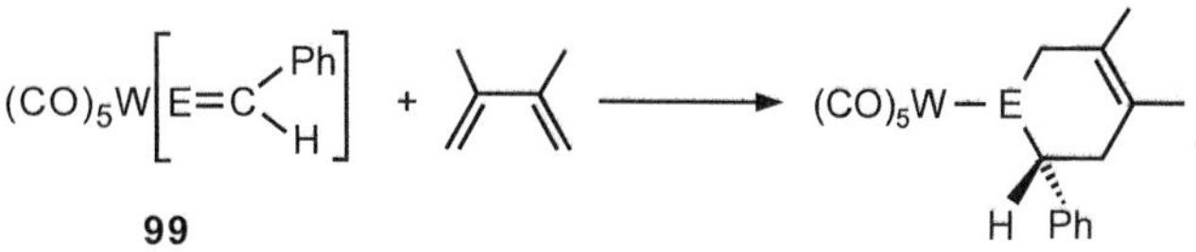

$$(CO)_5W\left[E=C\begin{smallmatrix}Ph\\H\end{smallmatrix}\right] \; + \; \bigl\rangle\!\!=\!\!\bigl\langle \;\longrightarrow\; (CO)_5W-E \diagdown \qquad (27)$$

**99**                                                                H  Ph

E = S (**a**), Se (**b**), Te (**c**)

From the results of a detailed kinetic study of the reactions of $[Cr(CO)_5$

{Se=C(Ph)$_2$}], [W(CO)$_5${E=C(Ph)H}] [E = S (**99a**), Se (**99b**)], and [W(CO)$_5${Se=C(Ph)C$_6$H$_4$R-*p*}] (**43**) with several dienes and the dependence of the reaction rate on the solvent, the metal, the chalcogen atom E, and the substituents and the strongly negative activation entropy [about $-144 \ldots -147$ J/(molK)] an associative mechanism with an only weakly polar transition state was deduced.[230] The reaction of the aldehyde complexes was much faster than that of the ketone complexes, e.g., substitution of R = H in [W(CO)$_5${Se=C(Ph)R}] for R = Ph accelerated the reaction with cyclopentadiene by a factor of 73500. In contrast, the selenobenzaldehyde complex reacted only 19 times faster than the thiobenzaldehyde complex. Because a good positive correlation between the rate constants $k_2$ and the Hammett constants $\sigma_p$ of R ($\rho = 2.5$) was observed, these cycloadditions were classified as type I Diels–Alder reactions, which are controlled by the HOMO of the diene and the LUMO of the dienophile.[234] The LUMO in the complexes (essentially E=C-$\pi^*$ in character, see also Section III,A) is higher in energy than that in uncomplexed E=C(R$^1$)R$^2$ due to interaction with filled d orbitals of the M(CO)$_5$ fragment. Therefore, the reactions of the heterocarbonyl ligands are slower than those of the uncomplexed species.

When unsymmetrically substituted dienes were used, the [4+2] cycloadditions were regioselective. Coordination of E=C(R$^1$)R$^2$ to M(CO)$_5$ enhanced the regioselectivity that is obviously determined by electronic factors.[37,231]

When cyclic 1,3-dienes were employed, the reactions were found to be *exo* selective [Eq. (28)]. In contrast, the Diels–Alder reactions of free thio- and selenoaldehydes with cyclopentadiene occur with a preference for the *endo* isomer.[235] Thus, coordination of thio- and selenoaldehydes to a pentacarbonylmetal fragment resulted in a reversal of the *endo* selectivity. The *exo/endo* product ratio of [W(CO)$_5${E=C(Ph)H}] was 7.3:1 (E = S) and 2.6:1 (E = Se), whereas the kinetically controlled *exo/endo* ratio of the cycloadducts obtained by trapping of the free heterobenzaldehyde with cyclopentadiene was reported to be 1:7[236] and 1:4[235,237] (E = S) and 1:2.6[238] and 1:4[239] (E = Se), respectively.

$$(CO)_5M\left[E{=}C\genfrac{}{}{0pt}{}{Ph}{H}\right] \xrightarrow{\phantom{xx}} \begin{array}{c}(CO)_5M\\E\\Ph\\H\end{array}\;\text{exo}\;+\;\begin{array}{c}(CO)_5M\\E\\H\\Ph\end{array}\;\text{endo} \quad (28)$$

M = Cr, W;  E = S, Se

It was suggested that the *endo* preference observed in the reaction of cyclopentadiene with heteroaldehydes is determined primarily by steric effects.[37] The formation of the *endo* isomer is favored by the less demanding interaction of the $CH_2$ bridge in cyclopentadiene with the hydrogen in the heteroaldehyde $E=C(R)H$ compared to that with the substituent R. The ame argument can be applied for the *exo* preference of the reactions of the heteroaldehyde complexes with cyclic 1,3-dienes, assuming that the $\eta^1$-E form (which is in rapid equilibrium with the $\eta^1$-Z and the $\eta^2$ isomer, see Scheme 1) reacts predominantly with the dienes. The bulky $M(CO)_5$ fragment at the heteroatom overrides the directing influence of the carbon substituent in determining the direction of the diene attack at the dienophile.[232,233]

In addition to pentacarbonyl complexes, other thio- and selenoaldehyde complexes were observed to react readily with conjugated dienes. From the thiobenzaldehyde ruthenium complexes $78^{47,66}$ [see Eq. (16)] and $63^{187}$ [see Eq. (14)] and cyclopentadiene, as well as from some complexes 78 with sterically less demanding phosphine ligands[47] and 2,3-dimethyl-1,3-butadiene, the corresponding ruthenium-coordinated thiacycles were obtained. When generated in the presence of pentamethylcyclopentadiene, the selenobenzaldehyde complex $[Mn(CO)_2\{\eta^2\text{-Se}=C(Ph)H\}(\eta^5\text{-}C_5H_5)]$ (42b) added the diene and a manganese-coordinated pentamethylselenanorbornene was formed.[172]

Although $\alpha,\beta$-unsaturated thio- and selenoaldehyde complexes also seemed to behave as dienophiles in Diels–Alder reactions with cyclopentadiene, the products were too unstable for characterization. Therefore, it remains unknown which double bond ($E=C$ or $C=C$) was involved in the cycloaddition.[64]

### 2. Reactions with Electron-Rich Alkynes

Cycloadditions to the $E=C$ functionality of metal-coordinated heteroaldehydes and -ketones are not confined to Diels–Alder reactions. The pentacarbonyl complexes $[M(CO)_5\{E=C(Ph)R\}]$ (M = Cr, W; E = S, Se; R = H, Aryl) (127) were found to react rapidly with the polar electron-rich alkyne 1-diethylaminoprop-1-yne to yield coordinated thio- and selenoacrylamide derivatives (129, R′ = Me) (Scheme 34).[240–244] The analogous reaction of 127 with the symmetrically substituted bis(diethylamino)ethyne gave the corresponding complexes 129 (R′ = $NEt_2$). Tungsten-coordinated telluroacrylamides were obtained from these alkynes and the tellurobenzophenone complex $[W(CO)_5\{Te=C(Ph)_2\}]$ (44).[245] The uncomplexed thio- and selenoacrylamides were obtained almost quantitatively when CO pressure was applied to solutions of the complexes 129 at 50°C.[240,242,244]

These reactions correspond to a formal insertion of the $C\equiv C$ bond of

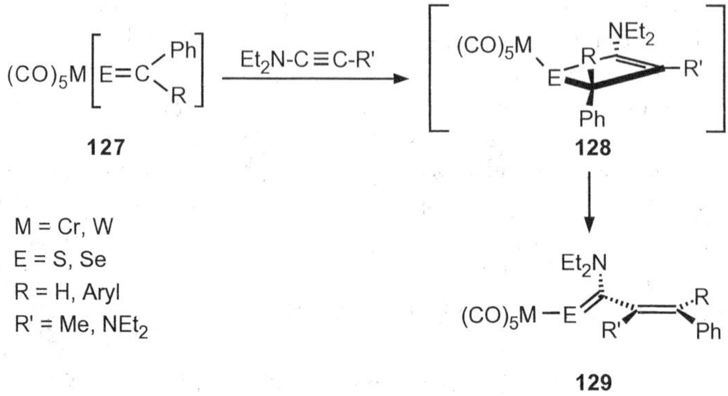

SCHEME 34

the alkyne into the $E=C$ bond of the coordinated heterocarbonyl. The reactions presumably proceed by a regiospecific formal $[2+2]$ cycloaddition of the alkyne to the $E=C$ bond to form metal-coordinated thietes and selenetes (128), and telluretes, respectively. Ring formation is followed by a rapid stereospecific electrocyclic opening of the four-membered heterocycle affording the complexes 129 (Scheme 34). The proposal is supported by the results of a kinetic study of the reaction of various thio- and selenoaldehyde and -ketone complexes of chromium and tungsten with 1-diethyl-aminoprop-1-yne and bis(diethylamino)ethyne.[240,242] The activation parameters and the influence of the polarity of the solvent, the metal, the heteroatom E, and the substituent R on the reaction rate are similar to those observed in the reactions of the same complexes with conjugated dienes. Therefore, the cyclo addition step in these reactions is rate limiting. The stereochemistry of the initially formed heteroacrylamide derivative is determined by the mode of the ring opening. This assumption is supported by the observation that in the reaction of $[W(CO)_5\{E=C(Ph)H\}]$ {99, E = S (a), Se (b)} with bis(diethylamino)ethyne first the isomer 130 (Ph and $NEt_2$ cis) was formed (Scheme 35). In a succeeding slower reaction the kinetically controlled product 130 isomerized to give 131 (Ph and $NEt_2$ trans) (Scheme 35). The structure of both thioacrylamide isomers was established by X-ray structural analyses.[244]

The stereoselectivity of the ring opening can be explained by steric arguments. Steric factors also play a major role in the selective ring opening of oxetenes formed from 1-diethylaminoprop-1-yne and benzaldehyde.[246]

The intermediacy of a thiete, selenete, or tellurete complex in the formation of the heteroacrylamide complexes is further supported by the isolation of several 2H-selenete complexes (132) from the reaction of $[W(CO)_5\{\eta^1-$

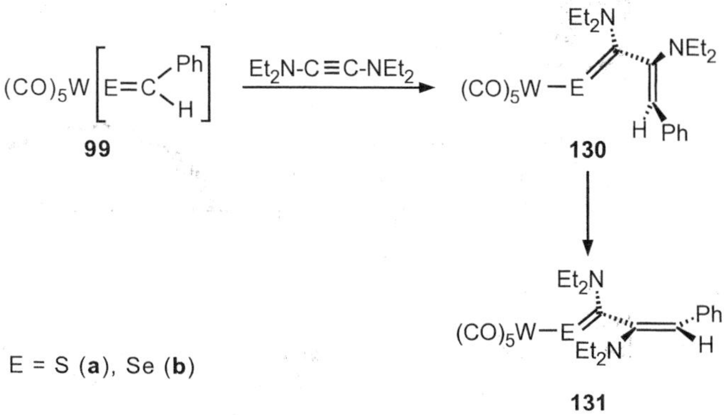

$$E = S \ (\mathbf{a}), \ Se \ (\mathbf{b})$$

SCHEME 35

Se$=$C(C$_6$H$_4$R-$p$)H}] (R = H, OMe, CF$_3$) with bis($t$-butylthio)ethyne. The structure of one 2$H$-selenete complex **132** (R = H) was confirmed by an X-ray structural analysis.[247] Based on spectra of these compounds, an equilibrium between the 2$H$-selenete complexes (**132**) and their thioselono-carboxylic ester isomers (**133**) was proposed (Scheme 36). The fraction of **133** in the eqilibrium was estimated to be about 6–10%. Decomplexation of the heterocyclic ligand in **132** (R = H) with [NEt$_4$]Br afforded the first uncoordinated 2$H$-selenete. In addition, a 3,4-dihydro-1,2-diselenine was obtained.[247]

The reactions of the selenobenzaldehyde complex **99b** with other organyl-

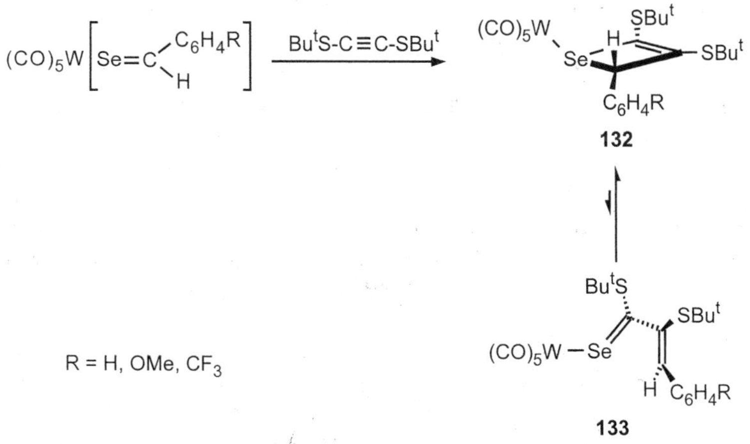

R = H, OMe, CF$_3$

SCHEME 36

thioalkynes turned out to be considerably substituent dependent. When bis($t$-butylthio)ethyne was replaced by other bis(organylthio)ethynes RS-C≡C-SR (R = Me, Ph, $C_6H_3Me_2$-$o,o'$) mixtures of the $E$ and $Z$ isomer (with respect to the C=C bond) of $\eta^1$-coordinated thioselonocarboxylic ester complexes related to **130b** and **131b** (SR instead of NEt$_2$) were isolated. Monitoring the reaction of **99b** with MeS-C≡C-SMe by $^1$H-NMR spectroscopy at $-53°$C revealed that the $Z$ isomer was the kinetically controlled reaction product. In the reaction of **99a** with Bu$^t$S-C≡C-SMe, only one isomer ($Z$) was detected, whereas again a mixture of the $E$ and $Z$(C=C) isomer was formed in the reaction of **99b** with MeS-C≡C-SMe. In addition to these thioselonocarboxylic ester complexes, two complexes containing selenacyclic ligands were obtained, **134** and **135**. Complex **134** is presumably formed by a 1,3-H-shift in the initially formed selenete complex and the binuclear dihydro-1,3-diselenine complex **135** by a [4+2] cycloaddition of **99b** to the thioselonocarboxylic ester complex. The reaction of **99b** with H-C≡C-OBu$^t$ finally also yielded a binuclear dihydrodiselenine complex (**136**), whatever the 1,2-isomer.[248]

134                              135                              136

The thiobenzaldehyde complex **99a** also reacted readily with polar electron-rich alkynes R-C≡C-XR' (R = H: XR' = OEt, OBu$^t$; R = Me: XR' = SMe, SeEt) by cycloaddition and subsequent electrocyclic ring opening to form highly regio- and stereoselectively the $E$ isomer of $\alpha,\beta$-unsaturated dithio, selenothiono, and carboxylic ester complexes. In contrast but similar to the reactions of the selenobenzaldehyde complex **99b,** those of **99a** with bis(organylthio)ethynes RS-C≡C-SR (R = Me, Bu$^t$) afforded $E/Z$ mixtures of the insertion products [M(CO)$_5$\{$\eta^1$-S=C(SR)-C(SR)=C(Ph)H\}]. The dithio- and thionoester ligands were cleaved intact from the metal by treatment with [NEt$_4$]Br, thus providing a simple route to these compounds.[249]

However, the insertion of alkynes into the E=C bond of [M(CO)$_5$\{$\eta^1$-E=C(R$^1$)R$^2$\}] is restricted to electron-rich alkynes. Nonactivated alkynes react with **99** by substitution of the alkyne for CO ligands. The monocarbonyl(heteroaldehyde) complexes [W(CO)(H-C≡C-Bu$^t$)$_2$\{E=C(Ph)H\}]

(E = S, Se) were obtained from H-C≡C-Bu$^t$ and **99**.[56] Obviously, the [2+2] cycloaddition is too slow and therefore cannot compete with CO substitution.

### 3. Reactions with Vinyl Ethers

The reaction of the thio- and selenoaldehyde complexes with vinyl ethers provides metal-coordinated substituted thietanes and selenetanes.[224,250–252] The cycloadditions are highly regio- and stereoselective.

The kinetically controlled product in the reaction of **99a,b** with a large excess of ethyl vinyl ether was the thermodynamically less stable diastereomer **137** (Ph and OR *trans*) (Scheme 37). Complex **137** was present in solution as two conformers that rapidly interconverted. At room temperature in CDCl$_3$ the complexes **137** isomerized completely within several hours to form the diastereomer **138.** In **138** the substituents OEt and Ph occupy mutual *cis* positions, thus minimizing steric interaction with the bulky W(CO)$_5$ fragment on the heteroatom.[250]

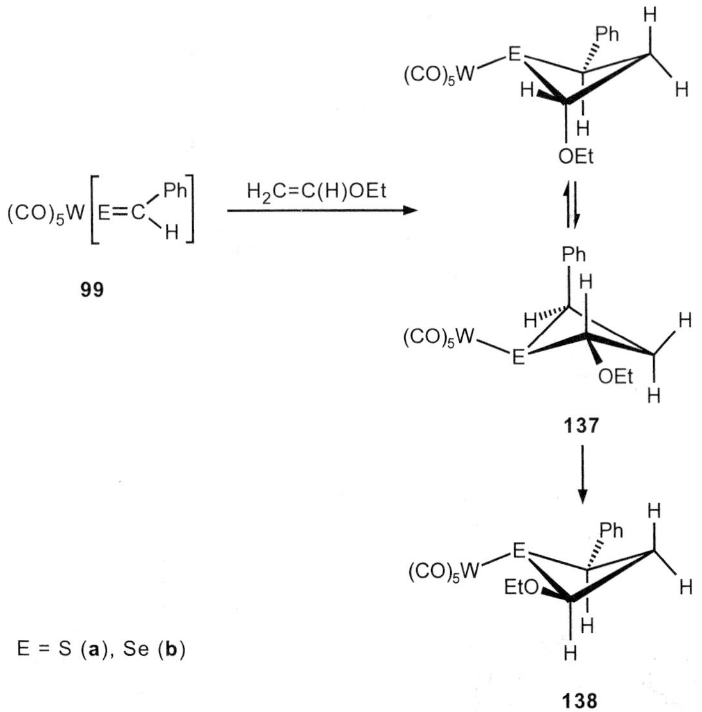

SCHEME 37

In addition to **99** and ethyl vinyl ether, other *para*-substituted thiobenzal-dehyde complexes $[W(CO)_5\{S=C(H)C_6H_4R\text{-}p\}]$ (R = H, Me, OMe, Cl, Br) and other open-chain vinyl ethers $R^1(R^2)C=C(R^3)OR^4$ ($R^1,R^2$ = H, Me; $R^3$ = H, Me, Ph, $C_6H_4Me\text{-}p$, $C_6H_4OMe\text{-}p$; $R^4$ = Me, Et, Bu$^t$) as well as cyclic vinyl ethers such as 2,3-dihydrofuran, 4,5-dihydro-2-methylfuran, and 3,4-dihydro-2*H*-pyran have also been employed successfully.[224,251] Thio- and selenoketone complexes did not react with vinyl ethers.

On addition the stereochemistry at the C=C bond of the vinyl ether remains constant. This was established by labeling studies and by the use of *cis* and *trans* ethyl propenyl ether. From the similarity of the activation parameters, the reaction characteristics, and the stereoselectivity of the cycloaddition step to those of the reactions of the same complexes with 1,3-dienes or ynamines an associative concerted nonsynchronous process was deduced.[224]

The isomerization **137** → **138** was found to be acid catalyzed. It very likely proceeds by protonation of OEt, subsequent elimination of EtOH and formation of a carbenium ion, and readdition of EtOH. This proposal is supported by the observation of an $OC_2D_5/OC_2H_5$ exchange in $C_2D_5OD/CF_3COOD$ as the solvent.[250,251]

The coordinated thietanes and selenetanes are reactive and could be used for the synthesis of various other heterocycles. The reaction of **137a** (E = S) and other thietane complexes with thiocyanate and $SiO_2/H_2O$ afforded tungsten-coordinated thiazinethione complexes (**139**) via insertion of the CN group into a S–C bond of the four-membered heterocycle, 1,3-migration of W(CO)$_5$, and protonation of the nitrogen atom. Thiazineselone complexes (**140**) were accessible from **137a** and [SeCN]$^-$ and $SiO_2/H_2O$, a selenazinethione complex (**141**) was finally obtained by reaction of **137b** with [SCN]$^-$ and $SiO_2/H_2O$.[252]

|            |          |         |
| ---------- | -------- | ------- |
| E = S      | X = S    | (**139**) |
| E = S      | X = Se   | (**140**) |
| E = Se     | X = S    | (**141**) |

In contrast, when **137a** was treated with [TeCN]$^-$ instead of a thiazinetellone complex, two isomeric complexes, each containing a thiatellurolane ligand, were isolated.[253] The reaction of **137b** with [TeCN]$^-$ afforded the analogous selenatellurolane. Both complexes are derived from **137a,b** by insertion of Te into a E–C bond. Related diselenolane complexes instead of selenazineselone complexes were formed when [SeCN]$^-$ was substituted for [SCN]⁻ in the reactions with selenete. Diselenolane complexes (**142**) were also obtained when **99b** was treated with only 1.5 equivalents of the vinyl ethers and under a CO pressure of 50 bar at −40°C [Eq. (29)].[252]

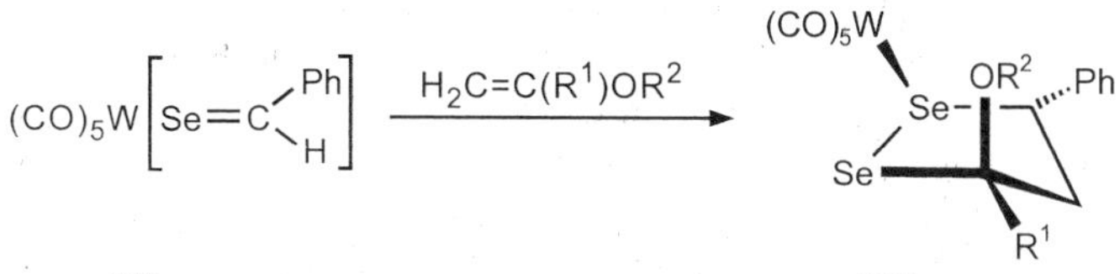

$$(CO)_5W \left[ Se=C \begin{array}{c} Ph \\ H \end{array} \right] \xrightarrow{H_2C=C(R^1)OR^2} \qquad (29)$$

**99b**                                **142**

R¹ = H: R² = Et;      R¹ = Me: R² = Me, Ph, C₆H₄Me-*p*

### 4. Reactions with Diazoalkanes

Three-membered rings should in principle be accessible by reaction of heteroaldehyde and -ketone complexes with C₁ sources. This was verified by the synthesis of the thiirane complex **143** from thiobenzaldehyde complex **99a** and diphenyl diazomethane [Eq. (30)]. The reaction was assumed to proceed via a 1,3,4-thiadiazolidine complex as the intermediate that loses dinitrogen to give **143**.[254]

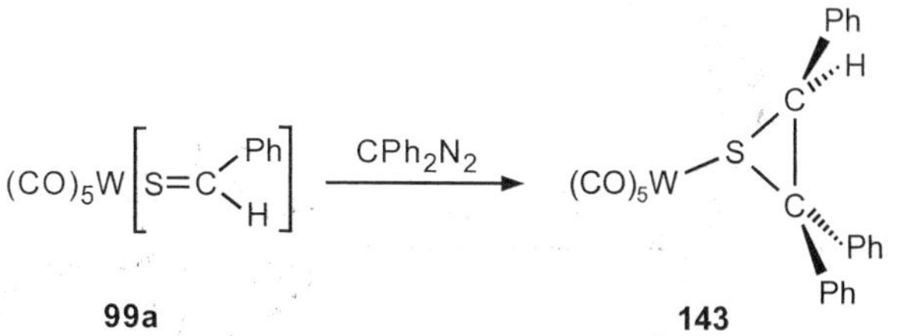

$$(CO)_5W \left[ S=C \begin{array}{c} Ph \\ H \end{array} \right] \xrightarrow{CPh_2N_2} \qquad (30)$$

**99a**                                **143**

Very likely, a selenirane complex was formed when the selenobenzophenone complex [W(CO)₅{Se=C(Ph)₂}] was treated with diphenyl diazomethane. It was not possible to spectroscopically detect the complex that decomposed rapidly by extrusion of tetraphenylethene. However, the intermediacy of a selenirane complex was supported by several observations. When the perdeuterated complex [W(CO)₅{Se=C(C₆D₅)₂}] was employed in the reaction with CPh₂N₂, a 1 : 4 : 3 mixture of the olefins (Ph)₂C=C(Ph)₂, (Ph)₂C=C(C₆D₅)₂, and (C₆D₅)₂C=C(C₆D₅)₂ was obtained. When the selenoaldehyde complex **99b** was used instead of the selenobenzophenone complex, the selenobenzophenone complex [W(CO)₅{Se=C(Ph)₂}] was formed along with stilbene H(Ph)C=C(Ph)H.[255]

E. *Insertion Reactions*

In contrast to $\eta^1$-heteroaldehyde and -ketone complexes, $\eta^2$-heteroaldehyde complexes are less prone to cyclo addition reactions. The reactivity of these complexes is characterized by the insertion of $C\equiv N$, $C=O$, and $C=N$ bonds into the M-C(heteroaldehyde) bond as has been demonstrated with the examples of zirconium and titanium complexes.

The $C\equiv C$ bond of several alkynes inserted into the Zr–C bond of the $\eta^2$-thioaldehyde zirconocene complexes **81** (see Scheme 21) to form the five-membered thiazirconacycles **144**. The imine metallacycle obtained from **81** and butyronitrile tautomerized under the reaction conditions to cleanly give the enamine metallacycle **145** (Scheme 38).[70]

In contrast, the related $\eta^2$-thioformaldehyde titanocene complexes **55** [see Eq. (11)] did not give clean insertion products with various triple ($C\equiv C$ and $C\equiv N$) and double ($C=O$, $C=N$, $C=S$) bonds.[67]

However, the reaction of the binuclear $\eta^2$-thioaldehyde titanocene complexes **112** (see Scheme 26) with valeronitrile, benzonitrile, and methyl thiocyanate proceeded in a similar fashion (Scheme 39).[69] Related metallacyclic compounds were also obtained from the reactions of **112** with benzophenone (Scheme 39), various imines, phenyl isothiocyanate, or dicyclohexyl carbodiimide via insertion of the $C=O$ and $C=N$ bond, respectively, into the Ti–C bond.[69,70]

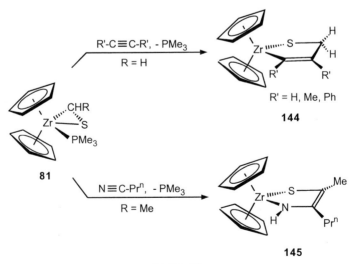

SCHEME 38

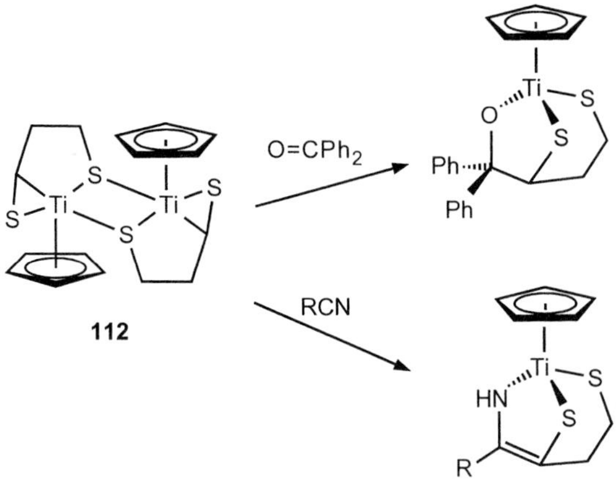

SCHEME 39

REFERENCES

(1) Duus, F., in *Comprehensive Organic Chemistry;* Jones, D. N., Ed.; Pergamon Press: Oxford, 1979; Vol. 3, Chapter 11.22; p. 373.

(2) Usov, V. A.; Timokhina, L. V.; Voronkov, M. G. *Russ. Chem. Rev.* **1990**, *59*, 378.

(3) McGregor, W. M.; Sherrington, D. C. *Chem. Soc. Rev.* **1993**, *22*, 199.

(4) Metzner, P. *Synthesis* **1992**, 1185.

(5) Whittingham, W. G. in *Comprehensive Organic Functional Group Transformations;* Katritzky, A. R., Meth-Cohn; O., Rees; C. W., Eds.; Pergamon Press: Oxford, 1995; Vol. 3; p. 329.

(6) Paulmier, C. *Selenium Reagents and Intermediates in Organic Chemistry;* Pergamon Press: Oxford, 1986; p. 58.

(7) Guziec, F. S., Jr. in *Organoselenium Chemistry;* Liotta, D.; Ed.; Wiley-Interscience: New York, 1987; p. 277.

(8) Guziec, F. S., Jr. in *The Chemistry of Organic Selenium and Tellurium Compounds;* Patai, S.; Ed.; Wiley: Chichester, 1987; Vol. 2, p. 215.

(9) Guziec, F. S., Jr.; Guziec, L. J. in *Comprehensive Organic Functional Group Transformations;* Katritzky, A. R.; Meth-Cohn, O.; Rees, C. W.; Eds.; Pergamon Press: Oxford, 1995; Vol. 3, p. 381.

(10) Giles, H. G.; Marty, R. A.; de Mayo, P. *Can. J. Chem.* **1976**, *54*, 537.

(11) Reid, D. H.; Webster, R. G.; McKenzie, S. *J. Chem. Soc. Perkin Trans. 1* **1979**, 2334.

(12) Mackie, R. K.; McKenzie, S.; Reid, D. H.; Webster, R. G. *J. Chem. Soc. Perkin Trans. 1* **1973**, 657.

(13) Muraoka, M.; Yamamoto, T.; Enomoto, K.; Takeshima, T. *J. Chem. Soc. Perkin Trans. 1* **1989**, 1241.

(14) Takeda, N.; Tokitoh, N.; Okazaki, R. *Chem. Eur. J.* **1997**, *3*, 62.

(15) Takeda, N.; Tokitoh, N.; Okazaki, R. *Angew. Chem.* **1996**, *108*, 714; *Angew. Chem. Int. Ed. Engl.* **1996**, *35*, 660.

(16) Takeda, N.; Tokitoh, N.; Okazaki, R. *Tetrahedron* **1997**, *53*, 12167.

(17) Okazaki, R.; Ishii, A.; Fukuda, N.; Oyama, H.; Inamoto, N. *J. Chem. Soc., Chem. Commun.* **1982**, 1187.

(18) Okazaki, R.; Kumon, N.; Inamoto, N. *J. Am. Chem. Soc.* **1989**, *111*, 5949.
(19) Okazaki, R.; Ishii, A.; Inamoto, N. *J. Am. Chem. Soc.* **1987**, *109*, 279.
(20) Ando, W.; Ohtaki, T.; Suzuki, T.; Kabe, Y. *J. Am. Chem. Soc.* **1991**, *113*, 7782.
(21) Vedejs, E.; Perry, D. A. *J. Am. Chem. Soc.* **1983**, *105*, 1683.
(22) Vedejs, E.; Perry, D. A.; Wilde, R. G. *J. Am. Chem. Soc.* **1986**, *108*, 2985.
(23) Erker, G.; Hock, R. *Angew. Chem.* **1989**, *101*, 181; *Angew. Chem. Int. Ed. Engl.* **1989**, *28*, 179.
(24) Segi, M.; Koyama, T.; Takata, Y.; Nakajima, T.; Suga, S. *J. Am. Chem. Soc.* **1989**, *111*, 8749.
(25) Back, T. G.; Dyck, B. P.; Parvez, M. *J. Org. Chem.* **1995**, *60*, 4657.
(26) Erker, G.; Hock, R.; Krüger, C.; Werner, S.; Klärner, F.-G., Artschwager-Perl, U. *Angew. Chem.* **1990**, *102*, 1082; *Angew. Chem. Int. Ed. Engl.* **1990**, *29*, 1067.
(27) Minoura, M.; Kawashima, T.; Okazaki, R. *J. Am. Chem. Soc.* **1993**, *115*, 7019.
(28) Minoura, M.; Kawashima, T.; Okazaki, R. *Tetrahedron Lett.* **1997**, *38*, 2501.
(29) Denifl, P.; Bildstein, B. *J. Organomet. Chem.* **1993**, *453*, 53.
(30) Linford, L.; Raubenheimer, H. G. *Adv. Organomet. Chem.* **1991**, *32*, 1.
(31) Linford, L.; Raubenheimer, H. G. *Comments Inorg. Chem.* **1991**, *12*, 113.
(32) Stumpf, R.; Fischer, H. *Trends Organomet. Chem.* **1994**, *1*, 465.
(33) Bock, H.; Aygen, S.; Rosmus, P.; Solouki, B.; Weißflog, E. *Chem. Ber.* **1984**, *117*, 187.
(34) Solouki, B.; Rosmus, P.; Bock, H. *J. Am. Chem. Soc.* **1976**, *98*, 6054.
(35) Bock, H.; Hirabayashi, T.; Mohmand, S. *Chem. Ber.* **1982**, *115*, 492.
(36) Collins, S.; Back, T. G.; Rauk, A. *J. Am. Chem. Soc.* **1985**, *107*, 6589.
(37) Vedejs, E.; Perry, D. A.; Houk, K. N.; Rondan, N. G. *J. Am. Chem. Soc.* **1983**, *105*, 6999.
(38) Adams, R. D.; Golembeski, N. M.; Selegue, J. P. *J. Am. Chem. Soc.* **1979**, *101*, 5862.
(39) Adams, R. D.; Golembeski, N. M.; Selegue, J. P. *J. Am. Chem. Soc.* **1981**, *103*, 546.
(40) Ziegler, T. *Inorg. Chem.* **1986**, *25*, 2721.
(41) Rosi, M.; Sgamellotti, A.; Tarantelli, F.; Floriani, C. *Inorg. Chem.* **1988**, *27*, 69.
(42) Delbecq, F., Sautet, P. *J. Am. Chem. Soc.* **1992**, *114*, 2446.
(43) Sakaki, S.; Kitaura, K.; Morokuma, K.; Ohkubo, K. *Inorg. Chem.* **1983**, *22*, 104.
(44) Fischer, H.; Zeuner, S.; Riede, J. *Angew. Chem.* **1984**, *96*, 707; *Angew. Chem. Int. Ed. Engl.* **1984**, *23*, 726.
(45) Fischer, H.; Zeuner, S.; Gerbing, U.; Riede, J.; Kreiter, C. G. *J. Organomet. Chem.* **1989**, *377*, 105.
(46) Fischer, H.; Zeuner, S. *Z. Naturforsch.* **1985**, *40b*, 954.
(47) Schenk, W. A.; Stur, T.; Dombrowski, E. *J. Organomet. Chem.* **1994**, *472*, 257.
(48) Quirós Méndez, N.; Arif, A. M.; Gladysz, J. A. *Angew. Chem.* **1990**, *102*, 1507; *Angew. Chem. Int. Ed. Engl.* **1990**, *29*, 1479.
(49) Quirós Méndez, N.; Mayne, C. L.; Gladysz, J. A. *Angew. Chem.* **1990**, *102*, 1509; *Angew. Chem. Int. Ed. Engl.,* **1990**, *29*, 1475.
(50) Fischer, H.; Märkl, R. *Chem. Ber.* **1982**, *115*, 1349.
(51) Gingerich, R. G. W., Angelici, R. J. *J. Organomet. Chem.* **1977**, *132*, 377.
(52) Bakhmutov, V. I.; Petrovskii, P. V.; Dolgova, S. P.; Setkina, V. N.; Fedin, E. J.; Kursanov, D. N. *J. Organomet. Chem.* **1983**, *246*, 177.
(53) Hill, A. F.; Roper, W. R.; Waters, J. M., Wright, A. H. *J. Am. Chem. Soc.* **1983**, *105*, 5939.
(54) Werner, H.; Paul, W.; Knaup, W.; Wolf, J., Müller, G.; Riede, J. *J. Organomet. Chem.* **1988**, *358*, 95.
(55) Werner, H.; Hofmann, L.; Wolf, J., Müller, G. *J. Organomet. Chem.* **1985**, *280*, C55.
(56) Fischer, H.; Gerbing, U.; Müller, G., Alt, H. G. *Chem. Ber.* **1987**, *120*, 1905.
(57) Allen, F. H., Kennard, O., Watson, D. G., Brammer, L., Orpen, A. G., Taylor, R. *J. Chem. Soc., Perkin Trans. II* **1987**, S1.

(58) Harmony, M. D.; Laurie, V. W.; Kuczkowski, R. L.; Schwendeman, R. H.; Ramsay, D. A.; Lovas, F. J.; Lafferty, W. J.; Maki, A. G. *J. Phys. Chem. Ref. Data* **1979**, *8*, 619.
(59) Johnson, D. R.; Powell, F. X.; Kirchhoff, W. H. *J. Mol. Spectrosc.* **1971**, *39*, 136.
(60) Kroto, H. W.; Landsberg, B. M. *J. Mol. Spectrosc.* **1976**, *62*, 346.
(61) Georgiou, K.; Kroto, H. W. *J. Mol. Spectrosc.* **1980**, *83*, 94.
(62) Kruger, G. J.; Linford, L.; Raubenheimer, H. G. *J. Chem. Soc., Dalton Trans.* **1984**, 2337.
(63) Fischer, H.; Fluck, K. H.; Troll, C. *Chem. Ber.* **1992**, *125*, 2675.
(64) Raubenheimer, H. G.; Kruger, G. J.; Linford, L.; Marais, C. F.; Otte, R.; Hattingh, J. T. Z.; Lombard, A. *J. Chem. Soc., Dalton Trans.* **1989**, 1565.
(65) Glidewell, C.; Liles, D. C.; Pogorzelec, P. J. *Acta Cryst.* **1983**, *C39*, 542.
(66) Schenk, W. A.; Stur, T.; Dombrowski, E. *Inorg. Chem.* **1992**, *31*, 723.
(67) Park, J. W.; Henling, L. M., Schaefer, W. P.; Grubbs, R. H. *Organometallics* **1990**, *9*, 1650.
(68) Huang, Y.; Nadasdi, T. T.; Stephen, D. W. *J. Am. Chem. Soc.* **1994**, *116*, 5483.
(69) Huang, Y.; Etkin, N.; Heyn, R. R.; Nadasdi, T. T.; Stephan, D. W. *Organometallics* **1996**, *15*, 2320.
(70) Buchwald, S. L.; Nielsen, R. B.; Dewan, J. C. *J. Am. Chem. Soc.* **1987**, *109*, 1590.
(71) Nelson, J. E.; Bercaw, J. E.; Marsh, R. E.; Henling, L. M. *Acta Cryst.* **1992**, *C48*, 1023.
(72) Mayr, A.; McDermott, G. A.; Dorries, A. M.; Holder, A. K.; Fultz, W. C.; Rheingold, A. L. *J. Am. Chem. Soc.* **1986**, *108*, 310.
(73) Takahashi, K.; Iwanami, M.; Tsai, A.; Chang, P. L.; Harlow, R. L.; Harris, L. E.; McCaskie, J. E.; Pfluger, C. E.; Dittmer, D. C. *J. Am. Chem. Soc.* **1973**, *95*, 6113.
(74) Harlow, R. L.; Pfluger, C. E. *Acta Cryst.* **1973**, *B29*, 2633.
(75) Buhro, W. E.; Patton, A. T.; Strouse, C. E.; Gladysz, J. A.; McCormick, F. B.; Etter, M. C. *J. Am. Chem. Soc.* **1983**, *105*, 1056.
(76) Burho, W. E.; Etter, M. C.; Georgiou, S.; Gladysz, J. A.; McCormick, F. B. *Organometallics* **1987**, *6*, 1150.
(77) Schenk, W. A.; Burzlaff, N.; Burzlaff, H. *Z. Naturforsch.* **1994**, *49b*, 1633.
(78) Bianchini, C.; Frediani, P.; Herrera, V.; Jiménez, M. V.; Meli, A.; Rincón, L.; Sánchez-Delgado, R.; Vizza, F. *J. Am. Chem. Soc.* **1995**, *117*, 4333.
(79) Chin, R. M.; Jones, W. D. *Angew. Chem.* **1992**, *104*, 340; *Angew. Chem. Int. Ed. Engl.* **1992**, *31*, 357.
(80) Brown, R. D.; Godfrey, P. D.; McNaughton, D.; Taylor, P. R. *J. Mol. Spectr.* **1986**, *120*, 292.
(81) Brown, R. D.; Godfrey, P. D.; McNaughton, D. *Chem. Phys. Letters* **1985**, *118*, 29.
(82) Hutchinson, M.; Kroto, H. W. *J. Mol. Spectr.* **1978**, *70*, 347.
(83) McCormick, F. B. *Organometallics* **1984**, *3*, 1924.
(84) Shin, J. H.; Parkin, G. *Organometallics* **1994**, *13*, 2147.
(85) Firth, A. V.; Stephan, D. W. *Organometallics* **1997**, *16*, 2183.
(86) Bernatis, P.; Haltiwanger, R. C.; DuBois, M. R. *Organometallics,* **1992**, *11*, 2435.
(87) Fischer, H.; Troll, C. Unpublished results.
(88) Mathur, P.; Manimaran, B.; Satyanarayana, C. V. V.; Varghese, B. *J. Organomet. Chem.* **1997**, *527*, 83.
(89) Bacchi, A.; Bianchini, C.; Herrera, V.; Jiménez, M. V.; Mealli, C.; Meli, A.; Moneti, S.; Peruzzini, M.; Sánchez-Delgado, R. A.; Vizza, F. *J. Chem. Soc., Chem. Commun.* **1995**, 921.
(90) Herrmann, W. A.; Weichmann, J.; Serrano, R.; Blechschmitt, K.; Pfisterer, H.: Ziegler, M. L. *Angew. Chem.* **1983**, *95*, 331; *Angew. Chem. Int. Ed. Engl.,* **1983**, *22*, 314; *Angew. Chem. Suppl.* **1983**, 363.
(91) Mathur, P.; Manimaran, B.; Trivedi, R.; Hossain, M. M.; Arabatti, M. *J. Organomet. Chem.* **1996**, *515*, 155.

(92) Fischer, H.; Früh, A.; Troll, C. Unpublished results.
(93) Mathur, P.; Manimaran, B.; Hossain, M. M.; Satyanarayana, C. V. V.; Puranik, V. G.; Tavale, S. S. *J. Organomet. Chem.* **1995**, *493*, 251.
(94) Rindorf, G.; Carlsen, L. *Acta Cryst.* **1979**, *35B*, 1179.
(95) Karcher, B. A.; Jacobson, R. A. *J. Organomet. Chem.* **1977**, *132*, 387.
(96) Raubenheimer, H. G.; Kruger, G. J.; Lombard, A. van A.; Linford, L.; Viljoen, J. C. *Organometallics* **1985**, *4*, 275.
(97) Barnes, J. C.; Bell, W.; Glidewell, C.; Howie, R. A. *J. Organomet. Chem.* **1990**, *385*, 369.
(98) Fischer, H.; Fluck, K.; Roth, G. Unpublished results.
(99) Andrianov, V. G.; Struchkov, Yu. T.; Petrovskii, P. V.; Fedin, E. I.; Kursanov, D. N.; Dolgova, S. P.; Setkina, V. N. *J. Organomet. Chem.* **1981**, *221*, 183.
(100) Dettlaf, G.; Behrens, U. Weiss, E. *J. Organomet. Chem.* **1978**, *152*, 95.
(101) Samb, A.; Demerseman, B.; Dixneuf, P. H.; Mealli, C. *Organometallics* **1988**, *7*, 26.
(102) Faraglia, G.; Graziani, R.; Guo, Z.; Casellato, U.; Sitran, S. *Inorg. Chim. Acta* **1992**, *192*, 17.
(103) Pasquali, M.; Leoni, P.; Floriani, C.; Chiesi-Villa, A.; Guastini, C. *Inorg. Chem.* **1983**, *22*, 841.
(104) Herbig, J.; Köhler, J.; Nuber, B.; Stammler, H. G.; Ziegler, M. L. *J. Organomet. Chem.* **1993**, *444*, 107.
(105) Skaugset, A. E.; Rauchfuss, T. B.; Wilson, S. R. *Organometallics* **1990**, *9*, 2875.
(106) Jones, P. G.; Meyer-Bäse, K.; Schroeder, T.; Sheldrick, G. M. *Acta Cryst.* **1987**, *C43*, 32.
(107) Jones, P. G.; Roesky, H. W.; Gries, T.; Meyer-Bäse, K.; Sheldrick, G. M. *Z. Anorg. Allg. Chem.* **1986**, *542*, 46.
(108) Minoura, M.; Kawashima, T.; Tokitoh, N.; Okazaki, R. *J. Chem. Soc., Chem. Commun.* **1996**, 123.
(109) Umland, H.; Behrens, U. *J. Organomet. Chem.* **1985**, *287*, 109.
(110) Daran; J.-C., Heim; B., Jeannin, Y. *J. Cluster Sci.* **1993**, *4*, 403.
(111) Pannell, K. H.; Mayr, A. J.; VanDerveer, D. *Organometallics* **1983**, *2*, 560.
(112) Kruger, G. J.; Lombard, A. van A., Raubenheimer, H. G. *J. Organomet. Chem.* **1987**, *311*, 247.
(113) Barrow, M. J.; Sim, G. A. *Acta Cryst.* **1979**, *35B*, 1223.
(114) Chen, J.; Day, C. L.; Jacobson, R. A.; Angelici, R. J. *J. Organomet. Chem.* **1996**, *522*, 21.
(115) Alper, H.; Silavwe, N. D.; Birnbaum, G. I., Ahmed, F. R. *J. Am. Chem. Soc.* **1979**, *101*, 6582.
(116) Adams, R. D.; Pompeo, M. P. *Organometallics* **1990**, *9*, 1718.
(117) Holden, H. D.; Johnson, B. F. G.; Lewis, J.; Raithby, P. R.; Uden, G. *Acta Cryst* **1983**, *C39*, 1197.
(118) Tokitoh; N.; Takeda, N.; Okazaki, R. *J. Am. Chem. Soc.* **1994**, *116*, 7907.
(119) Gladysz, J. A.; Avakian, R. W. *Syn. React. Inorg. Metalorg. Chem.* **1975**, *5*, 247.
(120) Alper, H.; Sachdeva, R. *J. Organomet. Chem.* **1979**, *169*, 63.
(121) Kursanov, D. N.; Setkina V. N.; Dolgova, S. P. *Bull. Acad. Sci. USSR Div. Chem. Sci.* **1979**, *28*, 814.
(122) Petrovskii, P. V.; Setkina, V. N.; Dolgova, S. P., Kursanov, D. N. *J. Organomet. Chem.* **1981**, *221*, 177.
(123) Kursanov, D. N.; Setkina, V. N.; Dolgova, S. P.; Nefedova, M. N. *Bull. Acad. Sci. USSR Div. Chem. Sci.* **1980**, *29*, 1360.
(124) Alper, H.; Silavwe, N. D. *J. Organomet. Chem.* **1978**, *161*, C8.
(125) Alper, H. *Inorg. Chem.* **1976**, *15*, 962.
(126) Alper, H., *J. Organomet. Chem.* **1974**, *73*, 359.
(127) Alper, H.; Chan, A. S. K. *J. Organomet. Chem.* **1973**, *61*, C59.

(128) Alper, H.; Paik, H.-N. *J. Organomet. Chem.* **1978**, *155*, 47.
(129) Ellis, J. E. *J. Organomet. Chem.* **1975**, *86*, 1.
(130) Alper, H. *J. Organomet. Chem.* **1973**, *61*, C62.
(131) Dupont, J.; Beydoun, N.; Pfeffer, M. *J. Chem. Soc., Dalton Trans.* **1989**, 1715.
(132) Davis, R. C., Grinter, T. J.; Leaver, D.; O'Neil, R. M.; Thomson, G. A. *J. Chem. Res.* *(S)* **1987**, *11*, 280.
(133) Siedle, A. R. *J. Organomet. Chem.* **1981**, *208*, 115.
(134) Faraglia, G.; Barbaro, F.; Sitran, S. *Transition Met. Chem.* **1990**, *15*, 242.
(135) Faraglia, G.; Guo, Z.; Sitran, S. *Polyhedron* **1991**, *10*, 351.
(136) Cabrino, R.; Biggi, G.; Pietra, F. *Inorg. Chem.* **1975**, *14*, 1213.
(137) Alper, H. *J. Organomet. Chem.* **1974**, *80*, C29.
(138) Okuda, J.; Herberich, G. E. *Organometallics* **1987**, *6*, 2331.
(139) Werner, H.; Kolb, O. *Chem. Ber.* **1985**, *118*, 880.
(140) Gambarotta, S.; Fiallo, M. L.; Floriani, C.; Chiesi-Villa, A.; Guastini, C. *Inorg. Chem.* **1984**, *23*, 3532.
(141) Mayer, R. in *Organosulfur Chemistry*; Janssen, M. J.; Ed.; Interscience: New York, 1967; p. 219.
(142) Campaigne, E. in *The Chemistry of the Carbonyl Group*; Patai, S.; Ed.; Interscience: New York, 1966; p. 917.
(143) Gingerich, R. G. W.; Angelici, R. *J. Am. Chem. Soc.* **1979**, *101*, 5604.
(144) Baird, M. C.; Wilkinson, G. *J. Chem. Soc. (A)* **1967**, 865.
(145) Browning, J.; Cundy, C. S.; Green, M.; Stone, F. G. A. *J. Chem. Soc. (A)* **1969**, 20.
(146) Commereuc, D.; Douek, I.; Wilkinson, G. *J. Chem. Soc. (A)* **1970**, 1771.
(147) Dittmer, D. C.; Takahashi, K.; Iwanami, M.; Tsai, A. I.; Chang, P. L.; Blidner, B. B.; Stamos, I. K. *J. Am. Chem. Soc.* **1976**, *98*, 2795.
(148) Patwardhan, B. H.; Parker, E. J.; Dittmer, D. C. *Phoshorus Sulfur* **1979**, *7*, 5.
(149) Pogorzelec, P. J.; Reid, D. H. *J. Chem. Soc., Chem. Commun.* **1983**, 289.
(150) Bogdanovic, B.; Krüger, C.; Locatelli, P. *Angew. Chem.* **1979**, *91*, 745; *Angew. Chem. Int. Ed. Engl.* **1979**, *18*, 684.
(151) Skaugset, A. E.; Rauchfuss, T. B.; Wilson, S. R. *J. Am. Chem. Soc.* **1992**, *114*, 8521.
(152) Jones, W. D.; Chin, R. M. *J. Am. Chem. Soc.* **1992**, *114*, 9851.
(153) Jones, W. D.; Chin, R. M. *J. Organomet. Chem.* **1994**, *472*, 311.
(154) Dailey, K. M. K.; Rauchfuss, T. R.; Rheingold, A. L.; Yap, G. P. A. *J. Am. Chem. Soc.* **1995**, *117*, 6396.
(155) Feng, Q.; Krautscheid, H.; Rauchfuss, T. B.; Skaugset, A. E.; Venturelli, A. *Organometallics* **1995**, *14*, 297.
(156) Feng, Q.; Rauchfuss, T. B.; Wilson, S. R. *Organometallics* **1995**, *14*, 2923.
(157) Krautscheid, H.; Feng, Q.; Rauchfuss, T. B. *Organometallics* **1993**, *12*, 3273.
(158) Dailey, K. K.; Rauchfuss, T. B. *Organometallics* **1997**, *16*, 858.
(159) Bianchini, C.; Meli, A.; Peruzzini, M.; Vizza, F.; Moneti, S.; Herrera, V.; Sánchez-Delgado, R. A. *J. Am. Chem. Soc.* **1994**, *116*, 4370.
(160) Bianchini, C.; Herrera, V.; Jimenez, M. V.; Meli, A.; Sánchez-Delgado, R.; Vizza, F. *J. Am. Chem. Soc.* **1995**, *117*, 8567.
(161) Bianchini, C.; Herrera, V.; Jiménez, M. V.; Laschi, F.; Meli, A.; Sánchez-Delgado, R.; Vizza, F.; Zanello, P. *Organometallics* **1995**, *14*, 4390.
(162) Bianchini, C.; Meli, A.; Peruzzini, M.; Vizza, F.; Frediani, P.; Herrera, V.; Sánchez-Delgado, R. A. *J. Am. Chem. Soc.* **1993**, *115*, 2731.
(163) Bennett, M. A.; Goh, L. Y.; Willis, A. C. *J. Am. Chem. Soc.* **1996**, *118*, 4984.
(164) Werner, H.; Feser, R.; Paul, W.; Hofmann, L. *J. Organomet. Chem.* **1981**, *219*, C29.

(165) Paul, W.; Werner, H. *Angew. Chem.* **1983**, *95*, 333; *Angew. Chem. Int. Ed. Engl.* **1983**, *22*, 316.

(166) Werner, H.; Paul, W. *Angew. Chem.* **1984**, *96*, 68; *Angew. Chem. Int. Ed. Engl.,* **1984**, *23*, 58.

(167) Headford, C. E. L.; Roper, W. R. *J. Organomet. Chem.* **1983**, *244*, C53.

(168) Clark, G. R.; Headford, C. E. L.; Marsden, K.; Roper, W. R. *J. Organomet. Chem.* **1982**, *231*, 335.

(169) Hofmann, L.; Werner, H. *J. Organomet. Chem.* **1983**, *255*, C41.

(170) Hofmann, L.; Werner, H. *Chem. Ber.* **1985**, *118*, 4229.

(171) Fischer, H. *J. Organomet. Chem.* **1981**, *219*, C34.

(172) Troll, C.; Fischer, H. Unpublished investigations.

(173) Fischer, H.; Zeuner, S.; Ackermann, K.; Schubert, U. *J. Organomet. Chem.* **1984**, *263*, 201.

(174) Fischer, H. *J. Organomet. Chem.* **1981**, *222*, 241.

(175) Fischer, H.; Zeuner, S. *Z. Naturforsch.* **1983**, *38B*, 1365.

(176) Fischer, H.; Zeuner, S. *J. Organomet. Chem.* **1983**, *252*, C63.

(177) Casey, C. P.; Polichnowski, S. W.; Tuinstra, H. E.; Albin, L. D.; Calabrese, J. C. *Inorg. Chem.* **1978**, *17*, 3045.

(178) Fischer, H.; Zeuner, S.; Ackermann, K. *J. Chem. Soc., Chem. Commun.* **1984**, 684.

(179) Fischer, H.; Reindl, D. *J. Organomet. Chem.* **1990**, *385*, 351.

(180) Fischer, H.; Früh, A.; Troll, C. *J. Organomet. Chem.* **1991**, *415*, 211.

(181) Raubenheimer, H. G.; Kruger, G. J.; Marais, C. F. *J. Chem. Soc., Chem. Commun.* **1984**, 634.

(182) Raubenheimer, H. G.; Kruger, G.; Marais, C.; Hattingh, J.; Otte, R.; Linford, L. in *Advances in Metal Carbene Chemistry;* Schubert, U.; Ed., Kluwer Academic Publishers: Dordrecht, 1989; p. 145.

(183) Shin, J. H.; Parkin, G. *Organometallics* **1995**, *14*, 1104.

(184) Köhler, J. U.; Lewis, J.; Raithby, R. P. *Angew. Chem.* **1996**, *108*, 1071; *Angew. Chem. Int. Ed. Engl.* **1996**, *35*, 993.

(185) Werner, H.; Wolf, J.; Zolk, R.; Schubert, U. *Angew. Chem.* **1983**, *95*, 1022; *Angew. Chem. Int. Ed. Engl.,* **1983**, *22*, 981.

(186) Wolf, J.; Zolk, R.; Schubert, U.; Werner, H. *J. Organomet. Chem.* **1988**, *340*, 161.

(187) Bianchini, C.; Glendenning, L.; Peruzzini, M.; Romerosa, A.; Zanobini, F. *J. Chem. Soc., Chem. Commun.* **1994**, 2219.

(188) Nelson, J. E.; Parkin, G.; Bercaw, J. E. *Organometallics* **1992**, *11*, 2181.

(189) Parkin, G.; Bunel, E.; Burger, B. J.; Trimmer, M. S.; van Asselt, A.; Bercaw, J. E. *J. Mol. Catal.* **1987**, *41*, 21.

(190) Belsky, K. A.; Asaro, M. F.; Chen, S. Y.; Mayr, A. *Organometallics* **1992**, *11*, 1926.

(191) Anderson, S.; Hill, A. F. *J. Chem. Soc., Dalton Trans.* **1993**, 587.

(192) Collins, T. J.; Roper, W. R. *J. Chem. Soc., Chem. Commun.* **1977**, 901.

(193) Collins, T. J.; Roper, W. R. *J. Organomet. Chem.* **1978**, *159*, 73.

(194) Broadhurst, P. V. *Polyhedron* **1985**, *4*, 1801.

(195) Collins, T. J.; Roper, W. R. *J. Organomet. Chem.* **1977**, *139*, C56.

(196) Roper, W. R.; Town, K. G. *J. Chem. Soc., Chem. Commun.* **1977**, 781.

(197) Buchwald, S. L.; Nielsen, R. B. *J. Am. Chem. Soc.* **1988**, *110*, 3171.

(198) Bleeke, J. R.; Ortwerth, M. F.; Rohde, A. M. *Organometallics* **1995**, *14*, 2813.

(199) Herrmann, W. A.; Rohrmann, J. *Chem. Ber.* **1986**, *119*, 1437.

(200) Touchard, D.; Dixneuf, P. H.; Adams, R. D.; Segmüller, B. E. *Organometallics* **1984**, *3*, 640.

(201) Touchard, D.; Fillaut, J.-L.; Dixneuf, P. H.; Adams, R. D.; Segmüller, B. E. *J. Organomet. Chem.* **1990**, *386*, 95.

(202) Khasnis, D. V.; Toupet, L.; Dixneuf, P. H. *J. Chem. Soc., Chem. Commun.* **1987**, 230.
(203) Khasnis, D. V.; Pirio, N.; Touchard, D.; Toupet, L.; Dixneuf, P. H. *Inorg. Chim. Acta* **1992**, *198–200*, 193.
(204) Fischer, H.; Zeuner, S.; Alt, H. G. *J. Organomet. Chem.* **1985**, *289*, C21.
(205) Fischer, H.; Kalbas, C.; Früh, A.; Fluck, K. Unpublished results.
(206) Fluck, K.; Fischer, H. Unpublished investigations.
(207) Adams, R. D.; Babin, J. E.; Tasi, M. *Organometallics* **1986**, *5*, 1920.
(208) Adams, R. D.; Babin, J. E.; Tasi, M. *Organometallics* **1987**, *6*, 1717.
(209) Bamber, M.; Froom, S. F. T.; Green, M.; Schulz, M.; Werner, H. *J. Organomet. Chem.* **1992**, *434*, C19.
(210) Umland, H.; Edelmann, F.; Wormsbächer, D.; Behrens, U. *Angew. Chem.* **1983**, *95*, 148; *Angew. Chem. Int. Ed. Engl.,* **1983**, *22*, 152; *Angew. Chem. Suppl.* **1983**, 156.
(211) Adams, R. D.; Golembeski, N. M. *J. Am. Chem. Soc.* **1979**, *101*, 1306.
(212) Davidson, J. L.; Sharp, D. W. A. *J. Chem. Soc., Dalton Trans.* **1975**, 2283.
(213) Herrmann, W. A.; Weichmann, J.; Serrano, R.; Blechschmitt, K.; Pfisterer, H.; Ziegler, M. J. *Angew. Chem.* **1983**, *95*, 331; *Angew. Chem. Int. Ed. Engl.* **1983**, *22*, 314; *Angew. Chem. Suppl.* **1983**, 363.
(214) Herberhold, M.; Ehrenreich, W.; Bühlmeyer, W. *Angew. Chem.* **1983**, *95*, 332, *Angew. Chem. Int. Ed. Engl.* **1983**, *22*, 315.
(215) Herrmann, W. A.; Weichmann, J.; Küsthardt, U.; Schäfer, A.; Hörlein, R.; Hecht, C.; Voss, E.; Serrano, R. *Angew. Chem.* **1983**, *95*, 1019; *Angew. Chem. Int. Ed. Engl.* **1983**, *22*, 979; *Angew. Chem. Suppl.* **1983**, 1543.
(216) Herrmann, W. A.; Hecht, C.; Ziegler, M. L.; Balbach, B. *J. Chem. Soc., Chem. Commun.* **1984**, 686.
(217) Frank, L.-R.; Evertz, K.; Zsolnai, L.; Huttner, G. *J. Organomet. Chem.* **1987**, *335*, 179.
(218) Herrmann, W. A.; Rohrmann, J.; Schäfer, A. *J. Organomet. Chem.* **1984**, *265*, C1.
(219) Herrmann, W. A.; Rohrmann, J.; Nöth, H.; Nanila, C. K.; Bernal, I.; Draux, M. *J. Organomet. Chem.* **1985**, *284*, 189.
(220) Herberhold, M.; Jellen, W.; Murray, H. H. *J. Organomet. Chem.* **1984**, *270*, 65.
(221) Parker, E. J.; Bodwell, J. R.; Sedergran, T. C.; Dittmer, D. C. *Organometallics,* **1982**, *1*, 517.
(222) Parker, E. J.; Dittmer, D. C. *Organometallics* **1982**, *1*, 522.
(223) Dittmer, D. C.; Parker, E. J.; Bodwell, J. R. *Org. Magn. Reson.* **1984**, *22*, 609.
(224) Fischer, H.; Ruchay, A.; Stumpf, R.; Kalbas, C. *J. Organomet. Chem.* **1993**, *459*, 249.
(225) Hartgerink, J.; Silavwe, N. D.; Alper, H. *Inorg. Chem.* **1980**, *19*, 2593.
(226) Herberhold, M.; Hill, A. F. *J. Organomet. Chem.* **1986**, *309*, C29.
(227) Herberhold, M.; Hill, A. F. *J. Organomet. Chem.* **1990**, *387*, 323.
(228) Fischer, H.; Gerbing, U. *J. Organomet. Chem.* **1986**, *299*, C7.
(229) Fischer, H.; Gerbing, U.; Riede, J.; Benn, R. *Angew. Chem.* **1986**, *98*, 80; *Angew. Chem. Int. Ed. Engl.* **1986**, *25*, 78.
(230) Fischer, H. *J. Organomet. Chem.* **1988**, *345*, 65.
(231) Fischer, H.; Gerbing, U.; Riede, J. *J. Organomet. Chem.* **1989**, *364*, 155.
(232) Fischer, H.; Treier, K.; Gerbing, U.; Hofmann, J. *J. Chem. Soc., Chem. Commun.* **1989**, 667.
(233) Fischer, H.; Gerbing, U.; Treier, K.; Hofmann, J. *Chem. Ber.* **1990**, *123*, 725.
(234) Sauer, J.; Sustmann, R. *Angew. Chem.* **1980**, *92*, 773; *Angew. Chem. Int. Ed. Engl.* **1980**, *19*, 779.
(235) Vedejs, E.; Stults, J. S.; Wilde, R. G. *J. Am. Chem. Soc.* **1988**, *110*, 5452.
(236) Krafft, G. A.; Meinke, P. T. *Tetrahedron Lett.* **1985**, 1947.

(237) Kirby, G. W.; Lochead, A. W.; Sheldrake, G. N. *J. Chem. Soc., Chem. Commun.* **1984**, 1469.

(238) Krafft, G. A.; Meinke, P. T. *J. Am. Chem. Soc.* **1986**, *108*, 1314.

(239) Segi, M.; Nakajima, T.; Suga, S.; Murai, S.; Ryu, I.; Ogawa, A.; Sonoda, N. *J. Am. Chem. Soc.* **1988**, *110*, 1976.

(240) Fischer, H.; Gerbing, U.; Tiriliomis, A. *J. Organomet. Chem.* **1987**, *332*, 105.

(241) Fischer, H.; Tiriliomis, A.; Gerbing, U.; Huber, B.; Müller, G. *J. Chem. Soc., Chem. Commun.* **1987**, 559.

(242) Fischer, H.; Hofmann, J.; Gerbing, U.; Tiriliomis, A. *J. Organomet. Chem.* **1988**, *358*, 229.

(243) Fischer, H.; Gerbing, U.; Tiriliomis, A.; Müller, G.; Huber, B.; Riede, J.; Hofmann, J.; Burger, P. *Chem. Ber.* **1988**, *121*, 2095.

(244) Fischer, H.; Treier, K.; Hofmann, J. *J. Organomet. Chem.* **1990**, *384*, 305.

(245) Fischer, H.; Pashalidis, I. *J. Organomet. Chem.* **1988**, *348*, C1.

(246) Fuks, R.; Viehe, H. G. *Chem. Ber.* **1970**, *103*, 564.

(247) Fischer, H.; Treier, K.; Troll, C.; Stumpf, R. *J. Chem. Soc., Chem. Commun.* **1995**, 2461.

(248) Fischer, H.; Treier, K.; Troll, C. *Chem. Ber.* **1995**, *128*, 1149.

(249) Fischer, H.; Treier, K.; Troll, C. *Chem. Ber.* **1995**, *128*, 883.

(250) Fischer, H.; Kalbas, C.; Gerbing, U. *J. Chem. Soc., Chem. Commun.* **1992**, 563.

(251) Fischer, H.; Kalbas, C.; Stumpf, R. *Chem. Ber.* **1996**, *129*, 1169.

(252) Fischer, H.; Kalbas, C.; Troll, C.; Fluck, K. H. *Z. Naturforsch.* **1993**, *48b*, 1613.

(253) Fischer, H.; Kalbas, C.; Hofmann, J. *J. Chem. Soc., Chem. Commun.* **1992**, 1050.

(254) Fischer, H.; Treier, K.; Gerbing, U. *J. Organomet. Chem.* **1992**, *433*, 127.

(255) Gerbing, U. Thesis, Technical University Munich, **1988**.

(256) Buchwald, S. L.; Fisher, R. A.; Davis, W. M. *Organometallics* **1989**, *8*, 2082.

ADVANCES IN ORGANOMETALLIC CHEMISTRY, VOL. 43

# Recent Progress in Transition Metal-Catalyzed Reactions of Silicon, Germanium, and Tin

JENNIFER A. REICHL and DONALD H. BERRY

*Department of Chemistry and Laboratory for Research on the Structure of Matter*
*University of Pennsylvania*
*Philadelphia, Pennsylvania 19104*

## I

## INTRODUCTION

An important focus of organometallic research has been the development and study of transition metal complexes as catalysts for transformations of organic compounds. These systems enable chemical reactions to occur under conditions that are often milder and more environmentally benign than more traditional routes, such as Lewis acid-catalyzed reactions, Grignard

197

additions, and thermal processes. Additionally, transition metal catalysts often provide a means of regio- and stereochemical control of reaction products. This research area has in recent years been coupled with increasing interest in the synthesis and reactivity of compounds that contain the Group 14 elements silicon, germanium, and tin. Many bond-forming reactions of these elements have been reported, resulting in the synthesis of new organic compounds and materials that are of value in a variety of applications. Major areas of research in the past decade include dehydrogenative coupling reactions, double additions of Group 14 elements to unsaturated substrates, and synthesis of oligomeric and polymeric species composed of extended chains of Group 14 elements.

Compounds of the Group 14 elements are employed in many synthetic processes and also possess unique features that make them valuable in materials applications. Organosilanes are used in organic synthesis as reducing agents or alkylating agents, for protection of carbonyl groups, and as potential controllers of chemo-, stereo-, and enantioselectivity in reactions.[1] Organosilanes are also extensively employed as coupling agents across interfaces. A coupling agent stabilizes the bond between an organic polymer and a mineral substrate by functioning as a surface modifier, primer, or adhesive, enhancing the adhesion of the two dissimilar materials. Additionally, the silane may improve wetting and rheological properties, and can be used to strengthen the organic and inorganic boundary layers.[2] Silicones are fluids, elastomers, or resins formed from the hydrolysis and condensation of methylchlorosilanes. These materials are characterized by high temperature stability, hydrophobicity, good dielectric properties, and high compressibility. Industrial uses include lubricants, coatings, pump oils, antifoaming agents, etc.[1c,3] Furthermore, certain organosilicon compounds have value as medicines or medicine modificants.[1c,4]

In contrast to the extensive employment of silicon-containing compounds, organogermanes are less frequently used. Compounds of germanium are more costly than those containing either silicon or tin, and therefore the other analogous Group 14 compounds are often preferentially employed. However, unique applications do exist for the germanium derivatives. Study of the biological activity of organogermanes has indicated potential as antitumor agents, biological antioxidants, and radioprotectants, among others. Additionally, germanium compounds are valuable in high-performance material applications.[5]

Organotin compounds are key intermediates in organic synthesis,[6] as they are useful in a variety of carbon–carbon bond-forming reactions. For example, lithium and cuprate reagents may be formed from organostannane precursors. The tin compounds are also used in catalytic reactions of diols (or triols) and diisocyanates for the manufacture of polyurethane. Because of the biocidal properties of organotin compounds, they are important components in antifouling paints for ships and in wood preservation applications. Unfortunately, tin biocides may negatively affect the environment. The most extensive use of organostannanes may be in the stabilization of PVC to prevent degradation from heat and light.[6a]

A unique characteristic of polymers composed of extended chains of Group 14 elements is the delocalization of electrons through the $\sigma$-bond framework of the polymer backbone.[7] These polymers are known to absorb in the ultraviolet, with absorption maxima dependent both on main chain substituents and on chain length. Several potential applications exist, such as photoconductors, photoresists in microelectronics, photoinitiators for radical reactions, and precursors to ceramic materials.

This review summarizes recent work in transition metal-catalyzed bond-forming reactions of silicon, germanium, and tin. The focus is on advances since the mid-1980s, although earlier work is included where applicable. We have chosen to organize the material by type of bond formation, rather than by class of catalyst or reactant, as this seems to best characterize the goals of research in this field. Areas of research that are not fully included in this review are the Direct Process[3,8] for copper-catalyzed conversion of elemental silicon to organosilicon compounds and the hydrosilylation of unsaturated substrates,[9] as each has been extensively reviewed elsewhere.

## II

## FORMATION OF E–C BONDS

### A. Additions of E–H Bonds to Unsaturated Substrates

#### 1. Hydrosilylation

It is likely that more silicon–carbon bonds are produced by the hydrosilylation of olefins than by any other method except the direct process. This deceptively simple addition of an Si–H bond to a C–C multiple bond can be promoted by a variety of means, but transition metal catalysis is by far the most significant. Two relatively old catalysts, $H_2PtCl_6$ ("Speier's catalyst") and $Pt_2(Me_2ViSiOSiMe_2Vi)_3$ ("Karstedt's catalyst"), remain the most effective, and the remarkable rates and turnover numbers observed in these systems are among the most impressive in all of organometallic chemistry. The bulk of the literature on hydrosilylation falls outside the scope of this review, and readers are directed to the comprehensive work on hydrosilylation edited by Marciniec.[9a]

Perhaps one of the more controversial issues surrounding hydrosilylation has been the suggestion by Lewis at General Electric that platinum colloids are responsible for the bulk of hydrosilylation catalysis.[10] However, more recent work has largely refuted this proposal. In particular, Lewis and Stein and co-workers examined Karstedt's catalyst *in situ* with extensive studies that probed the relation of reaction kinetics and product distribution to catalyst structure in solution (NMR, EXAFS) and the formation of metal colloids (TEM). The clear conclusions were that catalytic hydrosilylation is a homogeneous process and that formation of colloids under silane-rich/olefin-poor conditions is associated with deactivation of the system.[11]

### 2. Hydrogermylation and Hydrostannylation

Although hydrosilylation has been studied extensively, much less research has been directed toward the analogous hydrogermylation and -stannylation of unsaturated organic compounds. Both reactions are known to proceed in the presence of catalytic amounts of free radical initiators, such as AIBN (azobis[isobutyronitrile]) or $Et_3B$, but this technique may require long reaction times and can result in low yields of complex product mixtures.[12] Lewis acid-catalyzed routes have also been investigated.[13] Alternatively, transition metal catalysis has been shown to be an effective means of adding Sn–H or Ge–H bonds to a variety of substrates.

Transition metal-catalyzed hydrogermylation of unsaturated substrates has been known for many years,[14] but has generated little interest in comparison to hydrosilylation. However, a few instances of transition metal-catalyzed hydrogermylation have been reported, including reactions of alkynes,[15] allenes,[12d] and dienes.[16]

The addition of triphenylgermane to terminal acetylenes is catalyzed by $Pd(PPh_3)_4$ to yield alkenyl triphenylgermanes. Although both the $\alpha$- and the $\beta$-regioisomers are obtained, the reaction proceeds with high stereoselectivity to yield only the ($E$)-isomer of the $\beta$-adduct [Eq. (1)].[15c]

$$RC\equiv CH \ + \ Ph_3GeH \ \xrightarrow[\text{THF, 25 °C, 5h}]{Pd(PPh_3)_4} \ \underset{\substack{\alpha\text{-adduct}}}{\overset{\displaystyle R\diagdown_{\phantom{x}}\diagup H}{\underset{Ph_3Ge\diagup^{\phantom{x}}\diagdown H}{C=C}}} \ + \ \underset{\substack{\beta\text{-adduct}\\ E\text{-isomer}}}{\overset{\displaystyle R\diagdown_{\phantom{x}}\diagup H}{\underset{H\diagup^{\phantom{x}}\diagdown GePh_3}{C=C}}}$$

$$R = C_{10}H_{21}, \ Ph, \ PhCH_2OCH_2CH_2,$$
$$HOCH_2CH_2, \ Me_3Si$$

$$(1)$$

Other palladium complexes, such as $PdCl_2(PPh_3)_4$, $PdCl_2(CH_3CN)_2$, and $Pd(OAc)_2$, were tested for catalytic activity in the system, but only $Pd(PPh_3)_4$ proved effective.

Rhodium(I) complexes have also been investigated as catalysts for the hydrogermylation of phenylacetylene. Selective formation of the hydro-germylated $\alpha$-adduct is observed in reactions of tributylgermane and phenyl-acetylene catalyzed by either $Rh(acac)(C_2H_4)_2$ or $Rh(hfa)(C_2H_4)_2$. Other rhodium complexes, including $RhCl(CO)(PPh_3)_2$, $[RhCl(C_2H_4)_2]_2$, and $RhCl(cod)]_2$, catalyze the hydrogermylation reaction, but lead mainly to $\beta$-adduct. These results indicate that the regioselectivity of the reaction is dependent on the electronic and steric characteristics of the ligands around the Rh center.[15d]

Similarly, doubly bonded unsaturated compounds are good substrates for catalytic hydrogermylation reactions. Allylic triphenylgermanes are formed exclusively when $Pd(PPh_3)_4$ is used to catalyze the reaction of various allenes with $HGePh_3$. The germyl group always adds to one of the terminal carbon atoms of the allene.[12d] A system composed of a nickel complex, $Ni(CO)_4$, and a phosphine additive, $PPh_3$, catalyzes the hydrogermylation of butadiene in good yield.[16a] In addition, hydrogermylation of substituted 1,3-dienes is observed in the presence of catalytic amounts of an alternate nickel complex with a phosphine additive, $Ni(acac)_2/PPh_3$, resulting in high yields of the 1,4-cis addition products.[16b]

The intramolecular hydrogermylation of 3-hexenoxydipropylgermane to give a five-membered ring product is catalyzed by $H_2PtCl_6$.[17] 5-Propyl-1-germa-2-oxacyclopentane is the main product, along with minor amounts of a six-membered ring product. However, the overall yield of the reaction is quite low (28%). Interestingly, when triethylborane is employed to catalyze the reaction, a high selectivity for the six-membered product is observed.

The hydrostannylation of alkynes is catalyzed by a range of palladium, molybdenum, and rhodium complexes. High yields of vinyl stannanes are obtained from reaction of $Ph_3SnH$ or $^nBu_3SnH$ under mild conditions. Although the reaction is highly stereospecific, the regiochemistry of the reaction appears to be dependent on both the metal complex and the substrate employed. For example, the reaction of terminal and internal alkynes without bulky substituents is catalyzed by palladium complexes such as $Pd(PPh_3)_4$, $Pd(OAc)_2$, and $PdCl_2(PPh_3)_2$, to yield mixtures of the $\alpha$-adduct and the $(E)$-isomeric form of the $\beta$-adduct.[15c,18] Increased regiose-lectivity is seen in palladium-catalyzed reactions of terminal alkynes with bulky or heteroatom substituents, or those conjugated with electron-withdrawing groups.[18b,19] Similar regioselectivity is observed in reactions in which bulky ligands are present on the palldium metal center.[20] For example, in the hydrostannylation of propargyl alcohols with tributylstan-nane catalyzed by $PdCl_2(PPh_3)_2$, approximately equal mixtures of $\alpha$- and $\beta$-regioisomeric products were obtained. If a more bulky ligand is present at palladium, as in $PdCl_2(o\text{-}tol_3P)_2$, an increased selectivity is observed for $\beta$-vinyl stannanes [Eq. (2)].[19b]

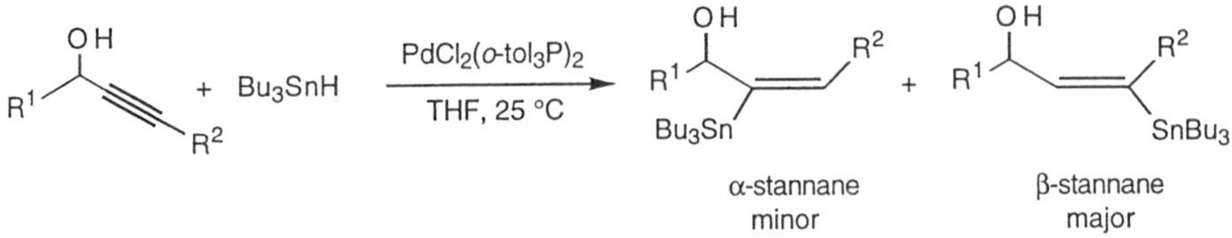

$$(2)$$

Rhodium[21] and molybdenum[18b] complexes are also effective catalysts for the hydrostannylation of terminal alkynes, interestingly resulting in preferential formation of the $\alpha$-regioisomer.

The hydrostannylation of allenes with tertiary stannanes is catalyzed by $Pd(PPh_3)_4$, giving allylstannanes in good yields.[12d,22] Three stannanes have been tested in the catalytic system, $Ph_3SnH$, $Me_3SnH$, and $Bu_3SnH$, and in all cases the reaction is highly regioselective. The reaction of aromatic allenes is highly stereospecific for production of only $(E)$-allylstannanes, although both $E$ and $Z$ isomeric products are obtained from aliphatic and alkoxyallene substrates. Stereoselectivity of the reaction of alkoxyallenes is dependent on steric considerations: for sterically bulky alkoxy groups or tin substituents, increased formation of the $Z$ isomer is observed. In contrast, the reaction is not stereoselective in the case of monosubstituted allenic ethers. Conjugated dienes react with regio- and stereoselectivity, giving 2-alkenylstannanes in good to excellent yields with the $E/Z$ ratio dependent on the substrate.[23] Sterically hindered dienes do not undergo hydrostannylation.

Although a number of acetylenic and allene compounds have been investigated for activity in the catalytic hydrostannylation system, very few reports exist that involve the reaction of simple alkenes. An ethylene organosilicon derivative, 2,6-dimethyl-2-vinyl-1,3-dioxa-6-aza-2-silacyclooctane, undergoes regioselective hydrostannylation to yield the $\beta$-adduct in the presence of catalytic amounts of $Rh(acac)(CO)_2$.[24] A palladium complex, $Pd(dba)_2/PPh_3$, catalyzes the hydrostannylation of oxabicyclic alkenes.[25] The regioselective addition of tributyltin hydride results in nearly quantitative yields of addition products with the $-SnBu_3$ moiety adding preferentially to the sterically less crowded carbon of the substrate. Interestingly, allyl alcohols give only very low yields when homogeneous palladium-catalyzed hydrostannylation is attempted, although high yields have been obtained with heterogeneous Pd/C systems.[26]

## B. *Dehydrogenative Silylation Reactions*

The well-established catalytic hydrosilylation of olefin substrates is in some instances accompanied by a dehydrogenative silylation reaction, yielding vinylsilane products in which the C–N double bond of the substrate has been preserved, as shown in Eq. (3).[27] Although the $E$ isomer of the product is predominantly formed, some $Z$-isomeric product is also observed in some cases. The hydrogen produced during the silylation is usually consumed in hydrogenation of a second equivalent of alkene.

$$2 \; R\diagup\!\!\!\diagdown \quad + \quad HSiR'_3 \quad \xrightarrow{\text{cat.}} \quad R\diagup\!\!\!\diagdown\!\!\!\diagup^{SiR'_3} \quad + \quad R\diagup^{CH_2\text{-}CH_3} \quad (3)$$

This type of dehydrogenative transformation was first reported in 1962 for the $Fe(CO)_5$-catalyzed reaction of hydrosilanes with various olefins.[28] In this work, it was shown that selectivity for the dehydrogenative silylation may be affected by the ratio of silane to alkene, as in reaction of ethylene with triethylsilane. Use of excess silane leads to high yields of only the saturated product, whereas excess alkene results in formation of only vinylsilane. In general, this silane to olefin ratio greatly affects the selectivity for dehydrogenative silylation, as excess substrate seems to be required to act as hydrogen acceptor. If an excess of alkene is not available, the result is often simple hydrosilylation. Many examples of the competitive hydrosilylation and dehydrogenative silylation of alkenes have been reported, involving a variety of transition metal catalysts.[29,30] The focus of recent work in this area has been to develop catalytic systems that display a high selectivity for dehydrogenative silylation.

Seki and Murai and co-workers have extensively studied the dehydrogenative silylation of alkenes in reactions catalyzed by metal carbonyl complexes. They found that $Ru_3(CO)_{12}$ is an effective catalyst for the highly selective synthesis of vinylsilanes from the reaction of hydrosilanes with aromatic alkenes, such as styrenes, substituted styrenes, and 2-vinylnapthalene.[31] A variety of trialkylsilanes can be used in the system, although reaction with dichloromethylsilane does not lead to vinylsilane formation. High yields of vinylsilanes are also obtained in reactions of certain aliphatic alkenes, e.g., 3,3-dimethylbut-1-ene. However, if an alkoxy or siloxy group is present at the allylic or vinylic position of the alkene, side reactions occur that involve C–O bond cleavage. Interestingly, if a hydrogen atom is present at the allylic position of the alkene, as in 1-hexene, a mixture of vinyl- and allylsilanes is obtained. Hydrosilylation is not generally observed in reactions catalyzed by $Ru_3(CO)_{12}$, although in some instances trace amounts of hydrosilylated products are formed.

In a more recent report from Seki and Murai, $Fe_3(CO)_{12}$ is shown to exhibit complete selectivity in the catalytic dehydrogenative silylation of styrenes.[32] No products resulting from hydrosilylation are observed with the iron complex catalyst, in comparison to the minor amounts of hydrosilylated product formed in reactions catalyzed by ruthenium and osmium carbonyl complexes. The iron catalyst is effective for various p-substituted styrenes, although other substrates, such as 1-hexene, give complex product mixtures that include only small amounts of vinyl- and alkylsilane.

The reaction of 1,5-dienes with hydrosilanes leads to dehydrogenative silylation when RhCl(PPh$_3$)$_3$ is used as catalyst.[33] The use of excess diene is important in obtaining the high selectivity for the dehydrogenated product. High yields of 1-silyl-1,5-dienes were obtained, along with hydrogenated alkene and, in some cases, small amounts of double-bond isomeric by-products. No simple hydrosilylation product was formed. The system appears to be specific to 1,5-dienes, as reaction of either 1,4- or 1,6-dienes does not result in exclusive formation of the analogous 1-silyl-1,$\omega$-dienes.

Seki and Murai and co-workers have also investigated the reactivity of $\alpha,\beta$-unsaturated esters with hydrosilanes in the presence of a number of transition metal complexes [Eq. (4)].[34]

$$(4)$$

With excess unsaturated ester substrate, Co$_2$(CO)$_8$ is an effective catalyst for the dehydrogenative silylation reaction, giving high product yields with high selectivity (ca. 90%). Various acrylic acid esters, methyl-, ethyl- and butylacrylate, are good substrates in this system, and hydrosilanes including HSiEt$_2$Me, HSiMe$_3$, and HSiMe$_2$Ph are all effective reagents. Interestingly, only those $\alpha,\beta$-unsaturated esters that do not have substituents at the $\beta$ position of the C=C double bond give products derived from dehydrogenative silylation, indicating that sterics influence the course of the reaction. A number of metal complexes known to be good hydrosilylation catalysts were investigated for reactivity under these conditions, i.e., with a methylacrylate : HSiEt$_2$Me ratio of 5 : 1. Both RhCl(PPh$_3$)$_3$ and RhCl(CO)(PPh$_3$)$_2$ give good yields of silylated products with good selectivity (ca. 70%) for the unsaturated product. Ruthenium and iridium complexes were found to be less active. The best catalysts were those with cobalt metal centers, Co$_2$(CO)$_8$ and the complex derived from reaction of Co$_2$(CO)$_8$ with hydrosilane, R$_3$SiCo(CO)$_4$.

Iridium complexes with O-donor ligand environments, Ir(triso)(ol)$_2$ [triso = tris(diphenyloxophosphoranyl)methanide; ol = C$_2$H$_4$, cyclooctene), catalyze the hydrosilylation and dehydrogenative silylation of ethylene with triphenylsilane.[35] Diphenylmethylsilane can also be used in the system, but a trialkylsilane, $Et_3SiH$, is inactive. Selectivity for the dehydrogenative silylation products, triphenylvinylsilane and ethane, is optimized at low ratios of silane to catalyst, as at higher ratios increased formation of the saturated product is observed. Interestingly, this iridium catalyst system is only effective in the case of ethylene; other alkenes, including styrene, methyl acrylate, 1-hepten, cyclooctene, and 3,3-dimethyl-1-butene, do not give silylated products from reaction with $Ph_3SiH$ even at elevated temperatures.

The highly selective dehydrogenative silylation of ethylene in the presence of $RuH_2(H_2)_2(PCy_3)_2$ has been reported.[36] Triethylsilane reacts with ethylene to give high yields of vinylsilane with approximately 97% selectivity. This high selectivity is obtained for the reaction in which ethylene is bubbled into a pentane solution of the catalyst before the addition of $Et_3SiH$. The reverse addition leads to lower selectivity (78%). This indicates the need for excess unsaturated substrate in the dehydrogenative reaction. The authors believe that the ruthenium bis(dihydrogen) complex is the catalyst precursor and that the active species is $RuH$ $[(\eta^3\text{-}C_6H_8)P(C_6H_{11})_2](C_2H_4)(PCy_3)$, derived from the dehydrogenation of a cyclohexyl ring of the initial complex.

Highly selective dehydrogenative silylation is also observed in the reaction of styrene with triethylsilane catalyzed by a cationic rhodium complex, $[Rh(COD)_2]BF_4$, with a phosphine cocatalyst.[37] The selectivity of the reaction was affected by four factors: ratio of styrene to silane, reaction temperature, hydrosilane, and added phosphine cocatalyst. Excess stryrene is again required to act as hydrogen acceptor, as hydrosilylation occurs at low ratios. At higher temperatures (70°C), the reaction is highly selective for the dehydrogenative silylation (92%) and a high overall yield is obtained (92%). In contrast, at room temperature the reaction gives a 19% overall yield of a product mixture of which only 17% is dehydrosilylated product. The selectivity of the reaction is also dependent on the hydrosilane used. In reactions with triethyl- or tri-$n$-propylsilane, dehydrogenative silylation is the predominant route (ca. 80% selectivity). Interestingly, dehydrosilylation is even more preferential in the case of bulkier silances, such as the 95% selectivity for reaction of $t$-butyldimethylsilane. The last factor in determining the selectivity of the reaction is the added phosphorous ligand. Dehydrogenative silylation is favored at higher ratios of phosphine to rhodium complex and also in the case of electron-withdrawing phosphine ligands, such as $PPh_2(C_6F_5)$.

Divinylsilane substrates undergo dehydrogenative silylation in the presence of $Rh(I)$ complexes, as shown in Eq. (5).[38] For the dehydrogenative reaction to occur, the substrate to silane ratio must be at least 2.

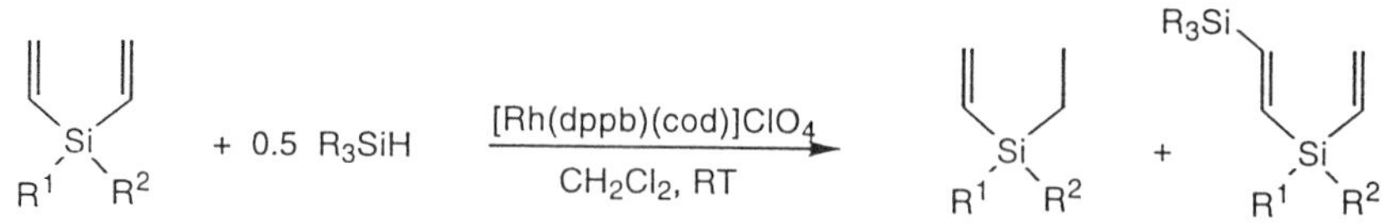

$$(5)$$

The reaction is successful for methyl($o$-tolyl)divinylsilane, dimethyldivinyl-silane, and diphenyldivinylsilane, giving the corresponding 1,2-disilylethene derivatives in good yield. Triethyl-, dimethylphenyl-, $t$-butyldimethyl-, and triethoxylsilane are all effective reagents in the system. Interestingly, prefer-ential activation of the allyl group vs the vinyl group of allydimethylvinyldis-ilane is observed. However, only hydrosilylation occurs in the reaction of diallyldimethylsilane or allyltrimethylsilane. A number of Rh(I) complexes are highly active in the system, including [Rh(dppb)(cod)]ClO$_4$, RhCl (PPh$_3$)$_3$, [RhCl(cod)]$_2$, and Rh(CO)$_2$(acac).

Allylsilanes are selectively synthesized in Ru(cod)(cot)-catalyzed dehy-drogenative silylations of olefins.[39] Reaction of hydrosilanes, HSiEt$_3$ or HSiEt$_2$Me, with alkenes such as 2-methyl-1-butene, 1-hexene, and ethyl ($E$)-2-butenoate results in good yields of allylsilanes. Only small amounts of vinylsilanes are detected in the product mixture. Although the overall mechanism is believed to be the same as in those dehydrogenative silylation reactions that lead to vinylsilanes, catalytic isomerization of the double bond of the starting alkene may occur before silylation, resulting in allylsilane formation. Silanes such as HSi(OMe)$_3$, HSi$^i$Pr$_3$, and HSiMe$_2$Ph undergo only slow, low-yield reactions.

An interesting variation of the dehydrogenative silylation system involves the platinum complex-catalyzed reaction of 1-alkenes with disilanes to pro-duce vinylsilanes.[40] In this system, one H atom and a silyl group are released by the reactants to yield the alkenylsilane product, rather than the two hydrogens released in reactions of hydrosilanes [Eq. (6)].

$$PhCH=CH_2 \;+\; FMe_2SiSiMe_2F \quad \xrightarrow[\text{PhLi}]{\substack{Pt(PPh_3)_4 \\ 150\ ^\circ C,\ C_6H_6}} \quad PhCH=CHSiMe_2Ph \;+\; Ph_2SiMe_2$$

$$(6)$$

Reactions of styrene and 1-hexene with 1,2-difluoro-1,1,2,2-tetramethyldisilane are effectively catalyzed by Pt(PPh$_3$)$_4$. However, only double silylation is observed for the reaction of ethylene with 1,2-diphenyl-1,1,2,2-tetramethyldisilane in the presence of Pt(PPh$_3$)$_4$. Changing the li-gand on the platinum center to one that is more electron donating and less sterically demanding, as in Pt(PMe$_3$)$_4$, leads to dehydrogenative silylation, although products from hydrosilylation and double silylation are also formed.

Dehydrogenative silylation has also been observed for terminal alkyne substrates. Doyle and co-workers reported in 1991 that a small amount (6%) of alkynylsilane was observed in the product mixture that results from reaction of phenylacetylene and $Et_3SiH$ catalyzed by $Rh_2(pfb)_4$.[41] The remaining components of the product mixture resulted from hydrosilylation. Crabtree and co-workers have found that in reaction of terminal alkynes with various tertiary silanes, dehydrogenative silylation can become the predominant route, depending on reaction conditions [Eq. (7)].[42]

$$R \!=\!\!\equiv\!\!=\! H \ + \ HSiR'_3 \ \xrightarrow{\ \ cat.\ \ } \ R \!=\!\!\equiv\!\!=\! SiR'_3 \ + \ RCH\!=\!CH_2$$

$$(7)$$

In the presence of catalytic amounts of $[IrH(H_2O)bqL_2]SbF_6$ (L = $PPh_3$; bq = 7,8-benzoquinolinato), high selectivity for dehydrogenative silylation is observed when the substituents on both the acetylene and the silane are sterically demanding. Additionally, dehydrogenative silylation is favored by a high alkyne : silane ratio, as the alkyne serves as the hydrogen acceptor in the reaction.

Conceptually related to the dehydrogenative silylation reaction is the general area of silyl group transfer or metathesis between olefins. Pioneered by Marciniec and co-workers, the basic reaction involves the silylation of a substrate such as styrene using a vinyl silane in the presence of ruthenium phosphine complexes, shown in Eq. (8c). The reaction of two olefins to yield two new olefins is highly reminiscent of the olefin metathesis process in which metal alkylidene complexes facilitate the cleavage and reassembly of the C–C doubly bonded framework. However, Marciniec has conclusively demonstrated that C–C redistribution is not occurring and that only a silyl group and a hydrogen are exchanged between olefins through a series of insertion and $\beta$-elimination processes [Eqs. (8a) and (8b)]. In a key mechanistic study, it was found that the reaction of $(C_6D_5)$ $CD\!=\!CD_2$ with $CH_2\!=\!CHSiMe_2Ph$ yields $E$-$(C_6D_5)CD\!=\!CD(SiMe_2Ph)$ and $CH_2\!=\!CHD$ during the initial stages of the reaction. Extensive isotopic scrambling occurs later in the reaction, presumably through reverse reactions with the initially produced ethylene-$d_1$. The extensive work on this topic is summarized in two reports.[43] In addition to providing a facile route for preparing new vinyl silanes, this work provides compelling evidence for the rapidity and reversibility of olefin insertion into metal–silyl bonds.

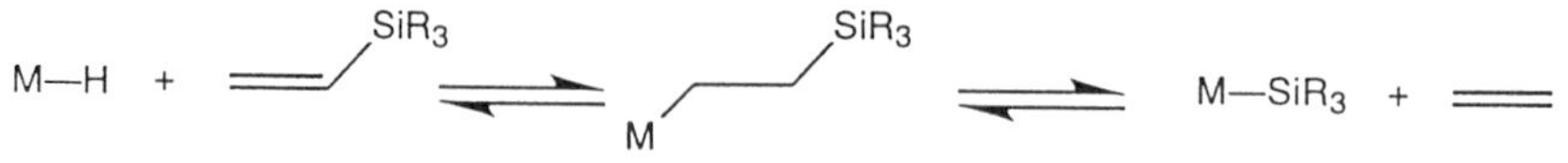

$$M{-}H \ + \ \underset{}{\overset{SiR_3}{=\!\!=\!\!=\!/}} \ \rightleftharpoons \ \underset{M}{\overset{SiR_3}{\diagup\!\!\diagup}} \ \rightleftharpoons \ M{-}SiR_3 \ + \ =\!\!=\!\!= \tag{8a}$$

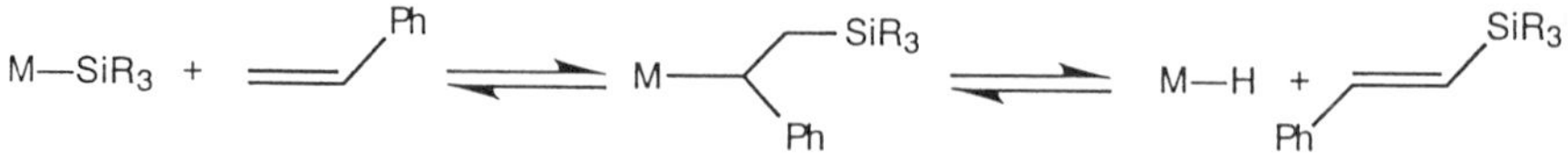

$$M{-}SiR_3 \ + \ \underset{}{\overset{Ph}{=\!\!=\!\!=\!/}} \ \rightleftharpoons \ M{-}\underset{Ph}{\overset{SiR_3}{\diagdown}} \ \rightleftharpoons \ M{-}H \ + \ \underset{Ph}{\overset{SiR_3}{/\!\!=\!\!=}} \tag{8b}$$

$$\underset{Ph}{/\!\!=\!\!=} \ + \ \underset{}{\overset{SiR_3}{=\!\!=\!\!/}} \ \overset{[M]}{\rightleftharpoons} \ \underset{Ph}{\overset{SiR_3}{/\!\!=\!\!/}} \ + \ =\!\!=\!\!= \tag{8c}$$

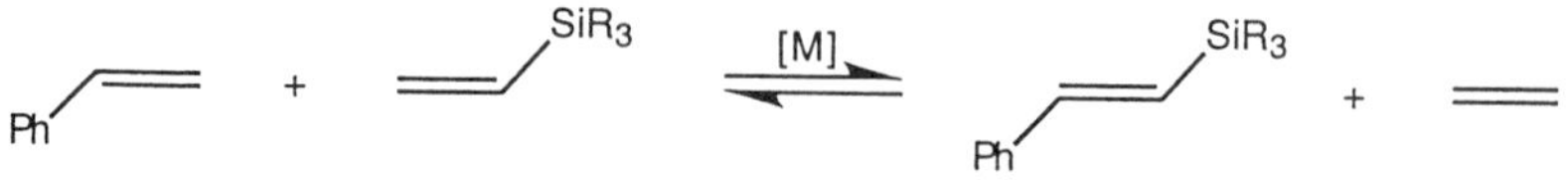

The most selective and active catalysts appear to be $RuCl(SiR_3)$ $(CO)(PPh_3)_2$ and $RuHCl(CO)(PPh_3)_3$. The former is by far the most active, and it has been shown that the latter, which is more readily available, must first lose a phosphine ligand by dissociation or oxidation before reaching optimum activity. The reaction of styrene with various vinyl silanes $R_3SiCH{=}CH_2$ [$R_3Si$ = $SiMe_3$, $SiMe_2Ph$, and $Si(OEt)_3$] at 110°C is regio- and stereoselective, producing 50–90% yields of the $E$-1-phenyl-2-silylethenes.

### C. Double Additions of Group 14 Atoms to Carbon–Carbon Multiply Bonded Substrates

#### 1. Double Silylation

The reactivity of compounds containing silicon–hydrogen bonds is known to be more analogous of that of dihydrogen than to that of similar compounds with carbon–hydrogen bonds. Ample precedent exists to support the casual statement that "silicon acts like a fat hydrogen," including similar catalytic additions of Si–H and H–H bonds to olefins and general trends in migratory aptitudes for organic and organometallic processes. Further extension of this analogy leads one to consider the addition of a Si–Si bond across an olefin, resulting in silicon–carbon bond formation. Indeed, the transition metal-catalyzed "double addition" of disilanes to unsaturated organic substrates is a highly efficient method of functionalization.[44] The first example of this double silylation reaction was reported by Kumada and co-workers in 1972, involving the nickel-catalyzed addition of a disilane to dienes.[45] A simplistic view of the process is presented in Eq. (9), although many variations have been reported.

$$R_3Si\text{-}SiR'_3 \ + \ R''C{\equiv}CR''' \ \xrightarrow{\text{[M]}} \ \underset{\underset{R'' \quad R'''}{|\quad\ \ |}}{R_3SiC{=}CSiR'_3} \qquad (9)$$

Several groups have screened a variety of transition metal complexes for activity in the double silylation system, but only compounds of nickel, palladium, and platinum appear to be viable catalysts. The key factor appears to be the involvement of a M(0) species, although certain M(II) complexes can also be used, presumably with *in situ* reduction to M(0). Generalizations regarding the activity of the different transition metal complexes are difficult, as many variables exist in each system. However, the most active complexes seem to combine palladium metal centers with dba, small basic phosphine, or isocyanate ligands.

The best substrates for double additions are unsaturated organic compounds such as dienes and alkynes. Alkenes are less reactive. In fact, only a few successful instances of double silylation have been demonstrated for alkenes not tethered to a disilane. The first concurrent addition of two silicons to a simple alkene such as ethylene was not reported until 1990.[46] Tanaka and co-workers investigated the reaction of ethylene with a series of disilanes, $XMe_2SiSiMe_2X$ (X = F, $OCH_3$, Cl, $CH_3$, $p\text{-}CF_3C_6H_4$, $C_6H_5$, $p\text{-}CH_3C_6H_4$), in the presence of $Pt(PPh_3)_4$. The yields of the 1,2-bis(silyl)ethanes closely follow the electronegativity trends: the disilane with the most electronegative substituent (X = F) gives nearly quantitative product yields, but a disilane substituted with a less electronegative group (X = $CH_3$) results in only an 18% yield. Interestingly, low yields (3.8–16.4%) are obtained for phenyl-substituted disilanes, despite the higher electronegativity of a phenyl group versus that of a methyl. In explanation of this result, the authors believe that the increased steric bulk of the phenyl substituent may lower the reactivity of the disilane. However, the yield of the phenyl derivative may be increased by altering the phosphine ligand of the platinum catalyst to one that is more electron donating and sterically less demanding, such as $PMe_3$. A variety of transition metal complexes were tested, but only those of type $Pt(PR_3)_4$ were found to catalyze the reaction. Even complexes known to be effective for the double silylation of acetylenes and dienes [e.g., $Pd(PPh_3)_4$, $PdCl_2(CN)_2$] do not give the desired product in the case of ethylene.

The reaction mechanism commonly accepted to account for the double silylation of unsaturated substrates involves three key steps. First, the disilane undergoes oxidative addition to the metal center, forming a transition metal–bis(silyl) complex. The unsaturated moiety inserts into the metal–silyl bond, followed by Si–C reductive elimination to give the double silylated product. Sakaki and co-workers have used ab initio MO calculations to determine the activation energies of individual steps in the model addition of $Si_2H_6$ to ethylene.[47] The most significant observation is an extremely high barrier for insertion of ethylene into a Pt–Si bond (45 kcal mol[-1]) arising from the strong *trans* influence of the silyl group. The authors believe this high activation energy for the rate-determining step explains the relatively low activity of simple alkenes toward catalytic double silylation. Other substrates, such as alkynes and dienes, may have lower barriers for the insertion step, which may explain the higher activity of these substrates.

Research in this field has mainly focused on disilane additions, although some studies have utilized tin and germanium compounds. Effective reagents include a range of Group 14 catenates. Interestingly, although peralkylstannanes are sufficiently reactive to undergo double additions, peralkylated disilanes have frequently given inferior results; one or more electronegative substituents are often required to activate the disilane. For example, in the palladium-catalyzed double silylation of bis(trimethylsilyl) butadiyne, reaction of $Me_3SiSiMe_3$ leads to only small amounts of the 1,2-addition products, whereas disilanes substituted with both halogen and alkyl groups lead to higher yields of either 1,4- or 1,2-double silylated products.[48] The usefulness of catalytic double addition would clearly be augmented by extension to high yield reactions for unactivated disilanes, and therefore, much work has been directed toward this goal. It has been found that functional groups such as phenyl and vinyl can enhance disilane activity. Cyclic disilanes are also more reactive, probably as a result of ring strain. Additionally, catalyst modification has proven to be successful in activating even peralkylated disilanes toward the double addition reaction.

Reports by Ishikawa *et al.*[49] indicate that a vinyl group serves to activate a disilane toward $Ni(PEt_3)_4$-catalyzed double silylation of conjugated dienes. In meistylene at high temperature ($\geq 150°C$), both 2,3-dimethylbutadiene and 1,3-cyclohexadiene react with 1,2,2,2-tetramethylphenylvinyldisilane to yield the 1,4-addition products in quantitative yield. The vinyl substituent on the disilane appears to be the key factor in this system, as reaction of pentamethylphenyldisilane under the same conditions does not result in adduct formation.

Although the phenyl group does not appear to provide sufficient activation when the nickel complex is used, it does activate the disilane in a system involving a palladium complex, as reported by Tsuji and co-workers in 1992.[50] Under carbon monoxide pressure, which is required to maintain catalytic activity, $Pt(CO)_2(PPh_3)_2$ is an effective catalyst for the double silylation of 1,3-dienes with a variety of phenyldisilanes. 1,4-Double-silylated products are obtained in high (51–91% isolated) yields, as shown in Eq. (10).

$$Me_2R^1Si\text{-}SiR^2Me_2$$

$$+ \quad \xrightarrow[\text{CO, 130 °C}]{Pt(CO)_2(PPh_3)_2} \quad Me_2R^1SiCH_2\overset{R^3}{\underset{|}{C}}{=}CHCH_2SiR^2Me_2 \quad (10)$$

$$\underset{H_2C{=}\overset{|}{C}\text{-}CH{=}CH_2}{R^3}$$

Contrasting results are obtained for the peralkylated disilanes $RMe_2SiSi-Me_2R$ (R = Me, $^nBu$, $^tBu$), which yield only trace amounts of the 1,4-addition products under the same reaction conditions. Attempts to increase double silylation yields by use of other platinum or palladium catalyst precursors under carbon monoxide pressure or inert atmosphere were also unsuccessful for these peralkylated disilanes. Additionally, the reaction of tetramethyl-1,2-divinyldisilane results in conversion to an intractable product mixture, with no incorporation of the 1,3-diene. Phenyl-substituted disilanes are also effective reagents in the $Pt(dba)_2$-catalyzed double silylation of phenylacetylene, but again, the alkylated disilanes and the vinyl-substituted disilane do not give double silylation products.

The authors propose that the influence of the phenyl group is not an electronic effect, as no product formation is observed with palladium catalysts known to be highly active in disilane systems substituted with electronegative elements. Rather, the phenyl group may allow for precoordination via a $\pi$-arene complex, which would accelerate the oxidative addition of an Si–Si bond to platinum, a key step in the proposed catalytic cycle.

Very different results were obtained by Tsuji and co-workers for the reaction of 1,3-dienes with alkyl, phenyl, and vinyl-substituted disilanes in the presence of $Pd(dba)_2$.[51] In this system, it appears that the solvent plays an important role in determining the course of the reaction. Double addition is now accompanied by the coupling of two diene molecules, a "dimerization-double silylation," previously observed in other systems by both the Sakurai[52a] and the West groups.[52b] In addition, the analogous dimerization-double stannylation has been investigated by Tsuji and co-workers (vide infra). The highly regio- and stereoselective reaction proceeds at room temperature with yields that are highly dependent on solvent, ranging from 19% in $CH_2Cl_2$ to 99% in dimethylformamide. No products resulting from 1,4-addition to a single diene are observed [Eq. (11)].

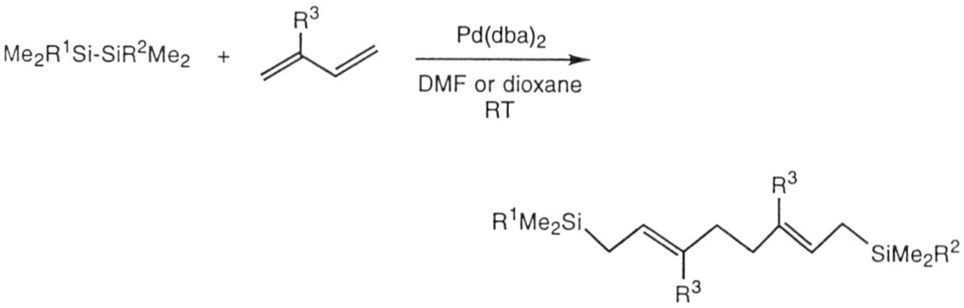

$$Me_2R^1Si\text{-}SiR^2Me_2 \quad + \quad \overset{R^3}{\diagup\!\!\!\diagdown} \quad \xrightarrow[\substack{\text{DMF or dioxane} \\ \text{RT}}]{Pd(dba)_2}$$

$$R^1Me_2Si\diagdown\diagup\diagdown\underset{R^3}{\diagup}\diagdown\diagup\overset{R^3}{\diagdown}\diagup\diagdown SiMe_2R^2$$

$$(11)$$

A total of eight other transition metal complexes were examined, but Pd(dba)$_2$ was found to be the most effective. Some of the other palladium complexes resulted in moderate yields, but no product was observed for any rhodium, ruthenium, or platinum complexes considered. Steric bulk of the 1,3-dienes also affects the reaction, as dienes with large substituents do not react even at elevated temperatures. This combination of Pd(dba)$_2$ catalyst precursor and appropriate solvent appears to greatly enhance the activity of the disilane, as those that were completely inactive in the earlier platinum-catalyzed system,[50] such as hexamethyldisilane, now give high yields of the dimerization-double silylation product. In addition, dimerization-double silylation does not occur with difluoro- and dichloro-tetramethyldisilanes, which would be expected to be highly active due to the presence of electronegative substituents.

Dimerization-double silylation has been extended to the synthesis of carbocycles from bis-diene substrates. Pd(dba)$_2$ is used to catalyze the regioselective reaction at room temperature, yielding carbocycles with allylic silane side chains [Eq. (12)].[53]

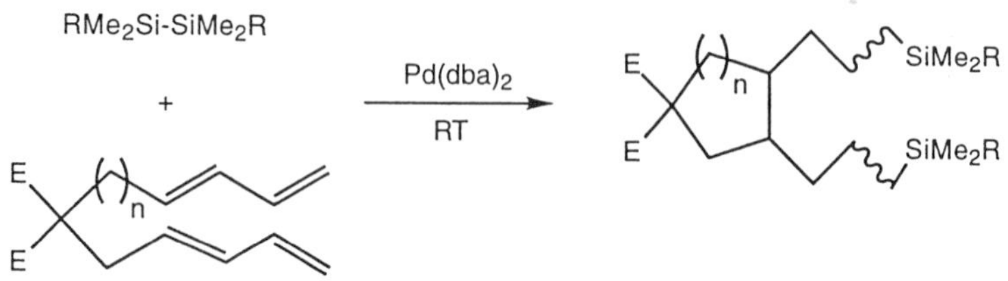

$$(12)$$

The success of the reaction, as well as its stereoselectivity, is dependent on both the disilane and the bis-diene employed. For example, the ethoxycarbonyl-substituted bis-diene reacts with either hexamethyldisilane or 1,2-diphenyltetramethyldisilane to give high yields of isomeric mixtures of the carbocycle products. In contrast, hexamethyldisilane does not have sufficient activity in reactions with the other bis-dienes studied. In addition, stereoselective formation of only one isomer is observed in some cases.

Disilanes connected via both the Si–Si bond and an organic or an organo-metallic linkage are activated toward reaction with unsaturated substrates to form cyclic bis(silyl) products. Reactions of 3,4-benzo-1,1,2,2-tetraethyl-1,2-disilacyclobutene with diphenylacetylene or benzaldehyde catalyzed by Ni(PEt$_3$)$_4$ proceed with addition across the multiple bond to form the ring-expanded product.[54] A second product is formed in a lesser amount in the case of diphenylacetylene, with insertion into the Si–C bond [Eq. (13)].

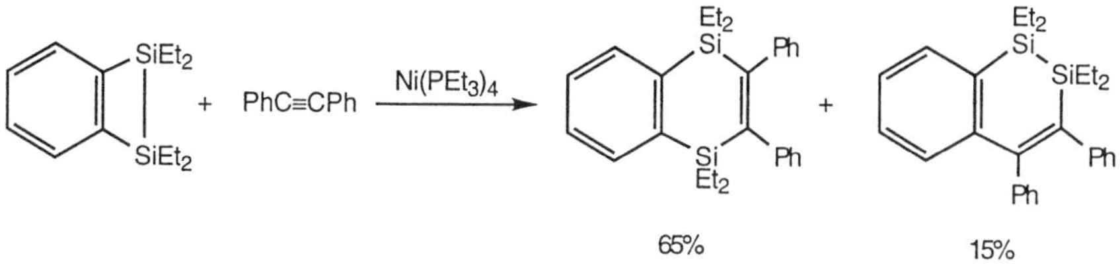

$$\tag{13}$$

Reactions of peralkylated 3,4-bis(alkylidene)-1,2-disilacyclobutane proceed with dependence on reactant silane and catalyst (Scheme 1).[55] The permethylated disilacyclobutane undergoes mono- and diinsertion of al-

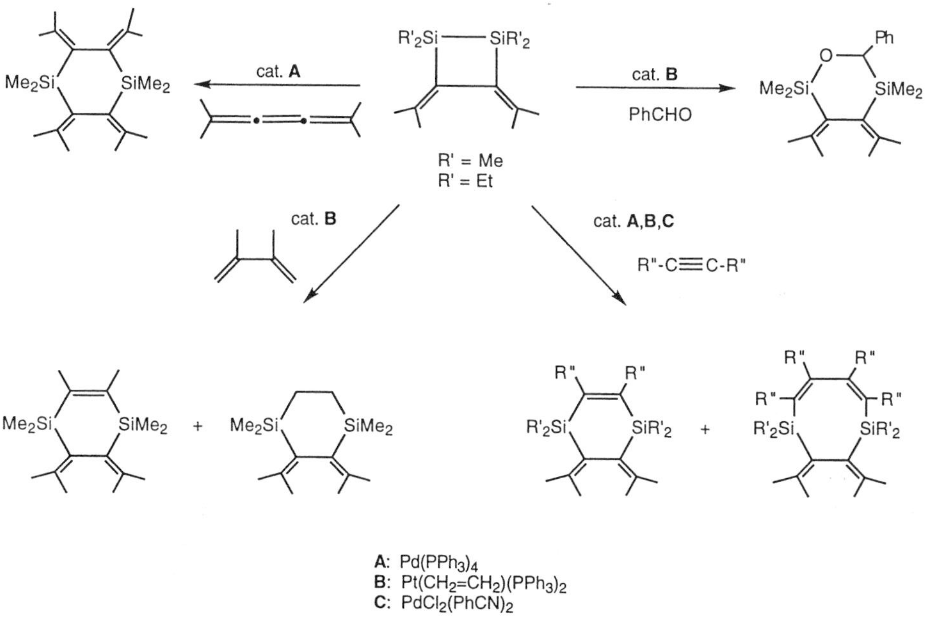

A: Pd(PPh$_3$)$_4$
B: Pt(CH$_2$=CH$_2$)(PPh$_3$)$_2$
C: PdCl$_2$(PhCN)$_2$

Scheme 1

kynes in the presence of $Pd(PPh_3)_4$ and $PdCl_2(PhCN)_2$, although use of $Pt(CH_2=CH_2)(PPh_3)_2$ only results in incorporation of one alkyne. Disilane metathesis can also occur in the presence of palladium catalysts, which may predominate over double silylation. For the perethylated disilane, both palladium and platinum complexes lead to the mono insertion product. Additions to butadiene or benzaldehyde proceed with moderate yield, but only the platinum complex is active. The minor product in the butadiene reaction mixture arises from insertion of ethylene derived from the catalyst. Tetramethylbutatriene selectively undergoes insertion at the central double bond to yield a 3:1 mixture of the twist and chair conformational isomers.

Manners and co-workers have demonstrated that ferrocene-containing silacarbocycles are readily obtained from the reaction of an ansa-ferrocenyl disilane with a number of terminal or activated internal alkynes and butadiene [Eq. (14)].[56]

1: R = R' = H
2: R = H, R' = Ph
3: R = R' = CO$_2$Me

(14)

Butadiene reacts to yield a dimerization-double silylation product that has been characterized by X-ray crystallography to reveal a *trans, trans* geometry. This is unusual, as transition metal-catalyzed insertion of unsaturated organic groups into Si–Si bonds regularly occurs with *cis* stereochemistry. However, $^{13}C$ and $^{29}Si$ nuclear magnetic resonance (NMR) point to the presence of a 7:3 mixture of the *trans, trans* and *cis, trans* isomers in solution. No reaction is observed for other dienes examined, including 1,3- and 1,5-cyclooctadiene and cyclopentadiene.

In the previous sections, double silylation was facilitated by using more reactive substrates such as alkynes and dienes with activated and strained disilanes. However, the use of palladium complexes with certain ligands allows successful reaction with the unactivated disilanes.

Prior to 1991, no high-yield double silylation had been reported using simple peralkyldisilanes such as hexamethyldisilane. In that year, Ito and Tanaka independently reported that palladium systems, with isocyanate and $P(OCH_2)_3CEt$ ligands, respectively, promote insertions into the Si–Si bond of unactivated disilanes.

Ito and co-workers obtained good to excellent yields of the double silylation products of terminal alkynes with hexamethyl- or 1,2-diphenyl-1,1,2,2-tetramethyldisilane when the Pd(OAc)$_2$/$t$-alkyl isocyanide system is used.[57] The reaction proceeds with high stereoselectivity, giving preferential formation of the $Z$ isomer. Internal alkynes, however, are unreactive. The choice of ligand is essential for catalysis; no reaction occurs when the isocyanide is not present. This system is also effective for the catalytic insertion of phenylacetylene into each Si–Si bond of permethylated trisilanes and tetrasilanes, giving regioisomeric products in good yields. In addition, the catalytic insertion of phenylacetylene into one or both Si–Si linkages of bis(disilanyl)alkanes occurs with use of the palladium/isocyanide catalyst [Eq. (15)].[58]

Me$_2$
Si—SiMe$_2$Ph
                                    Pd(OAc)$_2$          Me$_2$
        + Ph-C≡C-H                  $t$-OcNC            Si     Ph
                                    ———————▶                       +
                                    toluene, 90°C       Si    H
Si—SiMe$_2$Ph                                           Me$_2$
Me$_2$

                                                          **2**

                           SiMe$_2$Ph                         SiMe$_2$Ph
              Me$_2$                               Me$_2$
              Si         Ph                        Si        Ph
                                        +
              Si—SiMe$_2$Ph                        Si        Ph
              Me$_2$                               Me$_2$
                                                            SiMe$_2$Ph

                    **3**                               **4**

(15)

Reaction of bis(disilanyl)methane gives only trace amounts of the five-membered cyclic product, **2,** and higher yield (76%) of a mixture of the insertion products **3** and **4.** Substitution at the methylene tether leads to formation of **2** in good yield, with essentially no product from the insertion of two equivalents of phenylacetylene. Only mixtures of **3** and **4** are obtained from reaction of bis(disilanyl)alkanes tethered by longer chains.

Ito and co-workers have also used the Pd(OAc)$_2$/$t$-alkyl isocyanide catalyst to affect the double silylation of carbon–carbon multiple bonds in an *intra*molecular system to yield silacarbocycles.[59] Alkenes or alkynes that are tethered to a disilanyl group through a carbon chain, an ether linkage, or an amine functionality undergo intramolecular addition of the disilane moiety to the multiple bond. Activation of the disilane by the presence of electron-withdrawing groups on silicon is not necessary for the reaction to proceed. It is interesting to note that no competing intermolecular double silylation of the carbon–carbon multiple bond is observed, although independent intermolecular reactions do occur for carbon–carbon triple bonds.

A wide variety of disilanyl-substituted terminal alkenes undergo regio- and stereoselective intramolecular double silylation to yield cyclic bis(silyl) ring closure products [Eq. (16)].[59a,b,e]

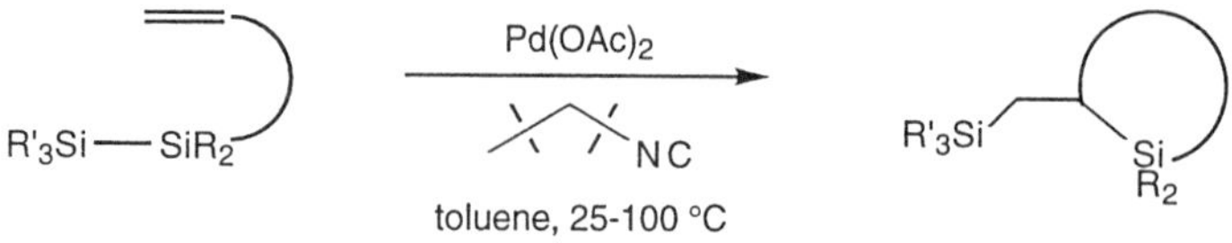

$$(16)$$

Only tethered terminal olefins are reactive, and ring junctures are always formed by coupling to the internal carbon of the multiple bond. If an asymmetric center is present in the tether, the reaction proceeds with high diasteroselectivity. Alkenes with substituents $\alpha$ to the double bond favor *trans* product formation, whereas $\beta$ substituents lead to *cis* products.

Similar results are observed in the intramolecular double silylation of alkynes.[59c,d] Both terminal and internal alkynes lead to exocyclic olefin formation. Interestingly, the reaction is not successful for internal alkynes tethered to a disilanyl functionality by a four atom chain, but is accomplished when four atoms link a disilanyl to a terminal alkyne. Internal alkynes with ester or olefin in conjugation with the C–C triple bond undergo chemoselective double silylation, and alkynes with substituents in the tether are also good substrates for the reaction.

Tanaka and co-workers also reported in 1991 that addition of the unactivated $Me_3Si$–$SiMe_3$ to terminal alkynes is catalyzed by the $Pd(dba)_2$ system in the presence of a particularly small basic phosphine ligand, $P(OCH_2)_3$-$CEt$.[60] Using this catalyst, it is also possible to insert acetylene into multiple Si–Si bonds in polysilanes and polycarbodisilanes [Eq. (17)].

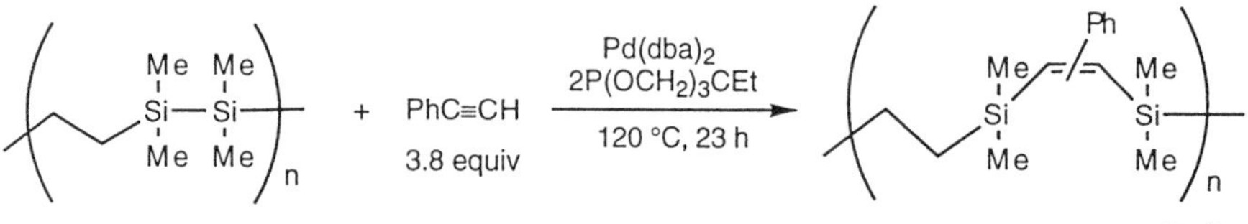

$$(17)$$

The extent of acetylene insertion can be limited by the control of reaction conditions, such as the ratio of reactant acetylene to the number of Si–Si bonds or reaction time. In addition, if an organic substrate with two alkyne functional groups is used, cross-linked polymers are formed. Other palladium catalysts and phosphine ligands investigated results in poor yields of the desired products. The authors attribute the enhanced activity of $Pd(dba)_2$-$2P(OCH_2)_3CEt$ to the reduced steric constraints of this phosphine ligand.

Another modification of the double silylation process reported by Tanaka and co-workers involves the use of a bis(hydrosilane) instead of a disilane as the reactant molecule.[61] This reaction can be described as a dehydrogenative double silylation, in that two Si–H bonds are activated rather than an Si–Si bond. The system is best catalyzed by $Pt(CH_2{=}CH_2)(PPh_3)_2$; other Pt, Pd, Ru, and Rh complexes give only very low yields of the double-silylated products. Alkynes, alkenes, and dienes undergo reaction with the bis(hydrosilane) with a range of results. Silicon–oxygen bonds and silicon–nitrogen bonds can also be formed by this method and are discussed in the appropriate sections later.

Nearly quantitative yields of the 1,2-double silylation product were obtained when internal alkynes were reacted with $o$-bis(dimethylsilyl)benzene in the presence of $Pt(CH_2{=}CH_2)(PPh_3)_2$ [(Eq. (18)].[61a]

$$\text{1a: } R^1 = R^2 = Ph$$
$$\text{b: } R^1 = R^2 = {}^nPr$$

2a: 98%
b:100%

$$(18)$$

Terminal alkynes give lower yields of the 1,2-addition product, and the 1,1-adduct is also formed. The length of the linkage between the two hydrosilane moieties is critical: of bis(hydrosilanes) with the general formula $HSiMe_2 ESiMe_2H$ [E = O, $(CH_2)_n$, n = 2, 3, or 4], only the compound with the -$CH_2CH_2$- linkage gave good yields of double addition to $^nPrC{\equiv}CPr^n$. The other bis(hydrosilanes) yield more simple hydrosilylation products.

The $Pt(CH_2{=}CH_2)(PPh_3)_2$-catalyzed dehydrogenative double silylation of olefins and dienes with $o$-bis(dimethylsilyl)benzene was also examined by Tanaka and co-workers.[61] The major product of the reaction with dienes, such as isoprene and penta-1,2-diene, is a result of 1,2-addition to the less substituted double bond. The reaction pathway for simple alkenes, shown in Eq. (19), appears to be dependent on the alkene substrate and, in some cases, on reaction temperature. Products resulting from 1,2-addition, **1,** and 1,1-addition, **2,** are detected for various substrates. In addition, hydrosilylation may occur to give the simple hydrosilylated product, **3,** or a by-product, **4,** derived from 1,4-migration of a methyl group in **3.**

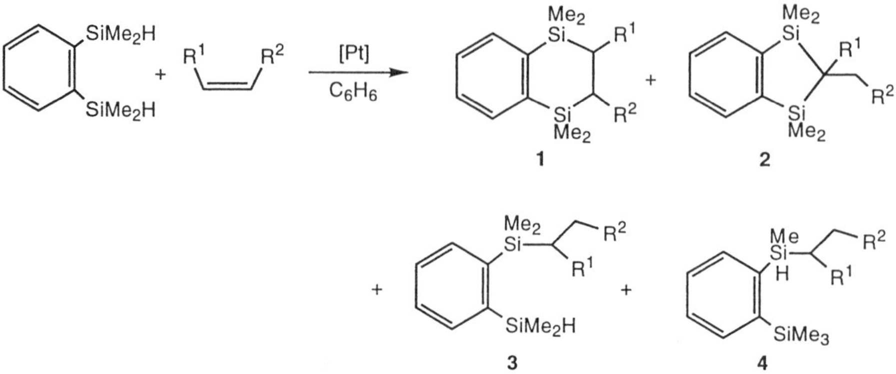

(19)

For example, the reaction of ethylene at 30°C yields both **1** and **2**, but the major portion of the product mixture is due to hydrosilylation. At 80°C, a higher selectivity for double silylation was observed. Similar temperature dependence was observed in the reaction of oct-1-ene, although formation of **1** is never observed. In contrast, the major pathway for reaction of styrene is 1,1-double silylation, independent of temperature. In general, internal olefins do not react with *o*-bis(dimethylsilyl)benzene even on heating of the reaction mixture, possibly as a result of steric hindrance. However, 1,2-double silylation does occur for the double bond of dimethyl maleate, which is presumably activated by the two electronegative ethoxy groups.

The dehydrogenative double silylation method is extended to the synthesis of fused cyclic polycarbosilanes by the condensation of various diynes with 1,2,4,5-tetrakis(dimethylsilyl)benzene.[62] Polymeric compounds are formed with elimination of $H_2$ [Eq. (20)].

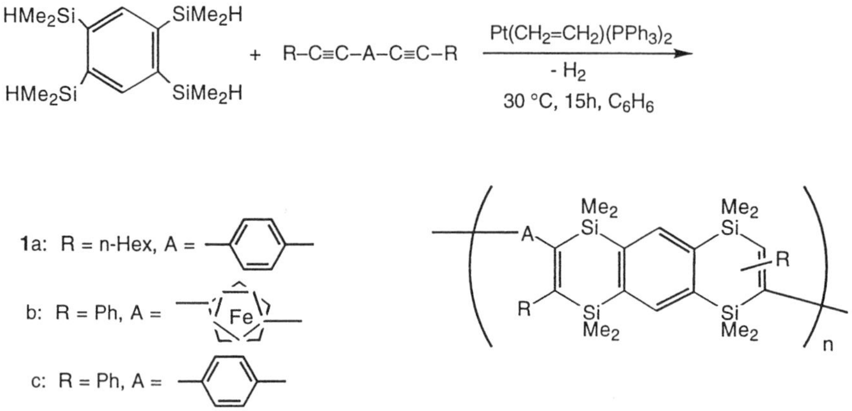

(20)

## 2. Double Germylation

Only a few examples of the double germylation of C–C multiple bonds have been reported in the past decade. The catalytic reaction of a digermirane with acetylenes results in the formation of ring-expanded digermacyclopentenes [Eq. 21)].[63]

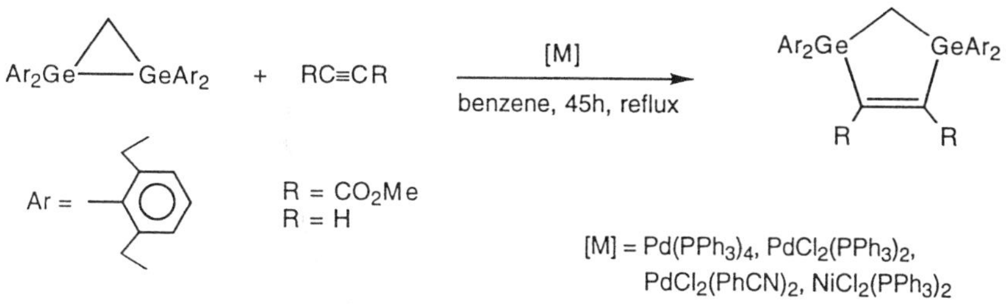

$$[M] = Pd(PPh_3)_4, PdCl_2(PPh_3)_2,$$
$$PdCl_2(PhCN)_2, NiCl_2(PPh_3)_2$$

(21)

Pd(0), Pd(II), and Ni(II) catalysts can all be used with equal success, although the active species is expected to be the M(0) complex generated by initial reduction of the metal center by the digermirane. Alternatively, the reaction of methyl acrylate gives a monogermylated ring-opened product. Acetylene insertion also occurs for digermiranes that possess a heteroatom such as sulfur or selenium in the ring. However, in this case the site of insertion is the Ge–X rather than the Ge–Ge bond [Eq. (22)].

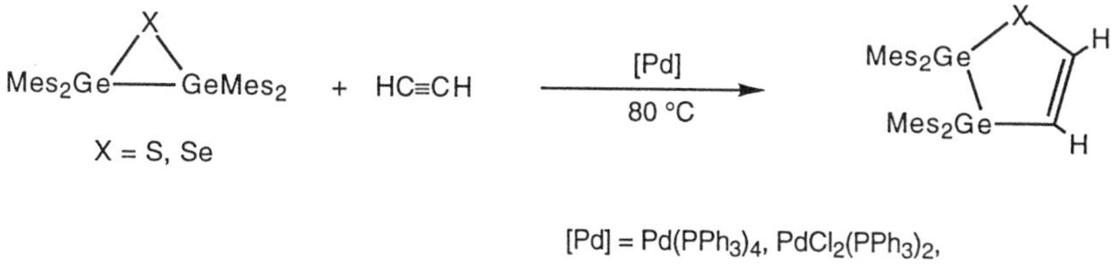

$$[Pd] = Pd(PPh_3)_4, PdCl_2(PPh_3)_2,$$

(22)

Tanaka and co-workers report that Pd(dba)$_2$-2P(OCH$_2$)$_3$CEt, highly active as a double silylation catalyst, is also effective for double germylation. High-yield reactions are observed for acetylenes and alkenes with 1,2-dichloro-1,1,2,2-tetramethyldigermane.[64] Although a variety of catalysts were investigated, the highest product yields were obtained with the Pd(dba)$_2$-2P(OCH$_2$)$_3$CEt complex. Complexes of general formula PdCl$_2$(PR$_3$)$_2$ were less active and showed dependence on the phosphine ligand. Interestingly, the selective formation of the Z or E olefin product from reaction of alkynes can be accomplished by the choice of catalyst. Use of Pd(dba)$_2$-2P(OCH$_2$)$_3$CEt gives only $E$ product in 72% yield, whereas Pt(PPh$_3$)$_4$ catalyzes formation of $Z$ isomer, but in low (20%) yield. Pd(dba)$_2$-2P(OCH$_2$)$_3$CEt also catalyzes the insertion of phenylacetylene into the Ge–Ge bonds of oligogermanes.[65]

### 3. Double Stannylation

In an interesting contrast to the analogous silicon and germanium systems, catalytic addition of the "unactivated" hexamethyldistannane to unsaturated substrate readily occurs with the common palladium catalysts Pd(PPh$_3$)$_4$ or Pd(dba)$_2$. In fact, prior to 1991 this was the only distannane investigated for addition reactions. The first example of transition metal-catalyzed double stannylation of unsaturated molecules, palladium-catalyzed $cis$ addition of hexamethyldistannane to 1-alkynes, was reported in 1983 by Mitchell and co-workers.[66] More recently, the system has been extended to include other substrates and both the hexaethyl- and the hexabutyldistannanes. All examples reported involve the use of a Pd(0) complex as the catalytic species.

The Pd(PPh$_3$)$_4$-catalyzed addition of Me$_3$SnSnMe$_3$ to allenes[67] proceeds with dependence on substrate and on reaction temperature. At lower temperatures, addition occurs across the more highly substituted double bond of the allene; at higher temperatures (ca. 85°C) the less congested product is formed. The reaction appears to be reversible, as decomposition to allene and distannane is observed. Double stannylation is also catalyzed by Pd(PPh$_3$)$_4$ in reactions of $\alpha,\beta$-acetylenic esters and $N,N$-dimethyl-$\alpha,\beta$-acetylenic amides with Me$_6$Sn$_2$, giving $Z$ and $E$ isomers of the addition products.[68] The $Z$ isomer, formed preferentially for $\alpha,\beta$-acetylenic ester substrate, is thermally unstable and cleanly rearranges to the $E$ isomer at 75–90°C. The palladium-catalyzed addition of Me$_6$Sn$_2$ to amides is slower and less efficient than that to esters, yielding a mixture of the $Z$ and $E$ double stannylation products.

Adding to their earlier work on the double stannylation of allene substrates with hexamethyldistannane, Mitchell and co-workers reported the double stannylation of allenes using hexaethyl- and hexabutyldistannane in 1991.[69] The reactions proceed in the same manner as for the methyl derivative, but require more forcing conditions. The reactivity of various cyclic allenes with R$_3$SnSnR$_3$ (R = Me, Et) was also examined [Eq. (23)].[70] The double stannylation proceeds in high yield to give the $cis$ and/or $trans$ isomers, depending on the length of the cyclic alkyl chain.

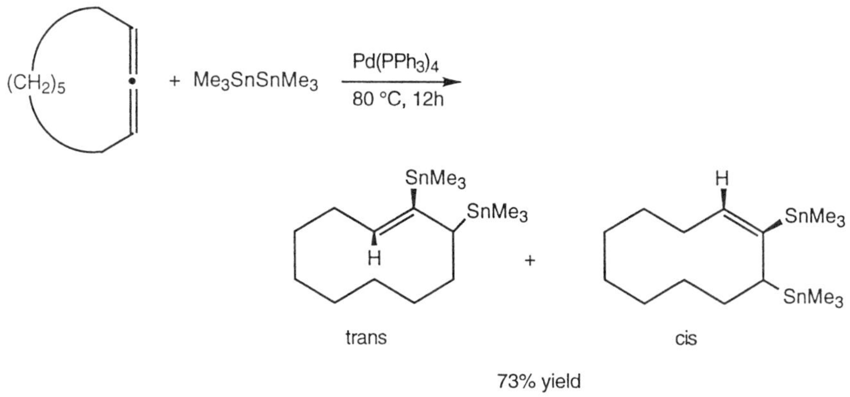

(23)

Two instances of the double stannylation of dienes have been reported by Tsuji and co-workers, both catalyzed by Pd(dba)$_2$. 1,3-Dienes undergo regio- and stereoselective dimerization-double stannylation with Me$_6$Sn$_2$ to form a single product.[71] Reactions do not occur for dienes with bulky substituents, such as 2-phenylbuta-1,3-diene. Steric bulk of the distannane substituents also appears to be important, as only hydrostannylation products are obtained with Bu$_6$Sn$_2$. In this case, a second molecule of diene may serve as the hydrogen source. In a later study, bis-dienes were shown to undergo regioselective double stannylation at the terminal positions. A single five-membered ring product with allylic stannane side chains is formed, having one $E$ allylic chain and one $Z$ allylic chain, rather than a mixture of products with $E,E$ or $Z,Z$ side chains as seen in analogous reactions of disilanes (vide infra).[53]

### 4. Silylstannation/Germylstannation

A logical extension of the addition of two Group 14 elements to an unsaturated substrate is the development of systems involving concurrent formation of bonds to two *different* elements. Research in this area has led to successful double additions of silylstannanes and germylstannanes, although it is interesting to note that no additions of silylgermane compounds have been described.

The silylstannylation of unsaturated organic moieties generally proceeds with high regioselectivity. With few exceptions, the silicon atom of the reactant molecule will always add to the less substituted carbon atom of an alkyne or to the central carbon atom of an allene. A new complication introduced with reactions of unsymmetrical Group 14 dimers is that disproportionation to the symmetrical dimers can also be catalyzed by the metal complex. This is not competitive with the addition of the mixed Group 14 dimers to highly active substrates, such as some alkynes, but may become a complication with bulky alkynes. Disproportionation appears to be a greater problem with $Pd(PPh_3)_4$ than it is with the $Pd(dba)_2/PR_3$ system.

Terminal alkynes undergo high-yield regioselective silylstannylation in the presence of $Pd(PPh_3)_4$, with addition of the stannyl group to the internal carbon atom to give regioisomer **1** in Eq. (24).[72] Only one exception to this regioselectivity has been reported: silylstannylation of methylpropiolate yields a 1:1 mixture of regioisomers **1** and **2** at temperatures between 25 and 115°C. However, the selectivity for **1** is greater than 90% if the reaction is run at 0°C.[72a]

$$RC{\equiv}CH \ + \ Me_3Si{-}SnMe_3 \ \xrightarrow[\text{60-70 °C}]{Pd(PPh_3)_4} \ \underset{\mathbf{1}}{\overset{R}{\underset{Me_3Sn}{>}}C{=}C\overset{H}{\underset{SiMe_3}{<}}} \ + \ \underset{\mathbf{2}}{\overset{R}{\underset{Me_3Si}{>}}C{=}C\overset{H}{\underset{SnMe_3}{<}}}$$

$$(24)$$

Silylstannylation generally proceeds with *cis* addition to acetylenes, yielding the *Z* olefins. Of many terminal alkynes examined, only two, acetylene and (trimethylsilyl)acetylene, give mixtures of *Z* and *E* olefins. The regiochemistry of the reaction is not affected by increasing the size of the R group on the stannyl moiety. In contrast, steric bulk does affect the comparative activity of the organic substrate, as alkynes with larger substituents are less reactive. A variety of substituents on the substrate are tolerated, but no adduct is formed if these substituents are in the propargyl position.[72c]

Silylstannylation is also observed in the $Pd(PPh_3)_4$-catalyzed reaction of alkynes with disilanyl stannanes, in which both Si–Si and Si–Sn bonds are present. The alkynes undergo insertion into the Si–Sn bond, to regio- and stereoselectively yield ($\beta$-disilanylalkenyl)stannanes [Eq. (25)].[73] With terminal alkynes, the stannyl group adds regioselectively to the internal carbon atom. Aliphatic alkynes are not reactive in this system.

$$R^1_3Sn{-}SiMe_2{-}SiMe_3 \ + \ R^2{-}C{\equiv}C{-}R^3 \ \xrightarrow[\text{40-60 °C, toluene}]{Pd(PPh_3)_4} \ \overset{R^1_3Sn}{\underset{R^2}{>}}C{=}C\overset{SiMe_2{-}SiMe_3}{\underset{R^3}{<}}$$

$$(25)$$

The initial $Pd(PPh_3)_4$-catalyzed addition of the Si–Sn bond of $Me_3Si$ $SnMe_3$ to 1,1-dimethylallene selectively places the silicon on the central allenic carbon, but no selectivity is observed for placement of the tin atom. However, heating the reaction mixture (90°C) in the presence of the same palladium catalyst results in an 80% preference for the regioisomer having the stannyl group on the less hindered carbon.[69,72a]

1,3-Dienes are good substrates for silylstannylation, although in this case Pd(0) complexes are found to be almost inactive. The best catalyst reported for reaction of dienes is $Pt(CO)_2(PPh_3)_2$.[74] The silylstannylation proceeds with high regio- and stereoselectivity, forming the $E$-1,4-product with allylic silane and stannane functionalities. Dienes with bulkier substituents result in lower yields. Bis-dienes react in an analogous fashion with silylstannanes as for distannanes and disilanes, resulting in carbocyclization and regioselective silylstannation [Eq. (12)].[53]

Tsuji and co-workers reported the only example of the silylstannylation of alkenes in 1993.[75] Both ethylene and norbornene undergo high yield insertions into the Si–Sn bond of a number of silylstannanes in the presence of catalytic amounts of $Pd(dba)_2$ in combination with a $PR_3$ ligand. The nature of the metal complex and cocatalyst employed in the system has an important effect on the outcome of the reaction. The combination of $Pd(dba)_2$ and $PBu_3$ gives high yields, but no reaction is observed if the phosphine ligand is absent from the system. Other less basic trialkylphosphine ligands lead to lower yields of the silylstannylated products. Although $Pd(PPh_3)_4$ shows high activity for the silylstannylation of alkynes (vide supra), only low-yield reactions occur in the case of alkenes. Two Pd(II) complexes, $PdCl_2$ and $PdCl_2(cod)_2$, show high catalytic activity in the system when combined with the $PBu_3$ cocatalyst. Steric bulk of the alkene substrate appears to affect the reaction, as no silylstannylation is observed for 1-hexene, styrene, cyclohexene, or cyclopentene. Additionally, silylstannanes with bulky alkyl substituents, $^tBuMe_2SiSnMe_3$ and $^tBuMe_2SiSnBu_3$, are not active in the system.

Only two studies of germylstannylation have been reported since the mid-1980s. The regioselective addition of $^nBu_3SnGeMe_3$ to $\alpha,\beta$-acetylenic esters is catalyzed by $Pd(PPh_3)_4$, with the formation of $E$ (major) and $Z$ (minor) isomers.[76] A later study probed the regio- and stereospecific palladium-catalyzed reaction of terminal alkynes with the same germylstannane.[77] Only a single product, the $E$ isomer, is formed in poor yield.

Germylstannanes react with allenes of general formula $RCH{=}C{=}CH_2$ to give products with regiochemistry dependent on the substituent on the organic substrate [Eq. (26)].[77]

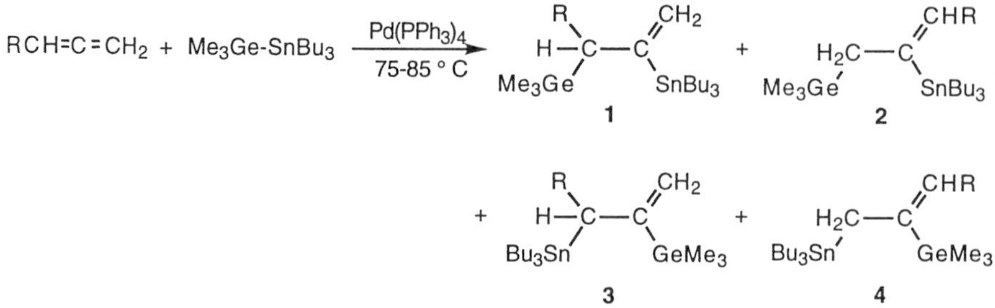

$$RCH=C=CH_2 + Me_3Ge\text{-}SnBu_3 \xrightarrow[75-85\,°C]{Pd(PPh_3)_4}$$

$$\begin{array}{ccc}
\underset{Me_3Ge}{\overset{R}{\underset{\underset{1}{}}{H-C-C}}}\overset{CH_2}{\underset{SnBu_3}{}} & + & \underset{Me_3Ge}{\overset{CHR}{\underset{\underset{2}{}}{H_2C-C}}}\overset{}{\underset{SnBu_3}{}}
\end{array}$$

$$+ \quad \underset{Bu_3Sn}{\overset{R}{\underset{\underset{3}{}}{H-C-C}}}\overset{CH_2}{\underset{GeMe_3}{}} \quad + \quad \underset{Bu_3Sn}{\overset{CHR}{\underset{\underset{4}{}}{H_2C-C}}}\overset{}{\underset{GeMe_3}{}}$$

$$(26)$$

For example, when R = OMe the regioisomeric vinyl tin compounds **1** and **2** are formed, but when R = *t*-Bu, only the allyl tin compounds **3** and **4** are generated. This is in contrast to the high regioselectivity observed in the analogous silylstannane reaction, in which the Si atom always adds to the central carbon atom. The regiochemistry may arise from steric influences on the intermediate in the catalytic cycle.

### 5. *1,1-Additions of E–E Bonds to Isonitriles*

Isonitriles react with oligo- and disilanes or silylstannanes to yield 1,1-addition products in which both Group 14 elements add to the same carbon atom of the substrate rather than across the C–N triple bond. Ito and co-workers report that palladium complexes, such as $Pd(PPh_3)_4$ and $Pd(OAc)_2$, effectively mediate these insertion reactions, an example of which is shown in Eq. (27).[78]

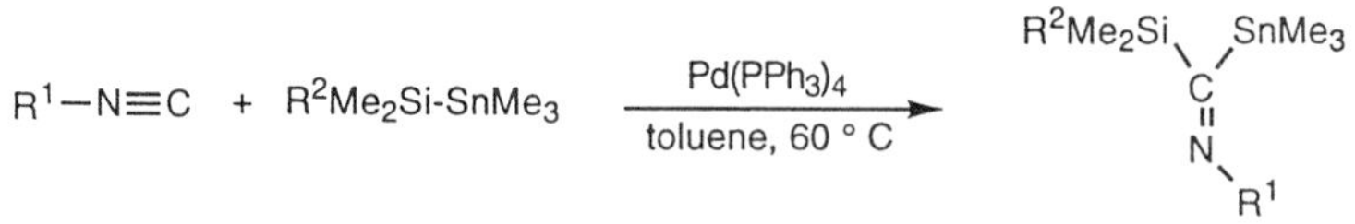

$$R^1-N\equiv C \; + \; R^2Me_2Si\text{-}SnMe_3 \xrightarrow[\text{toluene, 60 °C}]{Pd(PPh_3)_4} \underset{\underset{R^1}{\overset{\|}{N}}}{\overset{R^2Me_2Si \diagdown \diagup SnMe_3}{C}} \qquad (27)$$

In the case of oligosilanes, isonitrile insertion occurs at each Si–Si bond, giving moderate product yields. The reaction was attempted with a number of substituted isocyanides, with 2,6-disubstituted aryl isocyanides giving the most stable products. The degree of insertion into the Si–Si linkages of oligosilanes may be controlled by either varying the molar ratios of isocyanide to oligosilane or modifying the size of the substituents on the aryl ring of the isocyanide.

Interesting skeletal rearrangements are observed in the Pd-catalyzed reaction of tetrasilanes with 2,6-disubstituted aryl isocyanides [Eq. (28)].[78e,79]

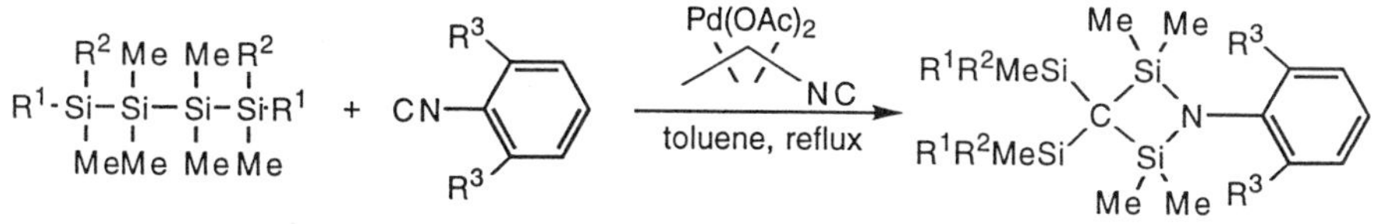

$$(28)$$

The reaction produces 3,3-disilyl-2,4-disila-1-azacyclobutane derivatives in moderate yields. Although the rearrangement is only observed for aryl isocyanides, addition of a *t*-alkyl isocyanide to the reaction mixture serves to increase product yields.

### D. *Reactions Producing Carbosilanes*

#### 1. *Dehydrogenative Coupling of Trialkylsilanes*

Procopio and Berry reported in 1991 that ruthenium phosphine complexes catalyze the formation of oligomeric carbosilanes by dehydrogenative coupling of alkyl silanes.[80] This appears to be the first example of a silane dehydrocoupling that produces Si–C rather than Si–Si bonds (cf. Section III,A). Although fairly stable to the catalytic conditions, $(PMe_3)_4RuH_2$ couples $HSiMe_3$ at a very slow rate ($<1$ turnover per day at 150°C). However, this reaction is of fundamental interest as it involves the catalytic activation and functionalization of an aliphatic C–H group.

Significantly, a new class of trimethylphosphine ruthenium complexes was prepared that exhibits appreciably greater activity ($>2$ turnovers per hour). These extremely stable seven-coordinate trihydride silyl complexes, $(Me_3P)_3Ru(H)_3(SiR_3)$, catalyze the reaction of trialkylsilanes to give oligomeric carbosilanes in $>90\%$ yield [Eq. (29)]. The oligomers contain up to ca. five silicon atoms and consist of mixtures of linear and branched isomers, indicating functionalization of more than one methyl group on a given silicon.[81]

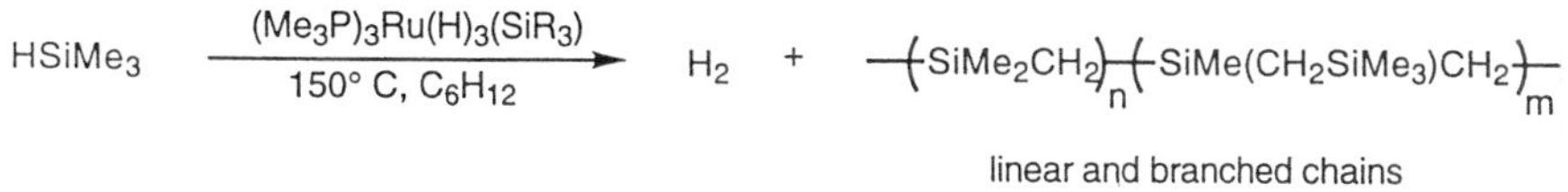

linear and branched chains

$$(29)$$

#### 2. *Transfer Dehydrogenative Coupling of Trialkylsilanes in the Presence of Olefins*

Preliminary studies of the mechanism of dehydrocoupling suggested that the principal turnover controlling process is *not* transfer of two hydrogen atoms from the alkyl silane to the ruthenium center, but rather subsequent elimination of $H_2$ from the metal. Furthermore, it was found that addition of a hydrogen acceptor such as $t$-butyl ethylene (3,3-dimethyl-1-butene) leads to an increase in the rate of dehydrocoupling of $HSiMe_3$ at temperatures as low as 80°C [Eq. (30)].[81]

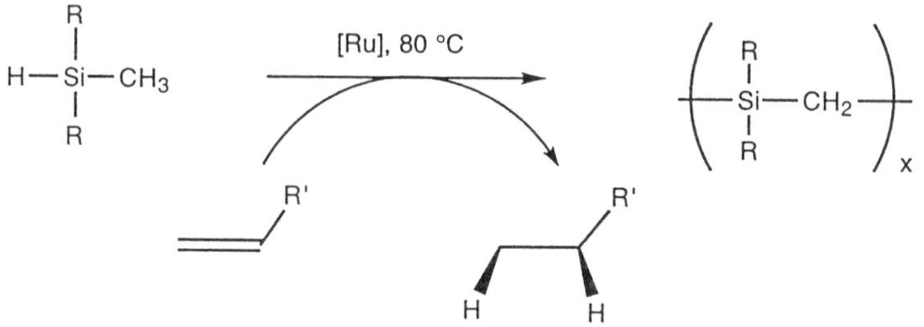

In addition to the lower reaction temperatures and greater rate, this process leads to appreciably larger oligomers with Mn = 700–800 Da and an average degree of polymerization of ca. 10. The dehydrocoupling catalyst, $(Me_3P)_3RuH_3(SiMe_3)$, does not competitively catalyze the hydrosilylation of $t$-butyl ethylene (<1%). Decomposition of the ruthenium complex appears to occur as absolute concentration of Si–H bonds drops during the reaction, ultimately limiting carbosilane molecular weight in this step polymerization process. However, appreciable turnover numbers (>1000) for production of the dimer, $HSiMe_2CH_2SiMe_3$, can be achieved by introducing multiple aliquots of $HSiMe_3$ and $t$-butyl ethylene as the reaction proceeds.

Ethylene coordinates too strongly to the ruthenium center to act as a hydrogen acceptor in this system, but slightly larger olefins do show activity. However, reaction selectivity is reduced compared with $t$-butyl ethylene. Thus, the use of $cis$ and $trans$ 2-pentene as hydrogen acceptor leads to products consisting of 90% carbosilane and 10% hydrosilylated olefin, and 1-hexene yields a carbosilane/hydrosilylation ratio of 4:6. In comparison, no hydrosilylation products are observed for dehydrocoupling in the presence of cyclohexene, but carbosilane formation is accompanied by disproportionation to benzene and cyclohexane as a side reaction.

Only silanes containing three alkyl or aryl groups undergo dehydrocoupling to produce carbosilanes. Secondary silanes such as $Me_2SiH_2$ are coupled in the presence of $t$-butyl ethylene to yield polysilanes containing ca. 25 silicon atoms.[82] In an interesting contrast, tertiary germanes do not undergo dehydrocoupling, but are instead coupled to polygermanes with loss of methane (cf. Section III,B).[83]

Two other systems have been discovered that exhibit activity for catalytic transfer dehydrogenation. The new class of compounds ($\eta^6$-arene) $Ru(H)_2(SiR_3)_2$ and the previously known compounds $Cp*Rh(H)_2(SiR_3)_2$ also catalyze the transfer dehydrocoupling of tertiary silanes using $t$-butylethylene as the hydrogen acceptor.[84,85] In particular, the rather hindered silane $HSiEt_3$ is coupled to the carbosilane dimer $HSiEt_2CHMeSiEt_3$ in reasonable yields [Eq. (31)]. The ruthenium and rhodium complexes also catalyze the heterocoupling of $HSiEt_3$ with arenes (vide infra), thus better yields of carbosilanes are obtained in nonaromatic solvents.

$$
\begin{array}{c}
\text{Et} \\
| \\
\text{H}-\text{Si}-\text{Et} \\
| \\
\text{Et}
\end{array}
\;+\; \underset{\underset{\displaystyle {}^t\text{Bu}}{\diagdown}}{\text{H}_2\text{C}=\text{CH}}
\;\xrightarrow[150\,°C]{[\text{Ru}]\ \text{or}\ [\text{Rh}]}\;
\begin{array}{c}
\text{Et}\;\;\text{Me}\;\;\text{Et} \\
|\quad|\quad| \\
\text{H}-\text{Si}-\text{C}-\text{Si}-\text{Et} \\
|\quad|\quad| \\
\text{Et}\;\;\text{H}\;\;\text{Et}
\end{array}
\;+\;
\underset{\underset{\displaystyle {}^t\text{Bu}}{\diagdown}}{\text{H}_3\text{C}-\text{CH}_2}
$$

(31)

Several mechanistic features are shared by the ruthenium-catalyzed silane dehydrocoupling process in the presence or absence of a hydrogen acceptor. Both appear to be initiated by the addition of an Si–H bond to the metal center, a step that renders subsequent (and more difficult) C–H bond activation steps *intramolecular* in nature. Second, the apparent ease of subsequent bond activations in metal silyl intermediates follows the order: $\alpha$-Si-H > $\beta$-SiC-H $\gg$ $\alpha$-Si-C. In contrast, the trend in the analogous reactions of germanes is inverted: $\alpha$-Ge-H > $\alpha$-Ge-C $\gg$ $\beta$-GeC-H.[83,86] In addition, the rate of the dehydrogenative reactions appears to be limited by the unfavorable equilibrium constant for loss of hydrogen. The concurrent removal of hydrogen with a hydrogen acceptor serves to increase the thermodynamic driving force for the desired reaction, effectively rendering the dehydrogenative reaction irreversible. Product distributions and isotopic labeling studies clearly indicate that the only C–H bonds activated are those adjacent to the silicon center in silanes that also contain an Si–H bond. This strongly suggests the mode of C–H activation is by $\beta$-hydrogen elimination to the metal in a silyl complex formed by prior Si–H addition. The $\beta$-H elimination from a metal–silyl to produce a silene complex or metallacycle is indirectly supported by the observation of selective H/D exchange processes in a related osmium phosphine system,[87] and in product analyses of rhodium and iridium complexes.[88] Formation of silene complexes by $\beta$-H elimination from isomeric silyl methyl complexes, $M\text{-}CH_2Si\text{-}Me_2H$, has been observed in several instances.[89] Direct evidence for $\beta$-H elimination from a ruthenium silyl has been obtained in a closely related model system in which the silene complex/metallacycle has been characterized *in situ* by multinuclear NMR techniques [Eq. (32)].[90]

$$(PMe_3)_4Ru(H)(SiMe_3) \;+\; BPh_3 \xrightarrow[\text{Me}_3\text{PBPh}_3 \downarrow]{\text{RT}}$$

$$
\begin{array}{c}
\text{Me}_3\text{P} \diagdown \quad \overset{\displaystyle \text{PMe}_3}{\underset{\displaystyle \text{PMe}_3}{|}} \quad \diagup \text{SiMe}_3 \\
\text{Ru} \\
| \quad \diagdown \text{H}
\end{array}
\;\; \rightleftharpoons \;\;
\begin{array}{c}
\text{Me} \\
\text{Me} \diagdown \overset{|}{\text{Si}} \\
\text{H} \diagdown \quad | \quad \diagup \text{CH}_2 \\
\text{H} - \text{Ru} \\
\text{Me}_3\text{P} \diagup \overset{|}{\underset{\text{PMe}_3}{}} \diagdown \text{PMe}_3
\end{array}
$$

$$(32)$$

Presumably, carbosilane formation in the catalytic cycle occurs via an analogous intermediate silene complex in which one of the two hydrides is replaced by a $SiMe_3$ group. Subsequent migration of the silyl group to the carbon of the silene would produce a new carbosilyl ligand.

### E. Reactions Forming Silicon–Arene Bonds

The ability of transition metal complexes to react with the C–H bonds of simple hydrocarbons has become well established over the past two decades. However, this knowledge has led to relatively few systems with the ability to catalytically functionalize C–H bonds.[91] This is unfortunate, as the potential use of C–H bonds in place of more reactive carbon–halide bonds would be particularly attractive in several industrial processes. The synthesis of arylsilanes, which are important intermediates in the silicone industry, is an excellent case in point. Although industrial needs for diphenyldichlorosilane and phenyltrichlorosilane are currently met through the Direct (Rochow) Process[3,8] and the Barry process,[92] most specialty arylsilanes and mixed aryl–alkyl silanes, including phenylmethyldichlorosilane, must be produced using Grignard or lithium reagents. The use of such reactive metal reagents introduces several limitations and expenses, including the growing costs of waste disposal. As a result, several research groups have focused on the development of new catalytic methods for the formation of arylsilanes.

In 1982, Curtis and co-workers reported that Vaska's complex promotes the formation of phenylsiloxanes from the reaction of hydridosiloxanes and benzene in a catalytic, albeit low-yield, process.[93] Catalytic arylsilane formation has also been reported by Tanaka and co-workers. Under photolytic conditions, $RhCl(CO)(PMe_3)_2$ catalyzes the C–H bond activation of arenes in reactions with hydrosilanes or disilanes, leading to the formation of new silicon–carbon bonds.[94] More recently, C–H bond activation of arenes resulting in arylsilane formation has been observed in the

$Pt_2(dba)_3$-catalyzed reaction of a series of substituted benzenes with $o$-bis(dimethylsilyl)benzene.[95] This intriguing process gives high yields of $o$-(aryldimethylsilyl)(dimethylsilyl)benzenes for a variety of arenes, but appears limited to this particular silane [Eq. (33)].

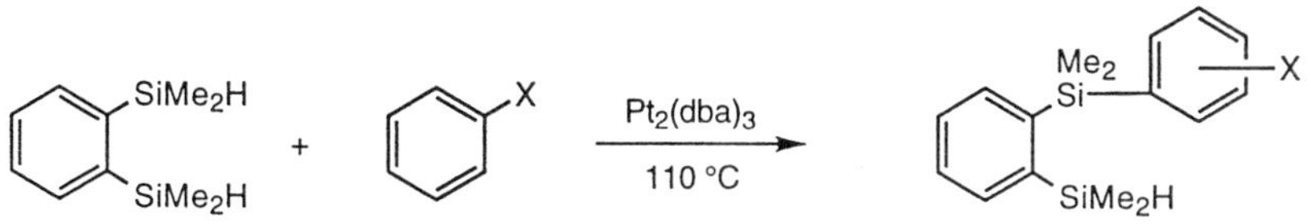

$$X = H, CH_3, Cl, OCH_3$$

(33)

Competition reactions between benzene and various substituted benzenes reveal that electron-withdrawing substituents on the arene increase reaction rates. All *ortho, meta,* and *para* forms of the arylsilanes are obtained. The major component of the arylsilane product formed in reactions of arylsilanes with electron-withdrawing groups, such as chlorobenzene, and anisole, is the *ortho* isomer. In contrast, toluene, which bears a less electronegative substituent, leads to *meta* isomer as major product.

In direct analogy to their work in the nickel-catalyzed double silylation of unsaturated substrates with 3,4-benzo-1,1,2,2-tetraethyl-1,2-disilacyclobutene, Ishikawa and co-workers have studied the catalytic formation of arene–silicon bonds. The $Ni(PEt_3)_4$ catalyst activates the C–H bond of the arene to result in net Si–H and Si–Ar bond formation [Eq. (34)].[96]

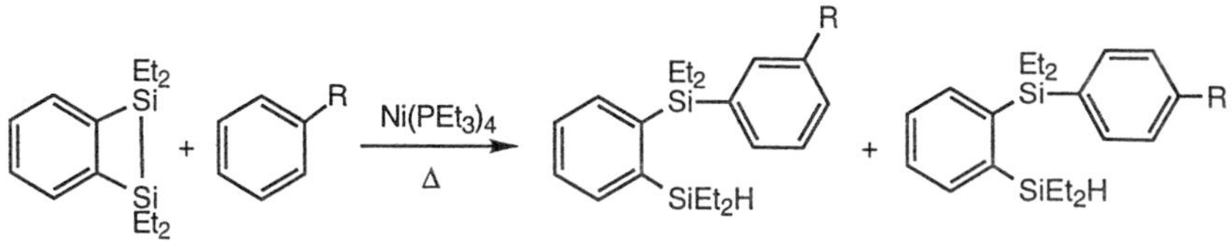

$$R = Me, CHMe_2$$

(34)

Perhaps the most industrially feasible approach has been developed by Rich and co-workers at General Electric, a palladium-catalyzed silylative decarbonylation reaction of aromatic acid chlorides with disilanes [Eq. (35)].[97] One of the silicon centers from the disilane is transferred to the arene whereas the other acts as a chloride acceptor to produce the chlorosilane. The basis for this process lies in previously discovered catalytic reactions of disilanes with chlorobenzene[98,99] or benzoyl chloride.[100] Although carbon monoxide is generally extruded during the reaction, the formation of varying amounts of acylsilanes is also observed depending on the disilane employed. Selective formation of phenylsilane vs acylsilane is observed in the reaction of aromatic acid chlorides with disilanes substituted with activating groups such as chlorides.

$$(35)$$

The catalyst system consists of $(PhCN)_2PdCl_2$ and a phosphine or amine cocatalyst. The cocatalyst is necessary for the reaction, except in the case of the most reactive acid chlorides, such as trimellitic anhydride acid chloride or $m$-nitrobenzoyl chloride. Although both phosphine and amine cocatalysts are effective in the system, greater reaction rates are observed with phosphines. The reaction does not proceed in the same manner for aliphatic acid chlorides.

The course of the reaction is affected by both aromatic and disilane substitution. In reactions with hexamethyldisilane, the ratio of arylsilane vs acylsilane product is dependent on the electronic characteristics of the arene substituent. Higher amounts of the desired product are obtained when highly electron-withdrawing substituents are present. An even greater effect on product formation is observed when methyl groups on the disilane are replaced by chlorines: if one or more chlorine atoms are present, no acylsilane is formed and the only silicon-containing aromatic product is arylsilane. In reactions with chlorinated disilanes, a good tolerance for substituents on the aromatic acid chloride is observed, with yields ranging from 15 to 80%. Acid chlorides possessing electron-withdrawing substituents have faster reaction rates and give higher yields. Unsymmetrically substituted chlorodisilanes react with preferential transfer of the least halogenated silicon atom to the arene.

As described in Section II,D, Berry and co-workers have reported catalytic activation and functionalization of alkylsilanes to produce oligomeric carbosilanes using homogeneous ruthenium and rhodium complexes, a process that involves the dehydrogenative coupling of C–H and Si–H bonds.[80,81,84b,87] In more recent work, this group has described an analogous catalytic transfer dehydrocoupling of arene C–H and silane Si–H bonds to produce arylsilanes [Eq. (36)].[85,101]

$$Et_3SiH + \overset{X}{\bigcirc} \xrightarrow[\underset{H \quad H}{\overset{^tBu}{\diagup\hspace{-0.3em}\diagup}}]{[cat.]} \overset{SiEt_3}{\underset{X}{\bigcirc}} + \underset{\substack{Et \quad CH_3 \quad Et}}{\overset{\substack{Et \quad \quad Et}}{H-Si-CH-Si-Et}}$$

(36)

This new C–H functionalization process is catalyzed by rhodium and ruthenium catalysts, $(\eta^5\text{-}C_5Me_5)Rh(H)_2(SiEt_3)_2$ and $(\eta^6\text{-arene})Ru(H)_2(SiEt_3)_2$, and their corresponding dimeric chloride precursors. In addition to the coupling of $Et_3SiH$ with arene, dimerization of $Et_3SiH$ results in the formation of a carbosilane, $Et_3Si\text{-}CHMe\text{-}SiEt_2H$. This material is the kinetic product of the coupling, whereas formation of arylsilane is thermodynamically preferred. The dehydrocoupling does not proceed in the absence of 3,3-dimethylbut-1-ene(*tert*-butylethylene) as hydrogen acceptor. Reaction rate and product selectivity are dependent on temperature (90–150°C) and arene substituents (H, CH$_3$, CF$_3$, F, Cl, Br). Arylsilane production is favored by increased reaction temperature and, in most cases, by electron-withdrawing groups on the arene. However, reactions with bromo- and chlorobenzene lead to competitive reduction of the substituted arene to benzene in addition to the formation of arylsilane. The steric bulk of the arene substituent plays a major role in determining the regiochemistry of C–H arene bond functionalization. Thus, silylation of toluene and $\alpha,\alpha,\alpha$-trifluorotoluene yields only the *meta* and *para* isomers, whereas the *ortho* isomer is also observed with fluorobenzene. The formation of carbosilane dimer and arylsilane is proposed to result from the competition between intramolecular activation of a silyl group C–H bond ($\beta$-H elimination to yield an $\eta^2$-silene) and intermolecular arene C–H bond activation. In the first case, silyl migration to the silene produces the Si–C bond of the carbosilane, and in the other instance the new Si–C bond is formed by reductive elimination of silyl and aryl ligands. Interestingly, although both ruthenium and rhodium complexes can be used as catalysts, a comparison of the two metal complexes reveals a higher selectivity for arylsilane when ruthenium is used. However, a slower reaction rate is also observed in this case.

In analogy to the catalytic reaction of disilanes with chlorobenzene[98] and arylbromides,[99] bromobenzene and iodobenzene undergo dehalogenative germylation with $ClMe_2GeGeMe_2Cl$ in the presence of $Pd(PPh_3)_4$.[102] The respective arylgermanes, $PhGeMe_2Cl$, are formed in fair yield. However,

the expected coproduct, $GeMe_2ClX$ ($X$ = Br, I), is not present or is observed in only trace amounts. Instead, both $PhGeMe_2X$ and $Me_2GeCl_2$ are
formed in significant yield. The authors have proposed various mechanistic
explanations for the appearance of these unexpected products, including
a possible germylene insertion pathway.

### F. Catalytic Bond-Forming Reactions with Concurrent Carbon Monoxide Incorporation

#### 1. Siloxymethylation

In 1977, Murai and co-workers described the catalytic addition of hydrosilane and carbon monoxide to an internal olefin to give enol silyl ethers in
which one molecule of CO is incorporated.[103–105] During the time period
covered by this review, the transition metal-catalyzed reaction of $HSiR_3$/
CO has been reported for many substrates. The catalytic system provides
a facile route to a number of materials that are valuable in organic synthesis.
The hydrosilane/CO system is very interesting, as different products can
be obtained depending on substrate, catalyst, and reaction conditions employed.

In 1991, the $Co_2(CO)_8$-catalyzed reaction of terminal olefins was reported
by Murai and co-workers. As two reaction sites are available in a terminal
alkene, four isomers of enol silyl ethers are obtained, each having one
additional carbon atom arising from CO incorporation [Eq. (37)].[106]

$$
R\diagup\diagdown \xrightarrow[\substack{50 \ (kg/cm^2) \\ C_6H_6, \ 140\ °C, \ 20 \ h}]{\substack{Co_2(CO)_8 \\ HSiEt_2Me \ / \ CO}} \quad \underset{Z \ and \ E \ isomers}{\overset{R}{\diagup\diagup}\text{OSiEt}_2\text{Me}} \quad + \quad \underset{Z \ and \ E \ isomers}{R\diagup\diagdown\diagup\diagdown\text{OSiEt}_2\text{Me}}
$$

$$(37)$$

Terminal dienes gave slightly different results. Both cyclic and linear enol
silyl ethers were produced in the reaction of 1,5-hexadiene, whereas for
1,7-octadiene only the straight chain $E$ and $Z$ isomers were produced. In
neither case is the branched product observed.[106]

When iridium complexes, $[IrCl(CO)_3]_n$ and $Ir_4(CO)_{12}$, are used to catalyze
the reaction of terminal alkenes with $HSiR_3$ and CO, good yields of enol
silyl ethers of acylsilanes are obtained. One molecule of CO and two molecules of silane are incorporated regioselectively at the terminal carbon atom
of the alkene to form a siloxy(silyl)methylene unit [Eq. (38)].[107]

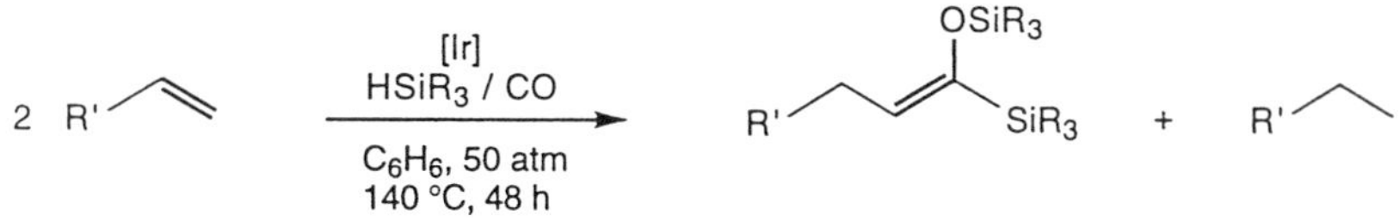

$$
2 \ R' \diagdown\diagup\diagdown \quad \xrightarrow[\substack{C_6H_6, \ 50 \ atm \\ 140 \ ^\circ C, \ 48 \ h}]{\substack{[Ir] \\ HSiR_3 \ / \ CO}} \quad R' \diagup\diagdown\diagup \overset{OSiR_3}{\underset{SiR_3}{|}} \quad + \quad R' \diagdown\diagup\diagdown
$$

$$(38)$$

This reaction is also a transfer dehydrogenative reaction, as two reactant hydrogen atoms are not incorporated into the enol silyl ether product but instead serve to hydrogenate another molecule of starting alkene. For example, in the reaction of vinylcyclohexane, ethylcyclohexane is obtained in equal amounts to the silylated product. Both iridium complexes effectively catalyze the reaction. Various silanes can be used, including diethylmethyl-, triethyl-, and dimethylphenylsilane. The reaction is successful for a range of terminal alkenes, even those bearing cyano, acetal, and epoxide functionalities. The $E$ isomer of the product is predominantly formed.

A variety of 1,6-diyne substrates have been investigated for activity in the $HSiR_3/CO$ system. A ruthenium complex, $Ru_3(CO)_{12,}$ with a tertiary phosphine additive catalyzes the formation of good yields of catechol products in which one molecule of diyne, two molecules of CO, and one or two molecules of silane are incorporated, as seen in Eq. (39).[108]

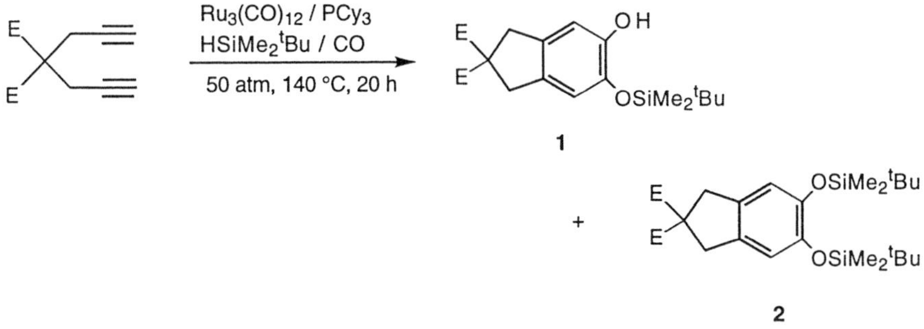

$$(39)$$

**1**

**2**

The product distribution depends on the reaction conditions. Selective formation of either **1** or **2** can be accomplished if the appropriate solvent and amount of silane are used. Functional groups such as ester, ketone, ether, and amide are well tolerated. The authors suggest that this system involves a different type of mechanism for CO incorporation than that seen in other reactions. In this case, the key catalytic step is proposed to involve an oxycarbyne complex as intermediate.

Siloxymethylation of cyclic ethers is accomplished using $Co_2(CO)_8$ under very mild reaction conditions, resulting in product formation in poor to excellent yields (14–96%) [Eq. (40)].[109]

$$
\begin{array}{c}
\text{R} \\
\triangle \\
\text{O}
\end{array}
\xrightarrow[\text{25 °C, 1 atm}]{\substack{\text{Co}_2(\text{CO})_8 \\ \text{HSiR}_3 \,/\, \text{CO}}}
\quad
\begin{array}{c}
\text{R} \\
\text{OSiR'}_3 \\
\text{OSiR'}_3
\end{array}
+
\begin{array}{c}
\text{R} \qquad \text{OSiR'}_3 \\
\text{OSiR'}_3
\end{array}
\qquad (40)
$$

The transformation proceeds with regioselectivity influenced by both steric and electronic attributes of the substrate. For most substituted compounds investigated, the presence of an electron-withdrawing group leads to ring opening mainly at the primary carbon center of the compound. Compounds possessing small substituents give a mixture of the primary and secondary ring-opening products. However, a substrate with a *tert*-butyl group yields only product from ring opening at the primary carbon, even at elevated temperatures. The reaction is tolerant of a variety of functional groups, although C=C moieties undergo hydrosilylation. Even germinally dialkyl-substituted oxiranes, in which tertiary carbon centers are present, will undergo silyformylation to give quaternary carbon centers. However, oxetanes with tertiary carbon centers only yield ring-opened silylated products, without CO incorporation.

The cobalt carbonyl complex is also an effective catalyst for the siloxy-methylation of aromatic aldehydes.[110] Arylethane-1,2-diol disilylethers are obtained in good yields, resulting from incorporation of one molecule of CO and two molecules of HSiR$_3$. Good selectivity for the siloxymethylation product is observed at 0°C in hexane. At 15°C, faster reaction rates are observed, but the selectivity for the CO-incorporated product is lower. In contrast, aliphatic aldehydes react under these conditions (1 atm CO, 0°C) to give only a small amount of CO-incorporated product, with a major product resulting from hydrosilylation.

Co$_2$(CO)$_8$-catalyzed reactions of benzylic acetates with trimethylsilane and CO proceed under mild reaction conditions to give trimethylsilylethers of β-phenethylalcohol in 43–76% yield. The highest yields are observed for benzyl acetates with electron-donating substituents.[111] Secondary alkyl acetates are also good substrates in the reaction system, yielding enol silyl ethers.[112] In addition, the cobalt complex is an effective catalyst for siloxy-methylation of five-membered cyclic *ortho* esters, as shown in Eq. (41).[113]

$$
\begin{array}{c}
\text{R'} \\
\text{O} \quad \text{O} \\
\text{OMe}
\end{array}
\xrightarrow[\text{25 °C, 1 atm, days}]{\substack{\text{Co}_2(\text{CO})_8 \\ \text{HSiR}_3 \,/\, \text{CO}}}
\quad
\begin{array}{c}
\text{R'} \\
\text{AcO} \quad \text{OSiR}_3
\end{array}
+
\begin{array}{c}
\text{R'} \\
\text{R}_3\text{SiO} \quad \text{OAc}
\end{array}
\qquad (41)
$$

The reaction is highly regioselective for substituted cyclic *ortho* esters, with ring opening mainly at the primary carbon center. However, reversed reactivity is observed for the phenyl-substituted derivative. Corresponding 1,4- and 1,5-diol derivatives are obtained in reaction of six- and seven-membered cyclic *ortho* esters.

Examples of the siloxymethylenation of nitrogen-containing compounds have also been reported by Murai and co-workers. The $[RhCl(CO)_2]_2$ complex catalyzes the reaction of enamines with $HSiEt_2Me$ under 50 atm of carbon monoxide. Regioselective incorporation of CO occurs at the $\alpha$-carbon of the enamines to yield $\alpha$-(siloxymethylene)amines, which can be subsequently hydrolyzed to $\alpha$-siloxy ketones. The highest yields are obtained for morpholine enamines, of which a number were investigated. Other rhodium complexes, $(Cp^*RhCl)_2$ and $Rh_6(CO)_{6,}$ also give good yields, but complexes of iron, cobalt, and ruthenium are ineffective.[114]

Rhodium-catalyzed reactions of N,N-acetals with $HSiR_3$ and CO were also reported in 1992.[115] Product formation is dependent on the structure of the substrate, revealing interesting reactivities for different acetals. For example, as seen in Eqs. (42) and (43), reaction of *N,N,N'N'*-tetramethylbenzylidenediamine, **1**, results in CO incorporation and cleavage of a C–N bond, but analogous reaction of *N,N,N'N'*-tetramethylmethylenediamine, **2**, yields a product resulting from incorporation of two CO molecules and transposition of an amino group. The expected by-product of the reaction of acetal **2**, $MeEt_2SiNMe_2$, appears to play an important role, and when $Me_3SiNMe_2$ is used as an additive the product yield improves to 75%.

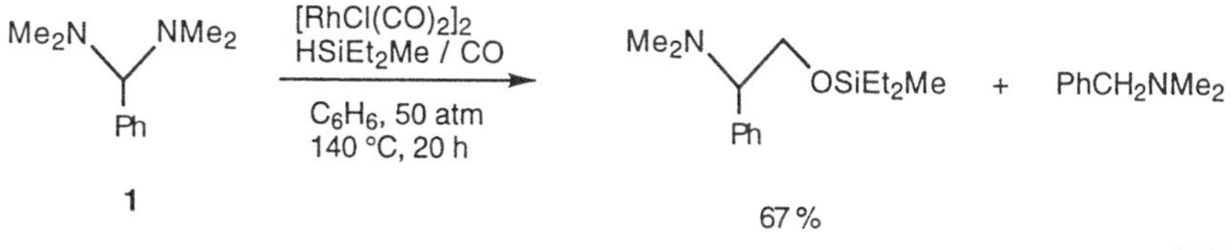

$$(42)$$

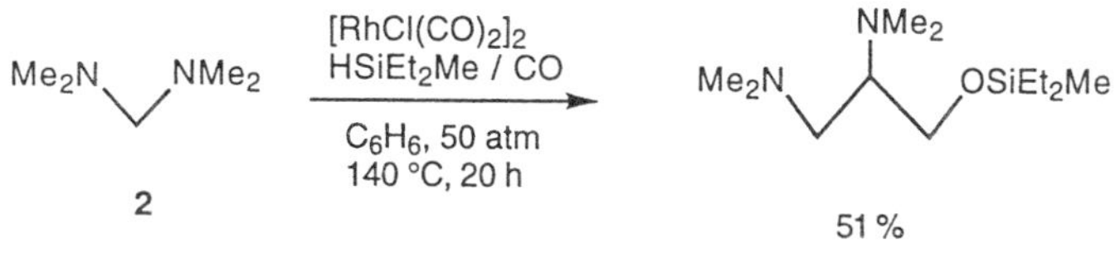

$$(43)$$

The reaction of five-membered cyclic N,N-acetals results in ring expansion with CO incorporation to give cyclic enediamines with no silyl moiety. No reaction is observed for alkyl-substituted cyclic N,N-acetals. Cyclic N,O-acetals give siloxymethylated ring-opened products, but corresponding O,O-acetals are unreactive.

## 2. Silyformylation and Germylformylation

Murai and co-workers reported the silylformylation of aliphatic aldehydes in 1979.[116] In this version of the transition metal-catalyzed reaction of HSiR$_3$ and CO with various substrates, a formyl moiety is always present in the final product of the reaction. Murai utilized the Co$_2$(CO)$_8$ complex with a triphenylphosphine cocatalyst to catalytically form $\alpha$-siloxy aldehydes from aliphatic aldehydes. An excess of reactant aldehyde is required to obtain the formyl products; if silane is in excess, 1,2-bis(siloxy)olefins are produced.[117]

More recently, Wright reported the [RhCl(COD)]$_2$-catalyzed silylformylation of aldehydes, with high yields of the $\alpha$-siloxy aldehydes under mild conditions [Eq. (44)].[118]

$$
\underset{R \quad H}{\overset{O}{\|}} \quad \xrightarrow[\text{thf, 23 °C, 24 h}]{\substack{[\text{RhCl(COD)}]_2 \\ \text{HSiMe}_2\text{Ph, CO}}} \quad R\text{—}\underset{CHO}{\overset{H}{|}}\text{—OSiMe}_2\text{Ph} \qquad (44)
$$

Reaction with dimethylphenylsilane is catalyzed at room temperature under 250 psi of carbon monoxide. Other silanes tested, triethyl- and triphenylsilane, are not effective reagents in this system. A variety of aldehydes are good substrates for the reaction, including benzaldehyde, substituted benzaldehydes, and heterocyclic aldehydes. Aliphatic aldehydes also yield $\alpha$-siloxy aldehyde products, but the reaction must be run at higher CO pressure (1000 psi) to avoid hydrosilylation. The reaction does not tolerate substrates bearing strong electron-withdrawing substituents, such as $p$-nitrobenzaldehyde.

Ring-opening silylformylation has been observed by Murai and co-workers in reactions of cyclic ethers. When the cobalt complex Co$_2$(CO)$_8$ is used as the catalyst in reactions of epoxides, an excess of substrate is required to prevent further reaction of the product siloxy aldehyde.[119a] Further investigation led to the discovery of [RhCl(CO)$_2$]$_2$/1-methylpyrazole as an effective catalyst combination for the reaction of oxiranes[119b] and oxetanes.[119c] For example, oxetane undergoes silylformylation to give 4-(dimethylphenylsiloxy)butanal in 81% yield [Eq. (45)].

$$
\underset{O}{\square} \quad \xrightarrow[\substack{\text{toluene} \\ \text{50 atm, 50 °C, 20 h}}]{\substack{[\text{RhCl}_2(\text{CO})_2]_2 \\ \text{1-methylpyrazole} \\ \text{HSiMe}_2\text{Ph, CO}}} \quad \text{PhMe}_2\text{SiO}\diagup\diagdown\diagup\diagdown\text{CHO} \qquad (45)
$$

Using the rhodium system, siloxy aldehydes are obtained as products in good yield even in the absence of excess substrate. 1-Methylpyrazole was found to be the best cocatalyst, as other amines do not lead to similar product yields. The authors suggest that the role of the amine is to accelerate the rate of carbon monoxide incorporation while suppressing possible side reactions, such as hydrosilylation.

Independent discovery of the silylformylation of alkynes was reported by the Matsuda and Ojima groups. The general reaction involves addition of both CO and tertiary hydrosilane to an alkyne to yield silyl alkenals, catalyzed by rhodium or rhodium–cobalt mixed metal clusters [Eq. (46)].

$$R^1 \!\!\!=\!\!\! R^2 \ + \ R_3SiH \quad \xrightarrow[\text{[Rh] or [Rh/Co]}]{\text{CO}} \quad \underset{OHC \qquad SiR_3}{\overset{R^1 \qquad R^2}{\diagup\!\!=\!\!\diagdown}} \ + \ \underset{R_3Si \qquad CHO}{\overset{R^1 \qquad R^2}{\diagup\!\!=\!\!\diagdown}}$$

$$(46)$$

As investigated by Matsuda and co-workers, $Rh_4(CO)_{12}$-catalyzed reactions of terminal or internal alkynes with CO and hydrosilane result in good to excellent yields of the Z and E isomers of the silyl alkenal products.[120] The reaction proceeds in the presence of 1 mol% of rhodium complex under CO pressure ($10–30$ kg/cm$^2$) at 100°C. Use of a base additive such as $Et_3N$ improves product yield and Z selectivity in most cases. Carbon monoxide incorporation is not observed when either $Co_2(CO)_8$ or $Ru_3(CO)_{12}$ is used. The reaction is regiospecific for the addition of the silicon group to the terminal carbon atom of 1-alkynes, producing 3-silyl-2-alkenals. In reactions of internal alkynes, the regioselectivity of the reaction appears to be affected by the steric bulk of the alkyne substituents: preferential formylation of the carbon atom of the triple bond that bears the bulkier substituent is observed, with the exception of those bearing strong electron-withdrawing groups. Silylformylation does not occur for some alkynes with sterically bulky substituents, such as 3,3-dimethyl-1-butyne, 1-phenyl-2-trimethylsilylethyne, and 1-trimethylsilylpropyne. Additionally, both silylformylation and hydrosilylation products are obtained in reactions of some internal alkynes with bulky substituents.

Substituted propargyl alcohols and propargyl amines are also good substrates for silylformylation catalyzed by $Rh_4(CO)_{12}$. Interestingly, by control of the reaction conditions the silylformylation can be modified to yield $\alpha$-(triorganosilyl)methylene-$\beta$-, $\gamma$-, and $\delta$-lactones or $\beta$-lactams via a cyclocarbonylation reaction [Eq. (47)].[121]

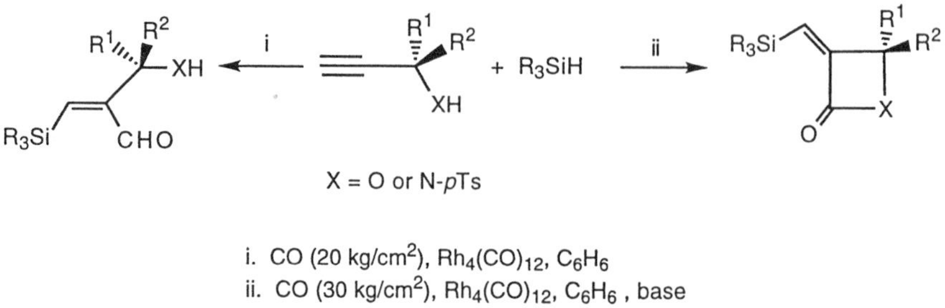
i
ii
XH
+ $R_3SiH$
$R_3Si$
CHO
O
X
X = O or N-pTs
i. CO (20 kg/cm$^2$), $Rh_4(CO)_{12}$, $C_6H_6$
ii. CO (30 kg/cm$^2$), $Rh_4(CO)_{12}$, $C_6H_6$ , base

(47)

Steric bulk of the substrates appears to play a role in determining which product is formed, but the key factor that favors formation of the cyclic product is the presence of a base cocatalyst, such as $Et_3N$ or DBU. Although silylformylation can occur for certain substrates when a base is part of the reaction mixture, in the absence of a base only silylformylated products are obtained. The silyl group always adds to the terminal carbon of the acetylene in the formation of either product.

Ojima and co-workers have also extensively studied the catalytic silylformylation of alkynes, employing a number of rhodium and rhodium–cobalt mixed metal complexes.[122] The reaction of 1-hexyne with hydrosilanes under either ambient pressure or 10 atm of CO is catalyzed by a variety of metal complexes to yield ($Z$)-1-silyl-2-formyl-1-hexene, **1,** and/or ($E$)-1-silyl-1-hexene, **2** [Eq. (48)].[122a,c]

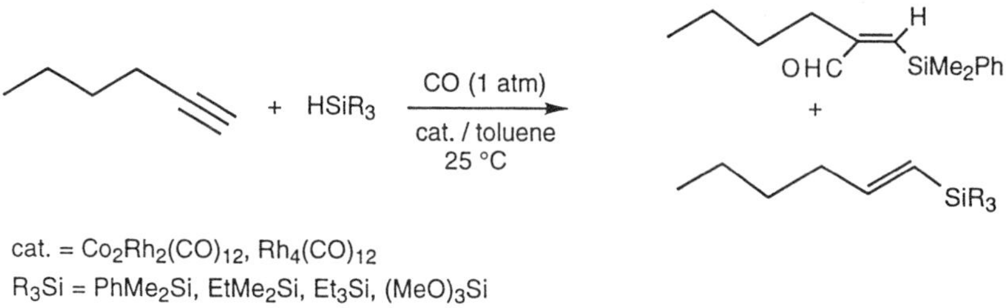
+ $HSiR_3$
CO (1 atm)
cat. / toluene
25 °C
H
OHC
$SiMe_2Ph$
+
$SiR_3$
cat. = $Co_2Rh_2(CO)_{12}$, $Rh_4(CO)_{12}$
$R_3Si$ = $PhMe_2Si$, $EtMe_2Si$, $Et_3Si$, $(MeO)_3Si$

(48)

The outcome of the reaction is affected by hydrosilane, CO pressure, and choice of catalyst. The hydrosilane substituents influence the selectivity of the reaction, i.e., relative ratio of silylformylation to hydrosilylation. For example, silanes with phenyl substituents, such as $Me_2PhSiH$, $Ph_2MeSiH$, and $Ph_3SiH$, lead exclusively to silylformylation products, but at least 80% of the product derived from trimethoxysilane is due to hydrosilylation. Trialkylsilanes generally give 40 : 60 mixtures of hydrosilylation and silylformylation products. The carbon monoxide pressure under which the reaction is run also effects the reaction. Although both $Rh_4(CO)_{12}$ and $Co_2Rh_2(CO)_{12}$ give similar results at 25°C and 1 atm of CO, with formation of substantial amounts of side products, superior results are obtained with the mixed metal catalyst at higher CO pressures, giving high yields of the silylformylation products with 93–100% selectivity (except in the case of trimethoxysilane). In contrast, large amounts of side products are still obtained at 10 atm CO pressure when $Rh_4(CO)_{12}$ is used, and selectivity for silylformylation is not quite as high. A comparison of six different rhodium and cobalt complexes indicates that the best silylformylation catalyst under these reaction conditions is $(^tBuNC)_4RhCo(CO)_4$, followed closely by $Rh(acac)(CO)_2$, $Co_2Rh_2(CO)_{12}$, and $Rh_4(CO)_{12}$. The complexes, $RhCl(PPh_3)_3$ and $Co_2(CO)_8$, are completely inactive.

The Ojima group has extended their studies of silylformylation to include more complex substrates, such as alkenyne, dialkyne, alkynyl nitrile, and ethynyl pyrrolidinone. Use of rhodium or rhodium–cobalt metal complexes catalyzes the silylformylation of these substrates with high chemoselectivity, as the other functionalities present are inert to the reaction.[122b,c,d]

A different rhodium complex, rhodium(II) perfluorobutyrate, was employed by Doyle and Shanklin in their work on the silylformylation reaction.[123] This is an effective catalyst for silylformylation, resulting in high yields of silylalkenal products (generally >70% following distillation) for a range of alkynes. Under mild conditions, the reaction proceeds with high stereoselectivity to give mainly the $Z$-isomeric form of the products ($ZE:\geq 10$). Functional groups such as alcohols, ethers, and esters are well tolerated.

Two studies of an intramolecular variation of the silylformylation reaction have been reported. In a series of substrates having the acetylene and the hydrosilane as part of the same molecule, silylformylation proceeds intramolecularly with high regio- and stereoselectivity to yield the exocyclic aldehydes in high yields [Eq. (49)].[124]

n = 1,2                        [Rh] = $Rh_4(CO)_{12}$ / $Et_3N$; (1,5-COD)$Rh^+(\eta^6\text{-}C_6H_6BPh_3)^-$
$R^1$ = H, $C_2H_5$, n-$C_4H_9$, Ph
$R_2$ = H, $CH_3$; R = $CH_3$; R' = $CH_3$, Ph

(49)

Both $Rh_4(CO)_{12}$ and the zwitterionic $(1,5\text{-COD})Rh^+(\eta^6\text{-PhBPh}_3)^-$ are effective as catalysts in this system. The reaction is run under 20 atm of CO at temperatures of 40–90°C. Two factors appear to affect the product yields for reactions of terminal alkynylsilanes: the substituents on the silicon atom and the size of the silacycloalkane framework. Higher yields are obtained for substrates containing -SiMePhH rather than -SiPh$_2$H groups, and for alkynylsilanes that result in six-membered versus five-membered ring products. Internal alkynylsilanes with three or four atom linkages separating the reactive centers also undergo the intramolecular silylformylation reaction to give product aldehydes in high yields. The reaction is highly regioselective to give only the *exo* cyclic isomer in all cases. Interestingly, the observed regiochemistry is opposite that of the intermolecular reaction, with formylation of the terminal carbon of the triple bond in terminal alkynylsilanes.

A second example of intramolecular silylformylation involves reactions of terminal and internal alkynes linked to a dimethylsiloxy group, as depicted in Eq. (50).[125]

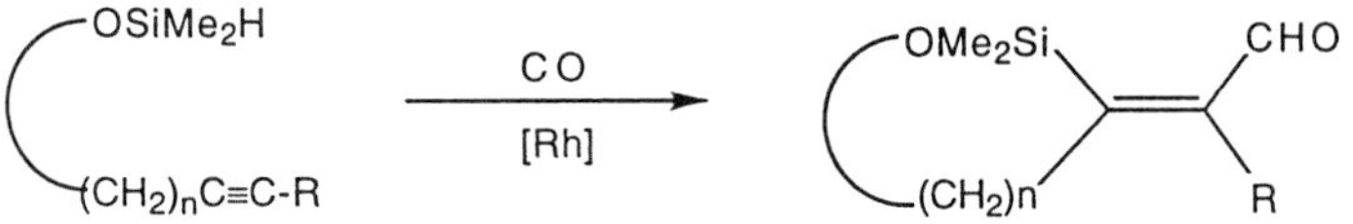

$$\text{(50)}$$

R = H, alkyl

[Rh] = ($^t$BuNC)$_4$RhCo(CO)$_4$, Rh$_2$Co$_2$(CO)$_{12}$, Rh(acac)(CO)$_2$

The $\omega$-siloxy-3-silyl-2-alken-1-al products are obtained in good to excellent yields, again displaying a reverse regioselectivity from the intermolecular version of the reaction. Cyclic molecules, such as $O$-(dimethylsilyl)-2-ethynyl-1-cyclohexanol, are also good substrates for the intramolecular reaction, giving high yields of the exocyclic products. The best catalyst of those investigated in this system is ($^t$BuNC)$_4$RhCo(CO)$_4$.

A surprising variation on the silylformylation reaction has been reported by Zhou and Alper. A zwitterionic rhodium(I) complex, $(1,5\text{-COD})Rh^+(\eta^6\text{-PhBPh}_3)^-$, catalyzes the silylformylation of alkynes under normal reaction conditions. However, if H$_2$ is added to the system, the reaction may proceed to yield silylalkenals of a different structure. Interestingly, although the H$_2$ must play a key role in the reaction, it is not incorporated in the product. At this time, the mechanistic role of the hydrogen remains unclear. The authors term this reaction a silylhydroformylation [Eq. (51)].[126]

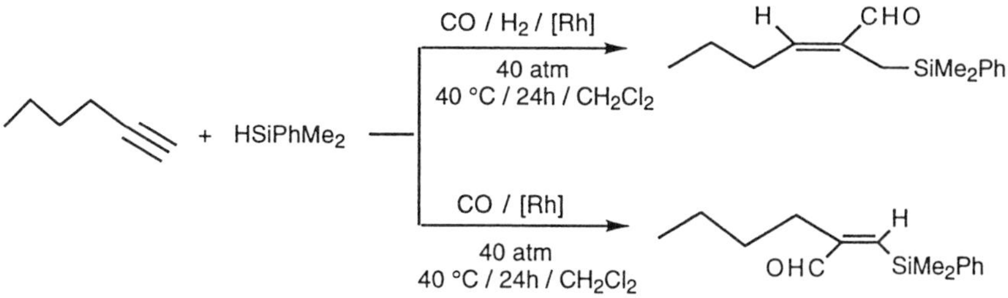

(51)

This appears to be the first report of the addition of $H_2$ to the silylformylation reaction mixture. Good yields are obtained when $Et_3SiH$ or $Ph_3SiH$ is used in the reaction of 1-hexyne or 4-phenyl-1-butyne. Although a variety of functionally substitued terminal alkynes have been studied, most lead only to the silylformylation product and do not appear to be affected by the presence of $H_2$ in the system. Other rhodium catalysts investigated, such as $[Rh(COD)(dppb)]^+BPh_4^-$ and $Rh_6(CO)_{16}$, catalyze the silylformylation reaction even under $H_2$ pressure and do not lead to any of the silylhydroformylated products.

The only report of catalytic germylformylation in the literature at this time is also effected by the zwitterionic rhodium(I) complex $(1,5\text{-COD})Rh^+(\eta^6PhBPh_3)^-$.[127] The synthesis of (Z)-3-germylalk-2-enals in good yield is always accompanied by a minor amount of hydrogermylation [Eq. (52)].

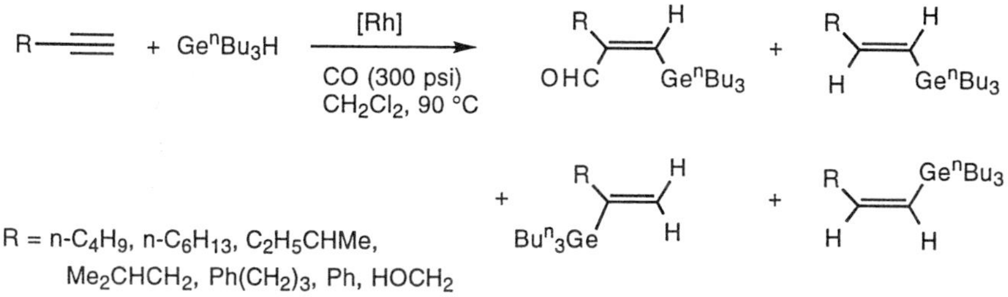

(52)

A series of internal alkynes has been investigated, revealing that the presence of an alkyl or aryl group does not appear to change the course of the reaction. Internal alkynes, however, do not undergo germylformylation in this system. In the case of tin, two reports exist of the stannylformylation of unsaturated carbon substrates, but both proceed by a free radical mechanism initiated by AIBN and do not require a transition metal catalyst.[128]

## 3. *Silacarbocyclization*

Catalytic reactions involving the hydrosilane/carbon monoxide system are in some cases quite sensitive to reaction conditions. For example, in the previously described rhodium-catalyzed silylformylations of propargyl alcohols and propargyl amines, alteration of reaction conditions by inclusion of a base cocatalyst may lead to a cyclocarbonylation reaction (vide supra).[121] In another study, formation of a small amount of 2,5-dibutyl-3-(dimethylphenylsilyl))-2-cyclopenten-1-one was observed in the reaction of 1-hexyne with $HSiMe_2Ph$, catalyzed by either $Co_2Rh_2(CO)_{12}$ or $Rh_4(CO)_{12}$.[122a] The reaction conditions can be optimized for the silacarbocyclization reaction: reaction of triethylsilane and 1-hexyne with $(^tBuNC)_4RhCo(CO)_4$ as catalyst at 60°C gives cyclopentenone as the major product in 54% yield [Eq. (53)].[129] Minor amounts of silylformylation and hydrosilylation products are formed in addition to the silacarbocycle.

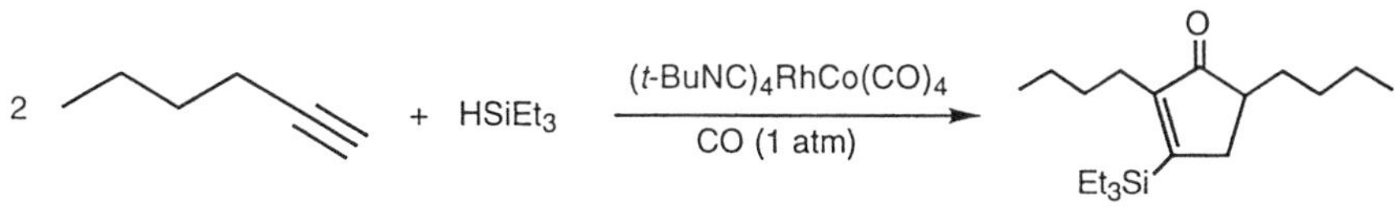

$$(53)$$

A silylcyclopentenone is also formed in the $Rh_4(CO)_{12}$-catalyzed reaction of phenylacetylene with $^tBuMe_2SiH$ and CO, although competitive silylformylation is always observed in this reaction.[130]

The silacarbocyclization of 1,6-diynes has been investigated by both the Ojima and Matsuda groups. 2-Silabicyclooctenones are formed in good to excellent yields in reaction of 1,6-diynes with $^tBuMe_2SiH$ and CO catalyzed by $Rh_4(CO)_{12}$,[131] $Rh(acac)(CO)_2$, or $Rh_2Co_2(CO)_{12}$.[130] Interestingly, when a ruthenium catalyst was used in the reaction of 1,6-diynes with $HSiR_3$ and CO, siloxymethylation occurs to give catechol derivatives (vide supra).[108] Similar dependence on reaction conditions is observed in the silacarbocyclization of 5-hexyn-1-al, with carbon monoxide pressure as the key factor.[122d] Ojima and co-workers have also extended the catalytic silacarbocyclization system to include reactions of enediynes[132] and alkynylamines.[133]

Murai and co-workers report that moderate yields of six- or seven-membered nitrogen-containing heterocycles result from the reaction of acetylene hydrazones with $HSiMe_2^tBu$ and CO in the presence of catalytic amounts of $Ir_4(CO)_{12}$ [Eq. (54)].[134]

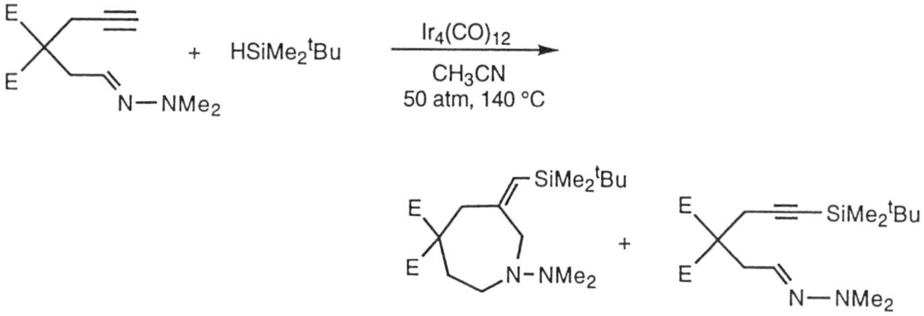

(54)

The incorporated CO is reduced to a methylene group, and the oxygen is consumed in the formation of a siloxane, $({}^tBuMe_2Si)_2O$. An additional product resulting from dehydrogenative silylation of the acetylene functionality was also formed in lower, although significant, yield. Two other iridium catalysts, $Ir(cod)_2BF_4$ and $[IrCl(cod)]_2$, were also found to be effective in the system. The solvent in which the reaction is run is crucial, as of nine solvents investigated, only use of $CH_3CN$ led to formation of the heterocycle. The analogous reaction is not observed for an acetylene hydrazone bearing a methyl substituent at the terminal carbon of the acetylene.

## III

## FORMATION OF E–E BONDS

### A. *Dehydrogenative Coupling*

Classical syntheses involving formation of silicon–silicon, germanium–germanium, or tin–tin bonds have inevitably required the reduction of element halide bonds, such as in the Wurtz-type coupling using alkali metals, or reactions of a element halide with a preformed magnesium or lithium element compound.[7d] These routes are often applicable to the formation of both small molecules and high polymers with Group 14 backbones, but suffer many synthetic limitations. Formation of element–element bonds in reactions catalyzed by transition metal complexes has been known for many years, but it was only with the advent of efficient dehydrocoupling catalysts in the mid-1980s that this avenue of research has gained widespread attention. Without question, efforts toward catalytic formation of silicon–silicon bonds have been spurred by an increased interest in polysilanes for their chemical and electronic properties and potential industrial applications.[7]

The discovery by Harrod, Samuel, and Aitken that dimethyltitanocene serves as a catalyst precursor for the dehydrogenative coupling of phenylsilane to moderate molecular weight poly(phenylsilane) [Eq. (55)] marked the beginning of a period of extensive investigation into dehydrocoupling as a route to E–E bond formation.[135] At this time, many research groups have reported studies on the dehydrogenative coupling system. As several reviews of progress in the field have already appeared,[136,27] only a cursory overview will be presented in this review.

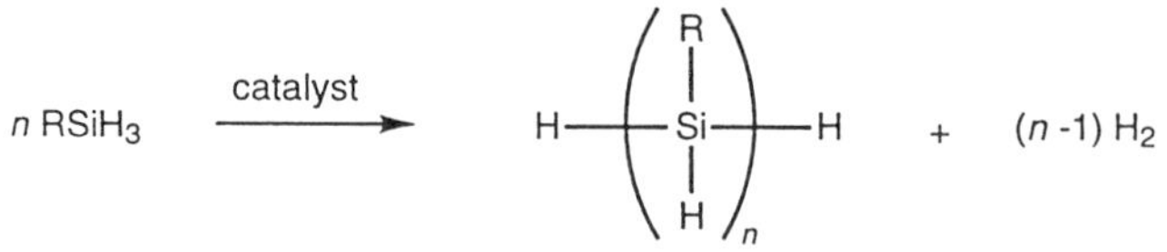

$$n\ RSiH_3 \quad \xrightarrow{\text{catalyst}} \quad H-\left(\begin{array}{c} R \\ | \\ -Si- \\ | \\ H \end{array}\right)_n-H \quad + \quad (n-1)\,H_2 \quad (55)$$

There seems to be a clear distinction between later metal catalysts,[137] which are relatively slow and produce substantial redistribution, and catalyst systems based on electrophilic early metal systems. The most successful early metal systems involve metallocene complexes, generally of titanium and zirconium. Lanthanide and actinide catalysts have also been reported, but do not seem to offer an advantage.[138] Significantly, the electrophilic catalysts are very sensitive to steric factors at the silane and are generally unsuccessful for the coupling of secondary and tertiary silanes to polymers. Tilley has studied various isolated zirconium and hafnium metallocene silyl complexes and provided substantial evidence that the key Si–H cleavage and Si–Si bond-forming steps proceed by $\sigma$-bond metathesis through four-centered transition states, as shown in Scheme 2.[136d,139] Although the Tilley mechanism or some variant is likely to be operative in most early metal catalyst systems, the complexity of some of the modified reaction conditions currently practiced makes it difficult to extrapolate mechanistic analogies with total confidence.

Although metallocene silyl and hydride complexes are the active species, many researchers have sought to develop more convenient precatalysts. The original dimethyltitanocene system reported by Harrod is reasonably easy to prepare, but Corey[140] and others have shown that *in situ* catalyst generation from metallocene dichlorides (Ti, Zr, and Hf) and $^n$BuLi is both simpler and equally effective. However, Harrod and Dioumaev have shown

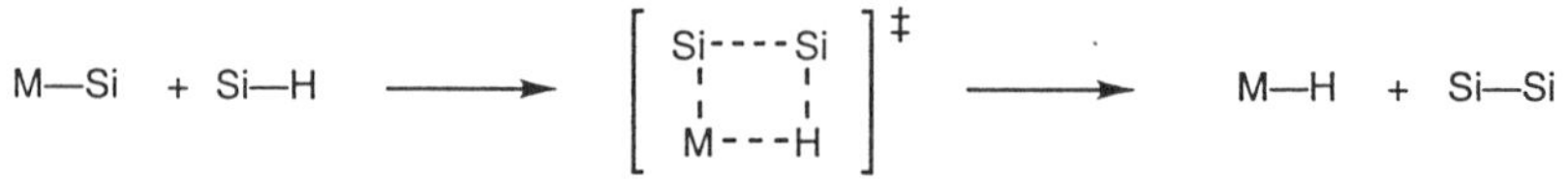

$$M-Si \quad + \quad Si-H \quad \longrightarrow \quad \left[\begin{array}{c} Si----Si \\ |\quad\quad| \\ M---H \end{array}\right]^{\ddagger} \quad \longrightarrow \quad M-H \quad + \quad Si-Si$$

SCHEME 2

that the $^n$BuLi reaction is mechanistically quite complicated and generates a plethora of Zr(III) and Zr(IV) species, several of which may be active in the catalytic cycle.[141] In addition, Corriu and Moreau and co-workers demonstrated that relatively stable titanocene and zirconocene phenoxides are reduced by $PhSiH_3$ in situ at 50°C to provide catalyst solutions that retain activity for hours at 20°C.[142]

Other promoters used to activate the metallocene dichlorides included Red-Al[143] and $^n$BuLi/$B(C_6F_5)_3$.[144] Both apparently act to reduce the metallocene and provide an additional Lewis acid component. In particular, the $CpCp^*ZrCl_2$/2n-BuLi / $B(C_6F_5)_3$ system yields somewhat higher molecular weight poly(phenylsilane) ($M_n$ ~7000) and lower amounts of the cyclic oligomers than the nonpromoted catalysts. The method of action of the strong Lewis acid promoters is thought to involve cationic zirconium centers and various redox steps, in addition to $\sigma$-bond metathesis.

Initial attempts to prepare poly(phenylgermanes) by the dehydrocoupling of phenylgermane with titanocene catalysts led to insoluble gels, although diphenylgermane is coupled to small oligomers.[145] Interestingly, it appears that Ti(IV) complexes are extremely inactive, and only on reduction to Ti(III) does coupling proceed beyond the dimer stage. In contrast, Tilley and co-workers have shown that dialkyltin hydrides undergo extremely facile dehydrocoupling to polystannanes of appreciable molecular weight ($M_n$ ~20,000) using a variety of zirconocene catalysts [Eq. (56)].[146]

$$R_2SnH_2 \xrightarrow[- H_2]{[Zr]} cyclo\text{-}(SnR_2)_m + H(SnR_2)_nH \qquad (56)$$

Sita and Babcock have also prepared high molecular weight poly(dibutylstannane) from dibutylstannane using $Rh(H)(CO)(PPh_3)_3$ as catalyst.[147] As is common with late metal catalysts, branching of the polymer chains is observed as a result of alkyl group redistribution.

The formation of simple disilanes by the dehydrocoupling of silanes is generally not feasible with early metal catalysts. Although primary silanes are undoubtedly coupled to tetrahydrodisilanes, subsequent polymerization is usually very fast. Secondary silanes are generally coupled to mixtures of dimers and trimers. Monohydridosilanes appear to be unreactive toward dehydrocoupling with both early and late metal systems, with the exception of the dehydrocoupling to carbosilanes (vida supra).[81,84] In contrast, early reports describe the coupling of $HGeR_3$ and $HSnR_3$ to produce the corresponding element–element-bonded dimers in the presence of iridium and rhodium complexes.[148] Although the yields of digermanes initially reported were low, a reinvestigation of the coupling of $HGeMe_3$ with Wilkinson's catalyst at 90°C demonstrated an 87% yield of $Ge_2Me_6$ under catalytic conditions [Eq. (57)].[86]

$$2 \ HGeMe_3 \quad \xrightarrow[\text{90 °C, } C_6H_6]{\text{RhCl(PPh}_3)_3} \quad Me_3GeGeMe_3 \ + \ H_2 \ + \ \underset{\text{(trace)}}{ClGeMe_3} \quad (57)$$

### B. Demethanative Coupling

Berry and co-workers have reported a new method for the synthesis of moderate to high molecular weight polygermanes in very high yields by the catalytic coupling of tertiary germanes.[83] This ruthenium-catalyzed process appears to be the first example of a catalytic *demethanative* coupling, in which element–element bonds are produced with the concurrent elimination of $CH_4$ [Eq. (58)].

$$HGeMe_2R \quad \xrightarrow[\text{25 °C, hours}]{\text{Ru(PMe}_3)_4(\text{GeMe}_3)_2} \quad CH_4 \ + \ H \underset{\substack{| \\ Me}}{\overset{\substack{R \\ |}}{\left( \! \! Ge \! \! \right)}}_{\!\! x} \! \! CH_3$$

R = Me, $p$-$C_6H_4X$ (X = H, $CH_3$, $CF_3$, F, OMe)          linear and branched chains

$$(58)$$

In contrast, previous attempts at catalytic dehydrocoupling of germanes have led to the formation of only low molecular weight oligomers.[145]

The catalyst, $Ru(PMe_3)_4(GeMe_3)_2$, is active at levels as low as 0.01 mol%, depending on monomer purity. As the catalyst is prepared from the reaction of $HGeMe_3$ with $Ru(PMe_3)_4Me_2$, this more readily obtainable ruthenium complex can be used directly in the polymerization as a convenient catalyst precursor. The properties of the polygermanes prepared with either complex appear to be identical.

Trimethylgermane is polymerized at 25°C in hydrocarbon solvents or as neat germane, with isolated polymer yields of 84–97%. It appears that alkyl substituents containing $\beta$ hydrogens (e.g., Et, Bu) lead to rapid deactivation of the catalytic system through olefin extrusion (e.g., $H_2C{=}CH_2$ and $H_2C{=}CHCH_2CH_3$) and result in the formation of only small oligomers.[149] However, aryldimethylgermanes, $HGeMe_2(p$-$XC_6H_4)$ [X = H, $CH_3$, $CF_3$, F, OMe], are readily polymerized.[150]

The demethanative coupling appears to be a step polymerization, as initial consumption of the germane monomer produces oligogermanes within the first few minutes, followed by further condensation to high polymer within hours at 25°C. Molecular weights of the permethylpolygermane (GPC, polystyrene calibration) range from $M_n \sim 1$–$4 \times 10^4$ and

$M_w \sim 2-7 \times 10^4$. The poly(arylmethylgermanes) exhibit somewhat lower molecular weights, ranging from $M_n \sim 2-7 \times 10^3$ and $M_w \sim 5-10 \times 10^3$. The GPC traces indicate a monomodal polymer distribution with no evidence of the cyclic oligomers commonly observed in catalytic dehydrocoupling reactions. Interestingly, on the basis of intrinsic viscosity, dynamic light scattering, and chemical analysis, the permethylpolygermane appears to be highly branched, but not cross-linked. In other words, the chains contain up to 40% of -(GeMeGeMe$_3$)- units, rather than only -(GeMe$_2$)- groups as would be expected in a linear polymer.

This unusual demethanative coupling reaction involves the cleavage of Ge–C bonds, and the formation of Ge–Ge bonds. Significantly, Ge$_2$Me$_6$ is not observed in the products and can be shown to be stable under the reaction conditions, indicating that simple dehydrocoupling does not occur. The authors purpose a catalytic cycle (Scheme 3) involving Ge–C cleavage by an $\alpha$-methyl migration from a germyl ligand to form a metal (methyl) (germylene) complex and Ge–Ge bond formation by subsequent germyl to germylene migration. Addition of another equivalent of HGeMe$_3$ would

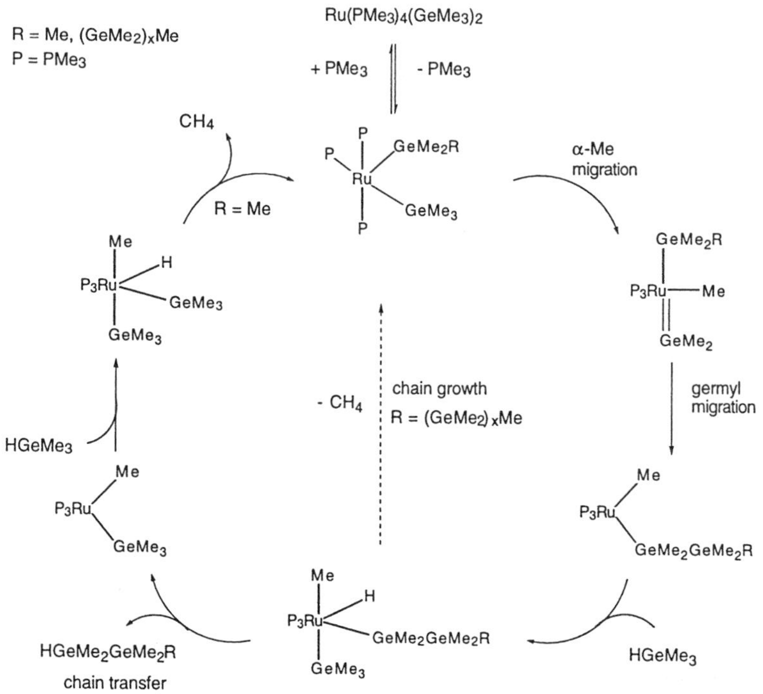

SCHEME 3

produce a seven-coordinate methyl hydride complex, from which $CH_4$ loss would regenerate a bis(germyl) species. Alternatively, exchange of $HGeMe_3$ for $HGeMe_2GeMe_2R$ would correspond to a chain transfer step.

# IV

## FORMATION OF Si–O BONDS

### A. Silane Alcoholysis

Transition metal complexes have been widely investigated as catalysts for the synthesis of alkoxysilanes via alcoholysis of hydrosilanes. The system provides a convenient method for the protection of hydroxy groups in organic synthesis and the synthesis of silyl ethers. The general reaction is shown in Eq. (59).

$$R_3SiH \ + \ HOR' \ \xrightarrow{\text{catalyst}} \ R_3SiOR' \ + \ H_2 \qquad (59)$$

A variety of transition metal complexes have been found to be effective catalysts in this system, including $Co_2(CO)_8$,[151] $RhCl(PPh_3)_3$,[152] $[FeH_2(PMePh_2)_4]$ and $[Fe(C_2H_4)(dppe)_2]$,[153] $BzCr(CO)_2(\eta^2\text{-}HSiHPh_2)$,[154] $(PMe_3)_2Ru(CO)_2Cl_2$[155] and $(PMe_3)_4Ru(H)_2$,[156] and a number of iridium complexes.[157] As the topic of silane alcoholysis has already been reviewed,[27,158] only a few examples will be described in detail. In general, rates of the alcoholysis reactions are most rapid for primary silanes and decrease with increasing silane substitution. In addition, reaction rates decrease with increasing alcohol chain length and degree of branching.

Doyle and co-workers have employed $Rh_2(pfb)_4$ as a highly selective catalyst for the room temperature synthesis of silyl ethers from alcohols and triethylsilane.[159] The selectivity of the catalyst is demonstrated by reactions of olefinic alcohols, in which hydrosilylation is not competitive with silane alcoholysis when equimolar amounts of silane and alcohol are employed. High yields (>85%) of triethylsilyl ethers are obtained from reactions of alcohols such as benzyl alcohol, 1-octanol, 3-buten-1-ol, cholesterol, and phenol. Tertiary alcohols are not active in this system.

The Corey group has reported that $Cp_2TiCl_2/^nBuLi$ effectively catalyzes the high yield alcoholysis of primary, secondary, and tertiary silanes with a variety of alcohols under mild reaction conditions.[160] All Si–H functionalities are replaced with alkoxy groups in reactions of ethanol and phenol, although increased temperatures and/or reaction times are required for highly substituted hydrosilanes. The system also appears to be influenced by the steric bulk of the alcohol, as *t*-butanol reacts with only one Si–H bond of secondary silanes and two Si–H bonds of primary silanes. Reactions of diols with secondary silanes were also investigated, leading to formation of five-, six-, or seven-membered 1,3-dioxa-2-silacycloalkanes [Eq. (60)]. The analogous reaction of primary silanes with diols leads to the formation of oligomeric products via rapid intermolecular condensation reactions.

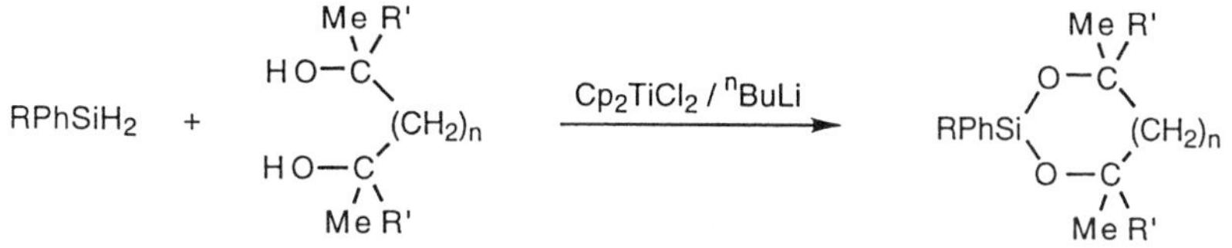

$$(60)$$

In 1992, Crabtree and co-workers reported the first nickel catalyst effective for silane alcoholysis.[161] The complex, $[Ni(tss)]_2$ (tss = salicylaldehyde thiosemicarbazone), bears a ligand that contains O and N donor groups and a semicarbazide sulfur. Alcoholysis of $Et_3SiH$ with ethanol or methanol occurs at room temperature in 50% dimethyl sulfoxide-benzene. However, the reaction is inhibited in the presence of strong donor ligands, $H_2$, or atmospheric pressure of CO.

Catalytic alcoholysis reactions between secondary silanes and allylic alcohols give allyloxyhydrosilane products, in which Si–H and C=C double bond functionalities are present [Eq. (61)].

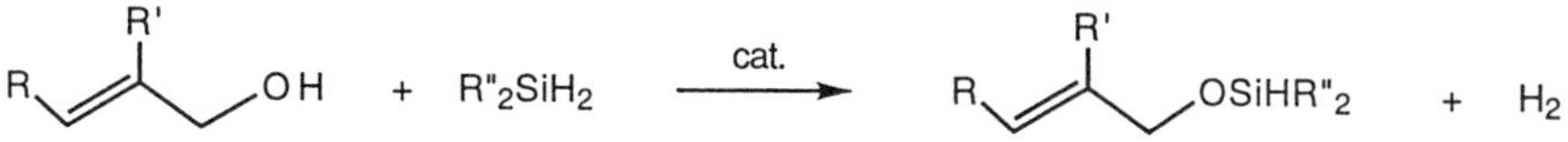

$$(61)$$

The allyloxyhydrosilanes are in turn good substrates for intramolecular hydrosilylation, yielding heterocycles containing both Si and O atoms.[162] Two groups have reported attempts toward a "one-pot" synthesis of the cyclic compounds, without isolation of the intermediate allyloxyhydrosilane.[163] Catalyst choice is crucial, as both the alcoholysis and the intramolecular hydrosilylation must be efficiently catalyzed. The overall reaction is shown in Eq. (62).

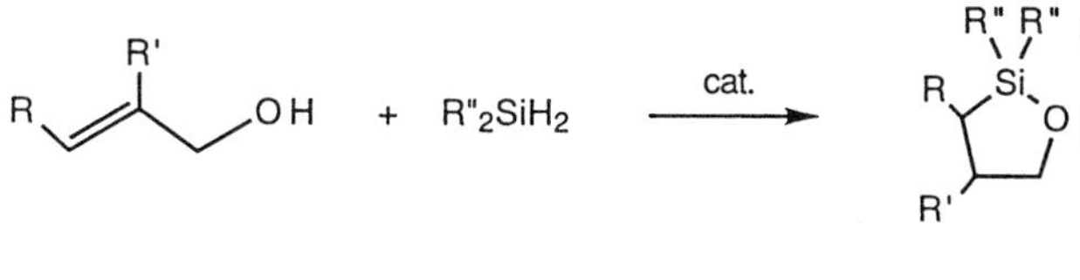

$$(62)$$

In reactions catalyzed by [Rh(diphos)(solvent)$_2$]ClO$_4$ (diphos = Ph$_2$ PCH$_2$CH$_2$PPh$_2$),[163b] allylic alcohols bearing Me, H, or Ph substituents react with diphenyl- or diethylsilane to give high yields of the cyclic products. In addition, allyloxysilane is obtained in lower (5–30%) yields. The system is also effective for reaction of $\alpha,\beta$-unsaturated aldehydes, giving a major product via 1,2-addition followed by cyclization, along with minor amounts of vinyl silyl ethers by 1,4-addition of the silane. Cyclic products and small amounts of disilylated diol are formed in the reaction of $\beta$-keto alcohols.

A number of alternate reaction pathways are available for the intermediate allyloxysilane, including reaction with another equivalent of allylic alcohol, redistribution to form a multisubstituted product and secondary silane, hydrogenation of olefin to give alkoxysilane, and intermolecular hydrosilylation to yield polymer. In reactions of 2-methylbut-1-en-4-ol, **1,** and 2-methylprop-1-en-3-ol, **2,** catalyzed by titanium catalysts, Cp$_2$TiMe$_2$ and *rac*-[EBTHI]TiMe$_2$ (EBTHI = ethylene-1,2-bis($\eta^5$-4,5,6,7-tetrahydroindenyl)), formation of cyclic products is accompanied by formation of products from the alternate pathways, in ratios dependent on substrate, catalyst, and silane employed.[163a] For example, in reactions of the allylic alcohols with PhMeSiH$_2$ catalyzed by dimethyltitanocene, the cyclic product is formed in greater yield from **2,** perhaps due to preferential formation of a five- vs a six-membered cyclic product. In contrast, reaction of **2** with PhMeSiH$_2$ catalyzed by *rac*-[EBTHI]TiMe$_2$ gives essentially no cyclic product. Analogous reactions of 2-sila-1,3-dioxanes from $\beta$- and $\gamma$-hydroxyketones were unsuccessful.

### B. *Hydrosilylation of Carbonyl-Containing Compounds*

Ojima and co-workers first reported the RhCl(PPh$_3$)$_3$-catalyzed hydrosilylation of carbonyl-containing compounds to silyl ethers in 1972.[164] Since that time, a number of transition metal complexes have been investigated for activity in the system, and transition metal catalysis is now a well-established route for the reduction of ketones and aldehydes.[9] Some of the advances in this area include the development of manganese,[165] molybdenum,[166] and ruthenium[167] complex catalysts, and work by the Buchwald and Cutler groups toward extension of the system to hydrosilylations of ester substrates.[168]

Additionally, the asymmetric catalytic hydrosilylation of ketones has attracted a great deal of attention in recent years. Chiral alcohols can be synthesized by enantioselective hydrosilylation of ketones to silyl ethers, followed by hydrolysis to give alcohols, as shown in Eq. (63). A number of complexes with chiral ligands may be employed to catalyze the hydrosily-lation, including those with rhodium,[169] titanium,[170] and iridium[171] metal centers. In general, the reaction displays moderate enantiospecificity (ca. 50–80% ee), although enantiomeric excesses greater than 90% have been reported. Further investigation is required for this system to be comparable to other highly enantiospecific processes such as hydrogenation.

$$
\underset{R^1 \quad R^2}{\overset{O}{\parallel}} \quad + \quad Ph_2SiH_2 \quad \xrightarrow[\text{2. } H_3O^+]{\text{1. catalyst}} \quad \underset{R^2}{\overset{OH}{\underset{R^1}{\bigwedge}}} H \tag{63}
$$

### C. Double Silylation of Carbonyl-Containing Compounds

The 1,4-disilylation of $\alpha,\beta$-unsaturated ketones has been reported by Ito and co-workers.[172] Unsymmetrically substituted disilanes, $PhCl_2SiSiMe_3$ and $Cl_3SiSiMe_3$, undergo reaction in the presence of catalytic amounts of palladium complexes with tertiary alkyl phosphine or bidentate phosphine ligands to yield $\beta$-silyl ketones [Eq. (64)].

$$
\underset{O}{\overset{R}{\diagdown}}\!\!\diagup\!\!\underset{}{\diagdown}R' \quad + \quad PhCl_2SiSiMe_3 \quad \xrightarrow[\substack{C_6H_6 \\ 40\text{-}80\,°C,\ 5\text{-}40\ h}]{Pd(PPh_3)_4} \quad \underset{PhCl_2Si \quad OSiMe_3}{\overset{R \quad R'}{\diagdown\!\diagup\!\diagdown}}
$$

R, R' = Me, Ph

$$\tag{64}$$

Regiospecific addition of the trimethylsilyl group of the disilane to the carbonyl oxygen and of the chlorinated silyl group to the $\beta$-carbon is observed. Reactions attempted with other unsymmetrically substituted disi-lanes, $(MeO)_3SiSiMe_3$ and $X_2MeSiSiMe_3$ (X = Cl, F), as well as with sym-metrically substituted disilanes, $XMe_2SiSiMe_2X$ (X = Me, Cl, F, Ph), were unsuccessful. A variety of ketones are good substrates, as demonstrated by moderate yield reactions of methyl vinyl ketone, 2-cyclohexenenone, and 2-cyclopentenone. When an optically active palladium complex, $PdCl_2$ [(+)-BINAP] (BINAP = 2,2'-bis(diphenylphosphino)-1,1'-binapthyl), is used as catalyst, highly enantioselective double silylation is observed. Enantioselectivity is also affected by solvent, reaction temperature, and ratio of disilane to substrate.

1,4-Double silylation is also observed in the reaction of $\alpha,\beta$-unsaturated ketones with a bis(disilanyl)dithiane, resulting in high yields of cyclic silyl enol ethers [Eq. (65)].[58] The catalyst for this reaction is a cyclic bis(silyl)pal-ladium(II) bis(tert-butyl isocyanide) complex. Analogous reactions of ester and nitrile substrates lead to Si–C bond formation by 1,2-addition in the presence of the cyclic palladium complex.

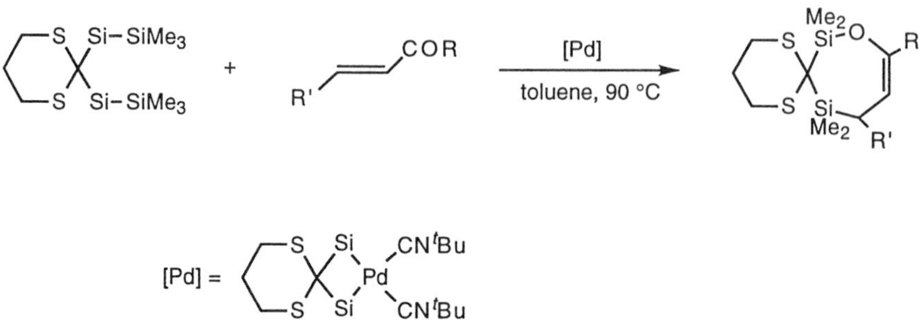

(65)

Tanaka and co-workers have investigated the dehydrogenative double silylation of carbonyl-containing compounds with *o*-bis(dimethylsilyl)benzene [Eqs. (66) and (67)].[173] High-yield 1,2-double silylation occurs in reactions of heptanal, benzaldehyde, and diphenylketene catalyzed by $Pt(CH_2=CH_2)(PPh_3)_2$ or $Pt(dba)_2$. In contrast, the 1,4-double silylation product is formed for $\alpha,\beta$-unsaturated aldehyde or $\alpha,\beta$-unsaturated ketone substrates, such as prop-2-enal and but-3-en-2-one. The system may also be affected by sterics: reaction of ($E$)-3-phenyl-2-propenal gives 1,2-adduct as the major product and only minor amounts of 1,4-adduct. Hydrosilylation products were not formed in any of the carbonyl systems studied.

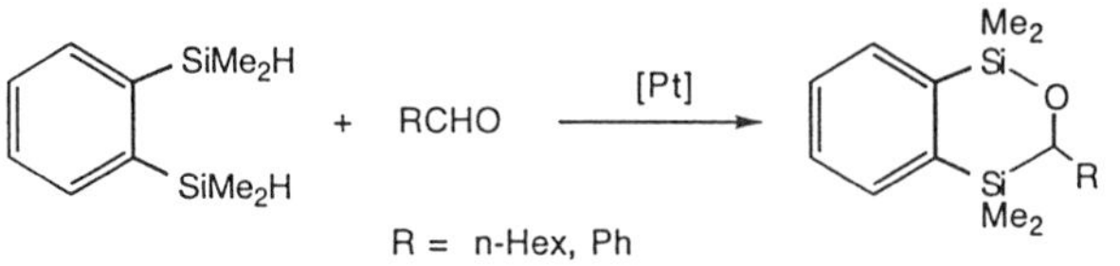

(66)

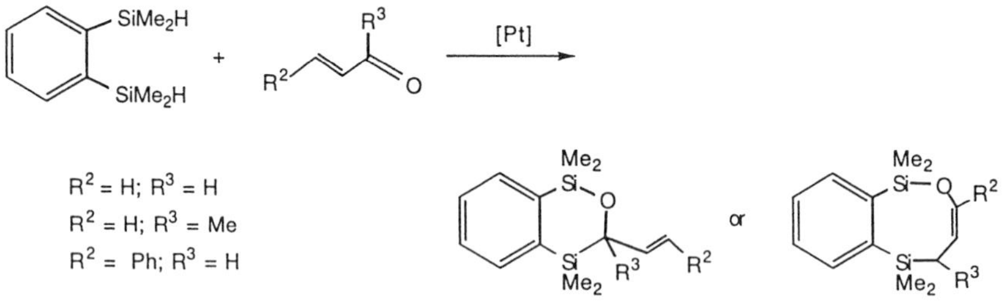

(67)

## D. *Ring-Expansion Reactions of Silacyclobutanes*

Tanaka and co-workers have reported two routes for the catalytic synthesis of cyclic silyl enol ethers from silacylobutanes. The strained silacarbocycle[174] can react directly with an acid chloride[175,176] or in a three-component reaction with an organic halide and carbon monoxide[177] to yield cyclic products that contain an Si–O bond [Eqs. (68) and (69)].

$$
\overset{}{\underset{}{\text{SiR}_2}} \;+\; \text{R'COCl} \quad \xrightarrow[\substack{-\text{Et}_3\text{NHCl} \\ 80\ ^\circ\text{C}}]{\substack{\text{PdCl}_2(\text{PhCN})_2 \\ \text{xs Et}_3\text{N}}} \quad \overset{}{\underset{}{\text{SiR}_2}} \qquad (68)
$$

$$
\overset{}{\underset{}{\text{SiMe}_2}} \;+\; \text{RX} \;+\; \text{CO} \quad \xrightarrow[\substack{-\text{Et}_3\text{NHX} \\ 80\ ^\circ\text{C}}]{\substack{\text{PdCl}_2(\text{dppf}) \\ \text{xs Et}_3\text{N}}} \quad \overset{}{\underset{}{\text{SiMe}_2}} \qquad (69)
$$

The reaction of silacyclobutane and acid chloride is best catalyzed by $PdCl_2$-$(PhCN)$; $PdCl_2(PPh_3)_2$ and $Pt(CH_2{=}CH_2)(PPh_3)_2$ can also be used but are not as active. For the three-component reaction, the most effective catalysts are those of general formula $PdCl_2L_2$, with less basic phosphine ligands such as $PPh_3$ or dppf. In the organic halide reactions, product yields are dependent on substrate. For example, no or low yields are obtained for chloro- and bromobenzene derivatives, but very high yields are observed in reactions of aromatic iodides. No reaction is observed for 1,1-dimethyl-1-silacyclopentane or -hexane, probably due to the lack of ring strain. In both systems, excess tertiary amine is required for the production of the cyclic silyl enol ethers, as only low yields are obtained when the amine is absent. Interestingly, addition of only a small amount (0.1 equiv) of triethylamine to the reaction of 1,1-dimethyl-1-silacyclobutane and benzoyl chloride leads to formation of 1-phenyl-4-(chlorodimethylsilyl)-1-butanone in good yield, as a result of Si–C, rather than Si–O, bond formation [Eq. (70)]. Therefore, γ-silylalkyl ketones can be formed by a modification of the system.

$$
\overset{}{\underset{}{\text{SiMe}_2}} \;+\; \text{PhCOCl} \quad \xrightarrow[\substack{\text{Et}_3\text{N (0.1 equiv)} \\ 80\ ^\circ\text{C}}]{\text{PdCl}_2(\text{PhCN})_2} \quad \text{Ph}\overset{O}{\underset{}{\|}}\!\!\sim\!\!\sim\!\!\text{SiMe}_2\text{Cl} \qquad (70)
$$

## V

## FORMATION OF Si–N BONDS

A less explored area of transition metal catalysis involves bond formation between Group 14 elements and nitrogen. In direct analogy to previously discussed areas of research, silicon–nitrogen bonds can be formed by dehydrocoupling, hydrosilylation, and dehydrogenative silylation. The compounds produced are valuable for use in organic synthesis or as polymer precursors to silicon nitride ceramics.

### A. Dehydrogenative Coupling Reactions

The coupling of a trialkylsilane and an amine with loss of $H_2$, catalyzed by palladium on carbon, was first reported by Sommer and Citron in 1967.[178] More recent work by Laine and Blum has involved the application of catalytic dehydrocoupling of compounds containing Si–H and N–H bonds to form aligo- and polysilazanes. These polymers, with silicon–nitrogen bonds in the backbone, are useful precursors to silicon nitride. In the presence of $Ru_3(CO)_{12}$, silicon–nitrogen bonds are cleaved and reformed [Eq. (71)].[179]

$$\underset{4}{\overline{[Me_2SiNH]}} + (Me_3Si)_2NH \xrightarrow[135\ ^\circ C]{Ru_3(CO)_{12}} Me_3SiNH-[Me_2SiNH]_x-SiMe_3$$

$$x = 1\text{-}12$$

$$(71)$$

The active catalytic species appears to involve a metal hydride complex, as added $H_2$ enhances the reaction rate. Additionally, $H_4Ru_4(CO)_{12}$ is observed during the course of the reaction and may also be used as an effective entry into the catalytic system. The oligomerization mechanism may proceed via a nucleophilic attack of the amine on a metal-activated Si–H bond, resulting in Si–N bond formation.

The Harrod group has investigated dehydrocoupling reactions of silanes with amines, hydrazines, and ammonia. In the presence of catalytic amounts of CuCl, hydrosilanes react with primary amines to give silazanes in high yields.[180] For example, the reaction of phenylmethylsilane with benzylamine leads to a mixture of three products, with dependence on initial molar ratio of silane to amine [Eq. (72)]. Equimolar silane to amine ratio leads to formation of **1** as major product, whereas **2** is the main product if a 2:1

ratio is employed. Conversely, a $1:2$ silane to amine ratio gives **3** as the major reaction product.

$$PhMeSiH_2 + BzNH_2 \xrightarrow[\text{105-145 °C}]{\text{CuCl}} \underset{\textbf{1}}{HPhMeSiNHBz} + \underset{\textbf{2}}{PhMeHSiN(Bz)SiHPhMe}$$

$$+ \underset{\textbf{3}}{BzNHSi(PhMe)NHBz}$$

$$(72)$$

Harrod and co-workers have employed dimethyltitanocene as a catalyst for dehydrocoupling reactions of ammonia and hydrosilanes.[181] A tertiary silane, $Ph_2SiMeH$, reacts with $NH_3$ to yield disilazane, with the best conversions observed at elevated temperatures (ca. 100°C). Reaction of a secondary silane, $PhMeSiH_2$, leads to initial formation of disilazane, followed by conversion at longer times to higher molecular weight oligomers. Interestingly, dehydrogenative couplings of primary silanes with ammonia proceed at slower rates than tertiary and secondary silanes due to competition with the homocoupling reaction. At long reaction times a polymeric product is obtained; a poly(aminosilane) structure is indicated by elemental analysis, NMR, and infrared.

Dimethyltitanocene is also a good catalyst for coupling reactions between silanes and hydrazines to give polysilazanes.[182] Although uncatalyzed reactions of various hydrazines with silanes are possible, the reaction rate is greatly enhanced by addition of the titanium catalyst. Phenylsilane undergoes reaction with methyl hydrazine and 1,1-dimethylhydrazine to produce silylhydrazine oligomers containing multiple structural units. However, analogous reaction with hydrazine leads to intractable gel products. Diphenylsilane is also active in this system, yielding oligomers from reactions of hydrazine, methyl hydrazine, and 1,1-dimethyl hydrazine. Methyl-substituted hydrazines display lower reaction rates, as indicated by the inactivity of 1,2-dimethylhydrazine toward diphenylsilane.

## B. *Hydrosilylation of Nitrogen-Containing Substrates*

Transition metal-catalyzed hydrosilylation, an established route for transformations of unsaturated carbon substrates, is also well known for reactions of nitriles,[183] imines,[184] and oximes.[185] Work in this area includes the $Co_2(CO)_8$-catalyzed addition of trialkylsilanes to aromatic, aliphatic, and $\alpha,\beta$-unsaturated nitriles, giving $N,N$-bis(silyl)amines and/or -enamines in fair to good yields [Eqs. (73) and (74)].[186]

$$RC{\equiv}N \quad + \quad 2\ HSiMe_3 \quad \xrightarrow[\substack{CO\ atm \\ 60\text{ - }100\ ^\circ C \\ toluene \\ PPh_3\ (for\ R=Ar)}]{Co_2(CO)_8} \quad RCH_2N(SiMe_3)_2 \tag{73}$$

R = aryl, alkyl

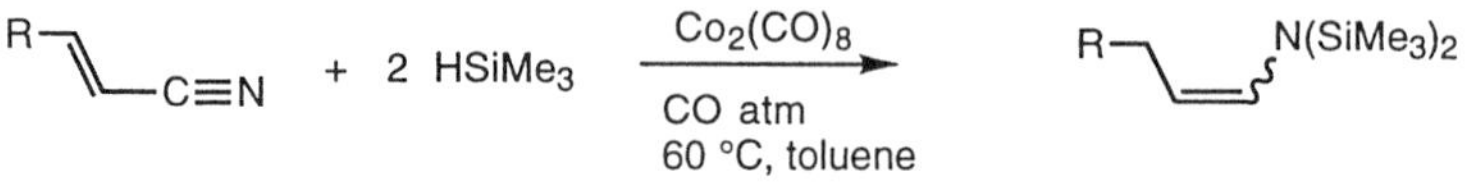

$$\text{R}\diagdown_{C{\equiv}N} \quad + \quad 2\ HSiMe_3 \quad \xrightarrow[\substack{CO\ atm \\ 60\ ^\circ C,\ toluene}]{Co_2(CO)_8} \quad \text{R}\diagdown\diagup N(SiMe_3)_2 \tag{74}$$

Aromatic and aliphatic nitriles react with trimethylsilane by Si–H addition across the C–N triple bond. A variety of functional groups can be tolerated, although reaction rates appear to be affected by electronic and steric characteristics of the substrates. For $\alpha,\beta$-unsaturated nitriles, consecutive 1,2- and 1,4-addition of two equivalents of $HSiMe_3$ occurs to selectively yield $Z$- or $E,N,N$-disilylenamines.

An interesting example of the highly enantioselective hydrosilylation of imines has been reported by Buchwald and co-workers. A titanium fluoride complex, $(S,S)$-$(EBTHI)TiF_2$, catalyzes the formation of silylamines in high yields, with good tolerance of functional groups [Eq. (75)].[187]

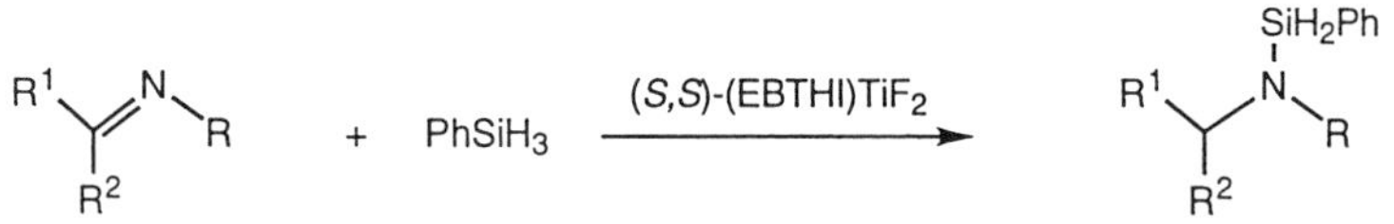

$$\begin{array}{c} R^1 \\ \diagdown \\ R^2 \end{array}{=}N{-}R \quad + \quad PhSiH_3 \quad \xrightarrow{(S,S)\text{-}(EBTHI)TiF_2} \quad \begin{array}{c} \quad\quad SiH_2Ph \\ R^1 \quad\ \ | \\ \diagdown\diagup^{N}{\diagdown}R \\ R^2 \end{array} \tag{75}$$

High enantioselectivity (ca. 95–99% ee) is observed in this system, better than that revealed in previous reports of the hydrosilylation of imines. The mechanism is as yet unclear; however, the authors propose that an active catalyst may be formed by cleavage of the Ti–F bond and generation of a Ti(III) hydride species. Insertion of an imine into the Ti–H bond, followed by a $\sigma$-bond metathesis with the silane in a four-centered transition state, may lead to the observed products. Another report on the activity of titanocene complexes as catalysts for the hydrosilylation of ald- and ketimines also indicates formation of a Ti–H species as catalyst.[188] Hydrosilylation proceeds to yield silylamines, with dependence on substitution at nitrogen and on the nature of the ligand bound to the metallocene precursor.

## C. Double Silylation of Nitriles

Tanaka and co-workers[189] have employed $Pt(CH_2{=}CH_2)(PPh_3)_2$ as catalyst for dehydrogenative double silylation across the C–N multiple bonds of nitriles with $o$-bis(dimethylsilyl)benzene. The type of heterocyclic compound formed is dependent on the starting nitrile [Eq. (76)].

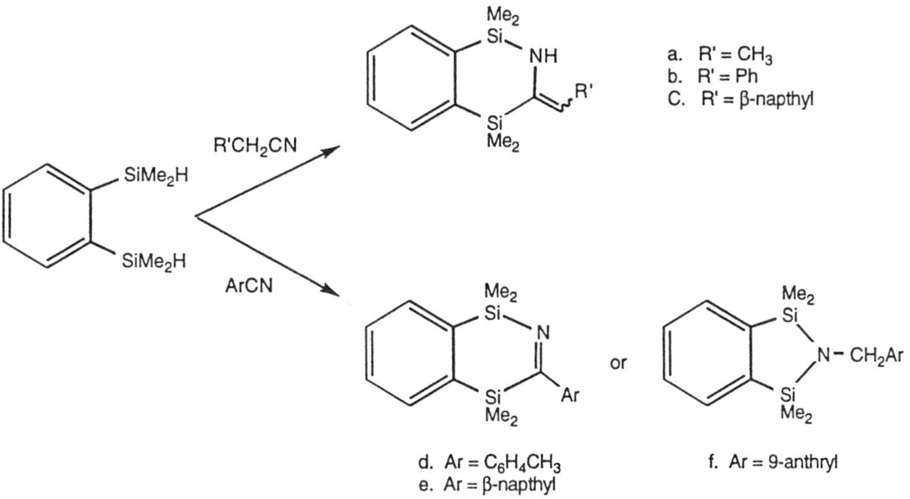

a. R' = CH$_3$
b. R' = Ph
C. R' = β-napthyl

d. Ar = C$_6$H$_4$CH$_3$
e. Ar = β-napthyl

f. Ar = 9-anthryl

(76)

In the case of nitriles containing $\alpha$-hydrogens, the reaction yields N-silylenamines. In contrast, for cyanoarenes where no $\alpha$-hydrogen is present, the corresponding imine is produced. Sequential hydrosilylation of the C–N multiple bond, yielding the $N,N$-bis(silyl)amine, is only observed for 9-anthroline of all nitriles investigated. In addition to these nitrile transformations, the platinum complex catalyzes the reaction of a C–N doubly bonded compound with $o$-bis(dimethylsilyl)benzene: 2-phenylazirine undergoes ring-opening dehydrogenative double silylation to give the $N,N$-bis(silyl) enamine in quantitative yield.

## VI

## CONCLUSIONS

It is clear from the enormous range of chemistries we have attempted to chronicle that research on catalytic transformations of the heavier Group 14 elements has been intense since the mid-1980s. Some reactions, such as double additions to olefins, previously known as curiosities have now been developed to a point where they possess significant synthetic utility to the nonspecialist chemist. Others have allowed ready access to new classes of polymers and materials. Substantial progress has also been made in establishing the diverse mechanisms that lead to the transformations of interest. However, it is also clear that there is a need for a deeper mechanistic understanding of many intriguing reactions in order to optimize product yields and selectivities. In addition, more exploratory work is necessary, particularly of reactions involving the heavier members of the group.

ACKNOWLEDGMENTS

The authors gratefully acknowledge the National Science Foundation MRSEC program (DMR-9632598) and Dow Corning for support of their research.

REFERENCES

(1) (a) *The Chemistry of Organosilicon Compounds*; Patai, S., Rappoport, Z., Eds.; Wiley: New York, 1989; Vol. 1; (b) *Frontiers of Organosilicon Chemistry*; Bassindale, A. R., Gaspar, P. P., Eds.; Proceedings of the IX[th] International Symposium on Organosilicon Chemistry; The Royal Society of Chemistry: Cambridge, **1991**; (c) Pawlenko, S. *Organosilicon Chemistry;* Walter de Gruyter: Berlin, **1986**; Chapter 4.
(2) Pleuddemann, E. P. *Silane Coupling Agents;* Plenum: New York, **1991**.
(3) *Catalyzed Direct Reactions of Silicon;* Lewis, K. M., Rethwisch, D. G., Eds.; Elsevier: Amsterdam, **1993**.
(4) Fessenden, R. J.; Fessenden, J. S. in *Advances in Organometallic Chemistry;* Stone, F. G. A., West, R., Eds.; Academic Press: New York, 1980; pp. 275–299.
(5) Riviere, P.; Riviere-Baudet, M.; Satge, J. in *Comprehensive Organometallic Chemistry II;* Abel, E. W., Stone, F. G. A., Wilkinson, G., Eds.; Pergamon: Oxford, **1995**; Vol. 2, pp. 137–216.
(6) (a) Davies, A. G. in *Comprehensive Organometallic Chemistry II;* Abel, E. W., Stone, F. G. A., Wilkinson, G., Eds.; Pergamon: Oxford, **1995**; Vol. 2, pp. 217–303; (b) Pereyre, M.; Quintard, J.-P.; Rahm, A. *Tin in Organic Synthesis;* Butterworth: London, 1987; (c) Mitchell, T. N. *J. Organomet. Chem.* **1986**, *304*, 1.
(7) For recent reviews of polysilanes, see: (a) West, R. in *Comprehensive Organometallic Chemistry II;* Abel, E. W., Stone, F. G. A., Wilkinson, G., Eds.; Pergamon: Oxford, **1995**; Vol. 2, pp. 77–110; (b) West, R.; Maxka, J. in *Inorganic and Organometallic Polymers;* Zeldin, M., Wynne, K. J., Allcock, H. R., Eds.; ACS Symposium Series 360; American Chemical Society: Washington, DC, **1988**; pp. 6–20; (c) Miller, R. D.; Michl, J. *Chem. Rev.* **1989**, *89*, 1359; (d) West, R. in *The Chemistry of Organic Silicon Compounds;* Patai, S., Rappoport, Z., Eds.; Wiley: New York, 1989; Chapter 19; (e) Zeigler, J. M. *Synth. Met.* **1989**, *28*, C581.
(8) Rochow, E. G.; Gilliam, W. F. *J. Am. Chem. Soc.* **1945**, *67*, 1772.
(9) For recent reviews, see: (a) *Comprehensive Handbook on Hydrosilylation;* Marciniec, B., Ed.; Pergamon: Oxford, 1992; (b) Ojima, I. in *The Chemistry of Organic Silicon Compounds;* Patai, S., Rappoport, Z., Eds.; Wiley: New York, **1989**; Chapter 25.
(10) (a) Lewis, L. N.; Lewis, N. *J. Am. Chem. Soc.* **1986**, *108*, 7228; (b) Lewis, L. N. *J. Am. Chem. Soc.* **1990**, *112*, 5998; (c) Lewis, L. N.; Uriarte, R. J.; Lewis, N. *J. Mol. Cat.* **1991**, *66*, 105; (d) Lewis, L. N.; Sy, K. G.; Bryant, G. L.; Donahue, P. E. *Organometallics* **1991**, *10*, 3750; (e) Lewis, L. N.; Uriarte, R. J. *Organometallics* **1990**, *9*, 621.
(11) (a) Lewis, L. N.; Stein, J.; Smith, K. A.; Messmer, R. P.; Legrand, D. G.; Scott, R. A. in *Progress in Organosilicon Chemistry;* Marciniec, B., Chojnowski, J., Eds.; Gordon

and Breach: Basel, 1995; pp. 263–285; (b) Stein, J.; Lewis, L. N.; Gao, Y. Submitted for publication in *J. Am. Chem. Soc.*

(12) (a) Sano, H.; Miyazaki, Y.; Okawara, M.; Ueno, Y. *Synthesis* **1986**, 776; (b) Ichinose, Y.; Nozaki, K.; Wakamatsu, K.; Oshima, K.; Utimoto, K. *Tetrahedron Lett.* **1987**, *28*, 3709; (c) Nozaki, K.; Oshima, K.; Utimoto, K. *J. Am. Chem. Soc.* **1987**, *109*, 2547; (d) Ichinose, Y.; Oshima, K.; Utimoto, K. *Bull. Chem. Soc. Jpn.* **1988**, *61*, 2693.

(13) (a) Asao, N.; Liu, J.-X.; Sudoh, T.; Yamamoto, Y. *J. Chem. Soc., Chem. Commun.* **1995**, 2405; (b) Asao, N.; Liu, J.-X.; Sudoh, T.; Yamamoto, Y. *J. Org. Chem.* **1996**, *61*, 4568; (c) Gevorgyan, V.; Liu, J.-X.; Yamamoto, Y. *J. Org. Chem.* **1997**, *62*, 2963.

(14) (a) Corriu, R. J. P.; Moreau, J. J. E. *J. Organomet. Chem.* **1972**, *40*, 55; (b) Corriu, R. J. P.; Moreau, J. J. E. *J. Organomet. Chem.* **1972**, *40*, 73.

(15) (a) Oda, H.; Morizawa, Y.; Oshima, K.; Nozaki, H. *Tetrahedron Lett.* **1984**, *25*, 3221; (b) Ichinose, Y.; Oda, H.; Oshima, K.; Utimoto, K. *Bull. Chem. Soc. Jpn.* **1987**, *60*, 3468; (c) Wada, F.; Abe, S.; Yonemaru, N.; Kikukawa, K.; Matsuda, T. *Bull. Chem. Soc. Jpn.* **1991**, *64*, 1701.

(16) (a) Salimgareeva, I. M.; Bogatova, N. G.; Panasenko, A. A.; Khalilov, L. M.; Purlei, I. I.; Mavrodiev, V. K.; Yur'ev, V. P. *Izv. Akad. Nauk SSSR* **1983**, 1605; (b) Bogatova, N. G.; Salimgareeva, I. M.; Yur'ev, V. P. *Izv. Akad. Nauk SSSR* **1984**, 930.

(17) Taniguchi, M.; Oshima, K.; Utimoto, K. *Chem. Lett.* **1993**, 1751.

(18) (a) Miyake, H.; Yamamura, K. *Chem. Lett.* **1989**, 981; (b) Zhang, H. X.; Guilbé, F.; Balavoine, G. *J. Org. Chem.* **1990**, *55*, 1857.

(19) (a) Casson, S.; Kocienski, P. *Synthesis* **1993**, 1133; (b) Greeves, N.; Torode, J. S. *Synlett* **1994**, 537.

(20) Boden, C. D. J.; Pattenden, G.; Ye, T. *J. Chem. Soc, Perkin Trans. 1* **1996**, 2417.

(21) Kikukawa, K.; Umekawa, H.; Wada, F.; Maatsuda, T. *Chem. Lett.* **1988**, 881.

(22) (a) Mitchell, T. N.; Schneider, U. *J. Organomet. Chem.* **1991**, *405*, 195; (b) Koerber, K.; Gore, J.; Vatele, J.-M. *Tetrahedron Lett.* **1991**, *32*, 1187; (c) Gevorgyan, V.; Liu, J.-X.; Yamamoto, Y. *J. Org. Chem.* **1997**, *62*, 2963.

(23) Miyake, H.; Yamamura, K. *Chem. Lett.* **1992**, 507.

(24) Voronkov, M. G.; Adamovich, S. N.; Khramtsova, S. Y.; Shternberg, B. Z.; Rakhlin, V. I.; Mirskov, R. G. *Izv. Akad. Nauk SSSR* **1987**, 1424.

(25) Lautens, M.; Klute, W. *Angew. Chem.* **1996**, *35*, 442.

(26) Lautens, M.; Kumanovic, S.; Meyer, C. *Angew. Chem.* **1996**, *35*, 1329.

(27) For a recent review on dehydrogenative silylation, see: Corey, J. Y. in *Advances in Silicon Chemistry;* Larson, G. L., Ed.; JAI: London; Vol. 1, pp. 327–387.

(28) Nesmeyanov, A. N.; Freidlina, R. K.; Chukovskaya, E. C.; Petrova, R. G.; Belyavsky, A. B. *Tetrahedron* **1962**, *17*, 61.

(29) (a) Millan, A.; Fernandez, M.-J.; Bentz, P.; Maitlis, P. M. *J. Mol. Cat.* **1984**, *26*, 89; (b) Onopchenko, A.; Sabourin, E. T.; Beach, D. L. *J. Org. Chem.* **1983**, *48*, 5101; (c) Ojima, I.; Fuchikami, T.; Yatabe, M. *J. Organomet. Chem.* **1984**, *260*, 335; (d) Duckett, S. B.; Perutz, R. N. *Organometallics* **1992**, *11*, 90; (e) Doyle, M. P.; Devora, G. A.; Nefedov, A. O.; High, K. G. *Organometallics* **1992**, *11*, 549; (f) Kesti, M. R.; Waymouth, R. M. *Organometallics* **1992**, *11*, 1095.

(30) For a recent mechanistic study on Pd(II)-catalyzed hydrosilylation and dehydrogenative silylation, see: LaPointe, A. M.; Rix, F. C.; Brookhart, M. *J. Am. Chem. Soc.* **1997**, *119*, 906.

(31) (a) Seki, Y.; Takeshita, K.; Kawamoto, K.; Murai, S.; Sonoda, N. *Angew. Chem. Int. Ed. Engl.* **1980**, *19*, 928; (b) Seki, Y.; Takeshita, K.; Kawamoto, K.; Murai, S.; Sonoda, N. *J. Org. Chem.* **1986**, *51*, 3890.

(32) Kakiuchi, F.; Tanaka, Y.; Chatani, N.; Murai, S. *J. Organomet. Chem.* **1993**, *456*, 45.

(33) Kakiuchi, F.; Nogami, K.; Chatani, N.; Seki, Y.; Murai, S. *Organometallics* **1993**, *12*, 4748.

(34) (a) Takeshita, K.; Seki, Y.; Kawamoto, K.; Murai, S.; Sonoda, N. *J. Chem. Soc., Chem. Commun.* **1983**, 1193; (b) Takeshita, K.; Seki, Y.; Kawamoto, K.; Murai, S.; Sonoda, N. *J. Org. Chem.* **1987**, *52*, 4864.

(35) Tanke, R. S.; Crabtree, R. H. *Organometallics* **1991**, *10*, 415.

(36) Christ, M. L.; Sabo-Etienne, S.; Chaudret, B. *Organometallics* **1995**, *14*, 1082.

(37) Takeuchi, R.; Yasue, H. *Organometallics* **1996**, *15*, 2098.

(38) Kawanami, Y.; Yamamoto, K. *Bull. Chem. Soc. Jpn.* **1996**, *69*, 1117.

(39) Hori, Y.; Mitsudo, T.; Watanabe, Y. *Bull. Chem. Soc. Jpn.* **1988**, *61*, 3011.

(40) Hayashi, T.; Kawamoto, A. M.; Kobayashi, T.; Tanaka, M. *J. Chem. Soc., Chem. Commun.* **1990**, 563.

(41) Doyle, M. P.; High, K. G.; Nesloney, C. L.; Clayton, T. W.; Lin, J. *Organometallics* **1991**, *10*, 1225.

(42) Jun, C.-H.; Crabtree, R. H. *J. Organomet. Chem.* **1993**, *447*, 177.

(43) (a) Marciniec, B.; Pietraszuk, C. *Organometallics* **1997**, *16*, 4320; (b) Marciniec, B. *New J. Chem.* **1997**, *21*, 815.

(44) For a review on pallidium-catalyzed reactions of silanes, see: Horn, K. A. *Chem. Rev.* **1995**, *95*, 1317.

(45) Okinoshima, H.; Yamamoto, K.; Kumada, M. *J. Am. Chem. Soc.* **1972**, *94*, 9263.

(46) Hayashi, T.; Kobayashi, T.; Kawamoto, A. M.; Yamashita, H.; Tanaka, M. *Organometallics* **1990**, *9*, 280.

(47) Sakaki, S.; Ogawa, M.; Musashi, Y. *J. Organomet. Chem.* **1997**, *535*, 25.

(48) (a) Kusumoto, T.; Hiyama, T. *Tetrahedron Lett.* **1987**, *28*, 1807; (b) Kusumoto, T.; Hiyama, T. *Bull. Chem. Soc. Jpn.* **1990**, *63*, 3103.

(49) Ishikawa, M.; Nishimura, Y.; Sakamoto, H.; Ono, T.; Ohshita, J. *Organometallics* **1992**, *11*, 483.

(50) Tsuji, Y.; Lago, R. M.; Tomohiro, S.; Tsuneishi, H. *Organometallics* **1992**, *11*, 2353.

(51) Obora, Y.; Tsuji, Y.; Kawamura, T. *Organometallics* **1993**, *12*, 2853.

(52) (a) Sakurai, H.; Eriyama, Y.; Kamiyama, Y.; Nakadaira, Y. *J. Organomet. Chem.* **1984**, *264*, 229; (b) Carlson, C. W.; West, R. *Organometallics* **1983**, *2*, 1801.

(53) Obora, Y.; Tsuji, Y.; Kakehi, T.; Kobayashi, M.; Shinkai, Y.; Ebihara, M.; Kawamura, T. *J. Chem. Soc., Perkin Trans. 1* **1995**, 599.

(54) Ishikawa, M.; Sakamoto, H.; Okazaki, S.; Naka, A. *J. Organomet. Chem.* **1992**, *439*, 19.

(55) Kusukawa, T.; Kabe, Y.; Ando, W. *Chem. Lett.* **1993**, 985.

(56) Finckh, W.; Tang, B.-Z.; Lough, A.; Manners, I. *Organometallics* **1992**, *11*, 2904.

(57) Ito, Y.; Suginome, M.; Murakami, M. *J. Org. Chem.* **1991**, *56*, 1948.

(58) Suginome, M.; Oike, H.; Ito, Y. *Organometallics* **1994**, *13*, 4148.

(59) (a) Murakami, M.; Andersson, P. G.; Suginome, M.; Ito, Y. *J. Am. Chem. Soc.* **1991**, *113*, 3987; (b) Murakami, M.; Suginome, M.; Fujimoto, K.; Nakamura, H.; Andersson, P. G.; Ito, Y. *J. Am. Chem. Soc.* **1993**, *115*, 6487; (c) Murakami, M.; Suginome, M.; Fujimoto, K.; Ito, Y. *Angew. Chem., Int. Ed. Engl.* **1993**, *32*, 1473; (d) Murakami, M.; Oike, H.; Sugawara, M.; Suginome, M.; Ito, Y. *Tetrahedron* **1993**, *49*, 3933; (e) Suginome, M.; Yamamoto, Y.; Fujii, K.; Ito, Y. *J. Am. Chem. Soc.* **1995**, *117*, 9608.

(60) (a) Yamashita, H.; Catellani, M.; Tanaka, M. *Chem. Lett.* **1991**, 241; (b) Yamashita, H.; Tanaka, M. *Chem. Lett.* **1992**, 1547.

(61) (a) Tanaka, M.; Uchimaru, Y.; Lautenschlager, H.-J. *Organometallics* **1991**, *10*, 16; (b) Tanaka, M.; Uchimaru, Y.; Lautenschlager, H.-J. *J. Organomet. Chem.* **1992**, *428*, 1.

(62) (a) Uchimaru, Y.; Brandl, P.; Tanaka, M.; Goto, M. *J. Chem. Soc., Chem. Commun.* **1993**, 744; (b) Shimada, S.; Uchimaru, Y.; Tanaka, M. *Chem. Lett.* **1995**, 223.

(63) Tsumuraya, T.; Ando, W. *Organometallics* **1989**, *8*, 2286.

(64) Hayashi, T.; Yamashita, H.; Sakakura, T.; Uchimaru, Y.; Tanaka, M. *Chem. Lett.* **1991**, 245.

(65) Mochida, K.; Hodota, C.; Yamashita, H.; Tanaka, M. *Chem. Lett.* **1992**, 1635.

(66) Mitchell, T. N.; Amamria, A.; Killing, H.; Rutschow, D. *J. Organomet. Chem.* **1983**, *241*, C45.

(67) Killing, H.; Mitchell, T. N. *Organometallics* **1984**, *3*, 1318.

(68) Piers, E.; Skerlj, R. T. *J. Chem. Soc., Chem. Commun.* **1986**, 626.

(69) Mitchell, T. N.; Schneider, U. *J. Organomet. Chem.* **1991**, *407*, 319.

(70) Kwetkat, K.; Riches, B. H.; Rosset, J.-M.; Brecknell, D. J.; Byriel, K.; Kennard, C. H. L.; Young, D. J.; Schneider, U.; Mitchell, T. N.; Kitching, W. *J. Chem. Soc., Chem. Commun.* **1996**, 773.

(71) Tsuji, Y.; Kakehi, T. *J. Chem. Soc., Chem. Commun.* **1992**, 1000.

(72) (a) Mitchell, T. N.; Killing, H.; Dicke, R.; Wickenkamp, R. *J. Chem. Soc., Chem. Commun.* **1985**, 354; (b) Chenard, B. L.; Laganis, E. D.; Davidson, F.; RajanBabu, T. V. *J. Org. Chem.* **1985**, *50*, 3666; (c) Chenard, B. L.; Van Zyl, C. M. *J. Org. Chem.* **1986**, *51*, 3561; (d) Mitchell, T. N.; Wickenkamp, R.; Amamria, A.; Dicke, R.; Schneider, U. *J. Org. Chem.* **1987**, *52*, 4868.

(73) Murakami, M.; Morita, Y.; Ito, Y. *J. Chem. Soc., Chem. Commun.* **1990**, 428.

(74) Tsuji, Y.; Obora, Y. *J. Am. Chem. Soc.* **1991**, *113*, 9368.

(75) Obora, Y.; Tsuji, Y.; Asayama, M.; Kawamura, T. *Organometallics* **1993**, *12*, 4697.

(76) Piers, E.; Skerlj, R. T. *J. Chem. Soc., Chem. Commun.* **1987**, 1025.

(77) Mitchell, T. N.; Schneider, U.; Fröhling, B. *J. Organomet. Chem.* **1990**, *384*, C53.

(78) (a) Ito, Y.; Bando, T.; Matsuura, T.; Ishikawa, M. *J. Chem. Soc., Chem. Commun.* **1986**, 980; (b) Ito, Y.; Nishimura, S.; Ishikawa, M. *Tetrahedron Lett.* **1987**, *28*, 1293; (c) Ito, Y.; Matsuura, T.; Murakami, M. *J. Am. Chem. Soc.* **1988**, *110*, 3692; (d) Ito, Y. *Pure Appl. Chem.* **1990**, *62*, 583; (e) Ito, Y.; Murakami, M. *Synlett* **1990**, 245; (f) Ito, Y.; Suginome, M.; Matsuura, T.; Murakami, M. *J. Am. Chem. Soc.* **1991**, *113*, 8899.

(79) Ito, Y.; Suginome, M.; Murakami, M.; Shiro, M. *J. Chem. Soc., Chem. Commun.* **1989**, 1494.

(80) Procopio, L. J.; Berry, D. H. *J. Am. Chem. Soc.* **1991**, *113*, 4039.

(81) (a) Procopio, L. J.; Mayer, B.; Plössl, K.; Berry, D. H. *Polym. Prepr.* **1992**, *33*, 1241; (b) Procopio, L. J. Ph.D. Thesis, University of Pennsylvania, 1991.

(82) Hong, P. Ph.D. Thesis, University of Pennsylvania, 1995.

(83) Reichl, J. A.; Popoff, C. M.; Gallagher, L. A.; Remsen, E. E.; Berry, D. H. *J. Am. Chem. Soc.* **1996**, *118*, 9430.

(84) (a) Djurovich, P. I.; Carroll, P. J.; Berry, D. H. *Organometallics* **1994**, *13*, 2551; (b) Djurovich, P. I.; Dolich, A. R.; Berry, D. H. *J. Chem. Soc., Chem. Commun.* **1994**, 1897.

(85) Ezbiansky, K. A.; Djurovich, P. I.; LaForest, M.; Sinning, D. J.; Zayes, R.; Berry, D. H. *Organometallics* **1998**, *17*, 1455.

(86) Popoff, C. M. Ph.D. Thesis, University of Pennsylvania, 1995.

(87) Berry, D. H.; Procopio, L. J. *J. Am. Chem. Soc.* **1989**, *111*, 4099.

(88) (a) Zlota, A. A.; Frolow, F.; Milstein, D. *J. Chem. Soc., Chem. Commun.* **1989**, 1826; (b) Yamashita, H.; Kawamoto, A. M.; Tanaka, M.; Goto, M. *Chem. Lett.* **1990**, 2107.

(89) (a) Pannell, K. H. *J. Organomet. Chem.* **1970**, *21*, P17; (b) Randolph, C. L.; Wrighton, M. S. *Organometallics* **1987**, *6*, 365; (c) Campion, B. K.; Heyn, R. H.; Tilley, T. D. *J. Am. Chem. Soc.* **1988**, *110*, 7558; (d) Campion, B. K.; Heyn, R. H.; Tilley, T. D.; Rheingold, A. L. *J. Am. Chem. Soc.* **1993**, *115*, 5527.

(90) Dioumaev, V. K.; Plössl, K.; Carroll, P. J.; Berry, D. H. Manuscript in preparation.

(91) (a) Sakakura, T.; Tanaka, M. *Chem. Lett.* **1987**, *2*, 249; (b) Kunin, A. J.; Eisenberg, R.

Organometallics **1988**, 7, 2124; (c) Boese, W. T.; Goldman, A. S. Organometallics **1991**, 10, 782.

(92) Barry, A. J.; Gilkey, J. W.; Hook, D. E. Direct Process for the Preparation of Arylhalosilanes; ACS Monograph 23; American Chemical Society: Washington, DC, 1959; pp. 246–264.

(93) Gustavson, W. A.; Epstein, P. S.; Curtis, M. D. Organometallics **1982**, 1, 884.

(94) Sakakura, T.; Tokunaga, Y.; Sodeyama, T.; Tanaka, M. Chem. Lett. **1987**, 2375.

(95) Uchimaru, Y.; El Sayed, A. M. M.; Tanaka, M. Organometallics **1993**, 12, 2065.

(96) Ishikawa, M.; Okazaki, S.; Naka, A.; Sakamoto, H. Organometallics **1992**, 11, 4135.

(97) (a) Rich, J. D. J. Am. Chem. Soc. **1989**, 111, 5886; (b) Rich, J. D. Organometallics **1989**, 8, 2609; (c) Rixh, J. D.; Krafft, T. E. Organometallics **1990**, 9, 2040.

(98) (a) Atwell, W. H.; Bokerman, G. N. U.S. Pat. 3 772 447, 1971; (b) Atwell, W. H.; Bokerman, G. N. U.S. Pat. 3 746 732, 1971; (c) Matsumoto, H., Nagashima, S., Yoshihiro, K.; Nagai, Y. J. Organomet. Chem. **1975**, 85, C1; (d) Matsumoto, H.; Yoshihiro, K.; Nagashima, S.; Watanabe, H.; Nagai, Y. J. Organomet. Chem. **1977**, 128, 409; (e) Matsumoto, H.; Shono, K.; Nagai, Y. J. Organomet. Chem. **1981**, 208, 145.

(99) The reaction of hexamethyldisilane with varous functionalized aryl bromides has also been reported. Babin, P.; Bennetau, B.; Theurig, M.; Dunoguès, J. J. Organomet. Chem. **1993**, 446, 135.

(100) (a) Yamamoto, K.; Suzuki, S.; Tsuji, J. Tetrahedron Lett. **1980**, 21, 1653; (b) Eaborn, C.; Griffiths, R. W.; Pidcock, A. J. Organomet. Chem. **1982**, 225, 331.

(101) Berry, D. H.; Djurovich, P. I. U.S. Pat. 5 508 460, 1996.

(102) Reddy, N. P.; Hayashi, T.; Tanaka, M. Chem. Lett. **1991**, 677.

(103) For reviews on transition metal-catalyzed reactions of HSiR$_3$/CO: (a) Murai, S.; Sonoda, N. Angew. Chem. Int. Ed. Engl. **1979**, 18, 837; (b) Murai, S.; Seki, Y. J. Mol. Catal. **1987**, 41, 197; (c) Chatani, N.; Murai, T.; Ikeda, T.; Sano, T.; Kajikawa, Y.; Ohe, K.; Murai, S. Tetrahedron **1992**, 48, 2013; (d) Chatani, N.; Murai, S. Synlett **1996**, 414.

(104) Seki, Y.; Hidaka, A.; Murai, S.; Sonoda, N. Angew. Chem. Int. Ed. Engl. **1977**, 16, 174.

(105) Reaction of olefins with Et$_3$SiH and CO catalyzed by Co$_2$(CO)$_8$ was first reported in 1967. Although it appeared that CO and hydrosilane were incorporated, complicated product mixtures were obtained. (a) Chalk, A. J.; Harrod, J. F. J. Am. Chem. Soc. **1967**, 89, 1640; (b) Chalk, A. J.; Harrod, J. F. Adv. Organomet. Chem. **1968**, 6, 119.

(106) Seki, Y.; Kawamoto, K.; Chatani, N.; Hidaka, A.; Sonoda, N.; Ohe, K.; Kawasaki, Y.; Murai, S. J. Organomet. Chem. **1991**, 403, 73.

(107) Chatani, N.; Ikeda, S.; Ohe, K.; Murai, S. J. Am. Chem. Soc. **1992**, 114, 9710.

(108) Chatani, N.; Fukumoto, Y.; Ida, T.; Murai, S. J. Am. Chem. Soc. **1993**, 115, 11614. 11615.

(109) (a) Murai, T.; Yasui, E.; Kato, S.; Hatayama, Y.; Suzuki, S.; Yamasaki, Y.; Sonoda, N.; Kurosawa, H.; Kawasaki, Y.; Murai, S. J. Am. Chem. Soc. **1989**, 111, 7938; (b) Murai, T.; Kato, S.; Murai, S.; Hatayama, Y.; Sonoda, N. Tetrahedron Lett. **1985**, 26, 2683.

(110) Chatani, N.; Tokuhisa, H.; Kokubu, I.; Fujii, S.; Murai, S. J. Organomet. Chem. **1995**, 499, 193.

(111) Chatani, N.; Sano, T.; Ohe, K.; Kawasaki, Y.; Murai, S. J. Org. Chem. **1990**, 55, 5923.

(112) (a) Chatani, N.; Fujii, S.; Yamasaki, Y.; Murai, S.; Sonoda, N. J. Am. Chem. Soc. **1986**, 108, 7361; (b) Chatani, N.; Murai, S.; Sonoda, N. J. Am. Chem. Soc. **1983**, 105, 1370.

(113) Chatani, N.; Kajikawa, Y.; Nishimura, H.; Murai, S. Organometallics, **1991**, 10, 21.

(114) Ikeda, S.; Chatani, N.; Kajikawa, Y.; Ohe, K.; Murai, S. J. Org. Chem. **1992**, 57, 2.

(115) Ikeda, S.; Chatani, N.; Murai, S. Organometallics **1992**, 11, 3494.

(116) Murai, S.; Kato, T.; Sonoda, N.; Seki, Y.; Kawamoto, K. Angew. Chem. Int. Ed. Engl. **1979**, 18, 393.

(117) Seki, Y.; Murai, S.; Sonoda, N. Angew. Chem. Int. Ed. Engl. **1978**, 17, 119.

(118) Wright, M. E.; Cochran, B. B. *J. Am. Chem. Soc.* **1993**, *115*, 2059.
(119) (a) Seki, Y.; Murai, S.; Yamamoto, I.; Sonoda, N. *Angew. Chem. Int. Ed. Engl.* **1977**, *16*, 789; (b) Fukumoto, Y.; Chatani, N.; Murai, S. *J. Org. Chem.* **1993**, *58*, 4187; (c) Fukumoto, Y.; Yamaguchi, S.; Chatani, N.; Murai, S. *J. Organomet. Chem.* **1995**, *489*, 215.
(120) (a) Matsuda, I.; Ogiso, A.; Sato, S.; Izumi, Y. *J. Am. Chem. Soc.* **1989**, *111*, 2332. (b) Matsuda, I.; Fukuta, Y.; Tsuchihashi, T.; Nagashima, H.; Itoh, K. *Organometallics* **1997**, *16*, 4327.
(121) (a) Matsuda, I.; Ogiso, A.; Sato, S. *J. Am. Chem. Soc.* **1990**, *112*, 6120; (b) Matsuda, I.; Sakakibara, J.; Nagashima, H. *Tetrahedron Lett.* **1991**, *32*, 7431; (c) Matsuda, I.; Sakakibara, J.; Inoue, H.; Nagashima, H. *Tetrahedron Lett.* **1992**, *33*, 5799.
(122) (a) Ojima, I.; Ingallina, P.; Donovan, R. J.; Clos, N. *Organometallics* **1991**, *10*, 38; (b) Eguchi, M.; Zeng, Q.; Korda, A.; Ojima, I. *Tetrahedron Lett.* **1993**, *34*, 915; (c) Ojima, I.; Donovan, R. J.; Eguchi, M.; Shay, W. R.; Ingallina, P.; Korda, A.; Zeng, Q. *Tetrahedron* **1993**, *49*, 5431; (d) Ojima, I.; Tzamarioudaki, M.; Tsai, C.,-Y. *J. Am. Chem. Soc.* **1994**, *116*, 3643.
(123) (a) Doyle, M. P.; Shanklin, M. S. *Organometallics* **1993**, *12*, 11; (b) Doyle, M. P.; Shanklin, M. S. *Organometallics* **1994**, *13*, 1081.
(124) Monteil, F.; Matsuda, I.; Alper, H. *J. Am. Chem. Soc.* **1995**, *117*, 4419.
(125) Ojima, I.; Vidal, E.; Tzamarioudaki, M.; Matsuda, I. *J. Am. Chem. Soc.* **1995**, *117*, 6797.
(126) Zhou, J.-Q.; Apler, H. *Organometallics,* **1994**, *13*, 1586.
(127) Monteil, F.; Alper, H. *J. Chem. Soc., Chem. Commun.* **1995**, 1601.
(128) (a) Ryu, I.; Kurihara, A.; Muraoka, H.; Tsunoi, S.; Kambe, N.; Sonoda, N. *J. Org. Chem.* **1994**, *59*, 7570; (b) Tsunoi, S.; Ryu, I.; Muraoka, H.; Tanaka, M.; Komatsu, M.; Sonoda, N. *Tetrahedron Lett.* **1996**, *37*, 6729.
(129) Ojima, I.; Donovan, R. J.; Shay, W. R. *J. Am. Chem. Soc.* **1992**, *114*, 6580.
(130) Matsuda, I.; Ishibashi, H.; Ii, N. *Tetrahedron Lett.* **1995**, *36*, 241.
(131) Ojima, I.; Fracchiolla, D. A.; Donovan, R. J.; Banerji, P. *J. Org. Chem.* **1994**, *59*, 7594.
(132) Ojima, I.; McCullagh, J. V.; Shay, W. R. *J. Organomet. Chem.* **1996**, *521*, 421.
(133) Ojima, I.; Machnik, D.; Donovan, R. J.; Mneimne, O. *Inorg. Chim. Acta* **1996**, *521*, 299.
(134) Chatani, N.; Yamaguchi, S.; Fukumoto, Y.; Murai, S. *Organometallics* **1995**, *14*, 4418.
(135) Aitken, C.; Harrod, J. F.; Samuel, E. *J. Organomet. Chem.* **1985**, *279*, C11.
(136) (a) Mu, Y.; Harrod, J. F. in *Inorganic and Organometallic Oligomers and Polymers;* Harrod, J. F., Laine, R. M., Eds.; Kluwer Academic: Dordrecht, 1991; pp. 23–35; (b) Harrod, J. F. in *Inorganic and Organometallic Polymers;* Zeldin, M., Wynne, K. J., Allcock, H. R., Eds.; ACS Symposium Series 360; American Chemical Society: Washington, DC, 1988; pp. 89–100; (c) Harrod, J. F.; Mu, Y.; Samuel, E. *Polyhedron* **1991**, *10*, 1239; (d) Tilley, T. D. *Acc. Chem. Res.* **1993**, *26*, 22; (e) Gauvin, F.; Harrod, J. F.; Woo, H. G. in *Advances in Organometallic Chemistry;* Stone, F. G. A., West, R., Eds.; Academic Press: New York, 1998.
(137) (a) Ojima, I.; Inaba, S.-I.k Kogure, T.; Nagai, Y. *J. Organomet. Chem.* **1973**, *55*, C7. (b) Corey, J. Y.; Chang, L. S.; Corey, E. *Organometallics* **1987**, *6*, 1595.
(138) (a) Watson, P. L.; Tebbe, F. N. U.S. Patent 4 965 386, 1990; (b) Forsyth, C. M.; Nolan, S. P.; Marks, T. J. *Organometallics* **1991**, *10*, 2543; (c) Sakakura, T.; Lautenschlager, H.-J.; Nakajima, M.; Tanaka, M. *Chem. Lett.* **1991**, 913.
(139) Tilley, T. D.; Woo, H.-G. in *Inorganic and Organometallic Oligomers and Polymers,* Harrod, J. F., Laine, R. M., Eds.; Kluwer Academic: Dordrecht, 1991; pp. 3–11.
(140) Corey, J. Y.; Zhu, X.-H.; Bedard, T. C.; Lange, L. D. *Organometallics* **1991**, *10*, 924.
(141) Dioumaev, V. K.; Harrod, J. F. *Organometallics* **1997**, *16*, 1452.
(142) Bourg, S.; Corriu, R. J. P.; Enders, M.; Moreau, J. J. E. *Organometallics* **1995**, *14*, 564.

(143) Woo, H.-G.; Kim, S.-Y.; Han, M.-K.; Cho, E. J.; Jung, I. N. *Organometallics* **1995**, *14*, 2415.

(144) (a) Dioumaev, V. K.; Harrod, J. F. *J. Organomet. Chem.* **1996**, *521*, 133; (b) Dioumaev, V. K.; Harrod, J. F. *Organometallics* **1994**, *13*, 1548.

(145) Aitken, C.; Harrod, J. F.; Malek, A.; Samuel, E. *J. Organomet. Chem.* **1988**, *349*, 285.

(146) (a) Imori, T.; Tilley, T. D. *J. Chem. Soc., Chem. Commun.* **1993**, 1607; (b) Imori, T.; Lu, V.; Cai, H.; Tilley, T. D. *J. Am. Chem. Soc.* **1995**, *117*, 9931; (c) Lu, V.; Tilley, T. D. *Macromolecules* **1996**, *29*, 5763.

(147) Babcock, J. R.; Sita, L. R. *J. Am. Chem. Soc.* **1996**, *118*, 12481.

(148) (a) Glockling, F.; Wilbey, M. D. *J. Chem. Soc. (A)* **1970**, 1675; (b) Glockling, F.; Hill, G. C. *J. Chem. Soc. (A)* **1971**, 2137.

(149) Reichl, J. A.; Berry, D. H. Manuscript in preparation.

(150) Katz, S. M.; Reichl, J. A.; Berry, D. H. Manuscript in preparation.

(151) Chalk, A. J. *J. Chem. Soc., Chem. Commun.* **1970**, 847.

(152) (a) Corriu, R. J. P., Moreau, J. J. E. *J. Organomet. Chem.* **1976**, *114*, 135; (b) Corriu, R. J. P.; Moreau, J. J. E. *J. Organomet. Chem.* **1977**, *127*, 7.

(153) Haszeldine, R. N.; Parish, R. V.; Filey, B. F. *J. Chem. Soc., Dalton Trans.* **1980**, 705.

(154) Matarasso-Tchiroukhine, E. *J. Chem. Soc., Chem. Commun.* **1990**, 681.

(155) Oehmichen, U.; Singer, H. *J. Organomet. Chem.* **1983**, *243*, 199.

(156) Burn, M. J.; Bergman, R. G. *J. Organomet. Chem.* **1994**, *472*, 43.

(157) Blackburn, S. N.; Haszeldine, R. N.; Parish, R. V.; Setchfield, J. H. *J. Organomet. Chem.* **1980**, *192*, 329; (b) Luo, X.-L.; Crabtree, R. H. *J. Am. Chem. Soc.* **1989**, *111*, 2527.

(158) Lukevics, E.; Dzintara, M. *J. Organomet. Chem.* **1985**, *295*, 265.

(159) Doyle, M. P.; High, K. G.; Bagheri, V.; Pieters, R. J.; Lewis, P. J.; Pearson, M. M. *J. Org. Chem.* **1990**, *55*, 6082.

(160) Bedard, T. C.; Corey, J. Y. *J. Organomet. Chem.* **1992**, *428*, 315.

(161) Barber, D. E.; Lu, Z.; Richardson, T.; Crabtree, R. H. *Inorg. Chem.* **1992**, *31*, 4709.

(162) (a) Tamao, K.; Nakagawa, Y.; Arai, H.; Higuchi, N.; Ito, Y. *J. Am. Chem. Soc.* **1988**, *110*, 3712; (b) Bergens, S. H., Noheda, P.; Whelen, J.; Bosnich, B. *J. Am. Chem. Soc.* **1992**, *114*, 2121; (c) Denmark, S. E.; Forbes, D. C. *Tetrahedron Lett.* **1992**, *33*, 5037; (d) Burk, M. J.; Feaster, J. E. *Tetrahedron Lett.* **1992**, *33*, 2099.

(163) (a) Xin, S.; Harrod, J. F. *J. Organomet. Chem.* **1995**, *499*, 181; (b) Wang, X.; Ellis, W. W.; Bosnich, B. *J. Chem. Soc., Chem. Commun.* **1996**, 2561.

(164) (a) Ojima, I.; Nihonyanagi, M.; Nagai, Y. *J. Chem. Soc., Chem. Commun.* **1972**, 938; (b) Ojima, I.; Nihonyanagi, M.; Nagai, Y. *Bull. Chem. Soc. Jpn.* **1972**, 3722; (c) Ojima, I.; Nihonyanagi, M.; Kogure, T.; Kumagai, M.; Horiuchi, S.; Nakatsugawa, K.; Nagai, Y. *J. Organomet. Chem.* **1975**, *94*, 449.

(165) Cavanaugh, M. D.; Gregg, B. T.; Cutler, A. R. *Organometallics* **1996**, *15*, 2764.

(166) Schmidt, T. *Tetrahedron Lett.* **1994**, *35*, 3513.

(167) Wiles, J. A.; Lee, C. E.; McDonald, R.; Bergens, S. H. *Organometallics* **1996**, *15*, 3782.

(168) (a) Berk, S. C.; Kreutzer, K. A.; Buchwald, S. L. *J. Am. Chem. Soc.* **1991**, *113*, 5093; (b) Mao, Z.; Gregg, B. T.; Cutler, A. R. *J. Am. Chem. Soc.* **1995**, *117*, 10139.

(169) (a) Enders, D.; Gielen, H.; Breuer, K. *Tetrahedron: Asymmetry* **1997**, *8*, 3571; (b) Sudo, A.; Yoshida, H.; Saigo, K. *Tetrahedron: Asymmetry* **1997**, *8*, 3205; (c) Newman, L. M.; Williams, J. M. J.; McCague, R.; Potter, G. A. *Tetrahedron: Asymmetry* **1996**, *7*, 1597; (d) Langer, T.; Janssen, J.; Helmchen, G. *Tetrahedron: Asymmetry* **1996**, *7*, 1599; (e) Nishibayashi, Y.; Segawa, K.; Ohe, K.; Uemura, S. *Organometallics* **1995**, *14*, 5486; (f) Sawamura, M.; Kuwano, R.; Shirai, J.; Ito, Y. *Synlett* **1995**, 347; (g) Hayashi, T.; Hayashi, C.; Uozumi, Y. *Tetrahedron: Asymmetry* **1995**, *6*, 2503; (h) Nishibayashi, Y.; Singh, J.

D.; Segawa, K.; Fukuzawa, S.; Uemura, S. *J. Chem. Soc., Chem. Commun.* **1994**, 1375; (i) Brunner, H.; Kürzinger, A. *J. Organomet. Chem.* **1988**, *346*, 413.

(170) (a) Imma, H.; Mori, M.; Nakai, T. *Synlett* **1996**, 1229; (b) Xin, S.; Harrod, J. F. *Can. J. Chem.* **1995**, *73*, 999; (c) Halterman, R. L.; Ramsey, T. M.; Chen, Z. *J. Org. Chem.* **1994**, *59*, 2642; (d) Carter, M. B.; Schiøtt, B.; Gutiérrez, A.; Buchwald, S. L. *J. Am. Chem. Soc.* **1994**, *116*, 11667.

(171) (a) Nishibayashi, Y.; Segawa, K.; Takada, H.; Ohe, K.; Uemura, S. *J. Chem. Soc., Chem. Commun.* **1996**, 847; (b) Faller, J. W.; Chase, K. J. *Organometallics* **1994**, *13*, 989.

(172) (a) Hayashi, T.; Matsumoto, Y.; Ito, Y. *J. Am. Chem. Soc.* **1988**, *110*, 5579; (b) Hayashi, T.; Matsumoto, Y.; Ito, Y. *Tetrahedron Lett.* **1988**, *29*, 4147; (c) Matsumoto, Y.; Hayashi, T.; Ito, Y. *Tetrahedron* **1994**, *50*, 335.

(173) Uchimaru, Y.; Lautenschlager, H.-J.; Wynd, A. J.; Tanaka, M.; Goto, M. *Organometallics* **1992**, *11*, 2639.

(174) For palladium-catalyzed reactions of silacyclobutane with acetylenes, see: Takeyama, Y.; Nozaki, K.; Matsumoto, K.; Oshima, K.; Utimoto, K. *Bull. Chem. Soc. Jpn.* **1991**, *64*, 1461.

(175) Tanaka, Y.; Yamashita, H.; Tanaka, M. *Organometallics* **1996**, *15*, 1524.

(176) For related palladium-catalyzed reactions of acid chlorides, disilanes, and 1,3-dienes, see: (a) Obora, Y.; Tsuji, Y.; Kawamura, T. *J. Am. Chem. Soc.* **1993**, *115*, 10414; (b) Obora, Y.; Tsuji, Y.; Kawamura, T. *J. Am. Chem. Soc.* **1995**, *117*, 9814.

(177) Chauhan, B. P. S.; Tanaka, Y.; Yamashita, H.; Tanaka, M. *J. Chem. Soc., Chem. Commun.* **1996**, 1207.

(178) Sommer, L. H.; Citron, J. D. *J. Org. Chem.* **1967**, *32*, 2470.

(179) (a) Blum, Y.; Laine, R. M. *Organometallics* **1986**, *5*, 2081; (b) Biran, C.; Blum, Y. D.; Glaser, R.; Tse, D. S.; Youngdahl, K. A.; Laine, R. M. *J. Mol. Cat.* **1988**, *48*, 183; (c) Laine, R. M.; Blum, Y. D.; Tse, D.; Glaser, R. in *Inorganic and Organometallic Polymers;* Zeldin, M., Wynne, K. J., Allcock, H. R., Eds.; ACS Symposium Series 360; American Chemical Society: Washington, DC, 1988; pp. 124–142.

(180) Liu, H. Q.; Harrod, J. F. *Can. J. Chem.* **1992**, *70*, 107.

(181) Liu, H. Q.; Harrod, J. F. *Organometallics* **1992**, *11*, 822.

(182) He, J.; Liu, H. Q.; Harrod, J. F.; Hynes, R. *Organometallics* **1994**, *13*, 336.

(183) (a) Calas, R. *Pure Appl. Chem.* **1966**, *13*, 61; (b) Chalk, A. J. *J. Organomet. Chem.* **1970**, *21*, 207; (c) Ojima, I.; Kumagai, M. *Tetrahedron Lett.* **1974**, 4005; (d) Ojima, I.; Kumagai, M.; Nagai, Y. *J. Organomet. Chem.* **1976**, *111*, 43; (e) Corriu, R. J. P., Moreau, J. J. E.; Pataud-Sat, M. *J. Organomet. Chem.* **1982**, *228*, 301.

(184) (a) Ojima, I.; Kogure, T.; Nagai, Y. *Tetrahedron Lett.* **1973**, 2475; (b) Langlois, N.; Dang, T.-P.; Kagan, H. B. *Tetrahedron Lett.* **1973**, 4865; (c) Kagan, H. B.; Langlois, N.; Dang, T. P. *J. Organomet. Chem.* **1975**, *90*, 353; (c) Becker, R.; Brunner, H.; Mahboobi, S.; Wiegrebe, W. *Angew. Chem. Int. Ed. Engl.* **1985**, *24*, 995.

(185) (a) Brunner, H.; Becker, R.; Gauder, S. *Organometallics* **1986**, *5*, 739; (b) Zhorov, E. Y.; Pavlov, V. A.; Fedotova, O. A.; Shvedov, V. I.; Mistryukov, É. A.; Platonov, D. N.; Gorshkova, L. S.; Klabunovskii, E. I. *Izv. Akad. Nauk SSSR* **1991**, 763.

(186) (a) Murai, T.; Sakane, T.; Kato, S. *Tetrahedron Lett.* **1985**, *26*, 5145; (b) Murai, T.; Sakane, T.; Kato, S. *J. Org. Chem.* **1990**, *55*, 449.

(187) Verdaguer, X.; Lange, U. E. W.; Reding, M. T.; Buchwald, S. L. *J. Am. Chem. Soc.* **1996**, *118*, 6784.

(188) Tillack, A.; Lefeber, C.; Peulecke, N.; Thomas, D.; Rosenthal, U. *Tetrahedron Lett.* **1997**, *38*, 1533.

(189) Reddy, N. P.; Uchimaru, Y.; Lautenschlager, H.-J.; Tanaka, M. *Chem. Lett.* **1992**, 45.

ADVANCES IN ORGANOMETALLIC CHEMISTRY, VOL. 43

# Organometallic Compounds of the Heavier Alkali Metals

## J. DAVID SMITH

*School of Chemistry, Physics and Environmental Science*
*University of Sussex*
*Brighton BN1 9QJ, United Kingdom*

## I

## INTRODUCTION

This review covers developments in the organometallic chemistry of sodium, potassium, rubidium, and cesium since the previous accounts by Weiss,[1] Bock,[2] and Schleyer,[3,4] but earlier work is described where this is necessary to make sense of recent results. Further information can be found in *Comprehensive Organometallic Chemistry II*[5] and in the second edition of the *Dictionary of Organometallic Compounds*,[6] which is available both in hard copy and, brought up to date every 6 months, on CD-ROM.

There is widespread evidence from X-ray studies on crystalline samples, from multinuclear nuclear magnetic resonance (NMR) data on solutions, and from various kinds of computational investigations of gas-phase interactions that the bonding in organometallic compounds of the heavier alkali metals is strongly ionic, i.e., electronic charge is almost completely transferred from the metal to the organic group. Long before the development of modern spectroscopic and computational methods for the study of electronic structures, the compounds were recognized as carbanion sources in organic synthesis, and derivatization remains an important method for inferring the structures of very reactive intermediates. However, the widespread availability of X-ray diffraction studies for routine characterization has led to a mass of new knowledge about the crystalline organometallic compounds themselves. In the following discussion, the words cation and carbanion are used to describe metal and organic fragments, but this does not mean that the structures and chemical properties of cations and anions are independent.[7] It is almost impossible with the present state of knowledge to predict the structure of a compound isolated from a particular cation–base–carbanion mixture: there is a subtle balance both in the crystalline state and in the solution between interionic interactions and interactions between cations and amine or ether bases that are often used as solvents or added to obtain crystallizable solids. It is not easy to draw general conclusions from many of the structures described in the literature. More insight is gained from the study of closely related series of compounds in which the effects of incremental changes are more readily pinpointed. Even here, however, there are surprises. Small variations in reaction conditions can sometimes result in drastic changes in crystal structures.

The emphasis in this review is on compounds in which there are crystallographically recognized interactions between alkali metals and carbon, or transfer of charge from metal to a carbanionic species. The contents are classified into broad categories based on the nature of the organic fragment with compounds of the various alkali metals considered together, but there is inevitably some overlap between categories. Metal–carbon distances have been given in the text to provide a consistent thread throughout the discussion of a wide range of structural types. They are usually quoted as a range with the estimated standard deviations in parentheses to indicate the precision of individual bond lengths. It must be emphasized that data do not necessarily imply that interactions are localized at particular carbon centers. For lithium and sodium compounds the small size of the metal and the constraints of chelate rings may well draw the cation near to specific carbon atoms, but for potassium, rubidium, and cesium it is more usual for the crystallographically determined coordination number to be large, suggesting that the main interactions are between the cation and the negative charge widely delocalized over the anion. In some cases the cations are complexed by amines or ethers and are completely distinct from the anions so that the lattice is held together by long-range electrostatic forces.

The organometallic compounds of the heavier alkali metals are in general extremely air- and moisture-sensitive and many are highly pyrophoric. Great care is necessary in their manipulation.

## II

## SYNTHESIS

The most widely used method for the preparation of organometallic compounds of the heavier alkali metals is from the corresponding organo-lithium reagent and an alkoxide.

$$\text{LiR} + \text{MOR}' \rightarrow \text{MR} + \text{LiOR}' \ (M = \text{Na, K, Rb, Cs})\qquad(1)$$

The reaction is usually effected in hydrocarbon solvents containing some ether, and the group $R'$ is chosen to make it easy to separate the product $MR$ from the unwanted lithium alkoxide. For sodium and potassium derivatives, the group $R'$ is usually $^t$Bu, but for rubidium and cesium compounds 3-ethyl-3-hept-,[8,9] 2-methyl-2-pent-,[10] or 2-ethylhex-oxides[11,12,46] are commonly used. Mixtures of alkyl-lithium and sodium or potassium t-butoxides have long been used as "superbases" showing superior metallating power compared with alkyl-lithiums alone (see Refs. 9 and 29 for surveys of the literature) but there has been some controversy about the nature of the active species in solution. A multinuclear NMR study[9] has shown that the species in solutions obtained from LiCPh$_3$ and CsOC$_9$H$_{19}$, C$_9$H$_{19}$ = 3-ethyl-3-heptyl, are CsCPh$_3$ and LiOC$_9$H$_{19}$. Close contacts between Cs and CPh$_3$ are revealed by $^{133}$Cs, $^1$H HOESY (heteronuclear Overhauser effect spectroscopy), and the proximity of the Li and the alkoxo ions is confirmed by $^6$Li, $^1$H HOESY studies. No mixed aggregates were found within the NMR detection limit. Although this work implies that organometallic derivatives of the heavier Group 1 metals are probably the active species in solutions of superbases, the fact that mixed alkylalkoxides have been isolated and characterized in the solid state[13] suggests that this conclusion may not be valid in all circumstances. Computational studies indicate that the mixed alkylalkoxide dimers are more stable than other possible species.[14]

Organoalkali metal compounds may also be obtained from reactions between organomercury compounds and the elemental metals.

$$HgR_2 + 2M \rightarrow 2MR + Hg \quad (M = K, Cs) \quad (2)$$

This route has, for example, been used to make the butyl compounds.[15] The trimethylsilylmethyl derivatives have also been made[16,17]: they can be precipitated from cyclohexane, washed free from organomercury starting materials, and obtained without organometal-alkoxide contaminants. Alternatively, $KCH_2SiMe_3$ may be used *in situ,* e.g., for metallation of alkenes.[18]

Methyl, butyl, or trimethylsilylmethyl compounds of Na, K, Rb, or Cs can be used to generate other organometallic compounds by alkyl group exchange.

$$MR + R''H \rightarrow MR'' + RH \quad (3)$$

This reaction has been used to make benzyl derivatives without problems from coupling reactions,[16] alkenyl derivatives containing an activating group such as OMe,[17] and silicon-substituted organometallic compounds.[19,20] A wide range of alkenyl compounds RM have been obtained from the reaction of trimethylsilylmethyl-potassium or -cesium with alkenyltin compounds $RSnMe_3$ or $RSnBu_3$.[21]

$$RSnMe_3 + Me_3SiCH_2M \rightarrow RM + Me_3SiCH_2SnMe_3 \quad (M = K \text{ or } Cs)\,(4)$$

Compounds RH containing strongly acidic protons can be treated directly with alkali metals in the presence of strong amine or ether donors (L) to give soluble complexes of organometallic compounds.

$$2M + 2RH + 2L \rightarrow 2MR.L + H_2 \quad (5)$$

This reaction has been used to make crystalline triphenylmethyl derivatives that have been characterized by X-ray structure determinations,[22] and a technique has been developed for the reduction of hydrocarbons by liquid cesium activated by ultrasound irradiation in the presence of ethers such as diglyme.[23] The blue solutions obtained when cesium metal is dissolved in THF in the presence of [18]-crown-6 have been used to metallate a series of 1,4- or 1,5-hexadienes. The organocesium compounds have not been isolated but they have been identified by derivatization by carbonation and trimethylsilylation.[24] Substituted cyclopentadienyl derivatives of sodium have also been synthesized by this method.[25]

$$\overset{\text{THF/0°C}}{2\ C_5H_5R + 2Na \rightarrow \quad 2Na(C_5H_4R) + H_2} \quad (6)$$

(R = $COCH_2OMe$, $COCH_2CH_2SMe$, $CH_2COCH_3$, $CH_2COOEt$,
    $CH_2CH_2COOMe$, $CH_2CH_2OMe$, $COCH_2OMe$)

Benzoylcyclopentadienylsodium (R = COPh) can be made from $NaC_5H_5$ and ethyl benzoate.[26]

$$NaC_5H_5 + PhCOOEt \rightarrow NaC_5H_4COPh + EtOH \qquad (7)$$

Alkali metal hydrides can be used as metallating agents in the synthesis of base-free cyclopentadienylmetal derivatives.[27,28]

$$MH + RH \rightarrow MR + H_2 \qquad (8)$$

## III

## ALKYL DERIVATIVES

### A. Methyl Derivatives

The isolation and characterization of the methyl derivatives of Na, K, Rb, and Cs have been described previously,[1] and there is considerable experimental and theoretical evidence that these compounds are highly ionic. More recent work has been on compounds in which the hydrogen atoms of the methyl group are replaced by substituents that allow the ionic charge to be delocalized. A variety of novel structures can be isolated; many of these would be difficult or perhaps impossible to obtain with less bulky alkyl groups.

### B. Benzyl and Related Derivatives

Benzylpotassium **1** and benzylrubidium **2** have been isolated from toluene solution as PMDTA adducts.[29] No mixed alkyl-alkoxo derivative was obtained even in the presence of an excess of alkoxide. Both structures (Fig. 1) show polymeric zigzag chains in which the metal atoms are linked to two benzyl anions, one $\eta^6$ (K–C 315.0(2)–328.8(2) and Rb–C 314(2)–337(2) pm) and the other $\eta^3$(K–C 317.1(2)–329.7(2) pm and Rb–C 326(2)–330(2) pm). The configuration at the $CH_2$ group is planar like that in $NaCH_2Ph$. PMDTA.0.5 $C_6H_6$[30]; in contrast, the $CH_2$ group in $LiCH_2Ph$. THF.TMEDA is pyramidal.[31] The magnitude (151 Hz) of the C–H coupling constant of the sodium compound in THF-$d_8$ at $-20°C$ suggests that the planar configuration at the carbanionic $CH_2$ is preserved in solution. The crystallographic and spectroscopic results have been supported by computational studies, which explain the shift from $\eta^{1-3}$ interactions between the aromatic rings and cation in lithium or sodium derivatives to $\eta^6$ interactions in compounds of the heavier alkali metals.

The benzyl and o-xylyl compounds $[NaCH_2Ph.TMEDA]_4$[32] and $[NaCH_2-C_6H_4Me.TMEDA]_4$[30] form tetrameric units in the solid state. A tetrameric structure (Fig. 2) has also been found for the compound with stoichiometry

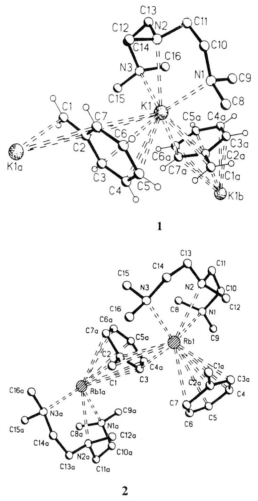

FIG. 1.  Molecular structure of benzylpotassium **1** and benzylsodium **2**. (Reprinted with permission from D. Hoffmann *et al., J. Am. Chem. Soc.* **1994**, *116*, 528. Copyright © 1994 American Chemical Society.)

Li$_{1.67}$ Na$_{2.33}$(CH$_2$Ph)$_4$.4TMEDA, **3,** isolated from the metallation of toluene with a butyl-lithium/butyl-sodium mixture.[33] The two metals are disordered over the four positions in the slightly puckered eight-membered ring with each edge of the metal square bisected with an approximately perpendicularly oriented CH$_2$Ph group with a trigonal bipyramidal carbon. The structure is very similar to that of the *all*-sodium analog.[32]

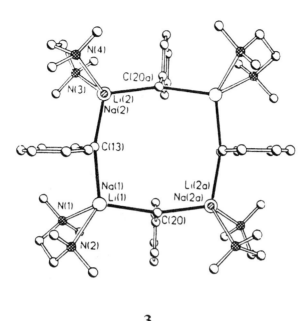

**3**

FIG. 2.   Molecular structure of $Li_{1.67}Na_{2.33}(CH_2Ph)_4 \cdot 4TMEDA$, **3**. (Reprinted with permission from D. R. Baker *et al.*, *Organometallics* **1994**, *13*, 4170. Copyright © 1994 American Chemical Society.)

Another organometallic compound, **4**, containing both lithium and sodium was obtained from the reaction between sodium methyl(4-methylbenzyl)amide and butyllithium in TMEDA. Both amide and carbanion centers were metallated. The $NaLi(4\text{-}MeC_6H_3CH_2NMe)$ units form tetramers (Fig. 3) with a $Li_4$ planar rhomboidal core and the *ipso* carbon atoms of two anion rings bridging $Li_3$ triangles on opposite sides. The sodium ions are more weakly bound. Each forms a bridge between one amide and two aryl fragments and is also solvated by TMEDA. There are two slightly different sodium coordination environments. In one the sodium, Na2, is asymmetrically bound in $\eta^6$ fashion to an aromatic ring (Na–C 290.4(6)–333.8(6)pm) and to the *ipso* and *ortho* carbon atoms of another (Na–C 304.9(6)–301.5(6) pm). In the other environment the sodium, Na1, is bound to the *ipso* and *ortho* carbon atoms of two rings (Na–C 266.0(6)–314.2(6) pm). The fact that the Na–C distance is longer than those in other phenylsodium derivatives (256.6–275.6 pm) is attributed to steric effects. The coordination number of seven for the *ipso* carbon atom C11 is remarkable[34] A similar structure is found for the organosodium–lithium alkoxide complex **5** isolated from the reaction between butyllithium and

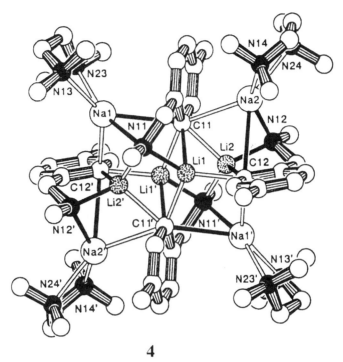

**4**

FIG. 3. Molecular structure of NaLi(4-MeC$_6$H$_3$CH$_2$NMe), **4**. (Reprinted with permission from S. Harder *et al.*, *Organometallics* **1995,** *14,* 2133. Copyright © 1995 American Chemical Society.)

sodium 2,4,6-trimethylphenoxide.[13] The core of the tetrameric structure (Fig. 4) consists of a Li$_4$O$_4$ cube with the sodium atoms coordinated to the carbanionic centers (Na–C 267.1(9) pm), through two further $\eta^3$ contacts to two aryl rings (Na–C 281.1(9)–358.3(9) pm), to one oxygen of the Li$_4$O$_4$ cube, and to TMEDA. Only half of the lithium atoms in the bis(pyridyl)-methyl compound LiCH(2-C$_5$H$_4$N)$_2$.2THF were displaced by NaO$^t$Bu at room temperature.

$$2\text{LiCH(2-C}_5\text{H}_4\text{N})_2.2\text{THF} + \text{NaO}^t\text{Bu} + 2\text{THF} \rightarrow$$
$$[\text{Na(THF)}_6][\text{Li}\{(\text{CH(2-C}_5\text{H}_4\text{N})_2\}_2] + \text{LiO}^t\text{Bu} \quad (9)$$

The product was formally a sodium dialkyllithate, but the crystal structure shows that the coordination is exclusively through the nitrogen atoms of the pyridyl groups with no Li–C bond.[35]

The structures of two adducts of 9,10-dihydroanthracenylsodium have been described.[36] The PMDTA derivative **6** (Fig. 5) forms monomeric ion

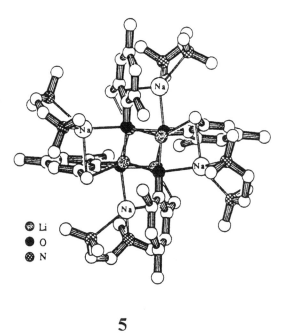

**5**

Fig. 4.    Molecular structure of $[LiOC_6H_2(2\text{-}CH_2Na)\text{-}4,6\text{-}Me_2]_2$, **5.** The tetrameric structure has approximately $S_4$ symmetry. (Reprinted with permission from S. Harder and A. Streitwieser, *Angew. Chem. Int. Ed. Engl.* **1993,** *32,* 1066. Copyright © 1993 Wiley VCH.)

pairs in the solid state with three normal Na–N bonds, one short Na–C distance (257.1(2) pm), and significant interaction with a hydrogen atom of the bridging $CH_2$ group. The TMEDA adduct **7** is similar but the sodium is coordinated to the two nitrogen atoms of the amine, one bridging carbon (Na–C 271.7(2) pm) and $\eta^2$ fashion to an aromatic ring of an adjacent ion pair, to give a linear polymeric structure.

### C. *Cumyl and Triphenylmethyl Derivatives*

A dark red solution containing cumylcesium, $CsCMe_2Ph$, has been made by the reaction of cesium metal with 2,3-dimethyl-2,3-diphenylbutane in THF and used in extensive measurements of carbon acidity.[37,38] Although the organocesium compound is thought to be basic enough to attack THF, steric hindrance slows the reaction sufficiently for the solution to be used in quantitative thermodynamic studies. An alternative organocesium compound for use in this area of research, $Cs_2C_2Ph_4$, is obtained from cesium

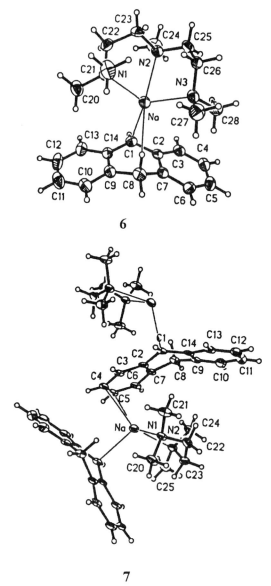

6

7

FIG. 5.   Molecular structures of 9,10-dihydroanthracenylsodium-PMDTA, **6,** and -TMEDA, **7.** (Reprinted with permission from H. Viebrock *et al., Chem. Ber.* **1994,** *127,* 1399. Copyright © 1994 Wiley VCH.)

metal and $Ph_2C=CPh_2$ in THF.[37] This is sufficiently stable in the solid state to be weighed out, a considerable advantage where it is necessary to make up solutions of accurately known concentration. Cumylpotassium (made from cumene, butyl-lithium, and potassium t-pentoxide) has found applications in synthesis. For example, reaction with cyclohexene oxide yields racemic *trans*- 2($\alpha$-cumyl)cyclohexanol from which the (+) isomer can be obtained by enzymatic resolution.[39]

The compounds $MCPh_3$.L (L = TMEDA, or PMDTA) are obtained readily from $LiCPh_3$, $MO^tBu$, and an excess of L. The lithium[40] and sodium[41] compounds consists of contact ion pairs in which there is clear interaction between the cation and the central carbon of the anion. The potassium compound $KCPh_3$.PMDTA.THF **8** (Fig. 6) also forms contact ion pairs,[42] but the $K^+$ is located nearly symmetrically above one phenyl ring with six K–C contacts ranging from 314.2(5) to 325.3(4) pm. The coordination sphere is completed by one K–O and three K–N contacts. There is significant shortening of the distance from the central carbon to the phenyl group which is coordinated to the metal. The configuration observed at the carbanionic center in the lithium derivative (and also in $LiCPh_3.2Et_2O^{43}$) is pyramidal but it is planar in the derivatives of the heavier alkali metals. In the compound $RbCPh_3$.PMDTA, **9,** each Rb is coordinated by two $\eta^6$ interactions to two different carbanions (Rb–C 335.1(4)–364.3(3) pm) linking the ion pairs into zigzag chains. The inner core of the central carbon C(1) and the three *ipso* carbons of the phenyl groups is planar and there is again no metal–C(1) interaction. The cesium derivative $CsCPh_3$.PMDTA, **10,** is also polymeric but the structure is different. The amine is coordinated unsymmetrically and the Cs atom makes no less than 13 additional interactions (Cs–C 334.8(4) to 382.0(4) pm) to neighboring carbon atoms: the phenyl ring of one carbanion is coordinated $\eta^6$ fashion and the neighboring carbanion is coordinated through the central carbon and the *ipso* and *ortho* carbon atoms of each of the three phenyl rings. It has been shown by $^{133}$Cs, $^1$H HOESY that the polymeric structure found in the solid is broken down in solution in THF-$d_8$ to give contact ion pairs with the Cs atoms close to the central carbon, the *ortho* carbon atoms of the anion, and the methyl groups of the triamine ligand. In another study, red derivatives $KCPh_3$.L (L = PMDTA, diglyme, or THF) were isolated from the reactions between $HCPh_3$, L and potassium metal in benzene.[22] The PMDTA adduct, **11** (Fig. 7), in contrast to **8** just described, was monomeric with the shortest K–C distance (293.1(3) pm) that to the central carbon C1, and significant interactions with *ipso* and *ortho* carbons (K–C 304.8(3)–338.6(3) pm). In the diglyme adduct, **12,** the shortest K–C distance (301.6(2) pm) is also that to the central carbon atom, but there are weaker interactions with a phenyl group from the same carbanion and one from

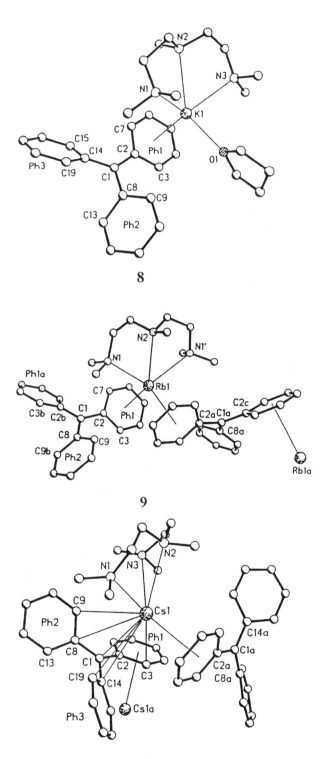

**8**

**9**

**10**

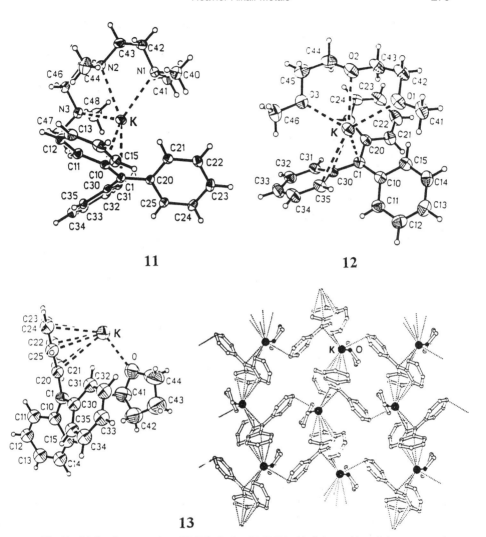

FIG. 7. Molecular structures of KCPh₃.L. L = PMDTA, **11,** diglyme, **12,** and the asymmetric unit and crystal packaging of KCPh₃.THF, **13.** (Reprinted with permission from H. Viebrock *et al., J. Organomet. Chem.* **1995,** *491,* 19. Copyright © 1995 Elsevier Science SA.)

FIG. 6. Molecular structures of the triphenylmethyl derivatives KCPh₃.PMDTA.THF **8,** RbCPh₃.PMDTA **9,** and CsCPh₃.PMDTA **10.** (Reprinted with permission from D. Hoffmann *et al., Organometallics* **1993,** *12,* 1193. Copyright © 1993 American Chemical Society.)

the neighboring ion pair, giving a chain structure. In the THF compound **13** there are interactions between the potassium and the central carbon (K–C 300.3(9) pm) of the ion pair and the neighboring *ipso* carbon atoms, $\eta^6$ coordination to a phenyl group of a neighboring carbanion to give infinite chains, and another series of short K–C contacts linking the chains into sheets.

Triphenylmethyl compounds have been compared with their diphenyl-pyridylmethyl analogues.[44] The coordination of the sodium in the compound $NaC(C_5H_4N)Ph_2$.3THF **14** (Fig. 8) is much less symmetrical than that in the corresponding $CPh_3$ derivative. Thus the distance from sodium to the central carbon is 349.6(10) pm (cf. 264.3(3) pm in $NaCPh_3$.TMEDA) and the shortest Na–C distances are 285.9(9) and 306.8(9) pm to the *ipso* and *ortho* carbon atoms of a phenyl ring. The strongest bonds to sodium are from the oxygen atoms of the three THF molecules and the nitrogen of the pyridyl ring. In the potassium compound $KC(C_5H_4N)Ph_2$. PMDTA.THF **15,** the potassium is strongly bound to the THF oxygen, the three nitrogen atoms of PMDTA, and the pyridyl nitrogen, but the coordination sphere is large enough for further significant interactions with three carbon atoms of a phenyl ring Ph1 (K–C 314.0(6), 320.1(6), and 345.5(7) pm). The coordination is only $\eta^3$; in contrast, the potassium in the triphenylmethyl analogue is almost over the center of the ring. The substitution of N for CH in the $CPh_3$ anion thus results in a shift of the alkali metal toward the nitrogen center where the negative charge is localized. Detailed comparisons of bond lengths in the anions have been made and extensive NMR data obtained.[44]

### D. *Silicon-Substituted Alkyl Derivatives*

Trimethylsilylmethyl derivatives of the heavier alkali metals have been used in the synthesis of other organometallic compounds.[6,16,17,45] but so far there do not appear to be any reports of structure determinations.

Bis(trimethylsilyl)methyl derivatives of Na, K, and Rb have been made from the organolithium compound and the alkoxides $NaO^tBu$, $KO^tBu$, $RbOC_6H_3{}^tBu_2$-2,6, or $CsOCH_2CHEtBu$.[46,47] The sodium and potassium compounds were precipitated from hexane, washed free from alkoxide, and crystallized as solvent-free solids. The structure of the sodium compound, **16** (Fig. 9), like that of the lithium analogue described previously, consists of chains of $Na^+$ cations alternating with anions having a planar $CHSi_2$ configuration (mean Na–C 255.5(10) pm) so that the overall coordination at carbon is trigonal bipyramidal.[48] The C–Na–C angle is 143(6)°, suggesting that the chain skeleton is distorted easily by weak interactions between the

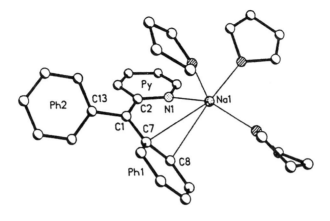

**14**

**15**

FIG. 8.   Molecular structures of $NaC(C_5H_4N)Ph_2.3THF$, **14,** $KC(C_5H_4N)Ph_2.PMDTA. THF,$ **15.** Compare the coordination in **15** with that in **8.** (Reprinted with permission from U. Pieper and D. Stalke, *Organometallics* **1993,** *12,* 1201. Copyright © 1993 American Chemical Society.)

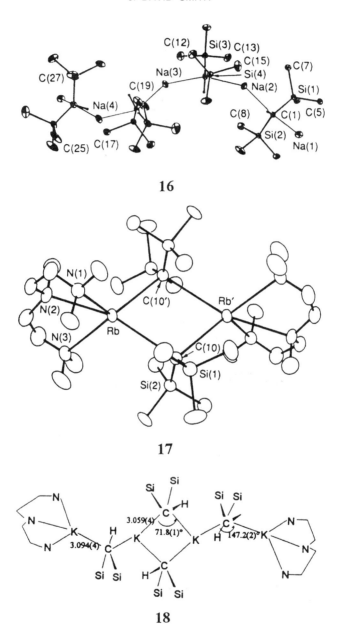

FIG. 9. Molecular structures of the bis(trimethylsily)methyl compounds $\{NaCH(SiMe_3)_2\}_n$, **16**, $[RbCH(SiMe_3)_2.PMDTA]_2$, **17**, and $(PMDTA)K(\mu\text{-}CHR_2)K(\mu CHR_2)_2$ $K(\mu\text{-}CHR_2)K$-(PMDTA), **18**, (R = $SiMe_3$). [Reprinted with permission from M. F. Lappert and D-S. Liu, *J. Organomet. Chem.* **1995**, *500*, 203. (Copyright © 1995 Elsevier Science SA) and P. B. Hitchcock *et al. J. Chem. Soc. Chem. Commun.* **1993**, 1386 (Copyright © 1993 Royal Society of Chemistry).]

metal and the methyl groups on the periphery of the anion. The potassium and rubidium compounds were isolated as PMDTA adducts. In crystals of the rubidium derivative there are electron-deficient dimers with Rb–C 336.1(10) and 348.5(8) pm. Although the structure resembles that of $Al_2Me_6$, it seems likely that the configuration at the bridging carbon is planar rather than tetrahedral. The Rb–C–Rb angle is 104.7(2) [cf. 75.7(1)° in $Al_2Me_6$]. The potassium compound (PMDTA)K($\mu$-CHR$_2$)K($\mu$-CHR$_2$)$_2$ K($\mu$-CHR$_2$)K(PMDTA), **18** (R = SiMe$_3$), shows both (CHR$_2$)$_2$ bridges (K–C 305.9(4) pm) as in the rubidium analog **17** and single CHR$_2$ bridges as found in the sodium compound or in KC(SiMe$_3$)$_3$, **20,** described later. The compound NaCH(SiMe$_3$)$_2$, like the lithium analog, is soluble in hexane and volatile, suggesting that the bonds within the chain are readily broken. Both the Na and the K compounds have proved to be useful reagents for further syntheses, e.g., for $\beta$-diketenimates or organometallic compounds of ytterbium such as Yb(CHR$_2$)$_2$(OEt$_2$)$_2$, NaYb(CHR$_2$)$_3$ or Yb(CHR$_2$)(OAr)(THF)$_3$ (R = SiMe$_3$ Ar = 2,6-$^t$Bu-4-MeC$_6$H$_2$).[49] Recent chemistry in this area has been summarized concisely.[46]

Organometallic compounds containing three silyl substituents at the carbanionic center show a variety of novel structures and interesting chemistry. Some of this is described in a review.[50] Solvent-free NaC(SiMe$_3$)$_3$, **19,** has not hitherto been isolated, but a TMEDA adduct can be obtained as colorless air- and moisture-sensitive crystals by metallation of a solution of HC(SiMe$_3$)$_3$ in Et$_2$O. The structure in the solid is that of a dialkylsodate,[19] in which the Na–C bond length (247.9(6) pm) is unusually short. The anion is isoelectronic with the neutral compound Mg[C(SiMe$_3$)$_3$]$_2$ and differs from it only in having a lower charge (+11) on the central nucleus instead of +12. As the charge is increased, electron density is drawn into the metal–carbon bond (Mg–C 211.6(2) pm) from the inner Si–C bonds, which increase in length from 181.7(8) in the sodate anion to 187.7(3) pm in the organomagnesium compound. In contrast, the potassium and rubidium compounds MC(SiMe$_3$)$_3$, **20** and **21,** respectively, crystallize, even from solvents containing strong amine or ether donors, as solvent-free species (Fig. 10). These have one-dimensional ionic structures in which cations alternate with carbanions having a planar CSi$_3$ core.[20] The mean distance from potassium to the central carbon is 309.7 pm, but there are six further K–C(Me) distances, three on either side, that do not differ significantly from the mean, 322.7 pm. This and the short inner Si–C distance (182.2(10) pm) suggest that the carbanionic charge is delocalized over the CSi$_3$ system and that the electrostatic interaction with the potassium is with the ion as a whole rather than through a specific K–C bond. The adduct KC(SiMe$_3$)$_3$.TMEDA, **22** (Fig. 11), has also been isolated.[12] It was found to be much more pyrophoric than the amine-free compound and conse-

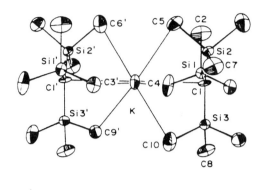

**20**

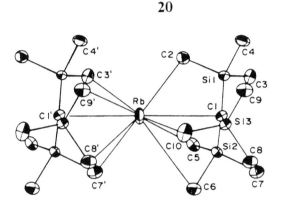

**21**

Fig. 10.  View across the chains in the structures of KC(SiMe₃)₃, **20**, and RbC(SiMe₃)₃, **21**. (Reprinted with permission from C. Eaborn *et al., J. Organomet. Chem.* **1995,** *500,* 89. Copyright © 1995 Elsevier Science SA.)

quently the quality of the X-ray data was much poorer, but the main features of the structure are clear. Each potassium is coordinated by two nitrogen atoms from the amine (K–N = 285(1) and 288(1) pm), the central carbon of one anion (K–C 292(1) pm), and a peripheral methyl group of another (K–C 321 pm). The extreme sensitivity of the compound is attributed to the low coordination number of the potassium, which is prevented by the bonds to the amine from making closer anion contacts.

In the rubidium compound RbC(SiMe₃)₃, **21**,[11] the mean Rb–C(1) distance is 328.9 pm and there are eight Rb–C(Me) distances, four on

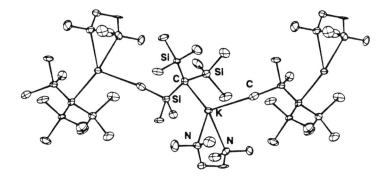

**22**

FIG. 11. Molecular structure of KC(SiMe₃)₃.TMEDA, **22**. (Reprinted with permission from C. Eaborn *et al., Organometallics* **1997,** *16,* 4728. Copyright © 1997 American Chemical Society.)

either side, in the range 327.8(9)–377.2(9) pm. Although the anion remains planar, there are significant distortions from $C_3$ symmetry induced by the need to accommodate additional Rb–C(Me) interactions required by the larger cation. The compound $CsC(SiMe_3)_3$ was expected to be similar to the K and Rb analogues, but it crystallized from benzene as a solvate $Cs(C_6H_6)_3C(SiMe_3)_3$, **23** (Fig. 12), with Cs–C 332.5(12) pm. It appears that $Cs-\eta^6-C_6H_6$ interactions are sufficiently strong to overcome the electrostatic interionic forces evident in the solid-state chain structures.[11,20]

The realization that the $C(SiMe_3)_3$ ligand is isoelectronic with N-$(SiMe_3)_2AlMe_3$ has led to the isolation of a number of compounds containing $N(SiMe_3)_2AlMe_3$ fragments in order to compare their structures with those of analogous compounds containing the $C(SiMe_3)_3$ group. Thus $NaN(SiMe_3)_2$ has been found to react with an equimolar amount of $AlMe_3$ to give the ionic compound $Na[Na\{N(SiMe_3)_2AlMe_3\}_2]$. The anion has a structure similar to that of the sodate $[Na(TMEDA)_2Et_2O][Na\{C(SiMe_3)_3\}_2]$. The $Na^+$ cation is surrounded by five methyl groups at the periphery of the anion [Na–C 273.3(5)-285.9(5) and one longer interaction of 310.9(5) pm].[51]

The isolation of solvent-free organometallic compounds of the alkali metals makes it possible to use them as ligand transfer reagents for the synthesis of compounds, which are unstable in ether solvents. Thus $KC(SiMe_3)_3$ has been used to make $Yb[C(SiMe_3)_3]_2$[52,53] and $Ca[C(SiMe_3)_3]_2$,[54] both of which are decomposed rapidly by $Et_2O$. Reactions with $LiC(SiMe_3)_3$

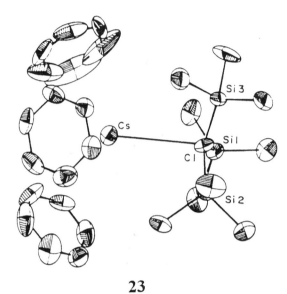

**23**

FIG. 12.   Molecular structure of Cs(C$_6$H$_6$)$_3$C(SiMe$_3$)$_3$, **23**. (Reprinted with permission from C. Eaborn *et al.*, *J. Organomet. Chem.* **1995,** *500,* 89. Copyright © 1995 Elsevier Science SA.)

in THF often lead to the isolation of ate complexes (for reviews, see Refs. 50 and 55), but new possibilities are opened up by the recent improved synthesis[56] of the solvent-free compound.[57]

A further series of organometallic compounds can be obtained by metallation of the compounds HC(SiMe$_3$)$_x$(SiMe$_2$Ph)$_{3-x}$ with NaMe or KMe.[12] The sodium derivatives NaC(SiMe$_3$)$_2$(SiMe$_2$Ph).TMEDA, **24,** and NaC (SiMe$_3$)(SiMe$_2$Ph)$_2$.TMEDA, **25,** have been isolated in the solid state and characterized by X-ray studies. Both adopt molecular structures (Fig. 13), with Na–C 256.6(3) and 256.5(5) pm, respectively, but the short inner Si–C distances and the near-planar configuration at carbon suggest that the species are best viewed as ion pairs with delocalization of negative charge from sodium to the CSi$_3$ skeleton. There are also short contacts of 288–303 pm to the *ipso* and *ortho* carbon atoms of one phenyl ring. There does not appear to be enough room, at least in the presence of TMEDA, for the sodium to coordinate to a second phenyl ring. The potassium compounds KC(SiMe$_3$)$_2$(SiMe$_2$Ph) and KC(SiMe$_3$)(SiMe$_2$Ph)$_2$ have been made similarly. The mono-substituted derivative was recrystallized from benzene and obtained as a dialkylpotassate, [K(C$_6$H$_6$)][K{C(SiMe$_3$)$_2$(SiMe$_2$Ph)}$_2$], **26** (Fig. 14).[58] The potassium cation is coordinated by one benzene molecule (K–C 328(1) pm) and phenyl groups from anions on either side (K–C

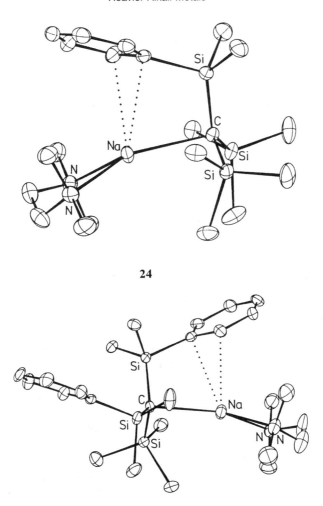

**24**

**25**

Fig. 13.   Molecular structures of NaC(SiMe$_3$)$_2$(SiMe$_2$Ph).TMEDA, **24,** and NaC(SiMe$_3$)-(SiMe$_2$Ph)$_2$.TMEDA, **25.** (Reprinted with permission from C. Eaborn *et al., Organometallics* **1997,** *16,* 4728 Copyright © 1997 American Chemical Society.)

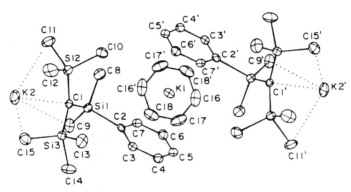

FIG. 14.  Molecular structure of the dialkylpotassate, $[K(C_6H_6)][K\{C(SiMe_3)_2(SiMe_2Ph)\}_2]$, **26**. (Reprinted with permission from C. Eaborn *et al., Angew. Chem. Int. Ed. Engl.* **1995**, *34*, 2679. Copyright © 1995 Wiley VCH.)

315.5(4)–319.0(5) pm). The other potassium is eight coordinate with significant interactions with the central carbon on either side (K–C 291.4(5) pm) and with three methyl groups (K–C 329.4(6)–344.8(5) pm). The environment of the potassium K2 thus resembles that in $KC(SiMe_3)_3$, but the fact that the K–C distances are significantly shorter is consistent with the formulation of the compound as a dialkylpotassate. The sum of the angles at the carbanionic center is 356.8°.

A similar potassate was obtained from the reaction between methylpotassium and $HC(SiMe_3)_2SiMe_2(CH{=}CH_2)$.[58] The fine crystals were too fragile for an X-ray structure determination but on standing in the supernatant solution much larger crystals separated. These were shown to consist of $[K(OSiMe_2)_7][K\{C(SiMe_3)_2SiMe_2(CH{=}CH_2)\}_2]$, but X-ray data were insufficiently accurate to justify extended discussion. The compound had been produced from the reaction between the initially formed product and silicone grease.

The compound $KC(SiMe_3)(SiMe_2Ph)_2$ has been isolated free from donor solvent, but the crystals obtained so far have not been suitable for single crystal diffraction studies.[12] In compounds containing the $C(SiMe_3)$ $(SiMe_2Ph)_2$ group (R), the methyl groups of the $SiMe_2Ph$ group are inequivalent and should therefore give two signals in [1]H- and [13]C-NMR spectra. In RH the signals are distinct at 300 MHz and 25°C, but in the alkali metal compounds separate signals are observed only when the samples are cooled, suggesting that inversion at the central carbon is occurring on the NMR time scale. The barrier to inversion decreases in the series Li > Na > K as the compounds become more ionic. The ionic character of the species in solution is also shown by the fact that the signals from the quaternary carbon atoms are shifted to low frequency and by the high values (typically 50–70 Hz) of the coupling constants $^1J(\text{CSi})$, which are larger than those in the silanes from which the alkali metal compounds are generated.

All five compounds $MC(\text{SiMe}_2\text{Ph})_3$ (M = Li, Na **27**, K, Rb **28**, or Cs **29**) have been characterized structurally. The lithium compound crystallizes with one mole of THF and forms discrete ion pairs with Li–C 212(2) pm and a short contact (240(2) pm) from the metal to the *ipso* carbon of the phenyl group.[59] The derivatives of the heavier alkali metals all separate, even from solvents containing amine or ether donors, as solvent-free yellow solids that are surprisingly insoluble in organic solvents and chemically unreactive.[12,19,20] The four compounds crystallize in different space groups, but in all cases the structures comprise chains of alternate cations and planar anions (Fig. 15). In the sodium, potassium, and rubidium compounds, each metal is coordinated to the central carbon, C1, of the nearest anion [C(19) for the Rb compound **28**], two attached phenyl groups and one phenyl group from the adjacent anion. In the cesium compound **29**, each metal ion is coordinated to the central carbon, one phenyl group from the nearest carbanion and two from the next anion. The M–C distances are Na–C(1) 275.4(14), Na–C(Ph) 280.9(13)–353.6(14); K–C(1) 336(2), K–C(Ph) 296(2)–358(2); Rb–C(19) 350.7(3), Rb–C(Ph) 317.4(2)–352.0(3); Cs–C(1) 365.5(7), Cs–C(Ph) 334.7(9)–362.0(6) pm. These data show that as the alkali metal ion becomes heavier and softer the principal interaction shifts from the central atom to the aromatic rings of the $C(\text{SiMe}_2\text{Ph})_3$ groups, the metal–ring interaction becomes more symmetrical, and the hapticity increases, viz. $\eta^1$ for Li, $\eta^3$ for Na, and $\eta^6$ for K–Cs.[12]

Several organometallic compounds of the alkali metals containing alkyl groups bearing donor substituents at silicon have been isolated.[60] For example, solvent-free $KC(\text{SiMe}_3)_2(\text{SiMe}_2\text{NMe}_2)$, **30**, and $KC(\text{SiMe}_2\text{NMe}_2)_3$, **31**, may be isolated from the metallation of the corresponding silanes $HC(\text{SiMe}_3)_n(\text{SiMe}_2\text{NMe}_2)_{3-n}$. Both form one-dimensional ionic solids with chains of alternate potassium cations and anions having planar $\text{CSi}_3$ skeletons (Fig. 16). In the mono-substituted derivative the potassium is coordinated to the central carbon of the adjacent anions (K–C 318.7(1) and 320(4) pm) and to the nitrogen from the attached $\text{SiMe}_2\text{NMe}_2$ group (K–N 284.8(1) pm). The coordination sphere is completed by short contacts to methyl groups in the next ion pair (K–C 311.3(2)–340.4(2) pm). In the tri-substituted compound the potassium is coordinated on one side to the central carbon (K–C 322.2(6) pm) and one $\text{NMe}_2$ group from the same carbanion and on the other side to two $\text{NMe}_2$ groups from the next ion pair (K–N 281.2(5)–282.3(5) pm). The structure of $KC(\text{SiMe}_3)_2(\text{SiMe}_2\text{NMe}_2)$ is therefore neatly intermediate between those of $KC(\text{SiMe}_3)_3$ and $KC(\text{SiMe}_2\text{NMe}_2)_3$. The compound $KC(\text{SiMe}_3)_2(\text{SiMe}_2\text{OMe})$ has been made

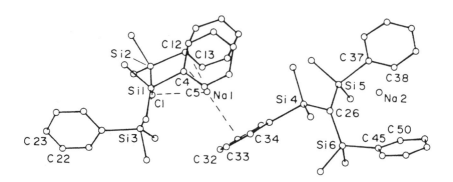

**27**

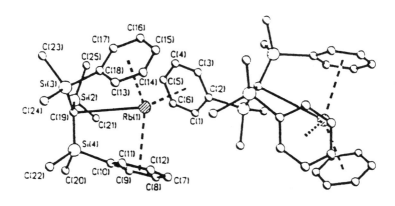

**28**

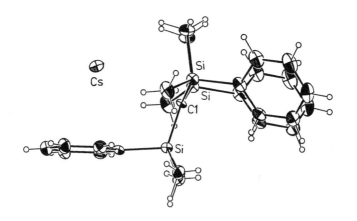

**29**

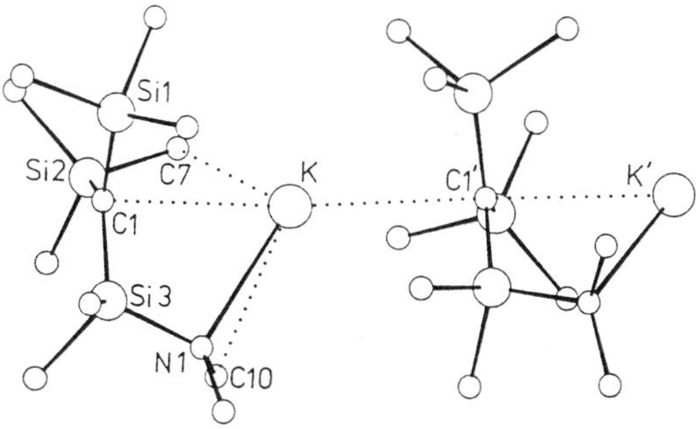

**30**

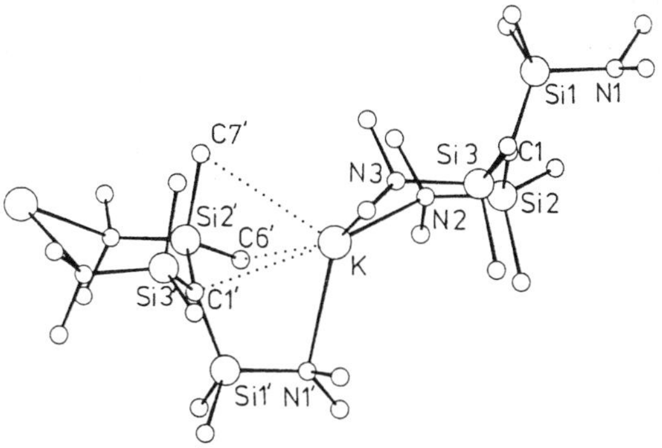

**31**

FIG. 16.   Molecular structures of KC(SiMe$_3$)$_2$(SiMe$_2$NMe$_2$), **30**, and KC(SiMe$_2$NMe$_2$)$_3$, **31**. Compare the coordination on the left of the potassium in **30** with that in **31** and the coordination on the right of the potassium in **30** with that in **20** (Fig. 10).

similarly and used in the synthesis of the first $\sigma$-bonded dialkylsamarium, isolated as a THF complex, Sm{C(SiMe$_3$)$_2$(SiMe$_2$OMe)}$_2$.THF.[61]

### E. *Alkylalkoxide Intermediates*

Sodium or potassium alkylalkoxides, formulated as [K([18]-crown-6)][OCH$_2$CH$_2$CH$_2$K] have been postulated as intermediates in the metalla-

FIG. 15.   Molecular structure of NaC(SiMe$_2$Ph)$_3$, **27**, RbC(SiMe$_2$Ph)$_3$, **28**, and CsC-(SiMe$_2$Ph)$_3$, **29**. [Reprinted with permission from C. Eaborn *et al.*, *Organometallics* **1997**, *16*, 4728. (Copyright © 1997 American Chemical Society) and S. S. Al-Juaid *et al.*, *Angew. Chem. Int. Ed. Engl.* **1994**, *33*, 1268 (Copyright © 1994 Wiley VCH).]

tion of anisole and triphenylmethane by oxetane in the presence of [18]-crown-6. The reaction is considered to involve formation of the oxetane radical anion, which cleaves at the C–O bond and reacts with the alkali metal to give the alkylalkoxide. No structural data for the intermediates have been reported.[62]

# IV

## ARYL DERIVATIVES

### A. *Phenyl Derivatives*

Although phenyl-sodium and -potassium may be made from reactions between phenyl-lithium and the heavier metal alkoxides, the structures of the donor-free solid compounds are apparently still unknown. The dimeric adducts [NaPh.PMDTA]$_2$ and [NaC$_6$H$_4$Ph.TMEDA]$_2$ have been characterized[1] and the structure of a solvent-free arylsodium [Na(C$_6$H$_3$Mes$_2$-2,6)]$_2$, **32** (Fig. 17), has been described.[63] The sodium interacts with the *ipso*

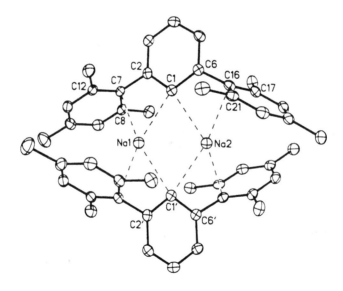

**32**

Fig. 17.   Molecular structure of [Na(C$_6$H$_3$Mes$_2$-2,6)]$_2$, **32.** (Reprinted with permission from M. Niemeyer and P. P. Power, *Organometallics* **1997,** *16,* 3258. Copyright © 1997 American Chemical Society.)

carbons of the central ring (Na–C 257.2(2), 260.9(2) pm) as well as with those of the substituted mesityl groups (Na–C 259.5(2), 266.3(2) pm, cf. 265.6(4), 268.2(6) pm in NaPh.PMDTA). A comparison between $[Na(C_6H_3Mes_2-2,6)]_2$ and $1,3-Mes_2C_6H_4$ shows that in the sodium compound the *ipso–ortho* C–C bond is lengthened from 139.1(6) pm to 142.0(3) and the angle at the *ipso* C is reduced from 122 to 113°. The potassium and cesium derivatives appear to be less stable in solution than the sodium derivative. In attempts to make them, the main product was $1,3-C_6H_4Mes_2$, although the orange potassium compound appeared to separate from hexane as an insoluble, presumably polymeric, solid below 0°C.

The colorless air- and moisture-sensitive 2,6-bis[(dimethylamino)methyl]phenyl derivative $NaC_6H_3(CH_2NMe_2)_2$, **33,** was obtained by the exchange reaction between the organolithium derivative and sodium 1,1-dimethylpropoxide.[64] An X-ray study (Fig. 18) showed that in the solid state it formed trimers in which the sodium was completely solvated intramolecularly, giving an overall propeller structure. The $Na_3C_3$ ring is slightly puckered and the aryl rings are at 66° to the least-squares plane. The Na–C distances

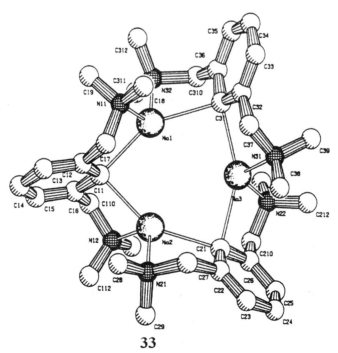

**33**

FIG. 18.    Molecular structure of $NaC_6H_3(CH_2NMe_2)_2$, **33.** (Reprinted with permission from R. den Besten *et al., J. Organomet. Chem.* **1995,** *498,* C6. Copyright © 1995 Elsevier Science SA.)

range from 249.5(5) to 255.4(5) pm. The compound is unstable in toluene above $-50°C$, but NMR spectra obtained from samples below this temperature confirm that the $CH_2NMe_2$ protons are diastereotopic, consistent with the preservation of the ring structure.

V

**CYCLOPENTADIENYL AND RELATED DERIVATIVES**

A. *Unsubstituted Cyclopentadienyl Derivatives*

Highlights in the chemistry of cyclopentadienyl compounds have been reviewed.[65] Trends in the metallation energies of the gas-phase cyclopentadienyl and methyl compounds of the alkali metals have been studied by ab initio pseudopotential calculations. Whereas there is a smooth increase in polarity of $M-(C_5H_5)$ bonds from Li to Cs, lithium appears to be less electronegative than sodium in methyl derivatives. The difference between $C_5H_5$ and $CH_3$ derivatives is attributed to differences in covalent contributions to the M–C bonds. In solution or in the solid state these trends may be masked by the effects of solvation or crystal packing.[66] The interaction between the alkali metal ions $Li^+-K^+$ and benzene has also been discussed.[67]

Although cyclopentadienylsodium and -potassium have been used widely in synthetic chemistry since the 1900s, the structures of these compounds in the solid state were described only in 1997. Because of difficulties in obtaining good crystals, single crystal diffraction methods could not be used, but the structures have been determined by powder techniques using high-resolution synchrotron radiation.[27] In the sodium compound $NaC_5H_5$, **34** (Fig. 19) ($LiC_5H_5$ is isostructural), the $C_5H_5$ anions lie on mirror planes perpendicular to the $c$ axis and the metal atoms are almost linearly coordinated by two $C_5H_5$ rings [Na–C 263.1(12)–267.1(6) pm and $c$–Na–$c$ = 177.7°]. Each sodium is surrounded by four $C_5H_5^-$ anions in the $ab$ plane.* The potassium compound, **35,** has a different structure. Instead of the "string of pearls" found in $NaC_5H_5$, the alternate cations and anions form a zigzag chain (K–C 295.5(5)–314.0(6) pm, K′–K–K″ 138°). On the exposed side, the potassium atoms interact in $\eta^2$ fashion with two further $C_5H_5^-$ anions (K–C 333.1(7)–366.3(8) pm). The coordination at potassium and the K–C distances are similar to those in $[KC_5H_4SiMe_3]_n$[68] and in $[KC_5H_5 \cdot OEt_2]$ (K–C 299.1(2)–303.2(2) pm).[69] In the latter case the K–$\eta^2$-$C_5H_5$ interaction between chains is replaced by interaction between the potassium and the lone pair of the oxygen in the ether.

---

* Throughout this review, for example, on pp. 294, 295, 301, 304, 319, 337, and 340, the centroids of $C_5H_5$, $C_6H_6$, and $C_8H_8$ rings are denoted by $c$. On pp. 294 and 295 the crystallographic axis of compound **34** is also denoted by $c$.

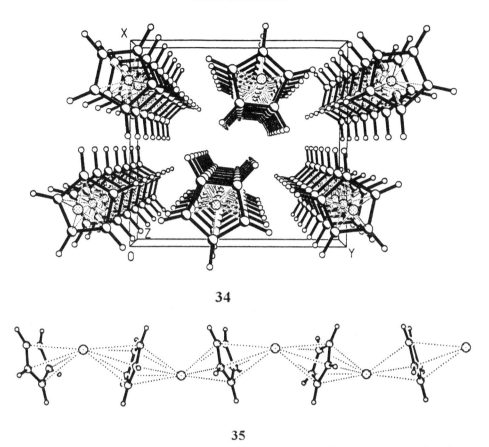

**34**

**35**

Fig. 19.   Molecular structure of $NaC_5H_5$, **34**, along the chains down the $c$ axis and that of $KC_5H_5$, **35**, viewed across the zigzag chains (K atoms are denoted by large circles). (Reprinted with permission from R. E. Dinnebier *et al.*, *Organometallics* **1997**, *16*, 3855. Copyright © 1997 American Chemical Society.)

   The rubidium analogue $RbC_5H_5$ was isolated as a white air-sensitive powder from the reaction between rubidium metal and $C_5H_6$ in THF.[70] An X-ray powder diffraction study showed that two phases crystallize simultaneously: in each there are polymeric chains with $\eta^5$-$C_5H_5$–Rb bonding (Rb–C 317.3–330.1 pm and $c$–Rb–$c$ 123.5 to 136.5°). The coordination sphere of the rubidium is completed by interactions with cyclopentadienyl rings in adjacent chains. The cesium compound **36**[71] (Fig. 20) is isostructural with one of the rubidium-containing phases [Cs–C 329.6(6)–338.9(6) pm and $c$–Cs–$c$ 129.7°]. Zigzag chains are also found in the adduct $NaC_5H_5$.TMEDA.[72]

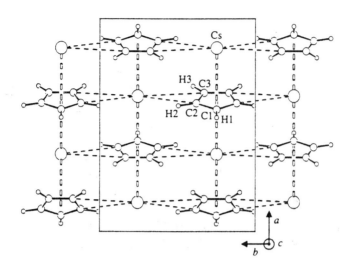

**36**

Fig. 20. Molecular structure of $CsC_5H_5$, **36**, viewed across the zigzag chains. (Reprinted with permission from R. E. Dinnebier *et al.*, *Acta Crystallogr. Sect. C*, **1997**, *53*, 699. Copyright © 1997 International Union of Crystallography.)

The reaction between $LiC_5H_5$ and $Ph_4PCl$ in a $2:1$ molar ratio gives a salt containing the "lithocene" anion.[73]

$$2LiC_5H_5 + Ph_4PCl \rightarrow [Ph_4P][Li(C_5H_5)_2] + LiCl \qquad (10)$$

However, the analogous reaction with $NaC_5H_5$ yielded dark red crystals of the salt $[Ph_4P][C_5H_5]$ containing discrete $[C_5H_5]^-$ anions. When a large excess of $NaC_5H_5$ was used the orange-brown compound $[Ph_4P]$ $[Na(C_5H_5)_2]$, containing the anion **37** was obtained.[74] This had a crystal structure very similar to that of the lithium analogue with the Na–C bond lengths in the narrow range $258.4(3)$–$267.0(3)$ pm. (In contrast, the mean Na–C distance in $NaC_5H_5 \cdot TMEDA$ is 292.4 pm.) The symmetry of the anion (Fig. 21) can be described as $D_{5d}$, but the high anisotropy in the thermal parameters shows the presence of rotational disorder and a small energy difference between staggered and eclipsed configurations. A compound containing an *ansa*-sodacene anion was obtained from the reaction between $NaC_5H_4CMe_2CMe_2C_5H_4Na$ and $Ph_4PCl$ in a $2:1$ molar ratio. The anion has Na–C distances to one $C_5H_5$ ring $265.0(4)$–$270.5(4)$ pm. (Bond lengths to the other ring are uncertain because of crystallographic disorders.) The $c$–Na–$c$ skeleton is not linear and a THF molecule is coordinated on the exposed side of the central sodium atom.[74]

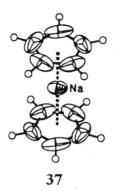

**37**

FIG. 21. Structure of the anion **37** in [Ph$_4$P][Na(C$_5$H$_5$)$_2$]. Reprinted with permission from S. Harder *et al.*, *Organometallics* **1996,** *15,* 118. Copyright © 1996 American Chemical Society.)

The ions [M(C$_5$H$_5$)$_2$]$^-$ (M = Li, Na, or K) may also be isolated in the [(Me$_2$N)$_3$S]$^+$ salts obtained by the reaction of MC$_5$H$_5$ with (Me$_2$N)$_3$S(C$_5$H$_5$) in MeCN and crystals suitable for X-ray studies grown by slow diffusion of Et$_2$O into the MeCN solution at $-30$ to $-40°$C. The structures of the red [(Me$_2$N)$_3$S][Na(C$_5$H$_5$)$_2$] and the violet [(Me$_2$N)$_3$S(C$_5$H$_5$)S(NMe$_2$)$_3$] [Na(C$_5$H$_5$)$_2$] have been determined. The [Na(C$_5$H$_5$)$_2$]$^-$ ion has almost ideal $D_{5d}$ symmetry with Na–C 259.6(3)–263.9(3) pm. These Na–C distances are some of the shortest yet measured. In contrast, the Li–C distances in [Li(C$_5$H$_5$)$_2$]$^-$ are long compared to those in organolithium compounds.[75] The difference between the sodium and the lithium compounds is attributed to reduced intramolecular repulsion between two cyclopentadienyl rings in the sodium compound in accord with ab initio calculations.[74]

The reaction between Cs(C$_5$H$_5$) and Ph$_4$PCl in a 2:1 mole ratio led to the formation of an orange-red compound. This was identified by an X-ray study as [Ph$_4$P][Cs$_2$(C$_5$H$_5$)$_3$], **38,** in which the anion lies across a center of symmetry and has a strongly bent structure ($c$–Cs–$c$ 115.6°).[76] The Cs–C bond lengths to the terminal (C$_5$H$_5$) ligands are from 331.3(6) to 338.5(5) pm. The bond distances to the central ring are in the same range but are determined less accurately as the ring is disordered. In addition to the bonds between a particular Cs atom and the adjacent bridging and terminal C$_5$H$_5$ rings there are interactions with the carbon atoms of one phenyl group and further $\eta^2$ interactions with two further C$_5$H$_5$ groups from neighboring ions (Fig. 22). The interionic Cs–C distances (369.9(5)–381.0(5) pm) are only 10% longer than the intraionic bond lengths.

The addition of NaC$_5$H$_5$ to Sn(C$_5$H$_5$)$_2$ in the presence of PMDTA gives the tris(cyclopentadienyl)stannate Na($\mu$-$\eta^5$,$\eta^5$-C$_5$H$_5$)Sn($\eta^5$-C$_5$H$_5$)$_2$. PMDTA, **39**[77] (Fig. 23). The tin center is almost planar. The sodium is

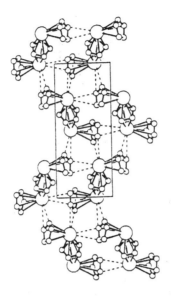

**38**

FIG. 22.    Packing of the $[Cs_2(C_5H_5)_3]^-$ anions in $[Ph_4P][Cs_2(C_5H_5)_3]$, **38,** showing intermolecular Cs–C contacts. (Reprinted with permission from S. Harder and M. H. Prosenc, *Angew. Chem. Int. Ed. Engl.* **1996,** *35,* 97. Copyright © 1996 Wiley VCH.)

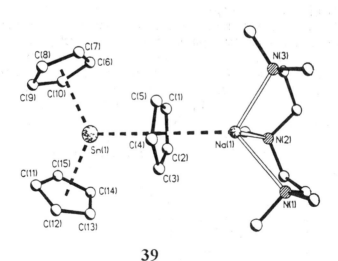

**39**

FIG. 23.    Molecular structure of $Na(\mu-\eta^5,\eta^5-C_5H_5)Sn(\eta^5-C_5H_5)_2$.PMDTA, **39.** (Reprinted with permission from M. G. Davidson *et al., Angew. Chem. Int. Ed. Engl.* **1992,** *31,* 1226. Copyright © 1992 Wiley VCH.)

coordinated to the three nitrogen atoms of the amine and $\eta^5$ coordinated to the bridging cyclopentadienyl ring: the Na distances are 278.2(2)–284.6(2) pm. The Sn–$c$–Na angle is 172.3°. The deviation from 180° is attributed to weak interactions between sodium in one molecular unit and the cyclopentadienyl ring in the next, which link the molecules into a polymeric array. The lead analogue is similar with Pb–$c$–Na 172.9° and Na–C 277.7(9)–288(1) pm. The structures of compounds have been examined by extensive model MO calculations and by solid-state NMR spectra. It is suggested that the unsolvated complexes Na($\mu$-$C_5H_5$)E($C_5H_5$)$_2$ (E = Sn or Pb) are best considered as loose contact complexes between Na($C_5H_5$) and E($C_5H_5$)$_2$.[78] Similar $\mu$-$\eta^5$,$\eta^5$ cyclopentadiene bridges are found in the compound NaYb($C_5H_5$)$_3$, **40** (Fig. 24), made by the reaction between NaC$_5$H$_5$ and Yb($C_5H_5$)$_2$ in THF and purified by sublimation under vacuum. The structure consists of a three-dimensional array of alternate sodium and ytterbium atoms with a triangular arrangement of ligands around both metals. The average Na–C distance is 283 pm.[79]

### B. *Substituted Cyclopentadienyl Derivatives*

In KC$_5$H$_4$SiMe$_3$, isolated as colorless, air-sensitive needles from the reaction between C$_5$H$_5$SiMe$_3$ and potassium metal or potassium hydride,

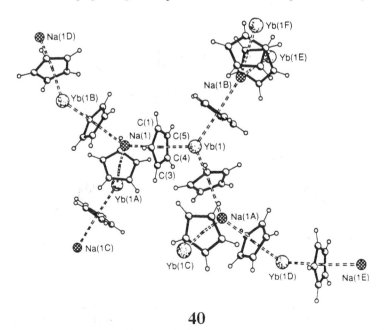

**40**

FIG. 24.   Structure of NaYb($C_5H_5$)$_3$, **40**. (Reprinted with permission from C. Apostolidis *et al.*, *Chem. Commun.* **1997,** 1047. Copyright © 1997 Royal Society of Chemistry.)

the molecules form a one-dimensional ionic solid with zigzag chains
(K–C 298.8(10)-307.4(10) and $c$–K–$c$ 150.7; cf. $KC_5H_5$ above).[68] There are
weaker $\eta^2$ interactions between the potassium ions of one chain and the
$C_5H_5$ rings attached to the adjacent chain (K–C ca. 367 pm). The related
compound $KC_5H_4CH_2CHMe_2$ has been made from potassium hydride and
isobutylcyclopentadiene. No structural data have been reported, but it has
been used in the synthesis of the volatile $Nd(C_5H_4CH_2CHMe_2)_2$. Derivatives with donor substituents $K(C_5H_4CH_2CH_2X)$ X = OMe or $NMe_2$ have
also been reported, but again no structural data have been given.[80] In
$NaC_5H_4COMe$, made by the reaction between $NaC_5H_5$ and methyl etha-
noate (acetate), the ethanoato-bridged dimers are linked by Na–$\eta^5$ interactions (Na–C 269(1)–294(1) pm).[81]

The compounds $M[(C_5HMe_4)_2Ti(\eta^1\text{-}C{\equiv}CSiMe_3)_2]$ (M = Na, K, Rb, Cs)
were made from reactions between the metals and the compound with
M = Li. In the potassium compound **41** (Fig. 25) the cyclopentadienyl
rings form $\eta^5,\eta^5$ bridges, although the K–C distances show a wide range
(307.8(5)–374.1(5), average 339.4(5) pm). There is also a strong tweezers-
like interaction with the alkyne systems (K–C 294.6(4)–302.1(4) pm).[82] The
pyridine adducts of pentamethylcyclopentadienylsodium and -potassium

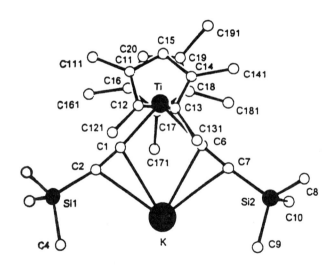

**41**

FIG. 25. Molecular structure of $K[(C_5HMe_4)_2Ti(\eta^1\text{-}C{\equiv}CSiMe_3)_2]$, **41**, showing the
tweezers-like interaction with alkyne ligands. (Reprinted with permission from V. Varga
*et al., J. Organomet. Chem.* **1996**, *514*, 219. Copyright © 1996 Elsevier Science SA.)

have also been isolated.[28] The sodium compound **42** is monomeric in the solid state with each sodium coordinated by three pyridine molecules and one $C_5Me_5$ ring (Na–C 265.8(5)–270.1(5) pm) (Fig. 26), and NMR data suggest that the monomeric structure is retained in pyridine solution. The potassium compound **43** crystallizes in zigzag chains ($c$–K–$c$ 138°), showing that the $\eta^5$–$C_5Me_5$ interactions (K–C 296.2(2)–310.4(2) pm) compete successfully with pyridine for sites in the coordination sphere of the potassium. Similar $\eta^5$–$C_5Me_5$ contacts are found in the green ate complexes isolated, for example, from the reaction between $KC_5Me_5$ and $C_5Me_5SmOAr$ (Ar = $C_6H_3{}^tBu_2$-2.6, $C_6H_2{}^tBu_2$-2,6-Me-4 and $C_6H_2{}^tBu_3$-2,4,6) in THF. The structures of the first two were found to consist of polymeric units in which the $C_5Me_5$ forms $\mu$-$\eta^5,\eta^5$ bridges between samarium and potassium (Fig. 27).[83] The pentabenzyl derivative $KC_5(CH_2Ph)_5$.3THF[84] has the same piano stool structure as $NaC_5Me_5$.3py. The bond distances from potassium to the $\eta^5$-$C_5(CH_2Ph)_5$ ring are in the range 296.8(5)–309.5(5) pm.

The bis(phenyl)-substituted cyclopentadienyl compound $Na(C_5H_3Ph_2)$ has been obtained in 81% yield from the reaction between 1,4-$C_5H_4Ph_2$ and $NaNH_2$.[85] It has been characterized in solution by NMR spectra and its reactions with InCl and $FeCl_2$ to give InR and $FeR_2$ (R = $C_5H_3Ph_2$). The compound $NaC_5HPh_4$.DME, **45,** obtained from $C_5H_2Ph_4$ and a sodium mirror in DME, has a structure (Fig. 28) like that of $NaC_5Me_5$.3py described earlier.[86] The sodium is coordinated on one side by the ether and on the other by the $[\eta^5\text{-}C_5HPh_4]^-$ anion, but the Na–C distances (273.5(2)–290.8(2) pm) are significantly longer than those in the pyridine adduct. In contrast, the reaction in the presence of [18]-crown-6 in hexane yield air- and moisture-sensitive crystals with the composition $\{[18]\text{-crown-6}\}_3Na(C_5HPh_4)_2$. An X-ray structure determination showed that this contained a "triple-decker cation" $[([18]\text{-crown-6})Na([18]\text{-crown-6})Na([18]\text{-crown-6})]^{2+}$ and two $[C_5HPh_4]^-$ anions. There is thus complete transfer of charge from sodium to the anion and no Na–C contact.[87]

The development of phospholide analogues of cyclopentadienide ions has opened up a wide range of further possibilities for the preparation of organometallic compounds. The reaction of the phosphole tetramer $(C_4PMe_2Ph)_4$ with sodium in DME resulted in cleavage of one of the P–P bonds (Fig. 29) to form the complex $[(C_4PMe_2Ph)_4Na_2(DME)_2]$, **46,** in which each sodium ion is bound in $\eta^5$ fashion to one phospholyl ring, $\sigma$ bound to a second ring, and coordinated to DME.[88] There may be a further weak interaction with the neutral phosphorus atom of another ring. The Na–C distances range from 280.7(2) to 293.7(2) pm and are longer than those in cyclopentadienyl compounds. A potassium derivative, containing the ion **47**, was made in crude form by treating the phosphole tetramer with four equivalents of potassium in DME. Crystals suitable for an X-ray

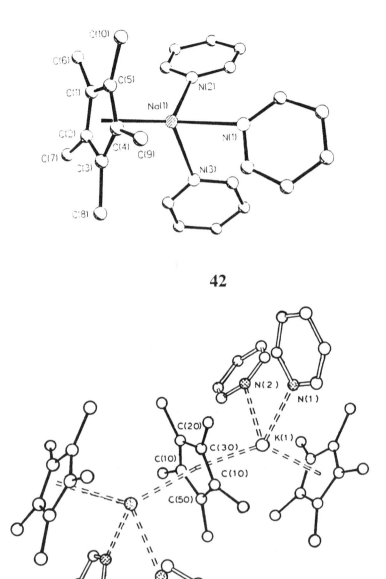

**42**

**43**

FIG. 26. Molecular structure of NaC$_5$Me$_5$, **42,** and KC$_5$Me$_5$, **43.** (Reprinted with permission from G. Rabe *et al., J. Organomet. Chem.* **1991,** *403,* 11. Copyright © 1991 Elsevier Science SA.)

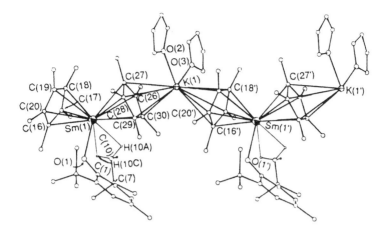

**44**

FIG. 27. Extended structure of [K(THF)$_2$Sm(C$_5$Me$_5$)$_2$OC$_6$H$_2$$^t$Bu$_2$-2,6-Me-4], **44.** (Reprinted with permission from Z. Hou *et al.*, *Organometallics* **1997,** *16*, 2963. Copyright © 1997 American Chemical Society.)

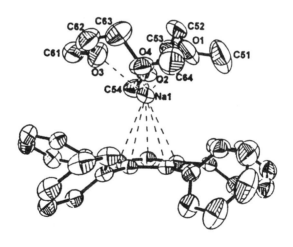

**45**

FIG. 28. Molecular structure of NaC$_5$HPh$_4$.DME, **45.** (Reprinted with permission from C. Näther *et al.*, *Acta Crystallogr. Sect. C* **1996,** *52*, 570. Copyright © 1996 International Union of Crystallography.)

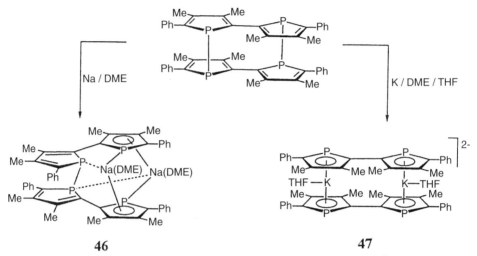

FIG. 29. Formation of the phosphole complex [(C$_4$PMe$_2$Ph)$_4$Na$_2$(DME)$_2$], **46**, and the anion in the potassium compound *trans*-[K([18]-crown-6)(THF)$_2$]$_2$[K(C$_4$PMe$_2$Ph)$_2$THF]$_2$. (Reprinted with permission from F. Paul *et al.*, *Angew. Chem. Int. Ed. Engl.* **1996**, *35*, 1125. Copyright © 1996 Wiley VCH.)

study were obtained by recrystallization from THF containing [18]-crown-6. The structure contained the anionic fragment [K(C$_4$PMe$_2$Ph)$_2$THF]$_2^{2-}$ related to the K(C$_5$H$_5$)$_2^-$ ion described earlier, showing that both P–P bonds of the phosphole tetramer had been cleaved. The mean K–C distance was 315.9 pm and the *c*–K–*c* angle was 137.9° with the THF coordinated on the open side of the central potassium atom.

### C. *Indenyl and Fluorenyl Derivatives*

The indenyl derivatives MC$_9$H$_7$.L (L = amine or ether donor) show a variety of structures that have been discussed previously.[1] Several further examples of this class have been described. The sodium compound NaC$_9$H$_7$.PMDTA is monomeric (from a poor quality of structure). The structure of the green compound KC$_9$H$_7$.TMEDA, **48** (Fig. 30), shows that the potassium is $\eta^5$ coordinated to two indenyl groups on one side [K–C distances vary widely from 295.2(2) to 315.6(2) pm] and to the amine ligand on the other to give polymeric zigzag chains.[69] The compound KC$_9$H$_7$.PMDTA, **49**, also forms polymeric zigzag chains but the structure is more complicated. There are two potassium atoms in the asymmetric unit; for each, one anion is $\eta^5$ coordinated (K–C 308.4–315.2 and 307.4–315.5 pm) and the other $\eta^3$ (K–C 329.4–330.5 and 319.5–329.8 pm).[69]

The crystal structures of several adducts of fluorenylsodium and -potassium have also been described.[1,89] The orange rubidium compound $RbC_{13}H_9$.PMDTA, **50**, crystallizes in what is described as a "square wave-like" polymeric chain (Fig. 31).[90] There are two crystallographically distinct Rb atoms. Both are coordinated $\eta^5$ [Rb–C 321.9(14)–341.1(15) and 321.3(13)–341.0(13) pm] and $\eta^3$ to two fluorenyl anions and to one PMDTA molecule. One rubidium cation, Rb1, coordinates in a $\eta^3$-benzylic fashion (Rb–C 315.6(13)–350.8(16) pm) to carbon atoms C21, C26, and C22 in the neighboring asymmetric unit and the other, Rb2, to carbon atoms C10, C9, and C13 (Rb–C 315.9(14)–352.4(15) pm). Crystals of $(C_{13}H_9Cs.THF)_n$ and $(C_{13}H_9Cs.PMDTA)_n$ were also obtained but X-ray data could not be refined. It was, however, clear that although the Cs was located above the five-membered ring in the PMDTA compound, it was above the six-membered ring in the THF adduct. The variety of structures found for the fluorene complexes has been attributed to the small difference in energy between $\eta^3$-, $\eta^5$-, and $\eta^6$-bonding modes. The fact that the rubidium compound crystallizes in a chain rather than as isolated ion pairs may be because the Coulomb energy for an infinite chain is higher than that for a finite cluster.

The potassium compounds obtained by the reaction of potassium with fluorene or 9-t-butylfluorene in the presence of TMEDA and extracting the product into $THF/Et_2O$ were identified as $KC_{13}H_9.(TMEDA)_2$, **51**, and $[K(THF)_2(\mu\text{-}C_{13}H_8{}^tBu)K(TMEDA)(\mu\text{-}C_{13}H_8{}^tBu)]$, **52**.[91] In the unsubstituted compound **51** (Fig. 32), the potassium interacts with the five-membered ring of one fluorenyl ligand (K–C 307.1(5)–332.0(4) pm) and with four nitrogen atoms from two TMEDA molecules. A second modification of this compound, with each potassium bound to two nitrogen atoms and the five-membered rings of two fluorenyl molecules, was described earlier.[92] The t-butyl-substituted compound **52** has a different structure. The potassium cations are coordinated alternatively to two THF or TMEDA molecules and to two six-membered rings of the fluorenyl anion so that the structure as a whole is polymeric. The coordination to the six-membered rings is unsymmetrical (K–C 309.1(6)–349.3(6) pm), but because there is no sharp discontinuity in the bond distances it is difficult to assign a specific hapticity. AM1 calculations show little difference in electronic structure between substituted and nonsubstituted hydrocarbons so the difference in the structures of the potassium derivatives is attributed to steric factors.

Similar structural principles are evident in compounds obtained by the metallation of 1,2-bis(dimethylamino)-1,2-di-9-fluorenyldiborane $(LH_2)$ by $MN(SiMe_3)_2$ (M = Na or K) or by cesium metal in hexane-THF (lithium, calcium, and barium derivatives have also been made).[93] The dianions were characterized by NMR spectroscopy, and the crystal structures of $K_2L.4THF$, **53**, and $K_2L.2DME$, **54**, were determined (Fig. 33). In each the

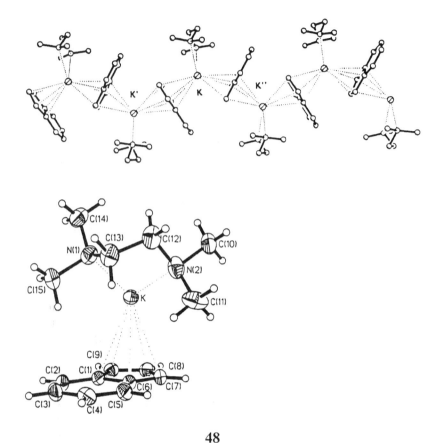

**48**

FIG. 30.   Molecular structure of KC$_9$H$_7$.TMEDA, **48,** and KC$_9$H$_7$.PMDTA, **49,** showing how molecules are linked into chains. (Reprinted with permission from V. Jordan *et al., J. Organomet. Chem.* **1996,** *517,* 81. Copyright © 1996 Elsevier Science SA.)

cations and anions form chains with two different K$^+$ environments. One, K1, is located between the fluorenyl groups of a particular anion with $\eta^5$ coordination to each. The other, K2, links two different anions and the bonding is $\eta^4$ and $\eta^6$. The DME complex is similar. One potassium ion, K1, is bound in $\eta^5$ fashion to two fluorenyl fragments in the same anion [K–C 298.1(5)–323.8(6) pm to one ring, and less symmetrically 298.7(5)–336.5(5) pm to the other]. The other, K2, links neighboring anions in $\eta^5$-$\eta^4$ fashion, with the metal to one side of the five-membered fluorene ring (K–C 306.9(6)–333.8(7) pm).

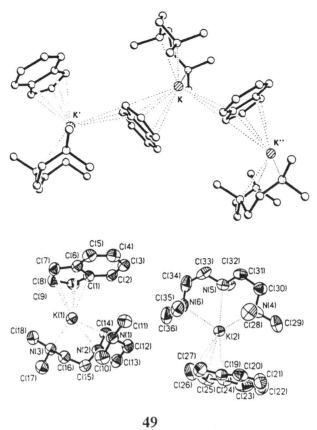

**49**

FIG. 30.   (*Continued*).

## D.  *Azulene and Pentalene Derivatives*

Azulene reacts with a sodium mirror at 150 K in diglyme to give colorless crystals that have been shown by an X-ray study to contain a dianion **55** in which azulene fragments are connected at the 6,6′ position.[94] Each sodium is unsymmetrically $\eta^5$ coordinated to the ring and to two diglyme molecules (Fig. 34). One of the diglyme oxygen atoms is bound to two sodium centers so that the contact triple ion pairs are linked into chains along the *a* axis. Cyclovoltammetric measurements and MNDO calculations suggest that the dianions are produced by disproportionation of the initially formed azulene radical anions.

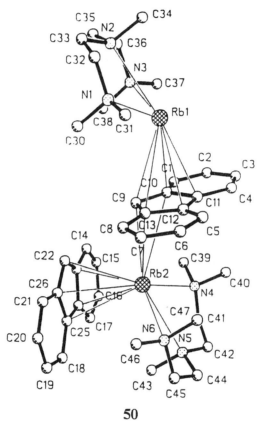

**50**

Fɪɢ. 31. Molecular structure of RbC₁₃H₉.PMDTA, **50**. (Reprinted with permission from D. Hoffmann *et al., J. Organomet. Chem.* **1993,** *456,* 13. Copyright © 1993 Elsevier Science SA.)

The potassium salt of the 1,5-triisopropylsilylpentalene dianion, **56** (Fig. 35), has been made in essentially quantitative yield by the reactions between the parent hydrocarbon and methylpotassium or potassium hydride in $Et_2O$. It reacts with $ThCl_4$ in THF at room temperature to give $Th[C_8H_4(Si^iPr_3)_2\text{-}1,5]_2$, a member of a new class of actinide sandwich compounds shown to consist of a mixture of staggered and eclipsed isomers, both in solution by NMR spectroscopy and in the solid state by an X-ray study.[95,96]

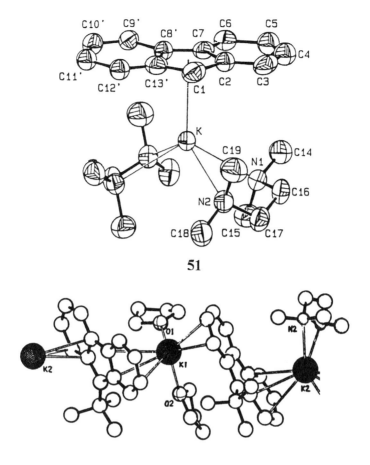

FIG. 32. Molecular structure of $KC_{13}H_9 \cdot (TMEDA)_2$, **51**, and $[K(THF)_2(\mu\text{-}C_{13}H_8{}^tBu)\text{-}K(TMEDA)(\mu\text{-}C_{13}H_8{}^tBu)]$, **52**. (Reprinted with permission from C. Janiak, *Chem. Ber.* **1993,** *126,* 1603. Copyright © 1993 Wiley VCH.)

# VI

## ALKENE AND ALKYNE DERIVATIVES

A. *Introduction*

Alkenes are deprotonated by the organometallic compounds of the heavier alkali metals to give allylmetal derivatives. As these are normally highly

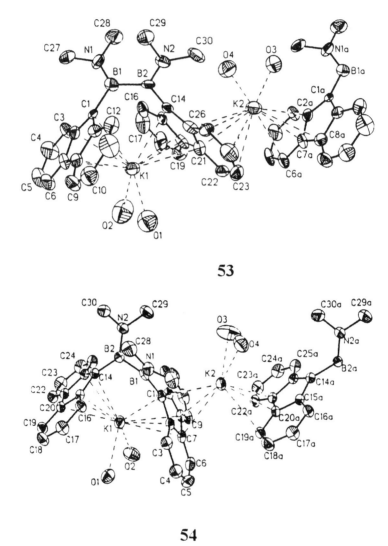

**53**

**54**

Fig. 33.    Molecular structures of $K_2L.4THF$, **53,** and $K_2L.2DME$, **54** ($LH_2$ = 1,2-bis(dimethylamino)-1,2-di-9-fluorenyldiborane). (Reprinted with permission from R. Littger *et al., Chem. Ber.* **1994,** *127,* 1901. Copyright © 1994 Wiley VCH.)

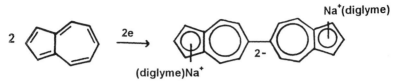

FIG. 34.  Formation of the dianion **55** by reduction of azulene.

reactive and undergo complex rearrangements, they are usually characterized by derivatization. However, where the anionic charge is delocalized, e.g., onto aryl or silicon-substituted groups, it is sometimes possible
under aprotic conditions to isolate ion pairs that can be characterized by
X-ray diffraction studies. In other cases the alkali metal donates an electron to the alkene to give a radical anion. This may then (a) dispropor-
tionate to give a dianion and a neutral alkene, as in the formation of
$Na_2[Ph_2CCPh_2].Et_2O$ from sodium and $Ph_2C=CPh_2$; (b) abstract a proton
from the medium to give an allyl anion, as in $Na[Ph_2CCHCPh_2]$ obtained
from sodium and $Ph_2C=C=CPh_2$ (the sodium is sandwiched by $\eta^6$ interactions to phenyl substituents); or (c) dimerize to give a new dianion with
formation of a C–C bond, as in the formation of $Na_2[Ph_2CCH_2CH_2C-$
$Ph_2].Et_2O$ from sodium and $Ph_2C=CH_2$.[2]

## B. *Allyl Derivatives*

Experimental investigations have been supported and stimulated by theoretical calculations giving comparative data for the whole series from lithium
to cesium. For example, ab initio studies of the allyl derivatives $MC_3H_5$
(M = Li, Na, K, Rb, or Cs) show that symmetrically bridged geometries are

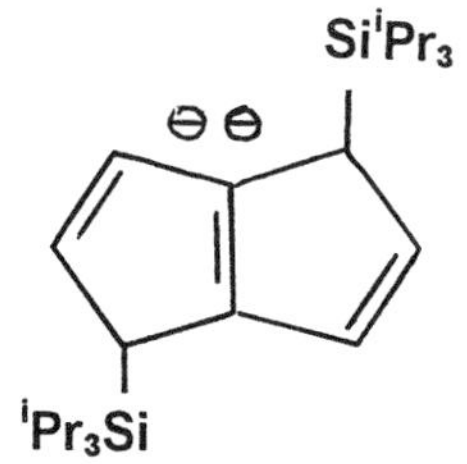

**56**

FIG. 35.  The 1,5-triisopropylsilylpentalene dianion, $K_2[C_8H_4(Si^iPr_3)_2-1,5)]$, **56.**

preferred, the bonding is highly ionic, and the barrier to rotation decreases systematically from Li to Cs, as found experimentally.[97]

Straight-chain derivatives of Na, K, Rb, and Cs have been obtained in solution by the standard preparative method [Eq. (1)]. On standing at $-52°C$ the 5-hexenyl derivatives undergo prototropic rearrangements as shown by quenching the reaction mixtures with MeOD to give 1-hexene, $(E)$- and $(Z)$-2-hexanes, and methylcyclopentane.[8] The lithium compound rearranges exclusively to give cyclopentylmethyllithium, but the heavier alkali metal derivatives cyclize only to a tiny extent. The formation of 2-hexenes with a high level of incorporation of deuterium shows that there are persistent propylallyl intermediates in solution. The small amount of cyclization observed for the heavy alkali metals suggests that these ions interact much less strongly than lithium with the alkene double bonds. In another study,[24] 1,4-, 2,4-, or 1,5-hexadienes have been metallated by cesium-([18]-crown-6) and 1,4- and 2,4-hexadienes by LiBu/CsO$^t$Bu in THF, and the geometries of the resulting anions have been derived from the products of trimethylsilylation or carbonation (Fig. 36). Thus $trans$-1,4-hexadiene gave mainly **57a** with **57b** as the minor product and $cis$-1,4-hexadiene gave **58a** with **58b** as a minor product. Reactions with $cis$–$trans$ and $trans$–$trans$ hexadienes were also studied. After storage at room temperature, an equilibrated mixture of anions was obtained and the main products were a mixture of the hexadienyltrimethylsilanes **57b, 58a,** and **58b.** Presumably, 1,4-dienes metallate more easily than 1,5-dienes because the metal is able to interact with the double bonds when these are closer together. The metallation can be achieved at lower temperature with cesium than with lithium so that rearrangements are restricted. This may be important in some organic syntheses: the work emphasizes the crucial role of the counterion in determining relative rates of reaction.

Reactions between KCH$_2$SiMe$_3$ and cyclohexene or methylcyclohexene gave white solids formulated as **59** or **60.** They were characterized by reaction with CH$_2$O and oxidized with (COCl)$_2$, but no structural data have been reported. Reactions with cyclooctene were similar.[45] The role of tertiary amine in the metallation of ethene, $n$-hexene, and $\alpha$-pinene has been studied.[15]

The reduction of the tetrasilacyclohepta-1,2-diene with a sodium mirror in Et$_2$O gave dark red crystals that were shown by an X-ray structure determination to be the sodium salt of the allyl anion **61** produced by a series of intramolecular rearrangements of the initially formed radical anion (Fig. 37). The Na–C distances are 265.6(5)–288.2(5) pm.[98]

The reactions of phenyl-substituted alkenes with sodium mirrors yield compounds containing anionic intermediates that may be contact ion pairs with significant Na–C contacts to the alkene skeleton or to phenyl substitu-

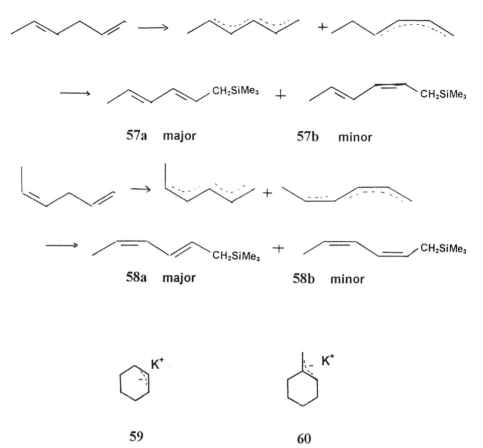

FIG. 36. Allyl intermediates from the reactions between alkali metals and hexadienes[24] and cyclohexenes.[45]

ents, or solvent-separated ion pairs with only long-range electrostatic inter-actions between cation and anion. Some of these have been obtained as crystalline compounds. Earlier work has been summarized.[2,99,100]

The tetrasilacyclobipentylidene $Me_2SiCH_2CH_2Me_2SiC = CSiMe_2CH_2CH_2Si-Me_2$, was readily reduced by lithium or sodium in THF to give dark red solutions containing the corresponding dianion, which was isolated in alkali metal salts with coordinated THF. The structures, determined by X-ray diffraction, showed that the cations were three coordinate (to C1, C2, and O1; Fig. 38). The sodium compound **62** has Na–C 248.7(4) and 250.1(4) pm, at the short end of the range for this kind of contact. The central C–C bond is stretched from 136.9(2) in the free alkene to 157.9(15) pm in the

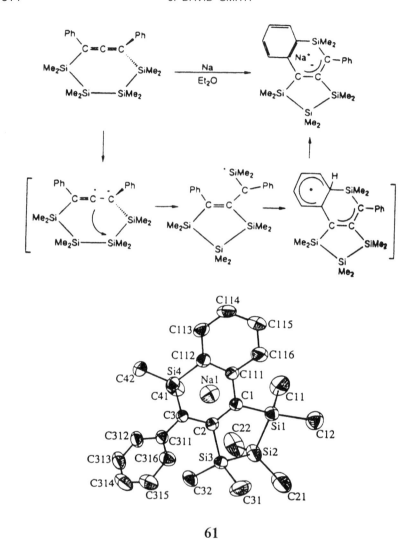

**61**

FIG. 37.  Formation and molecular structure of the sodium salt **61.** (Reprinted with permission from F. Hojo *et al., Organometallics* **1996,** *15,* 3480. Copyright © 1996 American Chemical Society.)

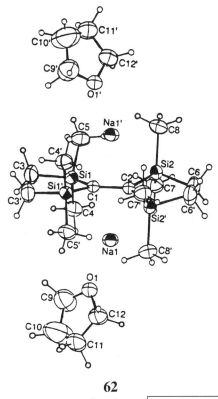

**62**

Fig. 38. Molecular structure of $Na_2(THF)_2[Me_2\overline{SiCH_2CH_2Me_2SiC}=\overline{CSiMe_2CH_2CH_2Si}$-$Me_2]$, **62**. (Reprinted with permission from A. Sekiguchi *et al., Organometallics* **1995,** *14,* 1092. Copyright © 1995 American Chemical Society.)

dianion and the C–Si bonds are shortened from 190.6(2) to 180.7(6) pm. The two halves of the dianion are twisted by 17.1°. It is not easy to see why this should be so since the anion in the lithium analogue is planar. Clearly both Coulombic and steric interactions are sensitive to the size of the cation.[101]

The reduction of tetraphenylbutadiene in the presence of DME yielded dark blue air- and moisture-sensitive crystals that were shown to contain contact triple ions $[Na(DME)_2]_2[Ph_2CCH=CHCPh_2]$, **63,** in which each sodium is coordinated by four ether oxygen atoms and three carbon atoms of the butadiene chain (Na–C 272, 273, and 292 pm).[2] In another preparation, apparently with very similar conditions, a different structure was obtained. The chains of contact triple ions found in the first structure were interleaved by chains of solvent-separated triple ions $[Na(DME)_3]_2[Ph_2CCH=CH$-

CPh₂], **64** (Fig. 39).[100] It seems that quite small changes in the preparative conditions can result in considerable modification to the crystal structure. The potassium analogue **65** was made from Na/K alloy and tetraphenylbuta-diene in diethyl ether. The structure of the black air- and moisture-sensitive crystals (Fig. 40) showed that each potassium was coordinated to one Et₂O,

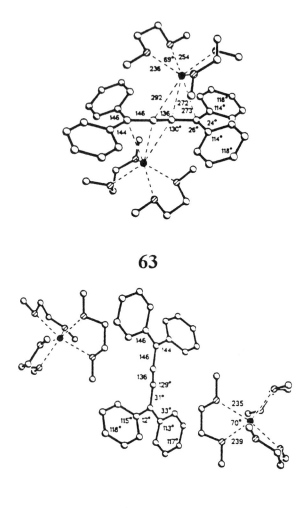

**63**

**64**

Fɪɢ. 39.   Structures of the contact triple ions **63** and the solvent-separated triple ions **64**, which are interleaved in one product from the reduction of Ph₂C=CHCH=CPh₂ with Na/DME. (Reprinted with permission from H. Bock *et al., J. Am. Chem. Soc.* **1992,** *114,* 6907. Copyright © 1992 American Chemical Society.)

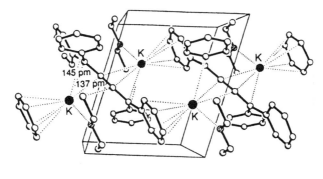

**65**

FIG. 40. Molecular structure of $[K(Et_2O)]_2[Ph_2CCH=CHCPh_2]$, **65**. (Reprinted with permission from H. Bock *et al., J. Chem. Soc. Chem. Commun.* **1992**, 765. Copyright © 1992 Royal Society of Chemistry.)

three carbons in the butadiene chain (K–C 298.7–303.7 pm), and to two phenyl rings, one $\eta^1$ (K–C 312.2 pm) in the same anion and the other $\eta^6$ in an adjacent anion. Thus the $K^+$ ion, larger and more weakly solvated than $Na^+$, makes up its coordination sphere by K–Ph interaction.[102]

Similar cesium compounds have also been isolated.[23] Reduction of $Ph_2C=CPh_2$ in diglyme gives black crystals of $[Cs_2(Ph_2C-CPh_2)(diglyme)_2]$, **66**, which has a structure (Fig. 41) in which each cesium interacts with four phenyl rings of two different tetraphenylethanediyl dianions (Cs–C 345(1)–358(1) pm and two slightly longer distances of 373(1) and 383(1) pm), as well as with three oxygen atoms of a diglyme molecule, giving an overall 15-fold coordination for the metal. The central C–C bond length is stretched from 136 pm in the neutral molecule to 151 pm in the dianion, and the $Ph_2C$ groups are twisted by 76°. Reduction of $Ph_2C=CHCH=CPh_2$ resulted in some attack on the diglyme solvent, and the violet-blue product was shown to be $[Cs_4(Ph_2C-CH=CH-CPh_2)(OCH_2CH_2OCH_3)_2(diglyme)_2]$, **67**. The short central C–C bond (138 pm) and the long outer C–C bonds (145 pm) show that the neutral $Ph_2C=CHCH=CPh_2$ is changed in the dianion into a but-2-ene-1,4-diyl system with the negative charge delocalized onto the phenyl rings. Each cesium is bound in $\eta^6$ or $\eta^7$ fashion to one dianion, to three oxygen atoms of a diglyme molecule, and to two oxygen atoms of 2-methoxyethanolate anions. There are short Cs–C distances (325–327 pm) to the carbon atoms bearing the two phenyl groups, i.e., to those with the highest negative charge. In contrast, the compound $[Na(DME)_3][Ph_2C=C=C=CPh_2]$ crystallizes in a lattice with discrete solvated cations and radical anions.[103]

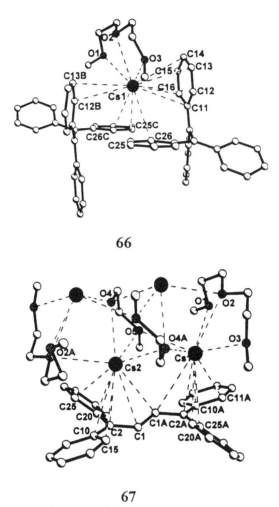

**66**

**67**

Fig. 41. Molecular structure of [Cs$_2$(Ph$_2$C—CPh$_2$)(diglyme)$_2$], **66,** and [Cs$_4$(Ph$_2$C—CH=CH—CPh$_2$)(OCH$_2$CH$_2$OCH$_3$)$_2$(diglyme)$_2$], **67.** (Reprinted with permission from H. Bock *et al., Organometallics* **1996,** 15, 1527. Copyright © 1996 American Chemical Society.)

Cyclooctatetraeneylpotassium is used as a reagent for the synthesis of transition metal complexes. The structure of the THF complex K$_2$(C$_8$H$_8$).3-THF, **68** (Fig. 42), shows that the potassium ions lie on opposite sides of a planar [C$_8$H$_8$]$^{2-}$ anion (K–C 293.6(6)–302.5(6) pm) linked into an extended structure by weakly bridging THF ligands. The open structure accounts for the extreme sensitivity of the compound toward air.[104] The reaction between sodium metal, [(C$_8$H$_8$)Ce($\mu$-Cl)(THF)$_2$]$_2$ and $^t$BuN=CHCH=N$^t$Bu, gave

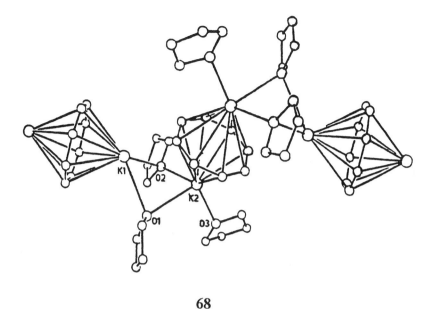

**68**

FIG. 42. Molecular structure of $K_2(C_8H_8) \cdot 3THF$, **68**. (Reprinted with permission from N. Hu *et al.*, *J. Organomet. Chem.* **1988**, *352*, 61. Copyright © 1988 Elsevier Science SA.)

brown crystals of $[Na(THF)_3(\mu\text{-}\eta^8,\eta^8\text{-}(C_8H_8)Ce(C_8H_8)]$, **69,** in which the planar cyclooctatetraene ring bridges sodium and cerium (Na–c 224.6(2) pm).[105] The structure (Fig. 43) is similar to that of $K(diglyme)(\mu\text{-}C_8H_8)\text{-}Ce(C_8H_8)$ described previously,[106] but the lithium analogue forms a solvent-separated ion pair $[Li(THF)_4][Ce(C_8H_8)_2]$ in the solid state.

### C. *Azaallyl Derivatives*

Although it is difficult to isolate allylmetal compounds for X-ray studies, azaallyl derivatives, in which CH is replaced by N, are obtained more readily and several have been characterized structurally.

The zincates $K[ZnR(^tBuNCH=CHN^tBu)]$ contain the diazabutadiene fragment. They have been made, via a series of radical-anion complexes, from the alkylation of $^tBuN=CHCH=N^tBu$ by potassium in the presence of organozinc compounds, and characterized as THF and $Et_2O$ adducts, **70** and **71,** respectively. Each forms linear coordination polymers comprising alternate potassium cations and zincate anions held together

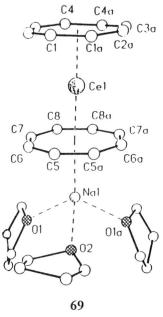

**69**

FIG. 43.   Molecular structure of $[Na(THF)_3(\mu-\eta^5,\eta^5-(C_8H_8)Ce(C_8H_8)]$, **69**. (Reprinted with permission from U. Kilimann *et al.*, *J. Organomet. Chem.* **1994**, *469*, C15. Copyright © 1994 Elsevier Science SA.)

by zinc diazabutadiene contacts with K–C 292.10(15)–304.0(7) and 289.2(15)–303.4(14) pm.[107]

The 1-azapentadienyl compound $K[\eta^4$-N(SiMe$_2^t$Bu)C$^t$Bu(CH)$_3$SiMe$_2^t$Bu], **72,** has been made from the reaction between $^t$BuCN and the allylpotassium compound $K[C(SiMe_2^tBu)CH{=}CHSiMe_2^tBu]$ (itself generated from $^t$BuMe$_2$SiCH$_2$CH$=$CHSiMe$_2^t$Bu by treatment with butyllithium in TMEDA and KO$^t$Bu and characterized by NMR spectroscopy as a pyridine adduct).[108] It was characterized in the solid state by an X-ray study (Fig. 45). The structure is polymeric with each $\eta^4$-azapentadienylanion interacting with two K$^+$ ions (K–N 287.8(3)–295.4(3) pm and K–C 294.5(4)–356.2(4) pm) with the shortest distance to C(1) and the longest to C(3).

A series of 2-azaallyl derivatives have also been made. The reaction between dibenzylamine and phenylsodium gave a red precipitate of dibenzylamidosodium that dissolved in PMDTA/toluene to give a much darker solution from which red-green dichroic crystals were isolated. These were shown to consist of the azaallyl compound Na[PhCHNCHPh]. PMDTA, **73,** formed the removal of one proton from each of the two benzylic CH$_2$ groups.[109] The crystal structure (Fig. 46) consists of ion pairs in which the

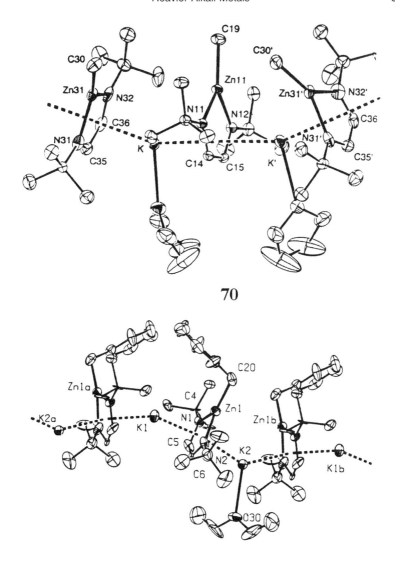

**70**

**71**

Fig. 44.   Molecular structure of **70** and **71**. (Reprinted with permission from E. Rijnberg *et al.*, *Organometallics* **1997,** *16*, 3158. Copyright © 1997 American Chemical Society.)

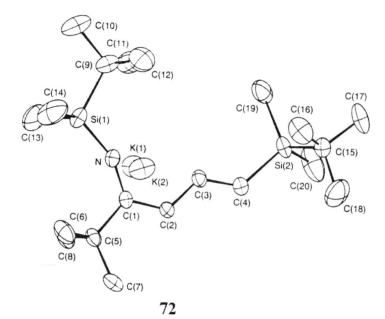

**72**

FIG. 45.   Molecular structure of K[$\eta^4$-N(SiMe$_2$$^t$Bu)C$^t$Bu(CH)$_3$SiMe$_2$$^t$Bu], **72,** showing the interaction between the anion and two neighboring K$^+$ ions. (Reprinted with permission from P. B. Hitchcock *et al., Chem. Commun.* **1996,** 1647. Copyright © 1996 Royal Society of Chemistry.)

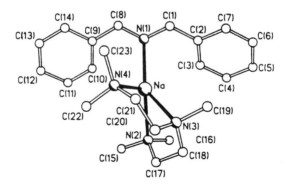

**73**

FIG. 46.   Molecular structure of Na[PhCHNCHPh]. PMDTA, **73.** (Reprinted with permission from P. C. Andrews *et al., J. Organometal. Chem.* **1990,** *386,* 287. Copyright © 1990 Elsevier Science SA.)

ring charge is delocalized over an extended planar system comprising the central CNC link and the phenyl groups, as shown by the C–C and C–N bond lengths. The sodium ion is coordinated by four nitrogens at the corners of a very distorted tetrahedron, but there are also short contacts between the metal and the *ortho* carbon of each phenyl ring (Na–C 313.3(4) and 316.6(4) pm). NMR data show that the structure in solution is the same as that in the solid, with restricted rotation about the *ipso* C–CH bond.

The reaction between dibenzylamine and butylpotassium gave a mixture of 1:1 and 1:2 PMDTA adducts of the potassium derivative K[PhCHNCHPh].[110]

$$(PhCH_2)_2NH + KBu + nPMDTA \rightarrow K[PhCHNCHPh]nPMDTA + \\ BuH + H_2 \quad (10)$$

The 1:2 compound was monomeric (from a crystal structure determination that was not of publishable standard) but the 1:1 compound, **74,** crystallized (Fig. 47) as an infinite zigzag chain of alternate cations and two distinct types of anions lying approximately perpendicular to each other, one azaallyl bound and the other allyl bound. Both types use delocalized $\pi$ systems to bridge pairs of $K^+$ cations. The azaallyl-bound ligand has K–N 299.6(2) and K–C 293.0(6) pm, and the allyl-bound ligand has K–C 336.6(7) and 326.8(5) pm, leaving the central nitrogen N(2) not bound to potassium. Azaallyls are the stronger bound ligands so that the ability of a second mole of PMDTA to displace the allyl-potassium bonds to form a 1:2 compound is readily understood. In contrast, the compound [K(PhCH= NCHPh)(18-crown-6)] is monomeric with the ligand $\eta^2$-C,N bonded to the

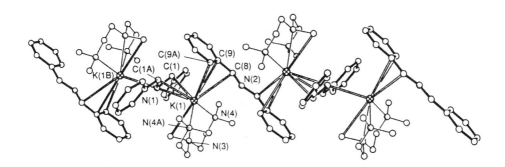

**74**

FIG. 47.   Crystal structure of K[PhCHNCHPh] PMDTA, **74.** (Reprinted with permission from P. C. Andrews *et al., Chem. Commun.* **1997,** 319. Copyright © 1997 Royal Society of Chemistry.)

cation (K–N 310.5(6), 301.7(7) and K–C 304.4(7)–364.7(8) pm).[111] The sodium analogue is also monomeric, but the bonds between cation and anion are through nitrogen only. The compound K[PhCHNCHPh) reacts with the imine $PhCH=NCH_2Ph$ and TMEDA in toluene to give benzylpotassium and the chiral compound $PhCH=NCHPhCHPhNHCH_2Ph$.[110] A detailed study by NMR spectroscopy and MO calculations shows that the transformation of the $[(PhCH_2)_2N]^-$ anion (characterized by X-ray diffraction in $[(PhCH_2)_2NNa.TMEDA]_2$ and $[(PhCH_2)_2NLi]_2.dioxane$) into $[PhCH=N=CHPh]^-$ involves $\beta$ elimination followed by metallation of the resulting imine [Eq. 11]. The steps in this sequence are influenced markedly by the particular alkali metal (M) and solvent (L) that are used.[112]

$$(PhCH_2)_2NH + MBu \rightarrow [(PhCH_2)_2NM]_n \rightarrow (PhCH_2)_2NM.L_m \rightarrow$$
$$\{PhCH_2N=CHPh.MHL_m\} \rightarrow [PhCH \cdots N \cdots CHPh]ML_m + H_2 \quad (11)$$

Reactions between $KCH(SiMe_3)_2$ and nitriles R'CN gave a variety of compounds depending on the group R'. With R' = $^tBu$ the product was the 1-azaallyl derivative $K[NSiMe_3C^tBuCHSiMe_3]$.[114] With R' = Ph, an unusual head-to-head coupling of the nitrile fragments was followed by $SiMe_3$ migration to nitrogen, giving the $\beta$-diketinimato compound $K[RN=CPhCH=CPhNR]$ (R = $SiMe_3$).[113]

$$KCHR_2 + PhCN \rightarrow R_2HCCPh=NK \rightarrow RCHKCPh=NR \quad R = SiMe_3$$
$$RCHKCPh=NR + PhCN \rightarrow K[RN=CPhCHRCPh=N] \rightarrow K[RN=CPhCH=CPhNR]$$
$$(12)$$

With the sterically hindered nitrile $2,5-Me_2C_6H_3CN$ the potassium $\beta$-diketinimato compound, which was almost insoluble in hexane, dissolved in the presence of an excess of the nitrile to give the 1,3-diazaallyl compound $[K\{NRCArNCArCHR\}(NCAr)]_2$ **75** (R = $SiMe_3$, Ar = $C_6H_3Me_2$-2,5), which was characterized by a crystal structure determination (Fig. 48) and shown to be dimeric. Each potassium is coordinated to five nitrogen atoms that form a square pyramid. The two N–C distances are the same (134(2) pm), showing that the electrons are delocalized over the NCN system. The K–C distances are 316(1)–319(2) pm.[115] The reaction of 1,3,5-triazine with $NaN(SiMe_3)_2$ resulted in ring opening and $\alpha,\omega$-$SiMe_3$ migration to give the 3-sodio-1,3,5,7,tetraazaheptatriene $[Na(Me_3SiNCHNCHNCHNSiMe_3)]_3$, **76** (Fig. 49). The X-ray structure shows that the three sodium ions are encapsulated by three tetraazaheptatriene anions and coordinated by nitrogen rather than by carbon atoms.[116,117] Likewise, in the $\beta$-diketiminato complex $Na[NRCPhCHCPhNR].THF$ and the 1,3-diazaallylsodium $[Na-(NRCPhNCPh=CHR)]_8$ (R = $SiMe_3$) the shortest contacts in the solid state are between sodium and nitrogen rather than between sodium and carbon.[117]

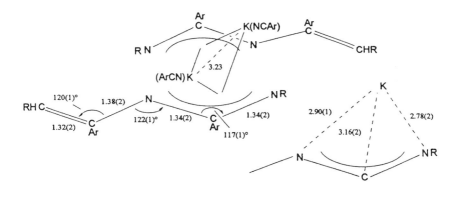

**75**

FIG. 48. Simplified bonding pattern in [K{NRCArNCArCHR}(NCAr)]₂, **75.** (Reprinted with permission from M. F. Lappert and D-S. Liu, *J. Organomet. Chem.* **1995,** *500,* 203. Distances in Å. Copyright © 1995 Elsevier Science SA.)

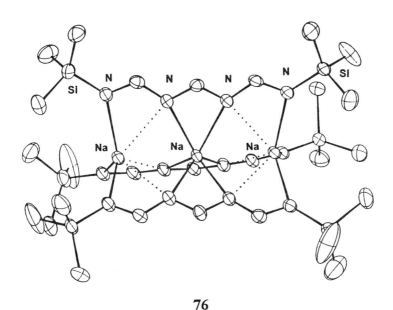

**76**

FIG. 49. Molecular structure of [Na(Me₃SiNCHNCHNCNSiMe₃)]₃, **76.**[117]

### D. Other Alkene Derivatives

In the open chain 2,4-dimethylpentadienylpotassium, anions and cations form zigzag chains with potassium coordinated to $[H_2C=CMeCHCMe=CH_2]^-$ anions on each side (K–C 306.9(7)–327.6(7) pm) as well as to TMEDA.[118]

Interactions between sodium atoms and $C=CH_2$ units (Na–C 270–2 287.3 pm) are apparent in the crystal structure of the enolate $Li_2Na_4(O-C(=CH_2)^tBu)_6\cdot2^iPr_2NH$ obtained from the reaction between lithium and sodium diisopropylamides and 3,3-dimethyl-2-butanone.[119]

### E. Alkyne Derivatives

The compounds $MC\equiv CCH_3$ (M = Na, K) were described earlier.[1] The rubidium and cesium derivatives have now been obtained by the reaction of the metals with propyne in liquid ammonia, and their crystal structures studied by X-ray powder diffraction. The compounds are isostructural with those isolated previously, and the $c/a$ ratio in the tetragonal unit cell decreases linearly with the increasing ionic radii of the alkali metals ions.[120]

## VII

## ARENE DERIVATIVES

### A. Compounds with Hydrocarbons

Compounds containing extensively delocalized aromatic systems have fascinated chemists for more than 100 years. The closeness in energy of the molecular orbitals means that they can both donate electrons to strong acceptors and accept electrons from donors such as the heavier alkali metals. It is sometimes possible to isolate compounds with identical arrangements of atoms but with different numbers of electrons and to examine in considerable detail how the geometry changes as electrons are added or taken away. The compounds provide examples for those wishing to test theoretical predictions for a range of computational techniques. The initial products of the reduction of aromatic hydrocarbons by sodium mirrors in the presence of donor solvents are radical anions that have been widely used as reducing agents in organic chemistry. Their formation depends on the hydrocarbon, the alkali metal donor, and the temperature. They are sometimes studied by quenching with proton donors such as methanol, but the final products are formed in a complex series of steps that depend critically on the conditions.[2,121] It is now, however, possible to obtain many of the radical anion salts in crystalline form. The structures may involve solvent-separated or contact ion pairs. In general, single electron transfer results in only small changes to the geometry of the organic fragment, but transfer of two or more electrons usually produces substantial structural changes. These points are illustrated by the examples below.

Calculations show that diglyme or crypt-221 is the best complexing agent for the sodium cation.[104] The product from sodium, diglyme, and naphthalene gives black crystals $[Na(diglyme)_2][C_{10}H_8]^{\bullet}$ in which essentially complete separation of the cation and radical anion is achieved, with the sodium coordinated by six oxygen atoms from the diglyme. The anthracene compound is similar.[122] Implications of detailed structural data for the study of extended hydrocarbon $\pi$ systems have been discussed. Similarly, the reduction of 9,10-diphenylanthracene by a sodium mirror in THF yields dark blue crystals that have been shown to be $[Na(THF)_6][C_{14}H_8Ph_2]^{\bullet}$ with separate cations and radical anions in the crystal lattice. The structure of the radical anion is similar to that of the neutral molecule with charge delocalized over the anthracene $\pi$ system and phenyl groups twisted 67° out of the anthracene plane. There is no significant Na–C contact: the ordered octahedrally coordinated $Na(THF)_6^+$ cation is unusual.[123] In another study[124] the sodium derivatives of the acenaphthylene and fluoranthene dianions were compared. The sodium salt, $(CH_2CH_2O)_5Na[C_{12}H_8]$-$Na(OCH_2CH_2)_5$ $(C_{12}H_8$ = acenaphthylene) in which sodium is solvated by [15]-crown-5, forms triple ions with 10-fold coordinated sodium ions above and below the five-membered ring (Na–C 268.5(2)–301.4(2) pm). In contrast, the solvated sodium salts of the fluoranthene dianion $L_2Na[C_{16}H_{10}]NaL_2$ $(C_{16}H_{10}$ = fluoranthene, L = DME or diglyme) form extended chains with 10-fold coordinated counterions above and below two different naphthalene six-membered rings (Na–C 267.8(2)–288.5(1), L = DME; 259.1(2)–305.1(2) pm L = diglyme). MNDO calculations show that the negative charge is associated mainly with the five-membered rings in the acenaphthylene anion but with the six-membered rings in the fluoranthene anion and provide a rationale for the crystallographic results.

The reduction of 1,2,4,5-tetrakis(trimethylsilyl)benzene under aprotic conditions yielded the sodium salt of the radical anion, and the red needles obtained from DME were shown by an X-ray investigation to have an ionic structure with no short Na–C contacts. There is a small but significant change from the neutral molecule ArH to the radical anion and the dianion.[125] As electrons are taken up by the aromatic system, the bond lengths between the substituted carbons are increased from 141 pm in ArH[126] to 147 pm in ArH$^{\bullet-}$ and 155 pm in ArH$^{2-}$. The reduction of 9,10-bistrimethylsilylanthracene is similar. The radical anion has been characterized by ESR spectroscopy, and the brown crystals isolated from DME were [Na(DME)$_3$][C$_{14}$H$_8$(SiMe$_3$)$_2$]$^\bullet$ with no short Na–C contacts.[127] Further examples of compounds that crystallize as ionic solids with distinct solvated sodium cations and radical anions are [Na(DME)$_3$][C$_{28}$H$_{18}$]$^\bullet$ (C$_{28}$H$_{18}$ = 9,9′-bianthryl),[128] Na(crypt 221) [C$_{13}$H$_9$]$^\bullet$ (C$_{13}$H$_9$ = fluorenyl),[104] [Na([18]-crown-6)][(O$_2$N)$_3$CNa([18]-crown-6)C(NO$_2$)$_3$]   (NaC(NO$_2$)$_3$.dioxane   is polymeric[129]), and Cs[(NC)$_2$C=C(CN)$_2$]$^{\bullet-}$, which crystallizes solvent free from DME-hexane (in contrast to sodium and potassium analogues).[130]

The radical anion from 2,3-diphenylquinoxaline has been isolated in the dark blue sodium or potassium salts, **77** and **78,** respectively.[131] The crystal structures show that the metal is coordinated (Fig. 50) to DME, to the nitrogen atoms of the heterocycle where the negative charge is concentrated, and to *ortho* and *ipso* carbon atoms of the adjacent phenyl rings (Na–C 302.2(6)–307.6(6) pm and K–C 312.1(3)–333.3(3) pm). The M–Ph interactions link the ion pairs into infinite chains. The formation and stability of the radical anion have been studied by ESR spectroscopy and cyclic voltammetry.

The reduction of pyrene by sodium in diethyl ether at 193 K yielded a dark red crystalline compound shown by an X-ray study to be NaC$_{16}$H$_{10}$.Et$_2$O, **79**

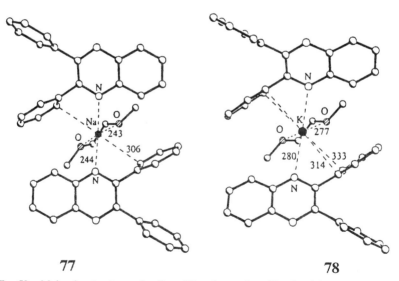

**77**                                        **78**

FIG. 50.   Molecular structures of sodium, **77,** and potassium, **78,** salts of the 2,3-diphenylquinoxaline radical anion. Distances in pm. (Reprinted with permission from H. Bock *et al., Helv. Chim. Acta* **1994,** *77,* 1505. Copyright © 1994 Verlag Helvetica Chimica Acta AG.)

(Fig. 51). The structure consisted of zigzag chains with the pyrene molecules tilted at 60° (**79a**). The Na–C distances to neighboring pyrene molecules range from 262(2) to 279(2) pm with the shortest contacts to the carbon atoms that bear the greatest negative charge.[132] A second polymorph (**79b**) has been isolated from a preparation at higher temperature. Again the structure consists of infinite chains with alternate sodium ions and pyrene radical anions. Each sodium is $\eta^6$ coordinated to one pyrene radical ion [Na–C 276.7(2)–300.3(2) pm with the shortest bonds to C17 and C18)] and $\eta^3$ coordinated to C1, C6, and C7 in the other pyrene radical anion (Na–C 263.1(2) to 296.2(2) pm). The hydrocarbon skeleton is planar in both polymorphs, showing that the charge is highly delocalized. The relative stabilities of the two polymorphs have been discussed.[133]

The effect of coordinating ethers on the solid-state structures adopted by a particular ion pair is shown by the mono-, di-, and tetraglyme complexes of disodiumperylene $Na_2[C_{20}H_{12}]$.[134] In all cases the planar dianion is only slightly distorted from the neutral molecule. In $Na_2[C_{20}H_{12}]$.4DME, **80** (Fig. 52), the sodium interacts significantly with the anion surface (Na–C 268–311 pm) and is bound to four oxygen atoms from the ether, but in $Na_2$-$[C_{20}H_{12}]$.2tetraglyme, **81**, the sodium makes only a peripheral contact with the anion (Na–C 279–309 pm). In $Na_2[C_{20}H_{12}]$.4diglyme, **82**, the coordination sphere of the $Na^+$ cations is occupied completely by oxygen atoms

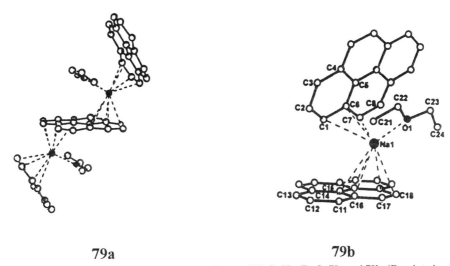

79a        79b

FIG. 51. Molecular structures of the two forms of $NaC_{16}H_{10}.Et_2O$, **79a** and **79b**. (Reprinted with permission from C. Näther *et al., Helv. Chim. Acta* **1996,** *79,* 84. Copyright © 1996 Helvetica Chimica Acta AG.)

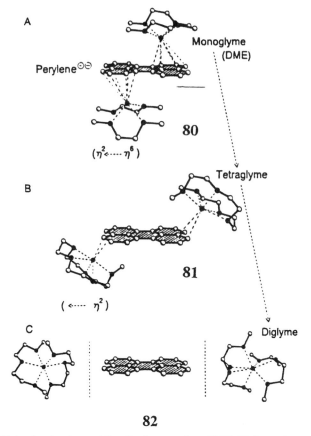

A

Monoglyme
(DME)

Perylene⊙⊖

**80**

$(\eta^2 \!\leftarrow\!\cdots\! \eta^6)$

Tetraglyme

B

**81**

$(\leftarrow\!\cdots\! \eta^2)$

C                                                            Diglyme

**82**

FIG. 52. Molecular structures of the disodium perylene Na[C$_{20}$H$_{12}$] complexes with mono-, di-, and tetraglyme showing the change from solvent-shared to solvent-separated ion pairs. (Reprinted with permission from H. Bock *et al.*, *J. Am. Chem. Soc.* **1995**, *117*, 3869. Copyright © 1995 American Chemical Society.)

from the ether and the metal is withdrawn from the surface of the electron-rich hydrocarbon anion.

Reduction of the hydrocarbon 5,6,11,12-tetraphenyltetracene (rubrene) with a sodium mirror in THF gave a dark green solution from which almost black crystals of the tetrakis-sodium salt **83** could be obtained.[135a] Two of the four sodium cations (each doubly solvated by THF) are located (Fig. 53) above and below the central tetracene skeleton and the other two are between pendant phenyl groups. The central sodium ions are 8-coordinate with Na–C 260–263 pm to the phenyl-substituted and 272 pm to the other

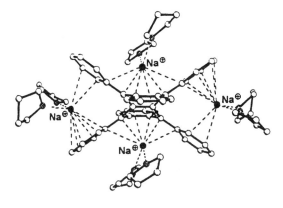

**83**

FIG. 53.   Molecular structure of Na$_4$(rubrene).8THF, **83**, showing the nonplanar structure of the tetraanion. (Reprinted with permission from H. Bock *et al.*, *Angew. Chem. Int. Ed. Engl.* **1996**, *35*, 631. Copyright © 1996 Wiley VCH.)

tetracene carbon atoms. The outer sodium ions are 10-coordinate. One phenyl group is $\eta^6$ (Na–C 277–293 pm) and the other $\eta^2$ (Na–C 275–284 pm). The hydrocarbon skeleton is distorted drastically compared to that of the neutral molecule in order to accommodate the four negative charges, e.g., it is bent by 43° at the phenyl-substituted carbons. The central C–C bond is shortened from 147 to 139 pm, the adjacent C–C bonds are lengthened from 142 to 148 pm, and there are numerous changes to bond angles. The related compound sodium tetraphenyl-*p*-benzosemiquinone crystallizes as two different tetrahydropyran adducts [Na$^+$(OC$_5$H$_{10}$)$_n$(C$_6$-Ph$_4$O$_2$$^{\bullet-}$)] ($n$ = 2 or 3): one with two and the other with three moles of coordinated ether.[135b,135c] The latter is obtained when the solvent contains TMEDA but the amine is not incorporated into the lattice. The structures have been rationalized in terms of Van der Waals interactions, and the species in solution have been identified by ESR studies as contact ion pairs M$^+$R$^{\bullet-}$ (M = Li, Na, Rb, or Cs; R$^{\bullet-}$ = C$_6$Ph$_4$O$_2$$^{\bullet-}$), contact triple radical cations [M$^+$$_2$R$^{\bullet-}$]$^+$ (M = Li, Na, or Cs), or [Li$^+$M$^+$R$^{\bullet-}$]$^+$ (M = Na, or Cs).

The coronene radical anion has been obtained in a half-sandwich contact ion pair **84**, (Fig. 54) with the [K(TMEDA)(THF)$_2$] cation.[136] The metal interacts with one of the outer rings of the hydrocarbon [K–C 318.5(4) to 345.5(4) and 313.3(4) to 354.8(4) pm in two crystallographically independent molecules]. There is almost no distortion of the anion compared with the neutral molecule, suggesting that the electron is well delocalized. UHF-

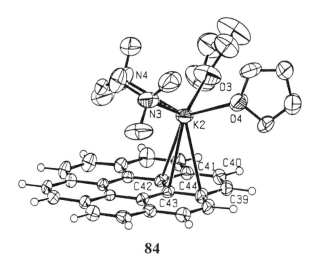

**84**

FIG. 54. Molecular structure of $[K(TMEDA)(THF)_2][C_{24}H_{12}]^{\bullet}$, **84** ($C_{24}H_{12}$ = coronene radical ion). (Reprinted with permission from C. Janiak and H. Hemling, *Chem. Ber.* **1994,** *127,* 1251. Copyright © 1994 Wiley VCH.)

PM3 and MNDO calculations confirm that the energy surface for the metal cation above the delocalized aromatic system is very flat and that the interaction is largely electrostatic.

A detailed study has been made of the reaction of LiBu/KOC-Me$_2$CH$_2$CH$_3$ with tribenzotriquinacene and its *centro*-alkyl(R)-substituted derivatives, **85.**[137] These undergo a series of deprotonations and elimination of RH (Fig. 55) to give the tribenzacepentalene dianion that can be trapped by various electrophiles to give 4,7-disubstituted tribenzoacepentalenes **86.** Evidence for intermediates has been obtained from NMR studies. The relative leaving tendencies of the alkyl groups as K$^+$R$^-$ correlate with the relative stabilities of the expelled anion. By restricting the amount of LiBu/KOCMe$_2$CH$_2$CH$_3$ it is possible to prepare the monopotassium salts, **87,** that can then be alkylated to give the disubstituted tribenzotriquinacenes, **88.** Treatment of the compound **88** (R = Me) with an excess of LiBu/KOCMe$_2$CH$_2$CH$_3$ followed by trapping with dimethyl sulfate gives the 1,4,7,10-tetramethyl tribenzotriquinacene, **89,** in 80% yield, thus corroborating the necessity for the formation of the trianions as key intermediates in the deprotonation–elimination process.

### B. *Other Compounds with Metal–Arene Interactions*

The compounds described in this section are not formally organometallic as they contain, in addition to organic groups, other electron-accepting

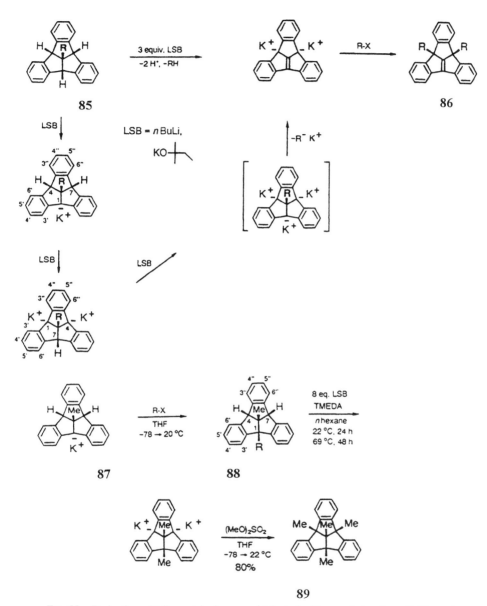

FIG. 55. Reduction of tribenzotriquinacene. (Adapted with permission from R. Haag *et al., J. Am. Chem. Soc.* **1995,** *117,* 10474. Copyright © 1995 American Chemical Society.)

substituents that can accommodate the charge transferred from the heavy alkali metal. In many cases, however, the charge is delocalized so that significant interactions are observed in crystal structures between the metal and the organic part of the anion.

The thiolates $[MSC_6H_3\text{-}2,6\text{-}Trip_2]$ (Trip $= (2,4,6\text{-}^iPr_3C_6H_2)$ were made

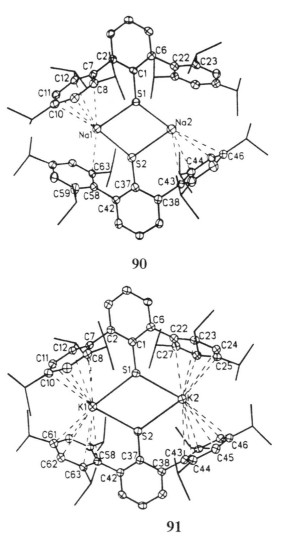

**90**

**91**

Fig. 56.   Molecular structures of $MSC_6H_3\text{-}2,6\text{-}(2,4,6\text{-}^iPr_3C_6H_2)_2$ M $=$ Na **90**, K **91**, and Cs **92**. (Reprinted with permission from M. Niemeyer and P. P. Power, *Inorg. Chem.* **1996**, *35*, 7264. Copyright © 1996 American Chemical Society.)

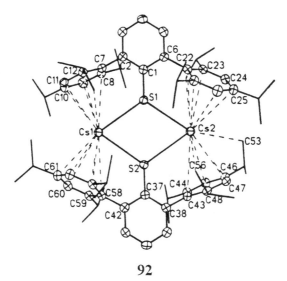

**92**

FIG. 56. (*Continued*).

by direct reaction between the metal M and the thiol in hexane.[138] The crystal structures (Fig. 56) all show an $M_2S_2$ core and further interactions between the metal and the rings of the Trip substituents. Several comparisons can be made in the series from Na to Cs: (i) the configuration at S becomes less pyramidal and is essentially planar for K–Cs; (ii) the $M_2S_2$ ring, which is almost planar in the Na, K, and Rb compounds, is folded in the Cs derivative; and (iii) interactions between the metal and the Trip substituents become more prevalent as the metal becomes heavier. Thus the sodium compound in the metal interacts strongly with two Trip rings and weakly with a third; in the K, Rb, and Cs compounds there are short $\eta^6$ contacts to all four Trip rings (Na–C 301.6(5)–310.8(5), 283.9(5)–324.9(5), and 310.8–326.6(5) pm; K–C 330.4(5)–356.4(5); Rb–C 334.4(7)–359.6(7); Cs–C 353.1(4)–382.8(4) pm).

Further examples of significant interactions between heavier alkali metal atoms and aromatic rings include the ate complexes $NaCrPh_5.3Et_2O.THF$ (Na–C = 263–292 pm),[139] $K[Nd(OC_6H_3{}^iPr_2\text{-}2,6)_4]$ (K–C 309.7(8)–347.3 (10) pm), **93**, [143b]$Cs[La(OC_6H_3Me_2\text{-}2,6)_4]$ **94**, (Cs–C 369.6(7)–384.7(7) pm)[140] (Fig. 57) (several similar compounds have been described), and $Cs_2[La(OC_6H_3{}^iPr_2\text{-}2,6)_5]$.[141] Although this last compound is an aryloxide, there are no Cs–O contacts less than 460 pm; the coordination sphere of each of the three independent cesium atoms in the complex structure is made up entirely of Cs–C contacts, 19 in all. For example, for one cesium

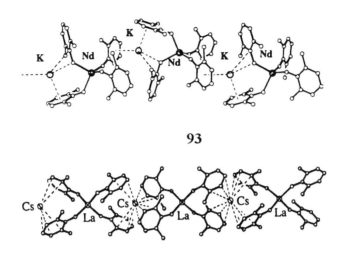

**93**

**94**

FIG. 57. Structures of KNd(OC$_6$H$_3^i$Pr$_2$-2,6)$_4$, **93**, and CsLa(OC$_6$H$_3$Me$_2$-2,6)$_4$, **94**. Some carbon atoms of $^i$Pr groups are omitted. (Reprinted with permission from D. L. Clark *et al.*, *Inorg. Chem.* **1996**, *35*, 667. Copyright © American Chemical Society.)

atom there are two $\eta^6$ interactions with arenes (Cs–C 332(2)–382(2) and 344(3)–357(3) pm), one $\eta^4$ to a third arene, two to methyl groups (Cs–C 381(2), 342(3) pm), and one to a methine carbon. The other cesium environments are similar. Similarly, in [K($\mu$-$\eta^5$,$\eta^5$-C$_5$H$_5$)$_2$Nd(OC$_6$H$_3$Me$_2$-2,6)$_2$], **95** (Fig. 58), the potassium ion makes no K–O contacts <440 pm.[142a] The coordination is to two C$_5$H$_5$ rings (K–C 301(2)-337(2) pm) and to two arene rings [K–C 317(2)–354(2) and 336(2)–393(2) pm with the shortest K–C on one side)] arranged in the form of a distorted tetrahedron. The K–aryl interactions link the ion pairs into a two-dimensional extended framework. They are also found in KSm($\mu$-OC$_6$H$_2^t$Bu$_2$-2,6-Me-4)$_3$. THF and K[Nd(OC$_6$H$_3^i$Pr$_2$)$_4$] (K–C 309.7(10)–347.3(11) and 322.4(11) to 331.3(11) pm) but there is also a significant K–O bond to the oxygen of the aryloxide.[142] In Na[Nd(OC$_6$H$_3$Ph$_2$-2,6)$_4$ the three phenyl groups interact intramolecularly in $\eta^1$ or $\eta^2$ fashion with the sodium (Na–C <300 pm).[143] Similarly, in [Na{N=Nb{N{C(CD$_3$)$_2$CH$_3$}(C$_6$H$_3$Me$_2$-3,5)}$_3$}]$_2$ $\eta^6$-aryl–sodium interactions are apparently strong enough to exclude THF from the coordination sphere of the niobium as the compound is obtained solvent free even from the THF solution.[144]

Alkali metal–phenyl interactions are also found in benzylgallates and -indates Cs[(PhCH$_2$)$_2$GaF$_2$] (Cs–C 349(2)–387(3) pm),[145] [Cs{(PhCH$_2$)$_3$-

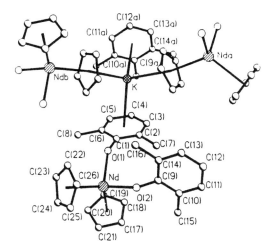

**95**

FIG. 58.   Portion of the pseudo-two-dimensional extended structure of [K($\mu$-$\eta^5$,$\eta^5$-C$_5$H$_5$)$_2$ Nd(OC$_6$H$_3$Me$_2$-2,6)$_2$], **95**. (Reprinted with permission from W. J. Evans *et al., Organometallics* **1995,** *14,* 558. Copyright © 1995 American Chemical Society.)

GaF}]$_2$ **96** (Cs–C 357–360 pm), [Cs{(C$_6$H$_2$Me$_3$-2,4,6)$_3$InF}.2MeCN]$_2$ **97**[146] (Fig. 59), Cs[(PhCH$_2$)$_3$GaN$_3$].0.5 toluene, Cs(PhCH$_2$)$_3$InCl.0.5toluene **98** (average Cs–C 362–436 pm), Cs$_2$[O{PhCH$_2$In(OCH$_2$Ph)$_2$}$_4$] (average Cs–C 385–446 pm), NaOPh.3H$_2$O (Na–*c* 306.7(2) pm), [{2,4,6-$^i$Pr$_3$C$_6$H$_2$}(F)SiPCs(Si$^i$Pr$_3$).0.5 THF,[148a] and in the aminophosphoranate Cs[Ph$_2$P($\mu$-NSiMe$_3$)$_2$], which forms infinite chains linked by Cs–Ph interactions (Cs–*c* = 338.5 pm) strong enough to exclude solvent from the coordination sphere. In contrast, the potassium and rubidium compounds have, respectively, 4 and 1 mol of solvated THF.[148b]

Crystal structures of solvates in which there are significant interactions between heavy alkali metal atoms and neutral arenes include the toluene solvate (K(C$_6$H$_5$Me)$_2$)[Lu{CH(SiMe$_3$)$_2$}$_3$ $\mu$-Cl] (K–*c* 300.3–312.8 pm),[149] the stannate(II) [K(C$_6$H$_5$Me)$_3$][Sn(CH$_2^t$Bu)$_3$],[150] and [(C$_6$H$_6$)KOSiMe$_2$Ph]$_4$ (K–C 323.3(14)–330.8(6) pm).[151]

Another series of compounds that illustrate the trends in coordination of the heavy alkali metals is obtained from deprotonation of the sulfinimi-namines Me$_3$SiN=S(R)NHPh (R = $^t$Bu or SiMe$_3$).[152] Although the sulfin-imidamide salts are not strictly organometallic, metal–phenyl interactions are clearly evident in the crystal structures. As the alkali metal atom becomes heavier, the $\eta^6$ coordination of the phenyl group becomes stronger,

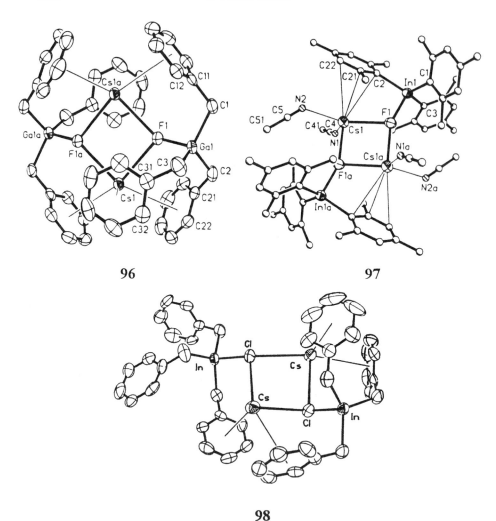

**96**                                              **97**

**98**

Fig. 59. Molecular structures of [Cs{(PhCH$_2$)$_3$GaF}]$_2$ **96,** [Cs{(C$_6$H$_2$Me$_3$-2,4,6)$_3$InF.2-MeCN}]$_2$ **97,** and Cs(PhCH$_2$)$_3$InCl.0.5toluene, **98.** (Reprinted by permission from B. Werner *et al., Organometallics,* **1996,** *15,* 3746, Copyright © 1996 American Chemical Society, and T. Kräuter *et al., Chem. Eur. J.* **1997,** *3,* 568, © 1997 Wiley VCH.)

as shown by a decrease in the metal-centroid distance, viz. Na 386.3, K 369.8, Rb 369.9, and Cs 355.1 pm.

Metal–arene interactions are also apparent in the triazinido derivative [Rb(THF)$_2$NCPh=NC$^t$Bu$_2$N=CPh]$_4$, **99,** which crystallizes as cyclic tetramers with $S_4$ symmetry. Each rubidium (Fig. 60) is coordinated by one nitrogen and one *ortho* carbon (Rb–C 335.2(3)–338.0(3) pm) from two triazinido ligands and other short Rb–C contacts of 345.0(3)–357.3(3) pm contribute to the metal-ligand bonding.[153]

Similar structural features have been found in the chemistry of the phenylhydrazides.[154] The compound PhHNN(SiMe$_3$)$_2$ is deprotonated smoothly in THF or benzene-THF to give PhNaNN(SiMe$_3$)$_2$, but in DME the product is [PhHNNPhNa]$_2$[PhNaNNPhNa].4DME

$$\text{PhNaNN(SiMe}_3)_2 \xleftarrow{\text{THF}} \text{PhHNN(SiMe}_3)_2 + \text{NaN(SiMe}_3)_2 \xrightarrow{\text{DME}}$$
$$[\text{PhHNNPhNa}]_2[\text{PhNaNNPhNa}] \quad (12)$$

Desilylation is also observed in the reaction between Ph$_2$NNHSiMe$_3$ and NaNH$_2$: the product Ph$_2$NNNaH crystallizes as a hexamer. The crystal structures of the adducts of PhNaNN(SiMe$_3$)$_2$ with 3THF, THF, Et$_2$O, and MeO$^t$Bu have been determined. In the THF adduct the sodium makes con-

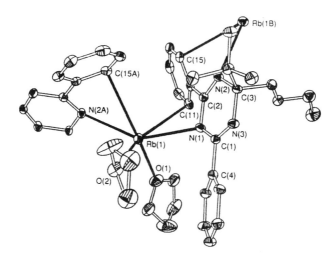

**99**

Fig. 60.    Part of the structure of the triazinido derivative [Rb(THF)$_2$NCPh=NC$^t$Bu$_2$N=CPh]$_4$, **99,** showing the coordination of Rb. (Reprinted with permission from W. Clegg *et al.*, *Chem. Commun.* **1997,** 1301. Copyright © 1997 Royal Society of Chemistry.)

tacts with a nitrogen atom of the hydrazine, an oxygen atom of the ether, an $\eta^3$-phenyl group (Na–C 270.1(6)–285.9(7) pm), and a methyl group attached to silicon. Similar features are apparent in the structures of the MeO$^t$Bu adduct, but the coordination to the phenyl ring is $\eta^2$(Na–C 272.7(8)–297.9(8) pm). There are also significant Na–Ph contacts in [PhHNNPhNa]$_2$[PhNaNN-PhNa] and [Ph$_2$NNNaH]$_6$. The structure of the cesium derivative shows that the cesium is coordinated to nitrogen and to two phenyl groups from adjacent anions (Cs–C 339.6(8)–377.5(7) pm). These interactions link the ion pairs into a three-dimensional lattice with long channels that may occlude loosely bound THF.[155] The related sodium derivative has been isolated as an ammonia adduct [Na[PhNN(SiMe$_3$)Ph}NH$_3$]$_2$: the structure shows that the Na$^+$ is coordinated to two nitrogen atoms from hydrazine and one from ammonia and also makes short contacts to *ipso* and *ortho* carbon atoms of phenyl rings (Na–C 303.6(8)–311.5(3) pm).

Short metal–arene distances have been observed in the structure of the disodium salt of 1,3-dimethyl ether *p*-t-butylcalix[4]arene, **100** (Fig. 61).[156] One of the two sodium atoms in the asymmetric unit, Na(2), bonds *exo* and bridges two calix[4]arene cones through Na–O contacts whereas the other, Na(1), bonds *endo* within the calix[4]arene cone with contacts to four *ipso* carbon atoms of the arene rings. The Na–C(*ipso*) distances are 279.2(3)–290.0(5) pm and further contacts are in the range 291.8(6)–305.6(6) pm. In [K$_2$(NCMe)$_3$OMo(NC$_6$H$_3$Me$_2$-2,6)(L)] (L = *p*-t-butyl-calix[4]arene), potassium–arene interactions (K–*c* 312.9(3) and 316.4(3) pm) distort the cone by drawing two opposite phenoxide rings together.[157] One of the remaining phenoxide rings is involved in $\pi$ interactions with a second potassium ion (K–*c* 286.2(3) pm) in a K$_2$O$_2$ ring. (The oxygen was probably derived from the THF solvent used for the synthesis.) Cation-$\pi$ interactions have also been detected by mass spectroscopy.[158]

As expected, there are also well-documented metal–arene interactions in the cesium calix[4]arenes. The monocesium derivative of *p*-tertiary-butylcalix[4]arene forms a 1:1 complex with MeCN in which the Cs–O distances are >400 pm and the shortest contacts are to the MeCN ligand(Cs–N 329 pm) and by unsymmetrical $\eta^6$ coordination to the four aromatic rings (Cs–C 354.5(3)–412.1(4) pm).[159] It has been suggested[160] that it may be possible to exploit metal–arene interaction in the development of selective sequestering agents for alkali metal cations. For example, a titration of 1,3-dimethoxycalix[4]arenecrown-6 with cesium picrate yielded a crystalline compound which was shown to have a structure in which each cesium is coordinated by eight oxygen atoms (six from the crown ether and two from the picrate) and $\eta^3$ fashion to one phenyl ring and $\eta^2$ to another. The Cs–C distances range from 335.4(7) to 368.4(8) pm. The cesium–arene interactions are important in changing the configuration of the ligand from that usually adopted in solution. By replacing the methoxo

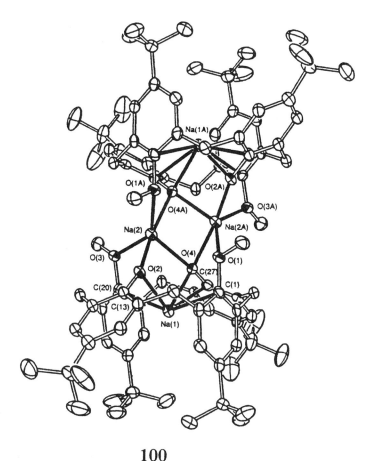

**100**

FIG. 61.   Molecular structure of the disodium salt of 1,3-dimethyl ether *p*-t-butylcalix[4]ar-ene, **100**. (Reprinted with permission from S. R. Dubberly *et al.*, *Chem. Commun.*, **1997,** 1603. Copyright © 1997 Royal Society of Chemistry.)

groups by OPr, $OC_8H_{17}$, or $O^iPr$, it was possible to synthesize new ligands with the same configuration by using cesium ions in the cyclization step. These proved to be highly selective sequestering agents with ratios of distribution coefficients $D(Cs)/D(Na)$ of about 30,000 with possible use for extracting traces of $^{137}Cs$ from solutions that are 4 *M* in $Na^+$.

## C. *Fullerides*

Alkali metal fullerides are not organometallic compounds in the normal sense, but their chemistry is in a number of respects similar to that of the alkali metal derivatives of aromatic hydrocarbons described earlier. Electrons are transferred from the metal to the fulleride framework and extensively delocalized. A notable feature of the chemistry of the fullerides is their superconductivity.[161] The critical temperature increases as the mass of the alkali metal increases and for $RbCs_2C_{60}$ or $Rb_2CsC_{60}$ is as high as 30 K. No attempt is made to give a full account of this rapidly expanding area, but further references can be found in other reviews.[161–165]

## VIII

## CONCLUSIONS

The work described in this review shows that the organometallic chemistry of the heavier alkali metals has ceased to be an exotic backwater. A good deal of information about the interaction between metal cations and carbanionic fragments has been obtained. Much of it has come from studies on crystalline solids, and although it may be reasonable to expect that the species found in the solid are also present in solution, this has to be established experimentally in each case. Some evidence has been obtained from multinuclear or multidimensional NMR spectroscopy, but so far there have been few studies using solid-state NMR to link structures found in the solid with those in solution. Even when the dominant species in solution is established, still more work is required to determine which species react fastest with particular substrates. The active species in a reaction may be present only in low concentration.

Recent work has confirmed earlier conclusions that the organometallic compounds are highly ionic. When the carbanion is small the negative charge is localized at a particular carbon center. In the simplest case, the $Me^-$ ion is, as expected, pyramidal.[1] When, however, the anion is large and such that the change can be extensively delocalized, the configuration at carbon is usually planar. The interaction with the metal is with the anion as a whole rather than with a particular carbon center. This is shown in Table I, which gives the range of M–C bond lengths based on data given earlier for individual compounds. For sodium the shortest bonds are normally those to central carbon atoms, but for rubidium and cesium the shortest bonds are often to peripheral phenyl groups. This point is well illustrated by results from series of closely related compounds,[12,29,42,152,166,167] and experimental work has been supported by computational studies. The coordination number of rubidium or cesium can be extraordinarily large: up to 20 neighbors are not uncommon.

Carbanionic ligands bearing donor atoms form compounds with sodium

TABLE I

METAL–CARBON DISTANCES (PM)

| | $\sigma^a$ | $\eta^{1b}$ | $\eta^{3b}$ | $\eta^{5b}$ | $\eta^{6b}$ | Me$^c$ |
|---|---|---|---|---|---|---|
| Na | 248–255 | 257–268 | 263–300 | 260–300 | 270–310 | 275–325 |
| K | 291–310 | 267–310 | 314–330 | 296–330 | 310–330 | 320–330 |
| Rb | 329–351 | | 325–330 | 320–340 | 314–360 | 330–370 |
| Cs | 366 | | 370–380 | 331–340 | 330–380 | |

$^a$ Distances to central carbon atoms in CR$_3$.
$^b$ Distances to carbon atoms in cyclopentadienyl or arene rings.
$^c$ Distances to methyl groups on the anion periphery.

in which the strongest bonds are to oxygen or nitrogen rather than carbon. For the heavier alkali metals, however, the bonds from the metal to aryl groups become increasingly strong and, in many cases, the oxygen or nitrogen is displaced from the coordination sphere. These points are illustrated by comparisons between CPh$_3$ and CPh$_2$(C$_5$H$_4$N) derivatives[42,44] and by the greater tendency in sodium than in cesium derivatives to form solvent-separated rather than solvent-shared ion pairs.

There have been few reports of the exploitation of the organometallic compounds of the heavier alkali metals in organic synthesis. In view of their greater reactivity compared with organolithium reagents (and therefore the possibility of effecting reactions at lower temperatures without interference from unwanted rearrangement of reactive intermediates), it is likely that many more applications will be found as reliable syntheses of the organometallic compounds become better known.

ACKNOWLEDGMENT

The author thanks the Engineering and Physical Sciences Research Council and the European Union for financial support.

REFERENCES

(1) Weiss, E. *Angew. Chem. Int. Ed. Engl.* **1993**, *32*, 1501.
(2) Bock, H.; Ruppert, K.; Näther, C.; Havlas, Z.; Herrmann, H-F.; Arad, C.; Göbel, I.; John, A.; Meuret, J.; Nick, S.; Rauschenbach, A.; Seitz, W.; Vaupel, T.; Solouki, B. *Angew. Chem. Int. Ed. Engl.* **1992**, *31*, 550.
(3) Schade, C.; Schleyer, P. v. R.; *Adv. Organomet. Chem.* **1987**, *27*, 169.
(4) Lambert C.; Schleyer, P. v. R. in *Methoden der Organische Chemie;* Vol. 19d; Thieme: Stuttgart, 1993.
(5) Beswick, M. A.; Wright, D. S. in *Comprehensive Organometallic Chemistry II;* Abel, E. W.; Stone, F. G. A.; Wilkinson, G., Eds.; Pergamon: Oxford, 1995; Vol. 1, pp. 1–34.

(6) *Dictionary of Organometallic Compounds,* Second Edition, Macintyre, J. E.; Hodgson, A. J. Eds.; Chapman and Hall: London, 1995.

(7) Lambert, C.; Schleyer, P. v. R. *Angew. Chem. Int. Ed. Engl.* **1994**, *33*, 1129.

(8) Bailey, W. F.; Punzalan, E. R. *J. Am. Chem. Soc.* **1994**, *116*, 6577.

(9) Bauer, W.; Lochmann, L. *J. Am. Chem. Soc.* **1992**, *114*, 7482.

(10) Weiss, E.; Köster, H. *Chem. Ber.* **1977**, *110*, 717.

(11) Eaborn, C.; Hitchcock, P.B.; Izod, K.; Smith, J. D. *Angew. Chem. Int. Ed. Engl.* **1995**, *34*, 687.

(12) Eaborn, C.; Clegg, W.; Hitchcock, P. B.; Hopman, M.; Izod, K.; O'Shaughnessy, P. N.; Smith, J. D. *Organometallics* **1997**, *16*, 4728.

(13) Harder, S.; Streitwieser, A. *Angew. Chem. Int. Ed. Engl.* **1993**, *32*, 1066.

(14) Kremer, T.; Harder, S.; Junge, M.; Schleyer, P. v. R. *Organometallics* **1996**, *15*, 585.

(15) Screttas, C. G.; Steele, B. R. *J. Organometal. Chem.* **1993**, *453*, 163.

(16) Hart, A. J.; O'Brien, D. H.; Russell, C. R. *J. Organomet. Chem.* **1974**, *72*, C19.

(17) Hartmann, J.; Schlosser, M. *Helv. Chim. Acta* **1976**, *59*, 453.

(18) Zaidlewicz, M. *J. Organomet. Chem.* **1985**, *293*, 139.

(19) Al-Juaid, S. S.; Eaborn, C.; Hitchcock, P. B.; Izod, K.; Mallien, M.; Smith, J. D. *Angew. Chem. Int. Ed. Engl.* **1994**, *33*, 1268.

(20) Eaborn, C.; Hitchcock, P. B.; Izod, K.; Jaggar, A. J.; Smith, J. D. *Organometallics* **1994**, *13*, 753.

(21) Desponds, O.; Schlosser, M. *J. Organomet. Chem.* **1991**, *409*, 93.

(22) Viebrock, H.; Panther, T.; Behrens, U.; Weiss, E. *J. Organomet. Chem.* **1995**, *491*, 19.

(23) Bock, H.; Hauck, T.; Näther C. *Organometallics* **1996**, *15*, 1527.

(24) Goel, S. C.; Grovenstein, E. *Organometallics* **1992**, *11*, 1565.

(25) Yeh, P.-H.; Pang, Z.; Johnston, R. F. *J. Organomet. Chem.* **1996**, *509*, 123.

(26) Jones, S. S.; Rausch, M. D.; Bitterwolf, T. E. *J. Organomet. Chem.* **1993**, *450*, 27.

(27) Dinnebier, R. E.; Behrens, U.; Olbrich, F. *Organometallics* **1997**, *16*, 3855.

(28) Rabe, G.; Roesky, H. W.; Stalke, D.; Pauer, F.; Sheldrick, G. M. *J. Organomet. Chem.* **1991**, *403*, 11.

(29) Hoffmann, D.; Bauer, W.; Hampel, F.; Eikema Hommes, N. J. R van; Schleyer, P. von R.; Otto, P.; Pieper, U.; Stalke, D.; Wright, D. S.; Snaith, R. *J. Am. Chem. Soc.* **1994**, *116*, 528.

(30) Corbelin, S.; Lorenzen, N. P.; Kopf, J.; Weiss, E. *J. Organomet. Chem.* **1991**, *415*, 293.

(31) Zarges, W.; Marsch, M.; Harms, K.; Boche, G. *Chem. Ber.* **1989**, *122*, 2303.

(32) Schade, C.; Schleyer, P. v. R.; Dietrich, H.; Mahdi, W. *J. Am. Chem. Soc.* **1986**, *108*, 2484

(33) Baker, D. R.; Clegg, W.; Horsburgh, L.; Mulvey, R. E. *Organometallics* **1994**, *13*, 4170

(34) Harder, S.; Lutz, M.; Kremer, T. *Organometallics* **1995**, *14*, 2133.

(35) Gornitzka, H.; Stalke, D. *Organometallics* **1994**, *13*, 4398.

(36) Viebrock, H.; Behrens, U.; Weiss, E. *Chem. Ber.* **1994**, *127*, 1399.

(37) Bors, D. A.; Kaufman, M. J.; Streitwieser, A. *J. Am. Chem. Soc.* **1985**, *107*, 6975.

(38) Xie, L.; Bors, D. A.; Streitwieser, A. *J. Org. Chem.* **1992**, *57*, 4986.

(39) Comins, D. L.; Salvador, J. M. *Tetrahedron Lett.* **1993**, *34*, 801.

(40) Brooks, J. J.; Stucky, G. D. *J. Am. Chem. Soc.* **1972**, *94*, 7333.

(41) Köster, H.; Weiss, E. *J. Organomet. Chem.* **1979**, *168*, 273.

(42) Hoffmann, D.; Bauer, W.; Schleyer, P. v. R.; Pieper, U.; Stalke, D. *Organometallics* **1993**, *12*, 1193.

(43) Bartlett, R. A.; Dias, H. V. R.; Power, P. P. *J. Organomet. Chem.* **1988**, *341*, 1.

(44) Pieper, U.; Stalke, D. *Organometallics* **1993**, *12*, 1201.

(45) Zaidlewicz, M. *J. Organomet. Chem.* **1991**, *409*, 103.

(46) Lappert, M. F.; Liu, D-S. *J. Organomet. Chem.* **1995**, *500*, 203.

(47) Tian, S.; Lappert, M. F. Personal communication.

(48) Hitchcock, P. B.; Lappert, M. F.; Leung, W-P.; Liu, D-S.; Shun, T. *J. Chem. Soc. Chem. Commun.* **1993**, 1386.

(49) Hitchcock, P. B.; Holmes, S. A.; Lappert, M. F.; Shun, T. *J. Chem. Soc. Chem. Commun.* **1994**, 2691.

(50) Eaborn, C.; Izod, K.; Smith, J. D. *J. Organomet. Chem.* **1995**, *500*, 89.

(51) Niemeyer, M.; Power, P. P. *Organometallics* **1996**, *15*, 4107.

(52) Eaborn, C.; Hitchcock, P. B.; Izod, K.; Smith, J. D. *J. Am. Chem. Soc.* **1994**, *116*, 12071.

(53) Eaborn, C.; Hitchcock, P. B.; Izod, K.; Lu, Z-R.; Smith, J. D. *Organometallics* **1996**, *15*, 4783.

(54) Eaborn, C.; Hawkes, S. A.; Hitchcock, P. B.; Smith, J. D. *Chem. Commun.* **1997**, 1961.

(55) Eaborn, C.; Smith, J. D. *Coord. Chem. Rev.* **1996**, *154*, 125.

(56) Schaller, F.; Schwarz, W.; Hausen, H-D.; Klinkhammer, K. W.; Weidlein, J. *Z. Anorg. Allg. Chem.* **1997**, *623*, 1455.

(57) Hiller, W.; Layh, M.; Uhl, W. *Angew. Chem. Int. Ed. Engl.* **1991**, *30*, 324.

(58) Eaborn, C.; Hitchcock, P. B.; Izod, K.; Smith, J. D. *Angew. Chem. Int. Ed. Engl.* **1995**, *34*, 2679.

(59) Eaborn, C.; Hitchcock, P. B.; Smith, J. D.; Sullivan, A. C. *J. Chem. Soc. Chem. Commun.* **1983**, 1390.

(60) Eaborn, C.; Clegg, W.; Farook, A.; Hitchcock, P. B.; Izod, K.; O'Shaunhessy, P. N.; Smith, J. D. Unpublished work.

(61) Clegg, W.; Eaborn, C.; Izod, K.; O'Shaughnessy, P. N.; Smith, J. D. *Angew. Chem. Int. Ed. Engl.* **1997**, *36*, 2815.

(62) Jedliński, Z.; Misiołek, A.; Jankowski, A.; Janeczek, H. *J. Organomet. Chem.* **1992**, *433*, 231.

(63) Niemeyer, M.; Power, P. P. *Organometallics* **1997**, *16*, 3258.

(64) Besten R. den.; Brandsma, L.; Spek, A. L.; Kanters, J. A.; Veldman, N. *J. Organomet. Chem.* **1995**, *498*, C6.

(65) Stalke, D. *Angew. Chem. Int. Ed. Engl.* **1994**, *33*, 2168.

(66) Lambert, C.; Kaupp, M.; Schleyer, P. v. R. *Organometallics* **1993**, *12*, 853.

(67) Caldwell, J. W.; Kollman, P. A. *J. Am. Chem. Soc.* **1995**, *117*, 4177.

(68) Jutzi, P.; Leffers, W.; Hampel, B.; Pohl, S.; Saak, W. *Angew. Chem. Int. Ed. Engl.* **1987**, *26*, 583.

(69) Jordan, V.; Behrens, U.; Olbrich, F.; Weiss, E. *J. Organomet. Chem.* **1996**, *517*, 81.

(70) Dinnebier, R. E.; Olbrich, F.; Smaalen, S. van; Stephens, P. W. *Acta Crystallogr. Sect. B.* **1997**, *53*, 153.

(71) Dinnebier, R. E.; Olbrich, F.; Bendele, G. M. *Acta Crystallogr. Sect. C.* **1997**, *53*, 699.

(72) Aoyagi, T.; Shearer, H. M. M.; Wade, K.; Whitehead, G. *J. Organomet. Chem.* **1979**, *175*, 21.

(73) Harder, S.; Prosenc, M. H. *Angew. Chem. Int. Ed. Engl.* **1994**, *33*, 1744.

(74) Harder, S.; Prosenc, M. H.; Rief, U. *Organometallics* **1996**, *15*, 118.

(75) Wessel, J.; Lork, E.; Mews, R. *Angew. Chem. Int. Ed. Engl.* **1995**, *34*, 2376.

(76) Harder, S.; Prosenc, M. H. *Angew. Chem. Int. Ed. Engl.* **1996**, *35*, 97.

(77) Davidson, M. G.; Stalke, D.; Wright, D. S. *Angew. Chem. Int. Ed. Engl.* **1992**, *31*, 1226.

(78) Armstrong, D. R.; Duer, M. J.; Davidson, M. G.; Moncrieff, D.; Russell, C. A.; Stourton, C.; Steiner, A.; Stalke, D.; Wright, D. S. *Organometallics* **1997**, *16*, 3340.

(79) Apostolidis, C.; Deacon, G. B.; Dornberger, E.; Edelmann, F. T.; Kanellakopulos, B.; MacKinnon, P.; Stalke, D. *Chem. Commun.* **1997**, 1047.

(80) Herrmann, W. A.; Anwander, R.; Munck, F. C.; Scherer, W. *Chem. Ber.* **1993**, *126*, 331.

346	J. DAVID SMITH

(81) Rogers, R. D.; Atwood, J. L.; Rausch, M. D.; Macomber, D. W.; Hart, W. P. *J. Organomet. Chem.* **1982**, *238*, 79.

(82) Varga, V.; Hiller, J.; Polášek, M.; Thewalt, U.; Mach, K. *J. Organomet. Chem.* **1996**, *514*, 219.

(83) Hou, Z.; Zhang, Y.; Yoshimura, T.; Wakatsuki, Y. *Organometallics* **1997**, *16*, 2963.

(84) Lorbeth, J.; Shin, S-H.; Wocadlo, S.; Massa, W. *Angew. Chem. Int. Ed. Engl.* **1989**, *28*, 735.

(85) Schumann, H.; Lentz, A.; Weimann, R. *J. Organomet. Chem.* **1995**, *487*, 245.

(86) Näther, C.; Hauck, T.; Bock, H. *Acta Crystallogr. Sect. C* **1996**, *52*, 570.

(87) Bock, H.; Hauck, T.; Näther, C.; Havlas, Z. *Angew. Chem. Int. Ed. Engl.* **1997**, *36*, 638.

(88) Paul, F.; Carmichael, D.; Ricard, L.; Mathey, F. *Angew. Chem. Int. Ed. Engl.* **1996**, *35*, 1125.

(89) Corbelin, S.; Kopf, J.; Weiss, E.; *Chem. Ber.* **1991**, *124*, 2417.

(90) Hoffmann, D.; Hampel, F.; Schleyer, P. v. R. *J. Organomet. Chem.* **1993**, *456*, 13.

(91) Janiak, C. *Chem. Ber.* **1993**, *126*, 1603.

(92) Zerger, R.; Rhine, W.; Stucky, G. D. *J. Am. Chem. Soc.* **1974**, *96*, 5441.

(93) Littger, R.; Metzler, N.; Nöth, H.; Wagner, M. *Chem. Ber.* **1994**, *127*, 1901

(94) Bock, H.; Arad, C.; Näther, C.; Göbel, I. *Helv. Chim. Acta* **1996**, *79*, 92.

(95) Cloke, F. G. N. Personal communication.

(96) Cloke, F. G. N.; Hitchcock, P. B. *J. Am. Chem. Soc.* **1997**, *119*, 7899.

(97) Eikema Hommes, N. J. R.; Bühl, M.; Schleyer, P. v. R. *J. Organomet. Chem.* **1991**, *409*, 307.

(98) Hojo, F.; Terashima, T.; Ando, W. *Organometallics* **1996**, *15*, 3480.

(99) Bock, H.; Ruppert, K.; Havlas, Z.; Bensch, W.; Hönle, W.; von Schnering, H. G. *Angew. Chem. Int. Ed. Engl.* **1991**, *30*, 1183.

(100) Bock, H.; Näther, C.; Ruppert, K.; Havlas, Z. *J. Am. Chem. Soc.* **1992**, *114*, 6907.

(101) Sekiguchi, A.; Ichinohe, M.; Kabuto, C.; Sakurai, H. *Organometallics* **1995**, *14*, 1092.

(102) Bock, H.; Näther C.; Ruppert, K. *J. Chem. Soc. Chem. Commun.* **1992**, 765.

(103) Bock, H.; Näther, C.; Havlas, Z.; John, A.; Arad, C. *Angew. Chem. Int. Ed. Engl.* **1994**, *33*, 875.

(104) Hu, N.; Gong, L.; Jin, Z.; Chen, W. *J. Organomet. Chem.* **1988**, *352*, 61.

(105) Kilimann, U.; Schäfer, M.; Herbst-Irmer, R.; Edelman, F. T. *J. Organomet. Chem.* **1994**, *469*, C15.

(106) Hodgson, K. O.; Raymond, K. N. *Inorg. Chem.* **1972**, *11*, 3030.

(107) Rijnberg, E.; Boersma, J.; Jastrzebski, J. T. B. H.; Lakin, M. T.; Spek, A. L.; van Koten, G. *Organometallics* **1997**, *16*, 3158.

(108) Hitchcock, P. B.; Lappert, M. F.; Wang, Z-X. *Chem. Commun.* **1996**, 1647.

(109) Andrews, P. C.; Mulvey, R. E.; Clegg, W.; Reed, D. *J. Organomet. Chem.* **1990**, *386*, 287.

(110) Andrews, P. C.; Armstrong, D. R.; Clegg, W.; Craig, F. J.; Dunbar, L.; Mulvey, R. E. *Chem. Commun.* **1997**, 319.

(111) Veya, P.; Floriani, C.; Chiesi-Villa, A.; Guastini, C. *J. Chem. Soc., Chem. Commun.* **1991**, 991.

(112) Andrews, P. C.; Armstrong, D. R.; Baker, D. R.; Mulvey, R. E.; Clegg, W.; Horsburgh, L.; O'Neil, P. A.; Reed, D. *Organometallics* **1995**, *14*, 427.

(113) Hitchcock, P. B.; Lappert, M. F.; Liu, D-S. *J. Chem. Soc. Chem. Commun.* **1994**, 1699.

(114) Hitchcock, P. B.; Lappert, M. F.; Liu, D-S. *J. Chem. Soc. Chem. Commun.* **1994**, 2637.

(115) Hitchcock, P. B.; Lappert, M. F.; Liu, D-S. *J. Organomet. Chem.* **1996**, *488*, 241.

(116) Boesveld, W. M.; Hitchcock, P. B.; Lappert, M. F. *Chem. Commun.* **1997**, 2091.

(117) Boesveld, W. M.; Hitchcock, P. B.; Lappert, M. F.; Sablong, R. Personal communication.

(118) Gong, L.; Hu, N.; Jin, Z.; Chen, W. *J. Organomet. Chem.* **1988**, *352*, 67.

(119) Henderson, K. W.; Willard, P. G.; Bernstein, P. R. *Angew. Chem. Int. Ed. Engl.* **1995**, *34*, 1117.

(120) Pulham, R. J.; Weston, D. P.; Salvesen, T. A.; Thatcher, J. J. *J. Chem. Res. (S)* **1995**, 254; Pulham, R. J.; Weston, D. P.; *J. Chem. Res. (S)* **1995**, 406.

(121) Screttas, C. G.; Ioannou, G. I.; Micha-Screttas, M. *J. Organomet. Chem.* **1996**, *511*, 217.

(122) Bock, H.; Arad, C.; Näther, C.; Havlas, Z. *J. Chem. Soc. Chem. Commun.* **1995**, 2393.

(123) Bock, H.; John, A.; Näther, C.; Havlas, Z.; Mihokova, E. *Helv. Chim. Acta* **1994**, *77*, 41.

(124) Bock, H.; Arad, C.; Näther, C. *J. Organomet. Chem.* **1996**, *520*, 1.

(125) Bock, H.; Ansari, M.; Nagel, N.; Havlas, Z. *J. Organomet. Chem.* **1995**, *499*, 63.

(126) Sekiguchi, A.; Ebata, K.; Kabuto, C.; Sakurai, H. *J. Am. Chem. Soc.* **1991**, *113*, 7081.

(127) Bock, H.; Ansari, M.; Nagel, N.; Claridge, R. F. C. *J. Organomet. Chem.* **1995**, *501*, 53.

(128) Bock, H.; John, A.; Näther, C.; Havlas, Z. *Z. Naturforsch. Teil B* **1994**, *49*, 1339.

(129) Bock, H.; Hauck, T.; Näther, C.; Havlas, Z. *Z. Naturforsch. Teil B* **1994**, *49*, 1012.

(130) Bock, H.; Ruppert, K.; *Inorg. Chem.* **1992**, *31*, 5094.

(131) Bock, H.; John, A.; Näther, C.; Ruppert, K. *Helv. Chim. Acta* **1994**, *77*, 1505.

(132) Jost, W.; Adam, M.; Enkelmann, V.; Müllen, K. *Angew. Chem. Int. Ed. Engl.* **1992**, *31*, 878.

(133) Näther, C.; Bock, H.; Claridge, R. F. C. *Helv. Chim. Acta* **1996**, *79*, 84.

(134) Bock, H.; Näther, C.; Havlas, Z. *J. Am. Chem. Soc.* **1995**, *117*, 3869.

(135) (a) Bock, H.; Gharagozloo-Hubmann, K.; Näther, C.; Nagel, N.; Havlas, Z. *Angew. Chem. Int. Ed. Engl.* **1996**, *35*, 631; (b) Bock, H.; John, A.; Näther, C. *J. Chem. Soc. Chem. Commun.* **1994**, 1939; (c) Bock, H.; John, A.; Kleine, M.; Näther C.; Bats, J. W. *Z. Naturforsch. Teil B* **1994**, *49*, 529.

(136) Janiak, C.; Hemling, H. *Chem. Ber.* **1994**, *127*, 1251.

(137) Haag, R.; Ohlhorst, B.; Noltemeyer, M.; Fleischer, R.; Stalke, D.; Schuster, A.; Kuck, D.; de Meijere, A. *J. Am. Chem. Soc.* **1995**, *117*, 10474.

(138) Niemeyer, M.; Power, P. P. *Inorg. Chem.* **1996**, *35*, 7264; Chadwick, S.; Englich, U.; Ruhlandt-Senge, K.; *Organometallics* **1997**, *16*, 5792.

(139) Müller, E.; Krause, J.; Schmiedeknect, K. *J. Organomet. Chem.* **1972**, *44*, 127.

(140) Clark, D. L.; Hollis, R. V.; Scott, B. L.; Watkin, J. G. *Inorg. Chem.* **1996**, *35*, 667.

(141) Clark, D. L.; Deacon, G. B.; Feng, T.; Hollis, R. V.; Scott, B. L.; Skelton, B. W.; Watkin, J. G.; White, A. H. *Chem. Commun.* **1996**, 1729.

(142) (a) Evans, W. J.; Ansari, M. A.; Khan, S. I. *Organometallics* **1995**, *14*, 558; (b) Evans, W. J.; Anwander, R.; Ansari, M. A.; Ziller, J. W. *Inorg. Chem.* **1995**, *34*, 5.

(143) (a) Clark, D. L.; Watkin, J. G.; Huffman, J. C. *Inorg. Chem.* **1992**, *31*, 1554; (b) Clark, D. L.; Gordon, J. C.; Huffman, J. C.; Vincent-Hollis, R. L.; Watkin, J. G.; Zwick, B. D. *Inorg. Chem.* **1994**, *33*, 5903.

(144) Fickes, M. G.; Odom, A. L.; Cummins, C. C. *Chem. Commun.* **1997**, 1993.

(145) Neumüller, B.; Gahlmann, F. *Chem. Ber.* **1993**, *126*, 1579.

(146) Werner, B.; Kräuter, T.; Neumüller, B. *Organometallics* **1996**, *15*, 3746; Kräuter, T.; Neumüller, B. *Chem. Eur. J.* **1997**, *3*, 568; Kopp, M. R.; Neumüller, B. *Organometallics* **1997**, *16*, 5623.

(147) Sieler, J.; Pink, M.; Zahn, G. *Z. Anorg. Allg. Chem.* **1994**, *620*, 743.

(148) (a) Driess, M.; Pritzkow, H.; Skipinski, M.; Winkler, U. *Organometallics* **1997**, *16*, 5108; (b) Steiner, A.; Stalke, D. *Inorg. Chem.* **1993**, *32*, 1977.

(149) Schaverien, C. J.; Mechelen, J. B. V. *Organometallics* **1991**, *10*, 1704.

(150) Hitchcock, P. B.; Lappert, M. F.; Lawless, G. A.; Royo, B. *J. Chem. Soc. Chem. Commun.* **1993**, 554.

(151) Fuentes, G. R.; Coan, P. S.; Streib, W. E.; Caulton, K. G. *Polyhedron,* **1991**, *10*, 2371.

(152) Pauer, F.; Stalke, D. *J. Organomet. Chem.* **1991**, *418*, 127.

(153) Clegg, W.; Drummond, A. M.; Mulvey, R. E.; O'Shaughnessy, P. N. *Chem. Commun.* **1997**, 1301.
(154) Gemünd, B.; Nöth, H.; Sachdev, H.; Schmidt, M. *Chem. Ber.* **1996**, *129*, 1335.
(155) Knizek, J.; Krossing, I.; Nöth, H.; Schwenk, H.; Seifert, T. *Chem. Ber.* **1997**, *130*, 1053.
(156) Dubberley, S. R.; Blake, A. J.; Mountford, P. *Chem. Commun.* **1997**, 1603.
(157) Gibson, V. C.; Redshaw, C.; Clegg, W.; Elsegood, M. R. J. *Chem. Commun.* **1997**, 1605.
(158) Inokuchi, F.; Miyahara, Y.; Inazu, T.; Shinkai, S. *Angew. Chem. Int. Ed. Engl.* **1995**, *34*, 1364.
(159) Harrowfield, J. M.; Ogden, M. I.; Richmond, W. R.; White, A. H. *J. Chem. Soc. Chem. Commun.* **1991**, 1159.
(160) Ungaro, R.; Casnati, A.; Ugozzoli, F.; Pochini, A.; Dozol, J-F.; Hill, C.; Rouquette, H. *Angew. Chem. Int. Ed. Engl.* **1994**, *33*, 1506.
(161) Gunnarsson, O. *Rev. Mod. Phys.* **1997**, *69*, 575.
(162) Fässler, T. F.; Spiekermann, A.; Spahr, M. E.; Nesper, R. *Angew. Chem. Int. Ed. Engl.* **1997**, *36*, 486.
(163) Prassides, K. *Curr. Opin. Solid State Mater. Sci.* **1997**, *2*, 433.
(164) Birkett, P. R.; Prassides, K. *Annu. Rep. Progr. Chem. Inorg. Chem.* **1996**, *92*, 539.
(165) Glarum, S. H.; Duclos, S. J.; Haddon, R. C. *J. Am. Chem. Soc.* **1992**, *114*, 1996.
(166) Klinkhammer, K. W.; Schwarz, W. *Z. Anorg. Allg. Chem.* **1993**, *619*, 1777.
(167) Klinkhammer, K. W. *Europ. Chem. J.* **1997**, *3*, 1418.

ADVANCES IN ORGANOMETALLIC CHEMISTRY, VOL. 43

# Organometallic Complexes in Nonlinear Optics II: Third-Order Nonlinearities and Optical Limiting Studies

## IAN R. WHITTALL, ANDREW M. McDONAGH, and MARK G. HUMPHREY

*Department of Chemistry*
*Australian National University, Canberra, ACT 0200, Australia*

## MAREK SAMOC

*Australian Photonics Cooperative Research Centre*
*Laser Physics Centre, Research School of Physical Sciences and Engineering*
*Australian National University, Canberra, ACT 0200, Australia*

## I

## INTRODUCTION

Interactions of electromagnetic fields (light) with matter change the nature of the incident light such that new field components with differing phase, frequency, amplitude, polarization, path, or other propagation characteristics are produced. These nonlinear optical (NLO) interactions are of technological importance in areas that utilize optical devices, with potential applications in optical signal processing, switching, and frequency generation. Materials that possess NLO properties may be used for optical data

349

storage, optical communication, optical switching, image processing, and, ultimately, optical computing. The technologically important effects may be classed as second order or third order. The first part of this review dealt with the second-order nonlinearities of organometallics.[1] Here, we focus on third-order nonlinearities.

Replacing electronic (electron-driven) devices by photonic (light-driven) devices in integrated optics will afford the possibility of higher bandwidths (because of higher carrier frequencies) and will result in significant speed enhancement; because photons travel much faster than electrons, optical computers could have processing speeds orders of magnitude greater than existing electronic-based computers. There are possible disadvantages with the introduction of photonic-based systems, though, including higher energy requirements and device size limitations related to the wavelength of light. The relatively small nonlinearity of common photonic materials (such as oxide glasses, particularly silica glass) necessitates very long interaction paths for devices utilizing nonlinear effects.

Semiconductors such as gallium arsenide or indium arsenide, particularly as reduced dimensionality species (quantum wells, quantum wires, or quantum dots), possess nonlinear optical effects that originate from saturable absorption,[2] with third-order NLO responses that are among the highest known.[3] A disadvantage of NLO processes based on resonant interactions (such as one-photon or two-photon absorption) is, however, that the speed of the NLO response may be relatively slow. Although nonresonant nonlinearities of some semiconductors such as $Al_xGa_{1-x}As$ are reasonably high, many organics [e.g., $\pi$-conjugated polymers such as polyacetylene, polydiacetylenes, and poly($p$-phenylenevinylenes)] offer nonlinearities that are of similar or higher magnitudes, with the inherent flexibility of their design being an additional attractive feature. Studies of the third-order responses of organic molecules have therefore commanded attention.

During the evolution of the study of organic molecules for nonlinear optics, experimental observation has enabled certain structure/NLO property relationships to be developed, from which useful insights may be gained. Computational investigations utilizing quantum theory have also afforded general qualitative rules. High polarizability of any order is associated with the existence of low energy molecular excited states which, because they are close in energy to the ground state, mix easily when the molecule is perturbed. One disadvantage is therefore a trade-off between nonlinear efficiency and optical transparency. Another possible disadvantage is lower thermal stability (especially in organic polymers). Sizable $\pi$-delocalization length (e.g., progressing from small molecules to $\pi$-conjugated polymers), the presence of donor and acceptor functional groups, chain orientation and packing density, conformation, and dimensionality (e.g., progressing from one-dimensional oligomers to two-dimensional porphyrins and phthalocyanines)[4] all impact favorably on third-order nonlinearity. Nonresonant NLO properties can be increased further by controlling bond-length alternation (the difference in length between $C-C$ single and double bonds in a $\pi$-conjugated system).[5] Changing the alternation may alter the charge distribution in the electronic ground state. For optimized alternation, enhancements of up to a factor of five may be observed. Even so, further improvement of the NLO properties and the reduction of linear and nonlinear absorption losses is necessary for practical device applications.

Like organic molecules, organometallic complexes can possess strong responses, fast response times, ease of fabrication, and integration into composites, as well as having the advantage over organic systems of much greater flexibility at the design state. Organometallic complexes can possess metal to ligand or ligand to metal charge transfer bands in the ultraviolet-visible region of the spectrum. Complexes with novel bonding patterns and coordination geometries allow for spatial arrangements of atoms that may not be easily accessible in other systems, increasing the design flexibility. Organometallic compounds are often strong oxidizing or reducing agents, as metal centers may be electron rich or poor depending on their oxidation state and ligand environment. Thus, the metal center may be an extremely strong donor or acceptor. Unusual and/or unstable organic fragments (e.g., carbenes) may be stabilized on metals, allowing the NLO properties of these species to be assessed. Variation in metal, oxidation state, ligand environment, and geometry can potentially permit NLO responses to be tuned in ways not possible for purely organic molecules. Studying the nonlinear properties of a variety of metals, oxidation states and ligands in systematic series of "families" of organometallic compounds can lead to an understanding of structure/property relationships. As with organics, organometallics can form polymers or be included in polymers, either as side chains or in the polymer backbone; this affords the possibility of introducing more polarizable atoms in a polymer chain than may be accessible in purely organic systems.

This review covers the theoretical background and some of the practical aspects of nonlinear optics, including a description of the origins of third-order nonlinearities, systems of units that are encountered, experimental techniques that have been used or may be used to probe the third-order NLO properties of organometallic complexes, and computational methods that have or could be used to calculate third-order NLO properties. Subsequent sections collect comprehensive data of organometallic complexes in tables categorized by complex type and discussions of the results of third-order NLO measurements and calculations performed on organometallic

complexes. Additionally, the optical limiting properties of a few organometallics are discussed. While it is intended that coverage of transition metal
complexes is comprehensive in scope, data for metalloidal systems (such
as silicon compounds) are indicative only.

# II

# BACKGROUND

## A. Microscopic and Macroscopic Third-Order Optical Nonlinearities

A more comprehensive discussion of the theoretical background can be
found in the first part of this review.[1] This necessarily more abbreviated
account focuses on those aspects relevant to third-order properties. As
discussed in the first part,[1] a convenient way to describe the nonlinear
optical properties of organic molecules is to consider the effect on the
molecular dipole moment $\mu$ of an external electric field:

$$\mu = \mu_0 + \alpha \mathbf{E}_{loc} + \beta \mathbf{E}_{loc}\mathbf{E}_{loc} + \gamma \mathbf{E}_{loc}\mathbf{E}_{loc}\mathbf{E}_{loc} + \ldots \qquad (1)$$

where $\mathbf{E}_{loc}$ is the local electric field, and the tensorial quantities $\alpha$, $\beta$, and
$\gamma$ are the linear polarizability, the second-order or quadratic hyperpolarizability, and the third-order or cubic hyperpolarizability, respectively. The
third-order hyperpolarizability $\gamma$ (also referred to as the second hyperpolarizability) is a fourth-rank tensor (or a $3 \times 3 \times 3 \times 3$ matrix). The electric
field of a light wave oscillating at the frequency $\omega$ can be expressed as

$$\mathbf{E}(t) = \mathbf{E}_0 \cos(\omega t) = \frac{\mathbf{E}_0}{2}\left[\exp(i\omega t) + \exp(-i\omega t)\right]$$

Therefore, for an arbitrary point in space, Eq. (1) can be written as

$$\begin{aligned}
\mu(t) &= \mu_0 + \alpha \mathbf{E}_0\cos(\omega t) + \beta \mathbf{E}_0^2\cos^2(\omega t) + \gamma \mathbf{E}_0^3\cos^3(\omega t) + \ldots \\
&= \mu_0 + \frac{1}{2}\alpha \mathbf{E}_0\exp(i\omega t) + \frac{1}{2}\beta \mathbf{E}_0^2 + \frac{1}{4}\beta \mathbf{E}_0^2\exp(2i\omega t) + \frac{3}{8}\gamma \mathbf{E}_0^3\exp(i\omega t) \\
&\quad + \frac{1}{8}\gamma \mathbf{E}_0^3\exp(3i\omega t) + \text{c.c.} + \ldots
\end{aligned} \qquad (2)$$

where c.c. represents complex conjugate terms. It is apparent that the
cubic term in Eq. (2) leads to various nonlinear optical effects, one being
oscillation of the induced dipoles at $3\omega$ (third harmonic generation). However, there is also a term containing $\gamma$ and a cube of the electric field
amplitude that oscillates at the unchanged fundamental frequency $\omega$. This
term is responsible for intensity-dependent refractive index, among other

effects. One way of visualizing this is by grouping the terms oscillating at $\omega$ in Eq. (2). The $\omega$ component of the oscillating dipole is equal to

$$\frac{1}{2}\alpha E_0 \exp(i\omega t) + \frac{3}{8}\gamma E_0^3 \exp(i\omega t) + \text{c.c.} = \frac{1}{2}[\alpha + \frac{3}{4}\gamma E_0^2]E_0 \exp(i\omega t) + \text{c.c.}$$

Thus, the linear polarizability $\alpha$ (responsible for the value of the refractive index $n$) can be treated as an electric field amplitude-dependent quantity, i.e., $\alpha_{eff} = \alpha + (3\gamma E_0^2)/4$. Remembering that the light intensity is proportional to the square of the field amplitude, this means that the third-order nonlinearity leads to the linear dependence of the refractive index on the light intensity and that, for example, the phase of the propagating beam is modified at high light intensities due to this dependence.

Equation (2) is, strictly speaking, not suitable for optical fields, which are rapidly varying in time. The damping of the oscillating dipole, and the resultant phase shift, is then conveniently expressed by treating the hyperpolarizabilities as complex, frequency-dependent quantities. For the cubic hyperpolarizability, the relation between the Fourier components of the electric field and the Fourier amplitude of the oscillation of the electric dipole gives

$$\Delta\mu^{(3)}(\omega_4) = \gamma(-\omega_4;\omega_1,\omega_2,\omega_3)E(\omega_1)E(\omega_2)E(\omega_3)$$

where the frequency $\omega_4 = \omega_1 + \omega_2 + \omega_3$ at which the dipole oscillates is the result of mixing of the input frequencies. Third-order NLO processes are often referred to as four-wave mixing processes, although all four fields do not have to be at different frequencies nor do they have to belong to geometrically separate light beams. An important process is that in which all the input fields have the same frequency and the mixing occurs in such a way that the output frequency is also the same (one of the input frequencies has a negative sign), i.e., that described by the degenerate hyperpolarizability $\gamma(-\omega;\omega,-\omega,\omega)$.

As has been discussed in part I of this review,[1] the microscopic hyperpolarizabilities of the $i$th order have their corresponding quantities at the macroscopic level in the form of nonlinear susceptibilities $\chi^{(i)}$. The macroscopic polarization is then given by

$$\mathbf{P} = \varepsilon_0(\chi^{(1)}\mathbf{E} + \chi^{(2)}\mathbf{E}^2 + \chi^{(3)}\mathbf{E}^3 + \ldots) \text{ in the SI system}$$

or

$$\mathbf{P} = \chi^{(1)}\mathbf{E} + \chi^{(2)}\mathbf{E}^2 + \chi^{(3)}\mathbf{E}^3 + \ldots \text{ in the cgs system}$$

analogous to the expansion in the molecular domain. The relation between the hyperpolarizabilities and the macroscopic susceptibilities is usually derived from the oriented gas model in which one performs a summation of appropriate tensor components of the molecules, taking into account the orientation of the molecules versus the external coordinate system (using directional cosines) and correcting for the difference between the external electric field and the local field (using local field factors). This, in general, may lead to quite complicated relations, but the situation is greatly simplified in the case of isotropic media (e.g., glasses or fluids). Due to statistical orientation of molecules, orientation averaging can be performed, which for fourth-rank tensors leads to substantial simplification; from symmetry considerations, the $\chi^{(3)}$ tensor for an isotropic medium can only have two independent components, namely $\chi^{(3)}_{1111}$ and $\chi^{(3)}_{1122}$. The component $\chi^{(3)}_{1111}$ can be related to components of the molecular hyperpolarizability tensor (those contributing to the so-called scalar part of the fourth-rank tensor) in the following way:

$$\chi^{(3)}_{1111}(-\omega_4;\omega_3,\omega_2,\omega_1) = L_{\omega_1}L_{\omega_2}L_{\omega_3}L_{\omega_4}N <\gamma(-\omega_4;\omega_3,\omega_2,\omega_1)> \qquad (3)$$

where $L_{\omega_i}$ is the local field factor at frequency $\omega_i$ (often approximated by the Lorenz factor $L_\omega = (n_\omega^2 + 2)/3$) and

$$\langle\gamma\rangle = \frac{1}{5}(\gamma_{1111} + \gamma_{2222} + \gamma_{3333} + 2\gamma_{1122} + 2\gamma_{1133} + 2\gamma_{2233})$$

The simplest case is that of an isotropic medium containing molecules with a single dominant component of $\gamma$, say $\gamma_{1111}$ (a reasonable approximation for rigid-rod $\pi$-conjugated molecules in which the hyperpolarizability component along the molecular axis is likely to be dominant); $\langle\gamma\rangle = \frac{1}{5}\gamma_{1111}$ is then a reasonable approximation.

## B. *Third-Order Nonlinear Optical Processes*

There are many possible third-order nonlinear processes, some of which are important as valuable tools for nonlinear spectroscopy, whereas others have technological significance. The presence of $\chi^{(3)}$ in any substance (even air) means that all materials exhibit third-harmonic generation of laser frequencies. The direct process of third-harmonic generation is, however, not usually exploited for the generation of short wavelength laser beams, a cascade of two second-order mixing processes ($\omega + \omega = 2\omega$ and $2\omega + \omega = 3\omega$) being preferred for generation of $3\omega$ from $\omega$ (one reason for this is that phase matching is virtually impossible to obtain for third-harmonic generation). From the technological point of view, the most interesting applications of $\chi^{(3)}$ are those that correspond to all-optical interactions of light beams. For interacting fields of the same frequency (the degenerate case), the frequency mixing scheme is $\omega - \omega + \omega \rightarrow \omega$, which means that the interaction of three fields of the same frequency generates a fourth field of the same frequency. A convenient way of looking at some of the degenerate interactions is by considering the refractive index changes involved in those processes. One can define the nonlinear refractive index $n_2$ by considering the change of the refractive index with the light intensity $I$:

$$n = n_0 + n_2 I$$

It can be shown that there is a relation between $n_2$ and the third-order nonlinear susceptibility, namely[6,7]:

$$n_2 = \frac{3}{4\varepsilon_0 n^2 c} \chi^{(3)} \left( -\omega;\omega,-\omega,\omega \right) \qquad (4)$$

where $\varepsilon_0$ is the free space permittivity. Note that one should be cautious here, as various relations between $n_2$ and $\chi^{(3)}$ are quoted in the literature, a result of many possible ways of defining $\chi^{(3)}$.[7] In addition, some authors use an alternative definition of the nonlinear refractive index, relating the change of the refractive index to the square of the amplitude of the electric field (i.e., $n = n_0 + n_2' E^2$) and not to the light intensity. In general, both $\chi^{(3)}$ and $n_2$ should be treated as complex quantities, with the real part (often called the refractive part) responsible for the changes in the refractive properties and the imaginary part related to changes in absorptive properties of a medium. The nonlinear refractive properties are of significant interest because of possible applications in *inter alia* all-optical switching of signals. The nonlinear absorptive properties can also be of interest. The simplest way of treating the nonlinearity of absorption processes is by considering the absorption equation:

$$\frac{dI}{dz} = - \alpha_{abs} I - \beta_2 I^2$$

where the linear absorption coefficient $\alpha_{abs}$ is replaced by the effective absorption coefficient $\alpha_{abs} + \beta_2 I$; the quadratic absorption coefficient $\beta_2$ is responsible for two-photon absorption, and is therefore most often referred to as the two-photon absorption coefficient. This equation can, however, also be used in situations where ordinary one-photon absorption is modified at high light intensities by some additional processes such as, for example, absorption saturation. The nonlinear (quadratic) absorption coefficient $\beta_2$ can then be called a photobleaching coefficient or a photodarkening coefficient (depending on the sign of $\beta_2$, i.e., on the direction of absorptivity changes with increasing light intensity). The imaginary part of $n_2$ or of the degenerate $\chi^{(3)}$ is related to nonlinear loss (or gain) phenomena: the nonlinear absorption coefficient $\beta_2$ is formally given by

$$\beta_2 = \frac{3\pi}{\varepsilon_0 n^2 c\lambda} \text{Im}\left[ \chi^{(3)} \left( -\omega;\omega,-\omega,\omega \right) \right]$$

The third-order nonlinear optical phenomena leading to changes in refractive and absorptive properties of media incorporate many physical mechanisms other than the electronic nonlinearity described by the molecular hyperpolarizabilities. Refractive index and absorption coefficient changes may be due to such factors as changes in population of ground state and excited state molecules (which may sometimes be equivalent to burning of spectral holes in absorption profiles), orientation of not fully symmetric molecules in the optical field, electrostriction, effects of electric fields produced by photogenerated space charges (the photorefractive effect), or thermal effects. The important feature of these processes is that the changes of the refractive and/or absorptive properties may take place over extended periods of time and relax with certain time constants, so time-dependent measurements of third-order nonlinear optical response are very useful in separating different mechanisms of nonlinearity.

### C. *Optical Limiting*

Two-photon absorption is an example of a process that causes transmission of light through a medium to be power dependent. In the absence of one-photon absorption, the transmission $T$ through a two-photon absorbing sample is

$$T = \frac{I}{I_0} = \frac{1}{1 + I_0 \beta_2 L} \tag{5}$$

where $L$ is the sample length. As the incident intensity $I_0$ increases, the transmitted intensity tends to saturate at the value of $(\beta_2 L)^{-1}$. Such behavior is referred to as optical power limiting.

There has been considerable interest in the application of optical power limiting, particularly for the protection of sensors from damage resulting from exposure to high energy laser pulses. In principle, the direct two-photon absorption process is suitable for optical limiting, but practical estimates show that power limiting properties of existing materials (even those with the largest two-photon absorption coefficients) are insufficient for the most important applications, namely the protection of sensors from laser pulses of duration of the order of nanoseconds. For this reason, other physical processes giving optical limiting behavior have been examined.

One important process for organics that affords optical limiting behavior is reverse saturable absorption (RSA). If a substantial proportion of the population of molecules is excited from the ground state to the excited state, then the absorption of the material is no longer the same as that of the population of ground state molecules. A common phenomenon is saturable absorption (absorption bleaching), i.e., increase of sample transmission as the ground state molecules are depleted. In order for reverse saturable absorption to take place, it is necessary that the excited state molecules exhibit a higher absorptivity at a given wavelength than the ground state molecules. The RSA phenomenon is thus a "photodarkening" effect.

A five-level molecular energy level diagram (Fig. 1) is useful to explain the processes involved. In this model, level 0 is the ground state (having a singlet character) and level 1 is the first singlet excited state, so the ground to first excited state transition is between levels 0 and 1 with an absorption cross section (absorptivity per molecule for the specific transition) of $\sigma_{01}$; more generally, transitions from state $x$ to state $y$ have a cross section $\sigma_{xy}$. The lifetime associated with an excited state $y$ before it returns to state $x$ is $\tau_{yx}$. Intersystem crossing associated with a spin flip occurs between singlet and triplet states.

If the excited state cross section is smaller than the ground state cross section (i.e., absorption in the excited state is less likely than in the ground state), then the transmission of the system will increase as the system is excited. This is saturable absorption (SA). If the situation is reversed, i.e., absorption in the excited state is more likely than in the ground state, then RSA will occur; the system will become less transmissive as excitation is increased.

The difference between the RSA process and two-photon absorption is that the two-photon absorption is virtually instantaneous whereas processes involving intermediate absorbing states exhibit certain kinetic behavior,

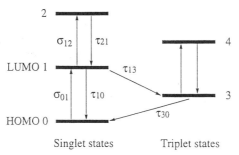

FIG. 1.   Five-level model for reverse saturable absorption. (Adapted with permission from G. R. Allan, D. R. Labergerie, S. J. Rychnovsky, T. F. Bogess, A. L. Smirl, and L. Tutt, *J. Phys. Chem.* **1992,** *96,* 6313. Copyright © 1992 American Chemical Society, Washington, DC.)

which is dependent on the lifetimes of the states involved. As with refractive third-order nonlinearity, time-resolved investigations of the changes of absorptive properties are necessary to evaluate the mechanism of power limiting in a given system.

### D. *Systems of Units*

The third-order nonlinear properties are specified in different ways by different authors and several systems of units are used. The conversions between different systems are not always obvious, as the numerical values of the conversion factors may depend on the definitions of particular properties. Table I lists some of the more important conversion factors and units. It should be noted that conversion of $n_2$ values to $\chi^{(3)}$ values can be performed using Eq. (4) in SI units. A frequently utilized conversion is that between $n_2$ values in SI units (cm$^2$ W$^{-1}$) and $\chi^{(3)}$ in cgs units (esu), namely $n_2 = (C_1\chi^{(3)})/n^2$, where $C_1$ is approximately 0.039.[7] Calculation of $\gamma$ values can be performed using Eq. (3). Reference 7 provides a discussion of the pitfalls that arise when applying conversion procedures between nonlinear properties defined in different ways.

It should be noted that polarizabilities of various orders can be defined in an alternative way in the SI system of units to that discussed previously. A quantity having the dimension of volume $\alpha' = \alpha/4\pi\varepsilon_0$ can be considered to be an SI analogue of the cgs polarizability. Analogously, $\gamma' = \gamma/4\pi\varepsilon_0$ (or $\gamma' = \gamma/\varepsilon_0$) can be used as the third-order hyperpolarizability in the SI system, with $\gamma'$ having the units of m$^5$ V$^{-3}$. The presence or absence of the factor of $4\pi$ in the definition of the hyperpolarizability is, unfortunately, not always obvious in literature data.

### E. *Experimental Techniques*

A variety of experimental techniques have been used to obtain information about the third-order optical nonlinearities and optical power limiting behavior of materials. This section includes descriptions of those techniques that have been used or have potential use with organometallics. For an excellent source of information about other techniques, the interested reader is directed to Ref. 6.

### 1. *Third Harmonic Generation*

Third harmonic generation is used to study the purely electronic molecular second hyperpolarizability of centrosymmetric materials; no other mechanism but the nonresonant electron cloud distortion can respond rapidly

## TABLE I
### Units and Conversion Factors for Important Properties

| Property | System of Units | | Conversion Factor |
|---|---|---|---|
| | SI | cgs | |
| Dipole moment | $\mu$ | C m | statC cm = statV cm$^2$ | $\mu_{SI} = \frac{1}{3} \times 10^{-11} \, \mu_{cgs}$ |
| Electric field | $E$ | V m$^{-1}$ | statV cm$^{-1}$ = (erg cm$^{-2}$)$^{1/2}$ | $E_{SI} = 3 \times 10^4 \, E_{cgs}$ |
| Linear polarizability | $\alpha$ | C m$^2$ V$^{-1}$ | cm$^3$ | $\alpha_{SI} = (\frac{1}{3})^2 \times 10^{-15} \, \alpha_{cgs}$ |
| Linear susceptibility | $\chi^{(1)}$ | Dimensionless | Dimensionless | $\chi^{(1)}_{SI} = 4\pi \, \chi^{(1)}_{cgs}$ |
| Second hyperpolarizability | $\gamma$ | C m$^4$ V$^{-3}$ | cm$^5$ statV$^{-2}$ = esu | $\gamma_{SI} = (\frac{1}{3})^4 \times 10^{-23} \, \gamma_{cgs}$ |
| Third-order susceptibility | $\chi^{(3)}$ | m$^2$ V$^{-2}$ | cm$^2$ statV$^{-2}$ = cm$^3$ erg$^{-1}$ = esu | $\chi^{(3)}_{SI} = (4\pi/3^2) \times 10^{-8} \, \chi^{(2)}_{cgs}$ |
| Nonlinear absorption coefficient | $\beta_2$ | m W$^{-1}$ | cm s erg$^{-1a}$ | |
| Nonlinear refractive index | $n_2$ | m$^2$ W$^{-1}$ | cm$^2$ s erg$^{-1b}$ | $n_{2\,SI} = 10^3 \times n_{2\,cgs}$ |
| Nonlinear refractive index related to field magnitude | $n'_2$ | m$^2$ V$^{-2}$ | cm$^2$ statV$^{-2}$ = cm$^3$ erg$^{-1}$ = esu | $n'_{2\,SI} = (\frac{1}{3})^2 \times 10^{-8} n'_{2\,cgs}$ |

[a] Although these are the cgs units, the alternative non-cgs units cm W$^{-1}$ or cm GW$^{-1}$ are often used.

[b] The alternative units cm$^2$ W$^{-1}$ are often used.

enough to produce a nonlinear polarization oscillating at the third harmonic.[6] It is technically difficult because all materials exhibit THG, including any glass used for a sample cell and even air. One technique that avoids
some of these problems involves placing the sample in a vacuum-sealed
cell inside a vacuum chamber. A simpler method involves using thick glass
windows that allow the contribution from air to be ignored; the third-order
susceptibility of the glass and solvent must be known. THG has been used
to study $\chi^{(3)}$ in many organic and organometallic molecules, particularly
those measured by EFISH for which an estimation of $\gamma$ is required to
extract an accurate value of $\beta$.[1]

### 2. Degenerate Four-Wave Mixing

In this technique, two coherent "pump" beams interact within a material
creating an interference pattern of light intensity. As the change in refractive
index of a third-order material depends on the intensity of the applied
field, a refractive index grating results, which in the simplest case can be
described by the dependence $\Delta n(r) = n_2 I(r)$. When a third beam is incident
on this grating, a fourth beam is generated, with the intensity of this beam
proportional to the product of all the input intensities and to the square
of the absolute value of the complex third-order susceptibility, i.e., $I_4 \sim
|\chi^{(3)}|^2 I_1 I_2 I_3$. Experimentally, one laser is used and the beam is split to provide
the pump beams and the probe beam. The so-called phase-conjugate geome-
try can be used for measurements, but with the use of very short subpicosecond laser pulses a BOXCARS geometry is more appropriate (Fig. 2).

Advantages of DFWM include (i) the ability (by using various combinations of polarizations for the four beams employed in the experiment) to
measure all of the independent $\chi^{(3)}$ tensor components of an isotropic
medium, (ii) the fact that absolute and relative measurements of $\chi^{(3)}$ are
possible,[6] and (iii) the fact that the time dependence of the nonlinear
response can be studied, permitting confirmation of its origin (off-resonance
electronic nonlinearities that show a practically instantaneous response can
be separated from slower processes giving contributions to the nonlinear
refractive index). A difficulty in using DFWM signals for measuring nonlinear optical properties is that it is necessary to distinguish between contributions from the real and imaginary part of the third-order susceptibility.
One way of doing this is to perform a series of measurements on solutions
of a compound with varying concentrations in a nonabsorbing solvent and
then interpret the concentration dependence of the DFWM signal as

$$I_{\text{DFWM}} \propto |\chi^{(3)}|^2 \propto [N_{\text{solvent}}\gamma_{\text{solvent}} + N_{\text{solute}} \, \text{Re}(\gamma_{\text{solute}})]^2 + [N_{\text{solute}} \, \text{Im}(\gamma_{\text{solute}})]^2$$

where it is assumed that the solvent contributes only to the real part of

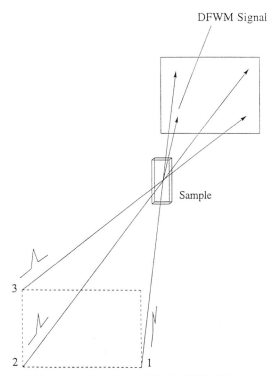

DFWM Signal

Sample

3

2                                    1

Fɪɢ. 2.    Geometry of degenerate four-wave mixing (BOXCARS geometry) for short-pulse, time-resolved measurements of the nonlinear response. Beams 1, 2, and 3 are derived from a single laser beam by the use of beam splitters and the beam paths are adjusted for the pulses to arrive simultaneously at the sample. By delaying one of the beams with respect to the others, the time-resolved measurements can be performed.

the solution suceptibility, whereas the solute can contribute to both the real (refractive) and the imaginary (absorptive) components.

Degenerate four-wave mixing has been widely used for the study of organometallics. At present, it forms a complementary technique to the technically less difficult Z-scan, in that is can be used to verify that the origin of the observed nonlinearity is electronic in nature.

### 3. *Z-Scan*

Z-scan[8] is a technique used to derive the nonlinear refractive index intensity coefficient $n_2$ (from which $\chi^{(3)}$ and $\gamma$ can be determined) by examining self-focusing or self-defocusing phenomena in a nonlinear material. Using a single Gaussian laser beam in a tight focus geometry (Fig. 3), the transmittance of a nonlinear medium through a fixed aperture in the far field is measured as the position of the material is varied through the $z$ direction. An example of a Z-scan trace is shown in Fig. 4 for a nonlinear material with a positive nonlinear refraction index. At the start (A) [and end (E)] of the scan the sample is far from the focal plane, the intensity of the beam is low and so lensing is not observed. As the material approaches the focal-plane (B), lensing causes the beam to focus earlier and hence reduces the measured transmittance. At the focal plane, $z = 0$ (C), there will be no change in transmittance as a thin lens at the focus will cause no change in the far field. After the focal plane (D), slight focusing of the beam by the lensing of the material causes an increase in the measured transmittance. The measured, normalized energy transmittance from a Z-scan experiment is numerically fitted to equations derived from theory and allows the determination of $n_2$, $\chi^{(3)}$, and $\gamma$.

The shape of the Z-scan curve can be modified if a nonlinear absorption or nonlinear transmission (absorption bleaching) takes place in the sample, e.g., due to the presence of an imaginary part of $\chi^{(3)}$ of the material. The curves then become asymmetrical due to increased absorption or transmission when the sample is close to the focal plane. By analyzing the shape of such a modified Z-scan curve, one can determine the nonlinear absorption coefficient $\beta_2$ or the related imaginary part of $\chi^{(3)}$. Alternatively, to determine the nonlinear absorption properties of a sample, the total transmission through the sample can be monitored, i.e., the total intensity of the transmitted beam can be measured without an aperture, as a function of the sample position with respect to the focal plane. Such an experiment is usually referred to as an "open aperture Z-scan." It is often used for the investiga-

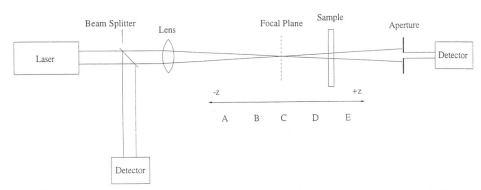

FIG. 3. Z-scan experimental. (Adapted with permission from M. Sheik-Bahae, A. A. Said, T. Wei, D. J. Hagan, and E. W. van Stryland, *IEEE J. Quant. Electr.* **1990**, *26*, 760. Copyright © 1990 Institute of Electrical and Electronic Engineers, New York.)

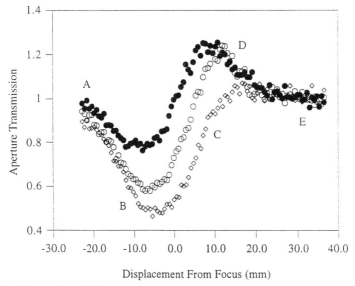

FIG. 4.   Comparison of closed aperture Z-scans for pure thf and solutions of Au(4,4'-C≡CC$_6$H$_4$C≡CC$_6$H$_4$NO$_2$)(PPh$_3$); ●, thf; °, 1.56 weight %; ◇, 3.08 weight %. Wavelength = 800 nm, pulse duration = 100 fs, beam waist = 50 μm, maximum light intensity = 93 GW cm$^{-2}$. The increasing asymmetry of the Z-scan indicates a strong two-photon absorption due to the imaginary part of the molecule nonlinearity. The real part of the nonlinearity of the solute is in this case positive (the same sign as the nonlinearity of the solvent).

tions of materials with potential optical limiting properties. For solutions, the changes of the nonlinearity with concentration of the solution can be determined and measurements performed in an absolute manner, or results can be referenced to a known standard (e.g., the nonlinear refractive index of silica equal to $n_2 = 3 \times 10^{-16}$ cm$^2$ W$^{-1}$ can be used).

Advantages of the Z-scan technique include (i) the ability to determine the sign and magnitude of the nonlinear refractive index, (ii) the ability to determine both real and imaginary parts of $\chi^{(3)}$, and (iii) simplicity (compared to DFWM) due to the single-beam configuration. Disadvantages of Z-scan include (i) the necessity for a high-quality Gaussian beam and good optical quality of samples, and (ii) the absence of information on the temporal behavior of the nonlinear response. The Z-scan technique has been used to determine the third-order nonlinear optical properties of organometallics as solutions and as thin films.

### 4. Optical Kerr Gate

In the optical Kerr gate experiment[6] (Fig. 5), the sample is subjected to a linearly polarized pump beam that induces optical birefringence (and

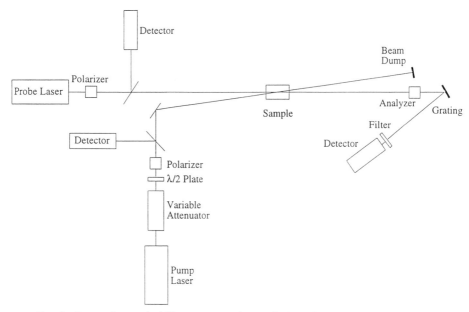

FIG. 5. Setup of an optical Kerr gate experiment. (Adapted with permission from R. L. Sutherland, *Handbook of Nonlinear Optics,* **1996**, p. 429. Copyright © 1996 Marcel Dekker, New York).

also, in some cases, optical dichroism, i.e., different absorption for different polarization sense). An almost colinear probe beam of known linear polarization (usually at 45° to that of the pump beam) is then allowed to pass through the material, and the resultant intensity of the light that passes through a crossed polarizer is measured. The Kerr gate transmittance is proportional to the square of the nonlinear phase shift between the slow and the fast axes of the induced birefringence, with the phase shift itself being proportional to $(\chi^{(3)}_{xxyy} + \chi^{(3)}_{xyyx})\ I_{pump}$. The quadratic dependence of the signal on the third-order susceptibility makes the Kerr gate experiment similar to DFWM: both real and imaginary parts of $\chi^{(3)}$ contribute to the signal. A slightly modified experiment, heterodyne Kerr gate, can be used to resolve these two contributions. It should be mentioned that, for electronic nonlinearity, the measured sum of the tensor components is equal to $\frac{2}{3}(\chi^{(3)}_{xxxx})$. Polarization ellipse rotation can be used as an auxiliary experiment to fully characterize the $\chi^{(3)}$ tensor.

Advantages of these experiments include (i) they are slightly simpler than DFWM (although not as simple as Z-scan) and (ii), as with Z-scan, both the real and the imaginary parts of $\chi^{(3)}$ can be measured. Temporal dependence of the nonlinear response can be followed in a Kerr gate experiment as in any pump-probe type technique. A disadvantage is the necessity to run two independent experiments to determine all the tensor components of $\chi^{(3)}$. These techniques were developed before DFWM and Z-scan and have largely been superceded as Z-scan and DFWM reveal more information from one experiment.

### 5. *Power-Dependent Transmission*

Information about nonlinear absorptive properties of a sample can be derived simply by measuring the sample transmission as a function of the incident light intensity. We note that Eq. (5) can therefore be inverted to read $T^{-1} = 1 + \beta_2 LI$; a linear dependence of the inverse transmission on the incident intensity can be used to determine the $\beta_2$ value. Because of the ease of determining the value of $\beta_2$ with open-aperture Z-scan, Z-scan is often the preferred method for quickly determining the nonlinear absorption coefficient. Point-by-point transmission measurements can be undertaken to verify the applicability of Eq. (5).

### 6. *Pump-Probe Experiments for Excited State Absorption Measurements*

Figure 6 shows an experimental setup to measure nonlinear transmittance by the pump-probe method. The pump is used to excite the sample and the transmission of the probe is measured. The probe beam may be delayed relative to the pump beam in order to perform time delay experiments. A continuum of frequencies may be used as the probe, thereby allowing

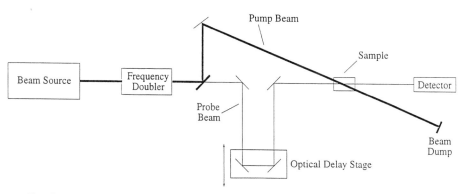

FIG. 6. Pump-probe/OPL experiment. (Adapted with permission from G. R. Allan, D. R. Labergerie, S. J. Rychnovsky, T. F. Bogess, A. L. Smirl, and L. Tutt, *J. Phys. Chem.* **1992,** *96,* 6313. Copyright © 1992 American Chemical Society, Washington, DC.)

a complete nonlinear absorption spectrum to be generated rather than investigate the change of absorption at a single frequency.[6] The pump-probe technique has advantages such as the capability of generating information on a single laser shot, as well as the possibility of observing absorption changes at varying intervals after excitation.

Pump-probe measurements of optical limiting have been carried out on very few organometallic complexes; results from these studies thus far are restricted to a range of tetrahedral metal clusters.

### F. *Comparisons between Results*

There are several problems in comparing results obtained by different groups using the multifarious experimental techniques that are available to investigate the NLO properties of molecules; the most important of these include dispersion, the measurement of different tensorial components, different physical processes contributing to nonlinearity, and solubility problems.

The dispersion of NLO properties is a major source of problems. Measurements are frequently available at one wavelength only, and the degree to which the results are influenced by material resonances close to the measurement wavelength is often difficult to quantify. It is possible to compensate for some of the dispersion effects in certain cases. However, unlike with second-order nonlinearity, the simple two-state model is generally considered insufficient for describing the dispersion of the third-order nonlinearity; at least two excited states have to be considered.

The tensorial character of the nonlinear polarizabilities is another experimental complication. Experimental techniques only provide access to specific tensorial components or combinations thereof, e.g., orientationally averaged $\gamma$. An especially challenging issue is the correct measurement and interpretation of the third-order nonlinear effects when the degenerate susceptibility $\chi^{(3)}(-\omega;\omega,-\omega,\omega)$ or the nonlinear refractive index $n_2$ is being investigated. Variations of the refractive index may be due to a plethora of physical processes, not just changes in the charge density distributions in molecules; the measured NLO response may therefore contain many contributions that need to be identified and properly separated. The use of very short laser pulses (usually in the subpicosecond range) and time-resolved techniques is helpful in resolving unclear cases.

A trivial, but important, issue in NLO measurements on many organometallics is that the most convenient technique for investigating the nonlinearity (as solutions of the molecules of interest in common solvents) is of little value if the solubility of the compounds is not sufficiently high. For example, if the nonlinearity of a solution can be measured with a 10% accuracy and the compound being investigated can only be dissolved at 1% concentration, the practical limit of detection is that the nonlinearity of the organometallic has to be at least 10 times that of the solvent.

Finally, a source of major frustration for anyone trying to compare third-order nonlinearity results obtained by different techniques and from different research groups is the use of varying definitions of the measured quantities and of varying measurement standards. This problem has been discussed in several monographs.[6,7,9] We note here that several different definitions of $\chi^{(3)}$ are possible, depending, for example, on the use or absence of the factor of $\frac{1}{2}$ before the Fourier components of the AC electric field and on the inclusion or exclusion in $\chi^{(3)}$ of so-called degeneracy factors that are dependent on the type of nonlinear process considered [as seen, e.g., in Eq. (2), different multipliers are present in front of the nonlinear terms responsible for different processes]. Thus, it is very helpful if $\chi^{(3)}$or $\gamma$ values for well-known standards (e.g., fused silica or solvents such as chloroform) used or determined in a given series of experiments are provided by the authors of original papers in which data on new compounds are published. In many cases, though, due to the factors mentioned earlier, direct comparison of numbers quoted in different papers is unfortunately not possible: trends observed for a series of compounds in a single set of experimental results are relatively reliable, though.

### G. *Computational Methodologies for Calculating Optical Nonlinearities*

The method of CHF-PT-EB CNDO has been used for several organometallic complexes. This method utilizes a coupled Hartree–Fock (CHF) scheme[10] applied through the perturbation theory (PT) of McWeeny,[11–13] and an extended basis (EB), complete neglect of differential overlap (CNDO/2) wavefunction. The exponents of the basis set are optimized with respect to experimental polarizabilities and second hyperpolarizabilities.[14,15] A detailed description of the CNDO/2 method may be found in Ref. 16.

Ab initio methods have thus far only been used to calculate the static third-order hyperpolarizabilities of ligated organometallic fragments. The CHF finite field (FF) method was utilized with MP2, MP3, MP4(SDTQ), QCISD(T), and CCSD(T) electron correlation corrections applied. The reader is directed elsewhere for a more detailed description of these electron correlation corrections.[10]

The calculated values of third-order NLO coefficients are tabulated in Section III,B, together with experimental comparison when possible.

## III

## ORGANOMETALLICS FOR THIRD-ORDER NONLINEAR OPTICS

A. *Third-Order Measurements*

The tables in this section contain data from approximately 35 research papers that have mostly employed THG, DFWM, or Z-scan techniques to measure the third-order hyperpolarizabilities of organometallic complexes.

The first investigation of the third-order nonlinearity of an organotransition metal complex was of ferrocene, and functionalized and polymeric metallocenes continue to attract interest (Table II). Metallocenes were studied originally due to the belief that interaction of an organic $\pi$-electron system with a metal would cause distortion of the $\pi$-electron system and influence the nonlinearity. With this idea in mind, the prototypical metallocene ferrocene has been measured a number of times. The earliest values obtained by optical power limiting (OPL) measurements[17] are very high compared to subsequent measurements performed by DFWM (and THG on the dibutylated ferrocene $[Fe(\eta^5\text{-}C_5H_4Bu^n)_2]^{23}$; these disparities in data are perhaps a function of the measurement technique, as OPL is carried out on a nanosecond time scale and the higher nonlinearity that was observed may include a contribution from thermal effects. Unlike the other class of complex studied intensively (acetylides: see later), results with the ferrocenyl complexes show no significant imaginary contribution, and the real part of $\gamma$ is positive[20] (Fig. 7).

Organic substitution on the ferrocene cyclopentadienyl rings invariably increases the observed nonlinearity. The styrylferrocene complex has a much larger $\gamma$ than the sum of its component molecules, ferrocene and styrene ($16.9 \pm 0.8 \times 10^{-36}$ esu),[20] suggesting that electronic communication between these components is important for optical nonlinearity. A significant increase in $\gamma$ is observed on formylation of the ring remote from the chromophore for ferrocenyl complexes, but permethylation of the remote ring in $[Ru(\eta^5\text{-}C_5R_5)(\eta^5\text{-}C_5H_4\text{-}(E)\text{-}CH{=}CHC_6H_4\text{-}4\text{-}NO_2)]$ or replacement of a vinylic proton by a cyano group leads only to minor or no variation in $\gamma$.

Large nonlinearities with monomeric ferrocenes have led some investigators to examine oligomeric and polymeric ferrocenes. However, the nonlinearity of the oligomer incorporating six to eight ferrocenyl groups is not significantly larger than that of the corresponding dimer. In this system, the increased $\pi$-electron conjugation in the oligomer has not significantly enhanced the nonlinearity, suggesting that the ferrocenyl linking groups are not effectively coupling the organic components. These results are similar to those seen in second-order nonlinear optical studies (see Ref.

## TABLE II

### Third-Order NLO Measurements of Ferrocenyl and Ruthenocenyl Complexes

| Complex | Solvent | $n_2$ | $\gamma$ | $\lambda_{max}$ (nm) | Technique | Fund. ($\mu$m) | Ref. |
|---|---|---|---|---|---|---|---|
| [FcH] | Molten | $1.1 \times 10^{-17}$ m² W⁻¹ | $4.4 \times 10^{-45}$ m⁵ V⁻² | — | OPL | 1.06 | 17 |
|  | Molten | $1.1 \times 10^{-17}$ m² W⁻¹ | $2.9 \times 10^{-45}$ m⁵ V⁻² | — | OPL | 1.06 | 18 |
|  | EtOH | $2.1 \times 10^{-19}$ m² W⁻¹ | $3.9 \times 10^{-45}$ m⁵ V⁻² | — | OPL | 1.06 | 17 |
|  | Molten | $1.48 \times 10^{-11}$ esu[a] | $3.9 \times 10^{-32}$ esu | — | OPL | 1.06 | 19 |
|  | EtOH | $2.6 \times 10^{-13}$ esu[a] | $1.2 \times 10^{-32}$ esu | — | OPL | 1.06 | 19 |
|  | thf |  | $(16.1 \pm 1.8) \times 10^{-36}$ esu | 439.6 | DFWM | 0.60 | 20 |
| [Fe($\eta^5$-C₅H₄SiMe₃)₂] | Neat | $1.3 \times 10^{-17}$ m² W⁻¹ | $5.5 \times 10^{-45}$ m⁵ V⁻² | — | OPL | 1.06 | 17,18 |
|  | Neat | $1.51 \times 10^{-11}$ esu[a] | $6.4 \times 10^{-32}$ esu | — | OPL | 1.06 | 19 |
| [FcCH=CHPh] | thf |  | $(85.5 \pm 19.8) \times 10^{-36}$ esu | 445 | DFWM | 0.60 | 20 |
| [Fe($\eta^5$-C₅H₄CHO)($\eta^5$-C₅H₄CH=CHPh)][b] | thf |  | $(305 \pm 36) \times 10^{-36}$ esu | 462.4 | DFWM | 0.60 | 20 |
| [Fe($\eta^5$-C₅H₄CH=CHPh)₂][b] | thf |  | $(270 \pm 26) \times 10^{-36}$ esu | 460.8 | DFWM | 0.60 | 20 |
| [FcCH=CHC₆H₄-4-CH=CHFc][b] | thf |  | $(504 \pm 52) \times 10^{-36}$ esu | 451.8 | DFWM | 0.60 | 20 |
| [FcCH=CHC₆H₄-4-CH=CH($\eta^5$-C₅H₄)Fe($\eta^5$-C₅H₄CHO)][b] | thf |  | $(925 \pm 86) \times 10^{-36}$ esu | 454.8 | DFWM | 0.60 | 20 |
| [Fe($\eta^5$-C₅H₄-CH=CHC₆H₄-4-CH=CH)($\eta^5$-C₅H₄)]ₙCHO n = 6-8[b] | thf |  | $(1550 \pm 270) \times 10^{-36}$ esu | 451.5 | DFWM | 0.60 | 20 |
| [FcCOCH₃] | p-dioxane |  | $(27 \pm 3) \times 10^{-36}$ esu | — | THG | 1.91 | 21,22 |
| [Fe($\eta^5$-C₅H₄Bu^n)₂] | Neat |  | $(25 \pm 4) \times 10^{-36}$ esu | — | THG | 1.91 | 23 |
| [FcC≡CC≡CC≡CFc] | CHCl₃ |  | $110 \times 10^{-36}$ esu | — | THG | 1.91 | 24 |
| [Ru($\eta^5$-C₅H₅)₂] | Molten | $3.3 \times 10^{-11}$ esu | $3.9 \times 10^{-45}$ m⁵ V⁻² | — | OPL | 1.06 | 19 |
|  | Molten | $2.2 \times 10^{-17}$ m² W⁻¹ | $3.9 \times 10^{-45}$ m⁵ V⁻² | — | OPL | 1.06 | 18 |
| [Ru($\eta^5$-C₅H₅)($\eta^5$-C₅H₄-(E)-CH=CHC₆H₄-4-NO₂)] | p-dioxane |  | $(114 \pm 11) \times 10^{-36}$ esu | 350, 390 | THG | 1.91 | 21,22 |
| [Ru($\eta^5$-C₅Me₅)($\eta^5$-C₅H₄-(E)-CH=CHC₆H₄-4-NO₂)] | p-dioxane |  | $(140 \pm 14) \times 10^{-36}$ esu | 370, 424 | THG | 1.91 | 21,22 |
| [Ru($\eta^5$-C₅Me₅)($\eta^5$-C₅H₄-(E)-CH=(CN)C₆H₄-4-NO₂)] | p-dioxane |  | $(105 \pm 11) \times 10^{-36}$ esu | 370, 443 | THG | 1.91 | 22 |

[a] $n_2$ related to the electric field; see Ref. 19.
[b] Stereochemistry not specified.

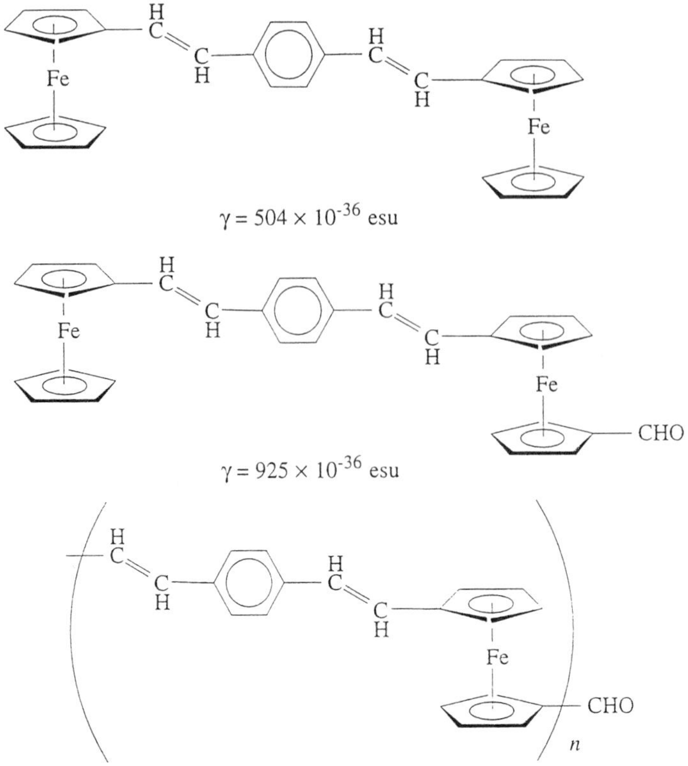

$\gamma = 504 \times 10^{-36}$ esu

$\gamma = 925 \times 10^{-36}$ esu

$\gamma = 1550 \times 10^{-36}$ esu

FIG. 7.   Ferrocenyl complexes with large $\gamma$.

1), which were rationalized as arising from poor coupling between the metal center and the substituent. The dominant contribution to the third-order nonlinearity in ferrocene comes not from the lowest lying $d$-$d$ transition, but from a combination of $d$-$\pi^*$ and $\pi$-$\pi^*$ levels[20]; the $\pi$-geometry of the metal–cyclopentadienyl interaction decreases the contribution of the metal to the nonlinearity.[25]

The effect of metal replacement has not been established with certainty. Direct comparison of ferrocene complexes with analogous ruthenocene complexes is only possible for the parent compounds, with ruthenocene having a slightly larger cubic nonlinearity.[18,19] This result needs to be treated cautiously, as the technique employed (OPL) is susceptible to thermal contributions (see earlier) and the low nonlinearities of the complexes induce substantial errors. The differing fundamental frequencies employed in the metallocenyl complex measurements further complicate the results, necessitating caution when carrying out comparisons.

The Group 4 metallocenes in Table III and Fig. 8 are formally $d^0$ 16 electron complexes; the most important contributor to optical nonlinearity is therefore the LMCT transition.[23] For halide complexes, the lowest energy transition is due to cyclopentadienyl to metal charge transfer. Compared to ferrocene, these complexes have very low nonlinearities (even when the cyclopentadienyl rings are methylated). Introduction of a metal-bound vinyl or acetylide ligand results in enhanced $\pi$-conjugation involving the cyclopentadienyl–metal bonding network and orbitals of like symmetry on the vinyl/acetylide ligands[23,26]; this has the effect of increasing the nonlinearity significantly. This is not simply an additive effect, as the resultant nonlinearities are greater than the sum of those of the precursor chlorides and terminal alkynes.[23] Similarly, the nonlinearity for $[Ti(C{\equiv}CPh)_2(\eta^5\text{-}C_5H_5)_2]$ is three times that of $[Ti(C{\equiv}CPh)Cl(\eta^5\text{-}C_5H_5)_2]$. The divinylbenzene-linked bimetallics $[(ZrCl(\eta^5\text{-}C_5H_5)_2)_2(\mu\text{-}(E)\text{-}CH{=}CHC_6H_4\text{-}n\text{-}(X)\text{-}CH{=}CH)]$ ($n = 3$, $X = Z; n = 4, X = E$) exhibit reasonably large nonlinearities; deconvolution of the contributions to enhancing $\gamma$ in progressing from $Z$ to $E$ stereochemistry or 3- to 4-aryl ring substitution in this system requires additional data.

The effect of metal replacement has been probed. Increased $\gamma$ is observed on progressing up the group, corresponding to increased electron-accepting ability of the metal[23]; for $d^0$ Group 4 metals, this has been rationalized by assuming mixing of titanium orbitals with those of the ligand is more effective than mixing of zirconium or hafnium with the ligand-based orbitals.[23]

Tricyclopentadienyl rare earth complexes ($[M(\eta^5\text{-}C_5H_5)_3]$; M = Er, Nd, Pr, Yb) have been measured by DFWM at a fundamental frequency corresponding to $\lambda = 1064$ nm.[29] They possess moderate nonlinearities despite having electronic transitions in the range 800–1550 nm, which might have been expected to result in dispersively enhanced values. Solution $\chi^{(3)}{}_{1111}$ for $[Yb(\eta^5\text{-}C_5H_5)_3]$ [the largest nonlinearity of the series, and the complex with $\lambda_{max}$ (1030 nm) closest to $\omega$] is only about half that of $CS_2$.

A range of thiophenyl and isobenzothiophenyl Group 10 metal complexes have been examined by four-wave mixing at 1064 nm (see Table IV).[30–32] The linear and nonlinear optical properties are highly dependent on the metal; both the wavelength of the optical transition and the magnitude of the second hyperpolarizability decrease on progressing down the group,[32] a trend which mirrors that observed in Group 10 metal alkynyl complexes (see later). Although no multiphoton absorption is observed, the magnitudes of the imaginary components are large; for conjugated systems where the degree of polarization is large, appreciable imaginary components resulting from electronic lattice coupling can be observed. Progressing from

## TABLE III
### Third-Order NLO Measurements of Other Metallocene Complexes and Tricyclopentadienyl Rare Earth Complexes

| Complex | Solvent | $n_2$ | $\chi^{(3)}$ ($10^{20}$ m$^2$ V$^{-2}$) | $\gamma$ | $\lambda_{max}$ (nm) | Technique | Fund. ($\mu$m) | Ref. |
|---|---|---|---|---|---|---|---|---|
| [HfCl$_2$($\eta^5$-C$_5$H$_5$)$_2$] | Molten | $1.3 \times 10^{-11}$ esu[a] | | — | — | OPL | 1.06 | 19 |
| [ZrCl$_2$($\eta^5$-C$_5$H$_5$)$_2$] | Molten | $1.0 \times 10^{-17}$ m$^2$ W$^{-1}$ | | $2.4 \times 10^{-45}$ V$^{-2}$ m$^5$ | — | OPL | 1.06 | 18 |
| | Molten | $1.0 \times 10^{-11}$ esu[a] | | | — | OPL | 1.06 | 19 |
| | Molten | $0.8 \times 10^{-17}$ m$^2$ W$^{-1}$ | | $1.9 \times 10^{-45}$ V$^{-2}$ m$^5$ | — | OPL | 1.06 | 18 |
| [TiF$_2$($\eta^5$-C$_5$H$_4$Me)$_2$] | CHCl$_3$ | | | $< 3 \times 10^{-36}$ esu | 284,[b]324,[b,c]414[d] | THG | 1.91 | 23,26,27 |
| [TiCl$_2$($\eta^5$-C$_5$H$_4$Me)$_2$] | CHCl$_3$ | | | $< 5 \times 10^{-36}$ esu | 274,[b]388,[b,c]394,[b,c]518[d] | THG | 1.91 | 23,26,27 |
| [TiBr$_2$($\eta^5$-C$_5$H$_4$Me)$_2$] | CHCl$_3$ | | | $< 5 \times 10^{-36}$ esu | 276,[b]328,[b]428,[b,c]568[d] | THG | 1.91 | 23,26,27 |
| [ZrCl$_2$($\eta^5$-C$_5$H$_5$)$_2$] | CHCl$_3$ | | | $< 5 \times 10^{-36}$ esu | 294,[b]334,[d]338,[b,c] | THG | 1.91 | 23,26,27 |
| [HfCl$_2$($\eta^5$-C$_5$H$_5$)$_2$] | CHCl$_3$ | | | $< 5 \times 10^{-36}$ esu | 306 | THG | 1.91 | 26 |
| [Ti(C≡CBu$^n$)$_2$($\eta^5$-C$_5$H$_4$Me)$_2$] | CHCl$_3$ | | | $(15 \pm 2) \times 10^{-36}$ esu | — | THG | 1.91 | 27 |
| [Ti(C≡CBu$^n$)$_2$($\eta^5$-C$_5$H$_5$)$_2$] | CHCl$_3$ | | | $(12 \pm 2) \times 10^{-36}$ esu | 390 | THG | 1.91 | 26 |
| [Ti(C≡CPh)$_2$($\eta^5$-C$_5$H$_5$)$_2$] | thf | | | $(92 \pm 14) \times 10^{-36}$ esu | 410,[c]416[d] | THG | 1.91 | 23,26 |
| [Zr(C≡CPh)$_2$($\eta^5$-C$_5$H$_5$)$_2$] | thf | | | $(58 \pm 9) \times 10^{-36}$ esu | 370,[d]390[c] | THG | 1.91 | 23,26 |
| [Hf(C≡CPh)$_2$($\eta^5$-C$_5$H$_5$)$_2$] | thf | | | $(51 \pm 8) \times 10^{-36}$ esu | 358,[d]390[c] | THG | 1.91 | 23,26 |
| [Ti(C≡CPh)Cl($\eta^5$-C$_5$H$_5$)$_2$] | thf | | | $(31 \pm 5) \times 10^{-36}$ esu | 402,510(sh) | THG | 1.91 | 23,26 |
| [Zr(C(Ph)=C(Ph)—C(Ph)=C(Ph)($\eta^5$-C$_5$H$_5$)] | thf | | | $(47 \pm 7) \times 10^{-36}$ esu | 370,474 | THG | 1.91 | 26 |
| [ZrCl(CH=CHC$_6$H$_4$Me)($\eta^5$-C$_5$H$_5$)$_2$][e] | thf | | | $(24 \pm 4) \times 10^{-36}$ esu | 356 | THG | 1.91 | 26 |
| [(ZrCl($\eta^5$-C$_5$H$_5$)$_2$)$_2$((E)-CH=CHC$_6$H$_4$-3-(Z)-CH=CH)] | thf | | | $(68 \pm 10) \times 10^{-36}$ esu | 356 | THG | 1.91 | 26 |
| [(ZrCl($\eta^5$-C$_5$H$_5$)$_2$)$_2$((E)-CH=CHC$_6$H$_4$-4-(E)-CH=CH)] | thf | | | $(154 \pm 23) \times 10^{-36}$ esu | 380 | THG | 1.91 | 26 |
| [ZrCl($\eta^5$-C$_5$H$_5$)$_2$)$_2$O] | thf | | | $(10 \pm 2) \times 10^{-36}$ esu | 282 | THG | 1.91 | 26 |
| [Yb($\eta^5$-C$_5$H$_5$)$_3$] | thf | $4.3$–$4.5 \times 10^{-11}$ esu | | | 1030 | OPL | 1.06 | 28 |
| [Yb($\eta^5$-C$_5$H$_5$)$_3$] | thf | | 1.3 | — | 1030 | DFWM | 1.06 | 29 |
| [Nd($\eta^5$-C$_5$H$_5$)$_3$] | thf | | 0.43 | — | 910 | DFWM | 1.06 | 29 |
| [Dy($\eta^5$-C$_5$H$_5$)$_3$] | thf | | 0.19 | | 1260 | DFWM | 1.06 | 29 |
| [Er($\eta^5$-C$_5$H$_5$)$_3$] | thf | | <0.1 | | 660 | DFWM | 1.06 | 29 |
| [Pr($\eta^5$-C$_5$H$_5$)$_3$] | thf | | <0.1 | | 1395 | DFWM | 1.06 | 29 |

[a] $n_2$ related to the electric field; see Ref. 19.
[b] Ref. 27.
[c] Ref. 23.
[d] Ref. 26.
[e] Geometry not specified.

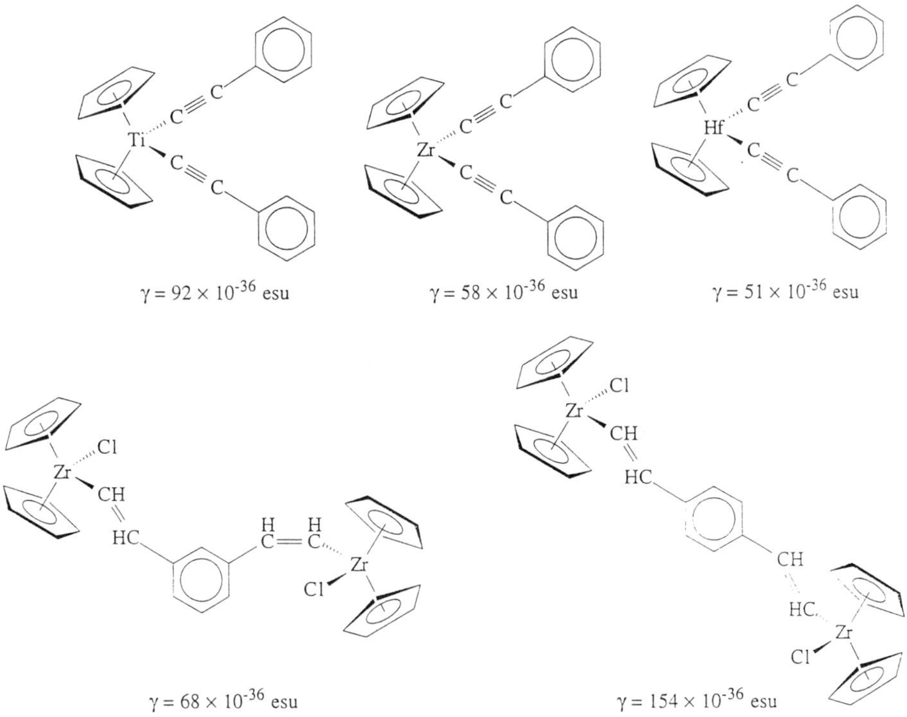
Ti
C
C
C
C
γ = 92 × 10^-36 esu
Zr
C
C
C
C
γ = 58 × 10^-36 esu
Hf
C
C
C
C
γ = 51 × 10^-36 esu
Zr
Cl
CH
HC
H H
C C
Zr
Cl
γ = 68 × 10^-36 esu
Cl
Zr
CH
HC
CH
HC
Zr
Cl
γ = 154 × 10^-36 esu

FIG. 8. Group 4 metallocene complexes with large $\gamma$.

thiophenyl to the related fused ring isobenzothiophenyl palladium and platinum complexes results in an increased third-order nonlinearity, suggesting that the response is dominated by the MLCT transition. The converse trend is observed with the nickel complexes. The presence of the ligated metal center is important in enhancing the response; the magnitudes of the nonlinearities are significantly larger than that for an organic analogue, terthiophene (Fig. 9).

Polymers are of interest for eventual device fabrication. Polymeric and oligomeric analogues of these *C*-coordinated thiophenyl Group 10 complexes show a substantial increase (one to two orders of magnitude) in nonlinearity over the corresponding monomers discussed earlier. Comparison to a related organic polymer, poly(3-butyl(thiophene)) ($|\gamma| = 10 \times 10^{-46}\ m^5\ V^{-2}$), reveals that the presence of the nickel enhances the nonlinearity despite a lower $\lambda_{max}$, a transparency/optical nonlinearity gain. The isobenzothiophene oligomer has a much smaller nonlinearity than the thiophene polymer, due to the decreased polymer chain length.

TABLE IV

THIRD-ORDER NLO MEASUREMENTS OF THIOPHENYL AND ISOBENZOTHIOPHENYL COMPLEXES

| Complex | Solvent | $\lambda_{max}$ (nm) | Re $\gamma$ ($\times 10^{-46}$ m$^5$ V$^{-2}$) | $\lvert$Im $\gamma\rvert$ ($\times 10^{-46}$ m$^5$ V$^{-2}$) | $\lvert\gamma\rvert$ ($\times 10^{-46}$ m$^5$ V$^{-2}$) | Technique | Fund. ($\mu$m) | Ref. |
|---|---|---|---|---|---|---|---|---|
| [Ni(2-C$_4$H$_3$S)$_2$(PEt$_3$)$_2$] | CHCl$_3$ | 374 | −5.1 | 4.6 | 6.8 | DFWM | 1.06 | 30–32 |
| [Pd(2-C$_4$H$_3$S)$_2$(PEt$_3$)$_2$] | CHCl$_3$ | 330 | −0.69 | 1.3 | 1.5 | DFWM | 1.06 | 30–32 |
| [Pt(2-C$_4$H$_3$S)$_2$(PEt$_3$)$_2$] | CHCl$_3$ | 320 | −0.25 | 0.60 | 0.65 | DFWM | 1.06 | 30–32 |
| [Ni(2-C$_8$H$_5$S)$_2$(PBu$^n_3$)$_2$] | CHCl$_3$ | 388 | −2−−1 | 3–10 | 3–10 | DFWM | 1.06 | 30–32 |
| [Ni(2-C$_8$H$_5$S)$_2$(PEt$_3$)$_2$] | CHCl$_3$ | 370 | −2.8 | 1.5 | 3.1 | DFWM | 1.06 | 31 |
| [Pd(2-C$_8$H$_5$S)$_2$(PEt$_3$)$_2$] | CHCl$_3$ | 370 | −2.1 | 0.34 | 2.1 | DFWM | 1.06 | 31 |
| [Pt(2-C$_8$H$_5$S)$_2$(PEt$_3$)$_2$] | CHCl$_3$ | 332 | −1.1 | 0.22 | 1.1 | DFWM | 1.06 | 31 |
| [Ni(2-C$_4$H$_2$S)$_2$(PBu$^n_3$)$_2$]$_n$ | CHCl$_3$ | 515 | −140−−170 | 170–200 | 220–260 | DFWM | 1.06 | 30–32 |
| [Ni(2-C$_8$H$_4$S)$_2$(PBu$^n_3$)$_2$]$_n$ $n_{av}$ = 9.5 | CHCl$_3$ | 580 | −20−−30 | 10–20 | 20–30 | DFWM | 1.06 | 30–32 |

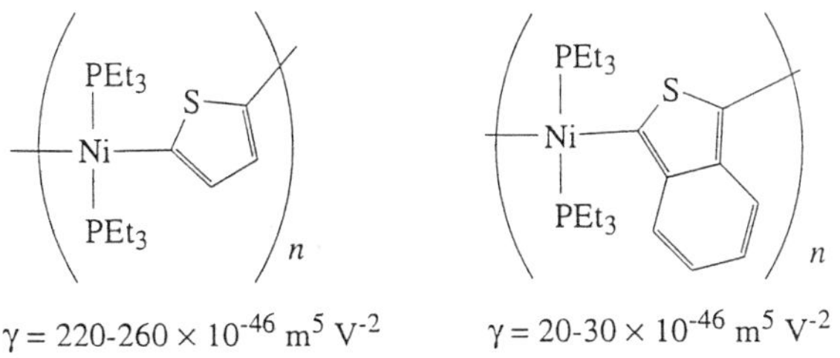

$$\gamma = 220\text{-}260 \times 10^{-46} \ m^5 \ V^{-2} \qquad \gamma = 20\text{-}30 \times 10^{-46} \ m^5 \ V^{-2}$$

FIG. 9.   Nickel thiophenyl and isobenzothiophenyl polymers with large $\gamma$.

The carbonyl complexes listed in Table V are of two types: tricarbonyl-chromium $\eta^6$-arene $\pi$-complexes, and pentacarbonyltungsten $\sigma$-pyridine complexes, with both complex types having relatively low $\gamma$. Nonlinearities increase on arene or pyridine $\pi$-system lengthening, and on proceeding from acceptor to donor substituent on the tricarbonylchromium-coordinated arene ring. Relative magnitudes and trends thus mirror those observed with quadratic nonlinearities of these complexes (see Ref. 1) (Fig. 11).

A systematically varied series of 16 electron-square planar arylpalladium and arylplatinum complexes has been examined by THG at 1.91 $\mu$m (see Table VI), with the focus on assessing the importance of halo or phosphine coligand, or of 4-aryl substituent; metal replacement for the complexes *trans*-[MI(C$_6$H$_4$-4-NO$_2$)(PEt$_3$)$_2$] (M = Pd, Pt) had no effect on nonlinearity. All complexes have significantly larger $\gamma$ than the carbonyl complexes described earlier, but values are still low in absolute terms (as expected

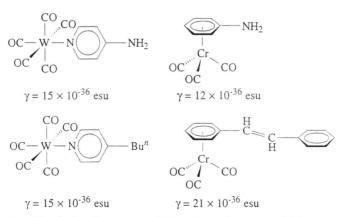

$$\gamma = 15 \times 10^{-36} \ esu \qquad\qquad \gamma = 12 \times 10^{-36} \ esu$$

$$\gamma = 15 \times 10^{-36} \ esu \qquad\qquad \gamma = 21 \times 10^{-36} \ esu$$

FIG. 10.   Carbonyl complexes of chromium and tungsten with large $\gamma$.

## TABLE V
### THIRD-ORDER NLO MEASUREMENTS OF CARBONYL COMPLEXES

| Complex | Solvent | $\gamma$ ($\times 10^{-36}$ esu) | $\lambda_{max}$ (nm) | Technique | Fund. ($\mu$m) | Ref. |
|---|---|---|---|---|---|---|
| [W(CO)$_5$(NC$_5$H$_4$-4-NH$_2$)] | DMSO | 15 | 290 | THG | 1.91 | 22 |
| [W(CO)$_5$(NC$_5$H$_4$-4-Bu$^n$)] | p-dioxane | 15 | 328 | THG | 1.91 | 22 |
| [W(CO)$_5$(NC$_5$H$_5$)] | Toluene | 8 | 332 | THG | 1.91 | 22 |
| [W(CO)$_5$(NC$_5$H$_4$-4-Ph)] | CHCl$_3$ | 12 | 330–340 | THG | 1.91 | 22 |
| [W(CO)$_5$(NC$_5$H$_4$-4-COMe)] | CHCl$_3$ | 14 | 420–440 | THG | 1.91 | 22 |
| [Cr(CO)$_3$($\eta^6$-C$_6$H$_6$)] | Toluene | 2 | 310 | THG | 1.91 | 22 |
| [Cr(CO)$_3$($\eta^6$-C$_6$H$_5$OMe)] | Toluene | 3 | 310 | THG | 1.91 | 22 |
| [Cr(CO)$_3$($\eta^6$-C$_6$H$_5$NH$_2$)] | p-dioxane | 12 | 313 | THG | 1.91 | 22 |
| [Cr(CO)$_3$($\eta^6$-C$_6$H$_5$NMe$_2$)] | Toluene | 10 | 318 | THG | 1.91 | 22 |
| [Cr(CO)$_3$($\eta^6$-C$_6$H$_5$COOMe)] | Toluene | 6 | 318 | THG | 1.91 | 22 |
| [Cr(CO)$_3$($\eta^6$-C$_6$H$_5$-(E)-CH=CHPh)] | p-dioxane | 21 | 410 | THG | 1.91 | 22 |

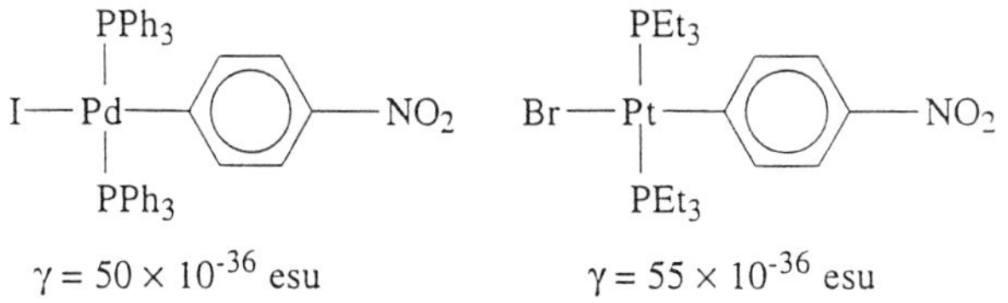

FIG. 11.   Square-planar platinum and palladium complexes with large $\gamma$.

for small molecules with little conjugation). Replacement of iodo by bromo results in a 50% increase in nonlinearity, and proceeding from triethylphosphine to triphenylphosphine coligand affords a similar increase in cubic NLO response, as does replacing a 4-formyl by 4-nitro substituent on the $\sigma$-bound aryl ring. Following this prescription, the most efficient complexes in this systematically varied series should be trans-[MBr($C_6H_4$-4-$NO_2$)($PPh_3$)$_2$] (M = Pd, Pt) but these were not examined (Fig. 11).

Most attention thus far has focused on metal acetylide complexes (see Table VII; like ferrocenyl systems, they are generally accessible by straightforward procedures in high yields, are thermally and oxidatively stable, and can be modified in a facile manner to afford polymeric analogues (Fig. 12).

Studies with both Group 4 and Group 8 metal acetylide complexes have revealed that the observed nonlinearities are not simply an additive effect of the molecular components, the values being substantially larger than the sum of those for the precursor chloride complexes and acetylides.[23,33,34] Increased $\gamma$ is observed on progressing up the group for Group 4 metal acetylide complexes, corresponding to increased electron accepting ability of the metal (see earlier)[23]; this is a further example of the possibility of tuning a nonlinearity by choice of metal.

Investigations of the third-order nonlinearities of a systematically varied series of ruthenium acetylides established the significance of (i) phosphine ligand substitution (replacement of $PPh_3$ by $PMe_3$ resulted in only a small change), (ii) variation in phenylacetylide substituent (replacing 4-H by 4-Br resulted in no change, but introduction of 4-$NO_2$ resulted in an appreciable increase), and (iii) chain lengthening (progression from one-ring to biphenyl to ene-linked and yne-linked two-ring to imino-linked two-ring resulted in increasing nonlinearity). Negative nonlinearities were observed for all the nitro-containing complexes; thermal lensing was rejected as the cause of negative $\gamma$, and two-photon dispersion deemed likely, but a negative static hyperpolarizability could not be ruled out.[33,34] trans-Bis(acetylide)bis(diphosphine)ruthenium complexes were investigated as monomeric models

TABLE VI

THIRD-ORDER NLO MEASUREMENTS OF SQUARE-PLANAR PLATINUM AND PALLADIUM COMPLEXES

| Complex | Solvent | $\gamma$ ($\times 10^{-36}$ esu) | $\lambda_{max}$ (nm) | Technique | Fund. ($\mu$m) | Ref. |
|---|---|---|---|---|---|---|
| trans-[PtBr($C_6H_4$-4-CHO)($PEt_3$)$_2$] | CHCl$_3$ | 37 | — | THG | 1.91 | 22 |
| trans-[PdI($C_6H_4$-4-NO$_2$)($PEt_3$)$_2$] | CHCl$_3$ | 36 | — | THG | 1.91 | 22 |
| trans-[PdI($C_6H_4$-4-NO$_2$)($PPh_3$)$_2$] | CHCl$_3$ | 50 | — | THG | 1.91 | 22 |
| trans-[PtI($C_6H_4$-4-NO$_2$)($PEt_3$)$_2$] | CHCl$_3$ | 36 | — | THG | 1.91 | 22 |
| trans-[PtBr($C_6H_4$-4-NO$_2$)($PEt_3$)$_2$] | CHCl$_3$ | 55 | — | THG | 1.91 | 22 |

of polymers; they have nonlinearities with very large imaginary components, implying significant two-photon absorption exists.

The same acetylide ligands were coupled to ligated gold centers. Values for some of the gold complexes are larger than those of their ruthenium analogues, the opposite trend to that observed with $\beta$, emphasizing that enhancing cubic optical nonlinearities of organometallic complexes does not simply involve increasing quadratic NLO merit. Cubic nonlinearities for $Au(C\equiv CC_6H_4\text{-}4\text{-}C\equiv CC_6H_4\text{-}4'\text{-}NO_2)(PPh_3)$ and $Au(C\equiv CC_6H_4\text{-}4\text{-}(E)\text{-}CH=CHC_6H_4\text{-}4'\text{-}NO_2)(PPh_3)$ are the largest values for any of the complexes in the ruthenium, gold, and nickel systems (and in fact the largest for monomeric organometallic acetylide complexes), despite the gold complexes having the lower valence electron count.[36] Data for the nickel complexes are, within the error margins, equivalent to those of the ruthenium complexes, although the ruthenium complexes possess a more easily oxidizable metal and greater delocalization possibilities with the extra phosphine coligand, indicating that these variables are not critical for enhancing nonlinearity in these complexes.[37] This metal replacement (ruthenium vs nickel) may have a more subtle impact on nonlinearity that is not apparent with the large errors. It is clear, though, that the later transition metal ruthenium, nickel, and gold complexes with the ligated metal acting as a donor group have significantly larger nonlinearities than earlier transition metal acetylide complexes with the metal acting as an acceptor moiety.

The nitro acceptor group is important in enhancing nonlinearity in these acetylide complexes. The mono- and dimetallic platinum and palladium acetylide complexes that lack nitro acceptor groups have much lower nonlinearities than the gold, ruthenium, and nickel complexes containing the acceptor nitro substituent, but have nonlinearities that are comparable in magnitude to those of the Group 4 metal, ruthenium, nickel, and gold acetylides lacking this group.

Measurements made by DFWM on Group 10 metal bis(acetylide) complexes are listed in Table VIII. Although results cannot be directly compared to those cited earlier, internal comparisons within the series are valid. These reveal that hyperpolarizability decreases progressing down the group for phenylacetylide examples (as observed with other acetylide complexes: see earlier), but that larger nonlinearities are observed for butadiynide complexes of the heavier metals. The complexes exhibit a high-order intensity dependence, characteristic of multiphoton resonant enhancement; for these complexes this is possibly due to three-photon resonant enhancement, as $\lambda_{max}$ is, in all cases, close to $3\omega$.

Data in Table IX are given as nonlinear refractive indices $n_2$; other experimental parameters are needed to derive $\gamma$ values that are required for comparison to the results given earlier. As with data in Table VIII,

TABLE VII

THIRD-ORDER NLO MEASUREMENTS OF ACETYLIDE AND BIS(ACETYLIDE) COMPLEXES

| Complex | Solvent | Re $\gamma$ ($10^{-36}$ esu) | Im $\gamma$ ($10^{-36}$ esu) | $\gamma$ ($10^{-36}$ esu) | $\lambda_{max}$ (nm) | Technique | Fund. ($\mu$m) | Ref. |
|---|---|---|---|---|---|---|---|---|
| [Ti(C≡CBu$^n$)$_2$($\eta^5$-C$_5$H$_4$Me)$_2$] | CHCl$_3$ | | | 15 ± 2 | — | THG | 1.91 | 27 |
| [Ti(C≡CBu$^n$)$_2$($\eta^5$-C$_5$H$_5$)$_2$] | CHCl$_3$ | | | 12 ± 2 | 390 | THG | 1.91 | 26 |
| [Ti(C≡CPh)$_2$($\eta^5$-C$_5$H$_5$)$_2$] | thf | | | 92 ± 14 | 410,[a]416[b] | THG | 1.91 | 23,26 |
| [Zr(C≡CPh)$_2$($\eta^5$-C$_5$H$_5$)$_2$] | thf | | | 58 ± 9 | 370,[b]390[a] | THG | 1.91 | 23,26 |
| [Hf(C≡CPh)$_2$($\eta^5$-C$_5$H$_5$)$_2$] | thf | | | 51 ± 8 | 358,[b]390[a] | THG | 1.91 | 23,26 |
| [Ti(C≡CPh)Cl($\eta^5$-C$_5$H$_5$)$_2$] | thf | | | 31 ± 5 | 402,510(sh) | THG | 1.91 | 23,26 |
| [Ru(C≡CPh)(PPh$_3$)$_2$($\eta^5$-C$_5$H$_5$)] | thf | ≤ 150 | — | | 311 | Z-scan | 0.80 | 33 |
| [Ru(C≡CC$_6$H$_4$-4-Br)(PPh$_3$)$_2$($\eta^5$-C$_5$H$_5$)] | thf | ≤ 150 | — | | 325 | Z-scan | 0.80 | 33 |
| [Ru(C≡CC$_6$H$_4$-4-NO$_2$)(PPh$_3$)$_2$($\eta^5$-C$_5$H$_5$)] | thf | -210 ± 50 | ≤ 10 | | 461 | Z-scan | 0.80 | 33 |
| | thf | -260 ± 60 | — | | 461 | DFWM | 0.80 | 33 |
| [Ru(C≡CC$_6$H$_4$-4-NO$_2$)(PMe$_3$)$_2$($\eta^5$-C$_5$H$_5$)] | thf | -230 ± 70 | 74 ± 30 | | 480 | Z-scan | 0.80 | 33 |
| [Ru(C≡CC$_6$H$_4$-4-C$_6$H$_4$-4-NO$_2$)(PPh$_3$)$_2$($\eta^5$-C$_5$H$_5$)] | thf | -380 ± 200 | 320 ± 160 | | 310,448 | Z-scan | 0.80 | 34 |
| [Ru(C≡CC$_6$H$_4$-(E)-CH=CHC$_6$H$_4$-4-NO$_2$)(PPh$_3$)$_2$($\eta^5$-C$_5$H$_5$)] | thf | -450 ± 100 | 210 ± 100 | | 346,476 | Z-scan | 0.80 | 33 |
| [Ru(C≡CC$_6$H$_4$-4-C≡CC$_6$H$_4$-4-NO$_2$)(PPh$_3$)$_2$($\eta^5$-C$_5$H$_5$)] | thf | -450 ± 100 | ≤ 20 | | 346,447 | Z-scan | 0.80 | 33 |
| [Ru(C≡CC$_6$H$_4$-N=CHC$_6$H$_4$-4-NO$_2$)(PPh$_3$)$_2$($\eta^5$-C$_5$H$_5$)] | thf | -850 ± 300 | 360 ± 200 | | 298,496 | Z-scan | 0.80 | 34 |
| trans-[RuCl(C≡CC$_6$H$_4$-4-NO$_2$)(dppm)$_2$] | CH$_2$Cl$_2$ | 170 ± 34 | 230 ± 46 | | 466 | Z-scan | 0.80 | 35 |
| trans-[RuCl(C≡CC$_6$H$_4$-4-C$_6$H$_4$-4-NO$_2$)(dppm)$_2$] | CH$_2$Cl$_2$ | 140 ± 28 | 64 ± 13 | | 448 | Z-scan | 0.80 | 35 |
| trans-[RuCl(C≡CC$_6$H$_4$-4-(E)-CH=CHC$_6$H$_4$-4-NO$_2$)(dppm)$_2$] | CH$_2$Cl$_2$ | 200 ± 40 | 1100 ± 220 | | 471 | Z-scan | 0.80 | 35 |
| trans-[Ru(C≡CC$_6$H$_4$-4-NO$_2$)$_2$(dppm)$_2$] | CH$_2$Cl$_2$ | 300 ± 60 | 490 ± 98 | | 474 | Z-scan | 0.80 | 35 |
| trans-[Ru(C≡CC$_6$H$_4$-4-C$_6$H$_4$-4-NO$_2$)$_2$(dppm)$_2$] | CH$_2$Cl$_2$ | ≤ 800 | 2500 ± 500 | | 453 | Z-scan | 0.80 | 35 |
| trans-[Ru(C≡CC$_6$H$_4$-4-(E)-CH=CHC$_6$H$_4$-4-NO$_2$)$_2$(dppm)$_2$] | CH$_2$Cl$_2$ | ≤ 1100 | 3400 ± 680 | | 367 | Z-scan | 0.80 | 35 |

| Complex | Solvent | | | | λmax (nm) | Technique | μm | Ref |
|---|---|---|---|---|---|---|---|---|
| $[Au(C{\equiv}CPh)(PPh_3)]$ | thf | 39 ± 20 | — | | 268,282,296 | Z-scan | 0.80 | 36 |
| $[Au(C{\equiv}CC_6H_4\text{-}4\text{-}NO_2)(PPh_3)]$ | thf | 120 ± 40 | 20 ± 15 | | 338 | Z-scan | 0.80 | 36 |
| $[Au(C{\equiv}CC_6H_4\text{-}4\text{-}C_6H_4\text{-}4\text{-}NO_2)(PPh_3)]$ | thf | 540 ± 150 | 120 ± 50 | | 274,287,350 | Z-scan | 0.80 | 36 |
| $[Au(C{\equiv}CC_6H_4\text{-}4\text{-}(E)\text{-}CH{=}CHC_6H_4\text{-}4\text{-}NO_2)(PPh_3)]$ | thf | 1200 ± 200 | 470 ± 150 | | 303,386 | Z-scan | 0.80 | 36 |
| $[Au(C{\equiv}CC_6H_4\text{-}4\text{-}(Z)\text{-}CH{=}CHC_6H_4\text{-}4\text{-}NO_2)(PPh_3)]$ | thf | 420 ± 150 | 92 ± 30 | | 298,362 | Z-scan | 0.80 | 36 |
| $[Au(C{\equiv}CC_6H_4\text{-}4\text{-}C{\equiv}CC_6H_4\text{-}4\text{-}NO_2)(PPh_3)]$ | thf | 1300 ± 400 | 560 ± 150 | | 301,362 | Z-scan | 0.80 | 36 |
| $[Au(C{\equiv}CC_6H_4\text{-}4\text{-}N{=}CHC_6H_4\text{-}4\text{-}NO_2)(PPh_3)]$ | thf | 130 ± 30 | 330 ± 60 | | 297,392 | Z-scan | 0.80 | 36 |
| $[Ni(C{\equiv}CPh)(PPh_3)(\eta^5\text{-}C_5H_5)]$ | thf | 15 ± 10 | < 10 | | 307 | Z-scan | 0.80 | 37 |
| $[Ni(C{\equiv}CC_6H_4\text{-}4\text{-}NO_2)(PPh_3)(\eta^5\text{-}C_5H_5)]$ | thf | -270 ± 100 | 70 ± 50 | | 368,439 | Z-scan | 0.80 | 37 |
| $[Ni(C{\equiv}CC_6H_4\text{-}4\text{-}C_6H_4\text{-}4\text{-}NO_2)(PPh_3)(\eta^5\text{-}C_5H_5)]$ | thf | -580 ± 200 | 300 ± 60 | | 263,310,413 | Z-scan | 0.80 | 37 |
| $[Ni(C{\equiv}CC_6H_4\text{-}4\text{-}(E)\text{-}CH{=}CHC_6H_4\text{-}4\text{-}NO_2)(PPh_3)(\eta^5\text{-}C_5H_5)]$ | thf | -420 ± 100 | 480 ± 150 | | 313,437 | Z-scan | 0.80 | 37 |
| $[Ni(C{\equiv}CC_6H_4\text{-}4\text{-}(Z)\text{-}CH{=}CHC_6H_4\text{-}4\text{-}NO_2)(PPh_3)(\eta^5\text{-}C_5H_5)]$ | thf | -230 ± 50 | 160 ± 80 | | 307,417 | Z-scan | 0.80 | 37 |
| $[Ni(C{\equiv}CC_6H_4\text{-}4\text{-}N{=}CHC_6H_4\text{-}4\text{-}NO_2)(PPh_3)(\eta^5\text{-}C_5H_5)]$ | thf | -640 ± 300 | 720 ± 300 | | 282,448 | Z-scan | 0.80 | 37 |
| *trans*-$[(C{\equiv}C_6H_4\text{-}4\text{-}C{\equiv}C)(PtCl(PBu^n_3)_2]$ | | < 120 | 360 ± 100 | 350ᶜ | — | FWM | 0.63 | 38 |
| *cis*-$[PtCl(PBu^n_3)_2(C{\equiv}CC_6H_4\text{-}4\text{-}C{\equiv}C)PtCl(PBu^n_3)_2]$ | thf | 11 ± 25% | 224 ± 25% | | — | OKG/IDS | 1.06/0.53 | 39 |
| *trans*-$[PtCl(PBu^n_3)_2(C{\equiv}CC_6H_4\text{-}4\text{-}C{\equiv}C)PtCl(PBu^n_3)_2]$ | thf | 19 ± 25% | 827 ± 25% | | — | OKG/IDS | 1.06/0.53 | 39 |
| *trans*-$[PtCl(PBu^n_3)_2(C{\equiv}CC_6H_4\text{-}4\text{-}C{\equiv}CC{\equiv}CC_6H_4\text{-}4\text{-}C{\equiv}C)PtCl(PBu^n_3)_2]$ | thf | 45 ± 25% | 1196 ± 25% | | — | OKG/IDS | 1.06/0.53 | 39 |
| *trans*-$[PtCl(PBu^n_3)_2(C{\equiv}CC_6H_4\text{-}4\text{-}C{\equiv}CC{\equiv}CC_6H_4\text{-}4\text{-}C{\equiv}C)PtCl(PBu^n_3)_2]$ | thf | 88 ± 25% | 2167 ± 25% | | — | OKG/IDS | 1.06/0.53 | 39 |
| *trans*-$[Pt(C{\equiv}CC_6H_4\text{-}4\text{-}C{\equiv}CH)_2(PBu^n_3)_2]$ | thf | 53 ± 25% | 759 ± 25% | | — | OKG/IDS | 1.06/0.53 | 39 |
| *trans*-$[Pt(C{\equiv}CC_6H_4\text{-}4\text{-}C{\equiv}CH)(PBu^n_3)_2(C{\equiv}CC_6H_4\text{-}4\text{-}C{\equiv}C)Pt(C{\equiv}CC_6H_4\text{-}4\text{-}C{\equiv}CH)(PBu^n_3)_2]$ | thf | 66 ± 25% | 1328 ± 25% | | — | OKG/IDS | 1.06/0.53 | 39 |
| *trans*-$[(C{\equiv}CC_6H_4\text{-}4\text{-}C{\equiv}C)(Pt(NCS)(PBu^n_3)_2)]$ | thf | 30 ± 25% | 1134 ± 25% | | — | OKG/IDS | 1.06/0.53 | 39 |
| *cis*-$[Pt(C{\equiv}CC_6H_4\text{-}4\text{-}C{\equiv}C)(CH)_2(PBu^n_3)_2]$ | | 230ᶜ | 260ᶜ | 290ᶜ | | FWM | 0.63 | 38 |
| *trans*-$[Pd(C{\equiv}CPh)_2(PBu^n_3)_2]$ | | — | — | 110ᶜ | | FWM | 0.63 | 38 |

ᵃ Ref. 23.

ᵇ Ref. 26.

ᶜ Error not quoted.

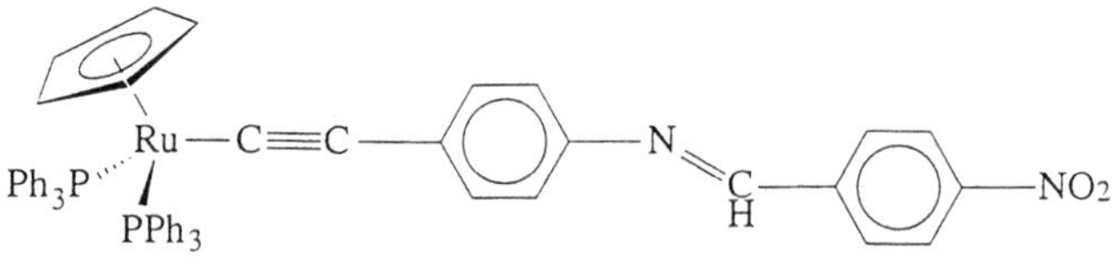

$Re\gamma = -850 \times 10^{-36}$ esu; $Im\gamma = 360 \times 10^{-36}$ esu

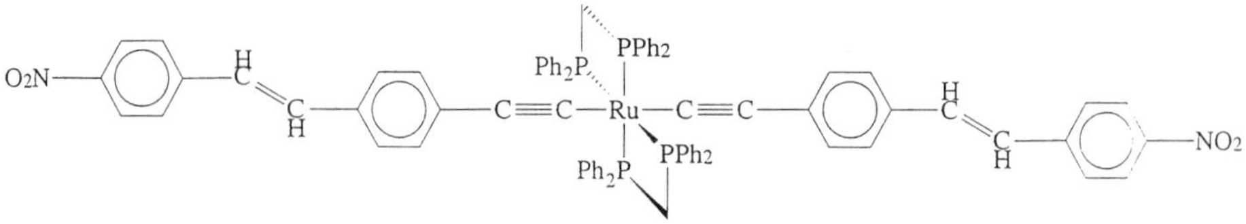

$Re\gamma$  $1100 \times 10^{-36}$ esu; $Im\gamma = 3400 \times 10^{-36}$ esu

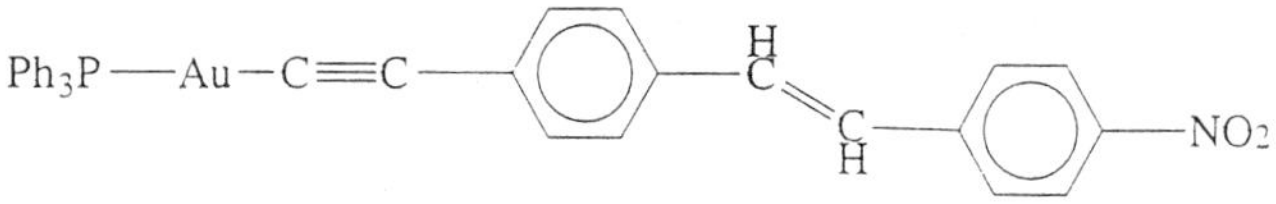

$Re\gamma = 1200 \times 10^{-36}$ esu; $Im\gamma = 470 \times 10^{-36}$ esu

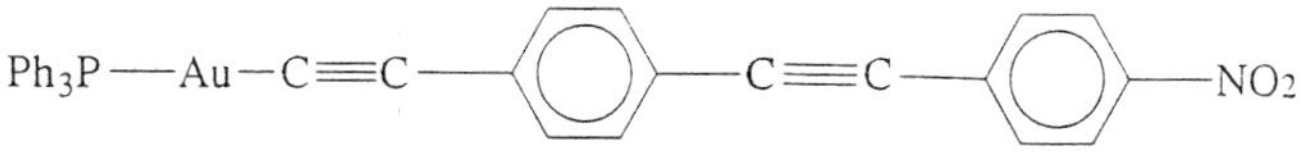

$Re\gamma = 1300 \times 10^{-36}$ esu; $Im\gamma = 560 \times 10^{-36}$ esu

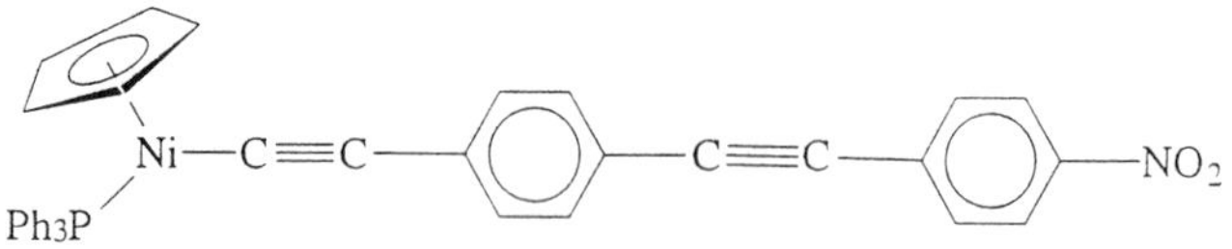

$Re\gamma = -640 \times 10^{-36}$ esu; $Im\gamma = 720 \times 10^{-36}$ esu

FIG. 12.    Acetylide and bis(acetylide) complexes with large $\gamma$.

results in Table IX permit internal conclusions to be drawn; the resultant trends can then be compared to those described earlier. Data here are consistent with increased nonlinearity on chromophore chain lengthening (as observed with acetylide complexes cited earlier) and with a metal efficiency series nickel > platinum > palladium.

In the acetylide polymers of square planar nickel, palladium, and platinum listed in Table X, the imaginary part of the nonlinearity is the major contributor, implying significant two-photon absorption. Some of the poly-

## TABLE VIII
### Third-Order NLO Measurements of Bis(acetylide) Complexes

| Complex | Solvent | Re $\gamma$ ($\times 10^{-44}$ m$^5$ V$^{-2}$) | $|\text{Im }\gamma|$ ($\times 10^{-44}$ m$^5$ V$^{-2}$) | $|\gamma|$ ($\times 10^{-44}$ m$^5$ V$^{-2}$) | $\lambda_{max}$ (nm) | Technique | Fund. ($\mu$m) | Ref. |
|---|---|---|---|---|---|---|---|---|
| trans-[Ni(C≡CPh)$_2$(PEt$_3$)$_2$] | CHCl$_3$ | −27.5 | 14.6 | 31.1 | 370 | DFWM | 1.06 | 40,41 |
|  | — | −0.028 | 0.015 | 0.031 | 370 | DFWM | 1.06 | 31 |
| trans-[Pd(C≡CPh)$_2$(PEt$_3$)$_2$] | CHCl$_3$ | −21.0 | 3.39 | 21.3 | 370 | DFWM | 1.06 | 40,41 |
|  | — | −0.021 | 0.0034 | 0.021 | 370 | DFWM | 1.06 | 31 |
| trans-[Pt(C≡CPh)$_2$(PEt$_3$)$_2$] | CHCl$_3$ | −11.2 | 2.15 | 11.4 | 332 | DFWM | 1.06 | 40,41 |
|  | — | −0.011 | 0.0022 | 0.011 | 332 | DFWM | 1.06 | 31 |
| trans-[Ni(C≡CC≡CH)$_2$(PEt$_3$)$_2$] | CHCl$_3$ | −7.87 | 17.2 | 18.9 | 336 | DFWM | 1.06 | 40,41 |
|  | — | −0.0079 | 0.017 | 0.019 | 336 | DFWM | 1.06 | 31 |
| trans-[Pd(C≡CC≡CH)$_2$(PEt$_3$)$_2$] | CHCl$_3$ | −3.85 | 0.919 | 0.396 | 290 | DFWM | 1.06 | 31,40,41 |
|  | — | −0.0039 | 0.00092 | 0.0040 | 290 | DFWM | 1.06 | 31 |
| trans-[Pt(C≡CC≡CH)$_2$(PEt$_3$)$_2$] | CHCl$_3$ | −1.93 | 0.771 | 2.08 | 318 | DFWM | 1.06 | 31,40,41 |
|  | — | −0.0019 | 0.00077 | 0.0021 | 318 | DFWM | 1.06 | 31 |

TABLE IX

THIRD-ORDER NLO MEASUREMENTS OF BIS(ACETYLIDE) COMPLEXES

| Complex | Solvent | $n_2$ ($\times 10^{-18}$ m$^2$ W$^{-1}$) | Technique | Fund. ($\mu$m) | Ref. |
|---|---|---|---|---|---|
| *trans*-[Pd(C≡CPh)$_2$(PBu$^n_3$)$_2$] | thf | $-0.5 \pm 0.1$ | Z-scan | 0.53 | 42 |
| *trans*-[Pt(C≡CPh)$_2$(PBu$^n_3$)$_2$] | thf | $-3.0 \pm 0.1$ | Z-scan | 0.53 | 42 |
| *trans*-[Ni(C≡CPh)$_2$(PBu$^n_3$)$_2$] | thf | $-16 \pm 5$ | Z-scan | 0.53 | 42 |
| *trans*-[Pd(C≡CC$_6$H$_4$-4-C≡CPh)$_2$(PBu$^n_3$)$_2$] | thf | $-25 \pm 3$ | Z-scan | 0.53 | 42 |
| *trans*-[Pt(C≡CC$_6$H$_4$-4-C≡CPh)$_2$(PBu$^n_3$)$_2$] | thf | $-209 \pm 27$ | Z-scan | 0.53 | 42 |

mers have nonlinearities that are significantly larger than related monometallic acetylide complexes. There does not seem to be a consistent trend
in nonlinearity on increasing polymer size. Although it is hard to compare
data across metal (because the polymers vary in length as well as composition), the platinum polymers are in many cases more efficient than the
analogous palladium polymers. The nonlinearities of these polymers do
not depend dramatically on aromatic ring substitution, but increasing the
number of diethynylarenes in the repeat unit from one to two increases
the nonlinearity[39] (Fig. 13).

Third-order nonlinearities of a range of boranes were obtained by THG
using a fundamental frequency of 1.91 $\mu$m (Table XI).[48] In these complexes
the dimesitylboranyl unit is acting as a $\pi$-acceptor due to its empty $\pi$-orbital.
Importantly, the steric demands of the mesityl groups cause them to be
rotated out of the plane of the $\pi$-orbital of boron so that observed nonlinearity should be due to charge transfer from the third "ligand."[48] As with
many other complexes investigated for third-order responses, these complexes were designed for second-order effects and $\gamma$ is not large. The
complex with largest $\gamma$ in this series [(4-Me$_2$NC$_6$H$_4$-($E$)-CH=CH)B(mes)$_2$]
has a donor-bridge-acceptor composition; it also has the largest $\beta$ value
(33 × 10$^{-30}$ esu) and the lowest absorption energy (405 nm) for this set of
complexes. Chain lengthening by yne linkage does not significantly increase
nonlinearity for [4-XC$_6$H$_4$B(mes)$_2$] proceeding to [4-XC$_6$H$_4$C≡CB(mes)$_2$]
but introduction of an $E$-ene linkage in proceeding to [4-XC$_6$H$_4$-($E$)-
CH=CHB(mes)$_2$] more than doubles the response. A range of donor
groups have been assayed; compounds with a diphenylphosphino donor
have low cubic nonlinearities, a similar result as observed with relative
quadratic NLO merit, emphasizing its ineffectiveness as a donor substituent.
The $E$-ene-linked examples [4-XC$_6$H$_4$-($E$)-CH=CHB(mes)$_2$] with four acceptor substituents not surprisingly have lower nonlinearities than most of
the four-donor-substituted analogues (Fig. 14).

Third-order measurements of fullerene derivatives are given in Table
XII. The organometallic fullerene derivative [($\eta^2$-C$_{60}$)Pt(PPh$_3$)$_2$] was measured by Z-scan using both circular and linear polarized light.[50] Different
selection rules allowing two-photon absorption exist when considering different polarizations of light. The two-photon absorption contribution to
nonlinearities is substantial here, confirmed by a significant difference in
the result obtained by using circular polarized light. These measurements
are complicated because of the comparatively large nonlinearity of the
solvent toluene (the complex was insufficiently soluble in thf). A broad
electron absorption band that tails out at ~800 nm renders extraction
of the off-resonance nonlinearity impossible because of the fundamental
frequencies employed. This contributes to making comparison of results

## TABLE X
### Third-Order NLO Measurements of Acetylide Polymer Complexes[a]

| Complex | Solvent | Re $\gamma$ | \|Im $\gamma$\| | \|$\gamma$\| | $\lambda_{max}$ (nm) | Technique | Fund. ($\mu$m) | Ref. |
|---|---|---|---|---|---|---|---|---|
| $[Ni(C\equiv CC\equiv C)(PBu^n_3)_2]_n$ | CHCl$_3$ | $-2.63 \times 10^{-42}$ m$^5$ V$^{-2}$ | $-2.41 \times 10^{-42}$ m$^5$ V$^{-2}$ | $3.57 \times 10^{-42}$ m$^5$ V$^{-2}$ | 412 | DFWM | 1.06 | 40,41,43 |
| — | — | $-2.63 \times 10^{-45}$ m$^5$ V$^{-2}$ | $-2.41 \times 10^{-45}$ m$^5$ V$^{-2}$ | $3.57 \times 10^{-45}$ m$^5$ V$^{-2}$ | 410 | DFWM | 1.06 | 31 |
| $[Pt(C\equiv CC\equiv C)(PBu^n_3)_2]_n$ | CHCl$_3$ | $-1.48 \times 10^{-42}$ m$^5$ V$^{-2}$ | $1.74 \times 10^{-42}$ m$^5$ V$^{-2}$ | $2.28 \times 10^{-42}$ m$^5$ V$^{-2}$ | 364 | DFWM | 1.06 | 40,41,43 |
| — | — | $-1.48 \times 10^{-45}$ m$^5$ V$^{-2}$ | $1.74 \times 10^{-45}$ m$^5$ V$^{-2}$ | $2.28 \times 10^{-45}$ m$^5$ V$^{-2}$ | 360 | DFWM | 1.06 | 31 |
| $[Ni(C\equiv CC_6H_4\text{-}4\text{-}C\equiv C)(PBu^n_3)_2]_n$ | CHCl$_3$ | $-10 \times 10^{-46}$ m$^5$ V$^{-2}$ | $20 \times 10^{-46}$ m$^5$ V$^{-2}$ | $20 \times 10^{-46}$ m$^5$ V$^{-2}$ | — | DFWM | 1.06 | 43 |
| $[Ni(C\equiv CC_6H_4\text{-}4\text{-}C\equiv C)(POc^n_3)_2]_n$ | CHCl$_3$ | $-40 \times 10^{-46}$ m$^5$ V$^{-2}$ | $100 \times 10^{-46}$ m$^5$ V$^{-2}$ | $100 \times 10^{-46}$ m$^5$ V$^{-2}$ | — | DFWM | 1.06 | 43 |
| $[Ni(C\equiv CC\equiv C)(Pt(C_8H_{17})_3)_2]_n$ | CHCl$_3$ | $-40 \times 10^{-46}$ m$^5$ V$^{-2}$ | $30 \times 10^{-46}$ m$^5$ V$^{-2}$ | $50 \times 10^{-46}$ m$^5$ V$^{-2}$ | — | DFWM | 1.06 | 43 |
| $[Pd_3(C\equiv CC_6H_4\text{-}4\text{-}C\equiv C)(\mu\text{-dppm})_2]_n$ | CHCl$_3$ | $-20 \times 10^{-46}$ m$^5$ V$^{-2}$ | $20 \times 10^{-46}$ m$^5$ V$^{-2}$ | $20 \times 10^{-46}$ m$^5$ V$^{-2}$ | — | DFWM | 1.06 | 43 |
| $[Pt(C\equiv CC_6H_4\text{-}4\text{-}C\equiv C)(PBu^n_3)_2]_n$ | | 890 | 130 | 1450 | | FWM | 0.63 | 38 |
| $[Pd(C\equiv CC_6H_4\text{-}4\text{-}C\equiv C)(PBu^n_3)_2]_n$ | | 390 | 380 | 490 | | FWM | 0.63 | 38 |
| $[Pd(C\equiv CC_6H_4\text{-}4\text{-}C\equiv C)(PBu^n_3)_2]_n$ n = 112 | thf | 102 | 3401 | | | OKG/IDA | 1.06/0.53 | 44 |
| $[Pt(C\equiv CC_6H_4\text{-}4\text{-}C\equiv C)(PBu^n_3)_2]_n$ ~32 000 amu | Benzene | | | 1470 | | THG | 1.06 | 45 |
| $[Pt(C\equiv CC_6H_4\text{-}4\text{-}C\equiv C)(PBu^n_3)_2]_n$ n = 112 | thf | 37 | 1906 | | | OKG/IDA | 1.06/0.53 | 39 |
| $[Pt(C\equiv CC_6H_3\text{-}2,5\text{-}Me_2\text{-}4\text{-}C\equiv C)(PBu^n_3)_2]_n$, n = 26 | thf | 29 | 1200 | | | OKG/IDA | 1.06/0.53 | 39 |
| | thf | 56 | 1199 | | | OKG/IDA | 1.06/0.53 | 46 |
| $[Pt(C\equiv CC_6H_3\text{-}2,5\text{-}Et_2\text{-}4\text{-}C\equiv C)(PBu^n_3)_2]_n$ n = 15 | thf | 43 | 956 | | | OKG/IDA | 1.06/0.53 | 39 |
| $[Pt(C\equiv CC_6H_3\text{-}3\text{-}F\text{-}4\text{-}C\equiv C)(PBu^n_3)_2]_n$ n = 18 | thf | 56 | 1260 | | | OKG/IDA | 1.06/0.53 | 39 |

| Compound | | | Solvent | Method | Wavelength (μm) | Ref. |
|---|---|---|---|---|---|---|
| $[Pt(C\equiv CC_6H_2\text{-}2,5\text{-}(OMe)_2\text{-}4\text{-}C\equiv C)(PBu^n_3)_2]_n$, n = 111, 105, 62 | 48,65,43 | 1724,1330,1586 | thf | OKG/IDA | 1.06/0.53 | 39 |
| $[Pt(C\equiv CC_6Me_4\text{-}4\text{-}C\equiv C)(PBu^n_3)_2]_n$, oligomer | 28 | 1324 | thf | OKG/IDA | 1.06/0.53 | 39 |
| $[Pt(C\equiv CC_6H_3\text{-}3\text{-}NH_2\text{-}4\text{-}C\equiv C)(PBu^n_3)_2]_n$, n = 76 | 18 | 1342 | thf | OKG/IDA | 1.06/0.53 | 39 |
| $[Pt(C\equiv CC_6H_3\text{-}3\text{-}CF_3\text{-}4\text{-}C\equiv C)(PBu^n_3)_2]_n$, n = 44 | 34 | 2148 | thf | OKG/IDA | 1.06/0.53 | 39 |
| $[Pt(C\equiv C(1\text{-}naphthyl)\text{-}4\text{-}C\equiv C)(PBu^n_3)_2]_n$, n = 62 | 19 | 2474 | thf | OKG/IDA | 1.06/0.53 | 39 |
| $[Pt(3\text{-}C\equiv C(C_5H_3N)\text{-}2\text{-}C\equiv C)(PBu^n_3)_2]_n$, n = 47, 35 | 33 | 2263 | thf | OKG/IDA | 1.06/0.53 | 39 |
| $[Pd(C\equiv CC_6H_2\text{-}2,5\text{-}Me_2\text{-}4\text{-}C\equiv C)(PBu^n_3)_2]_n$, n = 4 | 19 | 1169 | thf | OKG/IDA | 1.06/0.53 | 39 |
| $[Pd(C\equiv CC_6H_3\text{-}3\text{-}NH_2\text{-}4\text{-}C\equiv C)(PBu^n_3)_2]_n$, n = 12 | 15 | 1753 | thf | OKG/IDA | 1.06/0.53 | 39 |
| $[Pd(C\equiv CC_6H_2\text{-}2,5\text{-}(OMe)_2\text{-}4\text{-}C\equiv C)(PBu^n_3)_2]_n$ n = 67 | 22 | 2432 | thf | OKG/IDA | 1.06/0.53 | 39 |
| $[Pt(C\equiv CC_6H_4\text{-}4\text{-}C\equiv CC\equiv CC_6H_4\text{-}4\text{-}C\equiv C)(PBu^n_3)_2]_n$, n = 223, 97 | 90,121 | 4558,4025 | thf | OKG/IDA | 1.06/0.53 | 39 |
| $[Pt(C\equiv CC_6H_4\text{-}4\text{-}C\equiv CC\equiv CC_6H_4\text{-}4\text{-}C\equiv C)(PBu^n_3)_2]_n$, n > 144 | 856 | 3570 | thf | OKG/IDA | 1.06/0.53 | 44 |
| $[Pt(C\equiv CC_6H_2\text{-}2,5\text{-}Me_2\text{-}4\text{-}C\equiv CC\equiv CC_6H_2\text{-}2,5\text{-}Me_2\text{-}4\text{-}C\equiv C)(PBu^n_3)_2]_n$ n = 52,38 | 116 | 2432 | thf | OKG/IDA | 1.06/0.53 | 39 |
| $[Pt(C\equiv CC_6H_2\text{-}2,5\text{-}Me_2\text{-}4\text{-}C\equiv CC\equiv CC_6H_2\text{-}2,5\text{-}Me_2\text{-}4\text{-}C\equiv C)(PBu^n_3)_2]_n$ n = 52 | 181 | 4366 | thf | OKG/IDA | 1.06/0.53 | 44,47 |
| $[Pt(C\equiv CC_6H_2\text{-}2,5\text{-}Et_2\text{-}4\text{-}C\equiv CC\equiv CC_6H_2\text{-}2,5\text{-}Et_2\text{-}4\text{-}C\equiv C)(PBu^n_3)_2]_n$ n = 146 | 120 ± 30 | 5400 ± 500 | thf | OKG | 0.53 | 44 |
| | 79 | 4933 | thf | OKG/IDA | 1.06/0.53 | 39 |
| $[Pd(C\equiv CC_6H_4\text{-}4\text{-}C\equiv C)(PBu^n_3)_2]_n$, oligomer, 5, 2 | 66 | 2094 | thf | OKG/IDA | 1.06/0.53 | 39 |
| $[Pd(C\equiv CC_6H_2\text{-}2,5\text{-}Et_2\text{-}4\text{-}C\equiv CC\equiv CC_6H_2\text{-}2,5\text{-}Et_2\text{-}4\text{-}C\equiv C)(PBu^n_3)_2]_n$ n = 16 | 106 | 3490 | thf | OKG/IDA | 1.06/0.53 | 39 |
| $[Pt(C\equiv CC_6H_4\text{-}4\text{-}C\equiv C)(PBu^n_3)_2Pt(C\equiv CC_6H_4C_6H_4\text{-}4\text{-}C\equiv C)(PBu^n_3)_2]_n$ n = 66 | | 4466 | thf | OKG/IDA | 1.06/0.53 | 39 |

[a] Units of $10^{-36}$ esu except where indicated.

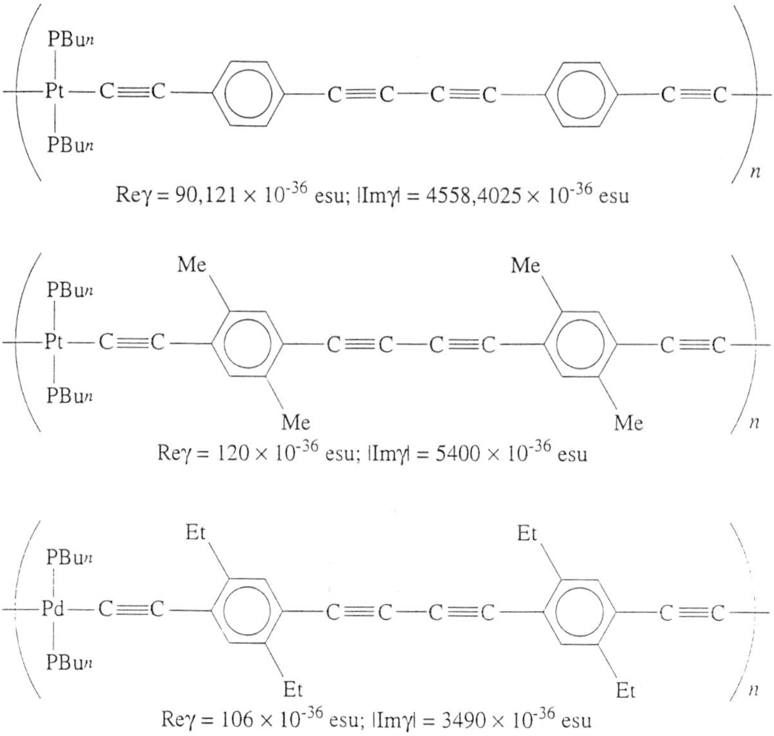
PBun
Pt
C≡C
C≡C—C≡C
C≡C
PBun
n
Reγ = 90,121 × 10^-36 esu; |Imγ| = 4558,4025 × 10^-36 esu
Me
Me
PBun
Pt
C≡C
C≡C—C≡C
C≡C
PBun
Me
Me
n
Reγ = 120 × 10^-36 esu; |Imγ| = 5400 × 10^-36 esu
Et
Et
PBun
Pd
C≡C
C≡C—C≡C
C≡C
PBun
Et
Et
n
Reγ = 106 × 10^-36 esu; |Imγ| = 3490 × 10^-36 esu

FIG. 13. Acetylide polymer complexes with large $\gamma$.

very difficult. Comparison to the free fullerene reveals that coordination to the ligated platinum center significantly increased nonlinear absorption (this is an order of magnitude greater in $[(Ph_3P)_2Pt(\eta^2\text{-}C_{60})]$ than in $C_{60}$) (Fig. 15).

The imaginary components of the third-order nonlinear susceptibility of palladium and iridium complexes of $C_{60}$ and $C_{70}$ were determined using saturation spectroscopy. In all cases, $\mathrm{Im}(\chi^{(3)})$ values are smaller than those of uncomplexed $C_{60}$ ($1.78 \times 10^{-16}$ m$^2$ V$^{-2}$) or $C_{70}$ ($7.55 \times 10^{-17}$ m$^2$ V$^{-2}$), a result explained by decreased conjugation in the molecule and consequent decreased electron delocalization, although differing photodynamics were not excluded.

Two (cyclopentadienyl)bis(phosphine)ruthenium chloride complexes have been investigated; the nonlinearities are low. These results have been used in conjunction with measured nonlinearities of acetylenes and acetylide complexes to demonstrate that values for the latter are not simply the sum of the molecular components (see earlier). An example of a metal

TABLE XI

THIRD-ORDER NLO MEASUREMENTS OF BORON COMPLEXES

| Complex | Solvent | $\lambda_{max}$ (nm) | $\gamma$ ($10^{-36}$ esu) | Technique | Fund. ($\mu$m) | Ref. |
|---|---|---|---|---|---|---|
| [(4-Me$_2$NC$_6$H$_4$)B(mes)$_2$] | CHCl$_3$ | 309,359 | 24 ± 5 | THG | 1.91 | 48,49 |
| [(4-MeSC$_6$H$_4$)B(mes)$_2$] | CHCl$_3$ | 249,335 | 32 ± 6 | THG | 1.91 | 48,49 |
| [(4-MeOC$_6$H$_4$)B(mes)$_2$] | CHCl$_3$ | 274,317 | 23 ± 5 | THG | 1.91 | 48,49 |
| [(4-BrC$_6$H$_4$)B(mes)$_2$] | CHCl$_3$ | 269,314 | 20 ± 4 | THG | 1.91 | 48,49 |
| [(4-IC$_6$H$_4$)B(mes)$_2$] | CHCl$_3$ | 315 | 27 ± 5 | THG | 1.91 | 48,49 |
| [(4-Me$_2$NC$_6$H$_4$)C≡CB(mes)$_2$] | CHCl$_3$ | 398 | 25 ± 8 | THG | 1.91 | 48,49 |
| [(4-MeSC$_6$H$_4$)C≡CB(mes)$_2$] | CHCl$_3$ | 368 | 33–75 ± 20% | THG | 1.91 | 48,49 |
| [(4-MeOC$_6$H$_4$)C≡CB(mes)$_2$] | CHCl$_3$ | 356 | 13–41 ± 30% | THG | 1.91 | 48,49 |
| [PhC≡CB(mes)$_2$] | CHCl$_3$ | 336 | 27 ± 8 | THG | 1.91 | 48,49 |
| [Ph$_2$PC≡CB(mes)$_2$] | CHCl$_3$ | 338 | 15 ± 3 | THG | 1.91 | 48 |
| [Ph-(E)-CH=CHB(mes)$_2$] | CHCl$_3$ | 290,330 | 37 ± 7 | THG | 1.91 | 48,49 |
| [(4-MeSC$_6$H$_4$-(E)-CH=CH)B(mes)$_2$] | CHCl$_3$ | 358 | 81 ± 16 | THG | 1.91 | 48,49 |
| [(4-MeOC$_6$H$_4$-(E)-CH=CH)B(mes)$_2$] | CHCl$_3$ | 347 | 54 ± 11 | THG | 1.91 | 48,49 |
| [(4-Me$_2$NC$_6$H$_4$-(E)-CH=CH)B(mes)$_2$] | CHCl$_3$ | 401 | 93 ± 19 | THG | 1.91 | 48 |
| [(4-H$_2$NC$_6$H$_4$-(E)-CH=CH)B(mes)$_2$] | CHCl$_3$ | 368 | 41 ± 8 | THG | 1.91 | 48,49 |
| [Ph$_2$P-(E)-CH=CHB(mes)$_2$] | CHCl$_3$ | 244,342 | 8 ± 2 | THG | 1.91 | 48,49 |
| [(4-NCC$_6$H$_4$-(E)-CH=CH)B(mes)$_2$] | CHCl$_3$ | 338 | 30 ± 6 | THG | 1.91 | 48,49 |
| [(4-O$_2$NC$_6$H$_4$-(E)-CH=CH)B(mes)$_2$] | CHCl$_3$ | 348 | 32 ± 6 | THG | 1.91 | 48,49 |
| [4,4'-Me$_2$NC$_6$H$_4$-(E)-CH=CHC$_6$H$_4$)B(mes)$_2$] | CHCl$_3$ | 405 | 230 ± 46 | THG | 1.91 | 48,49 |

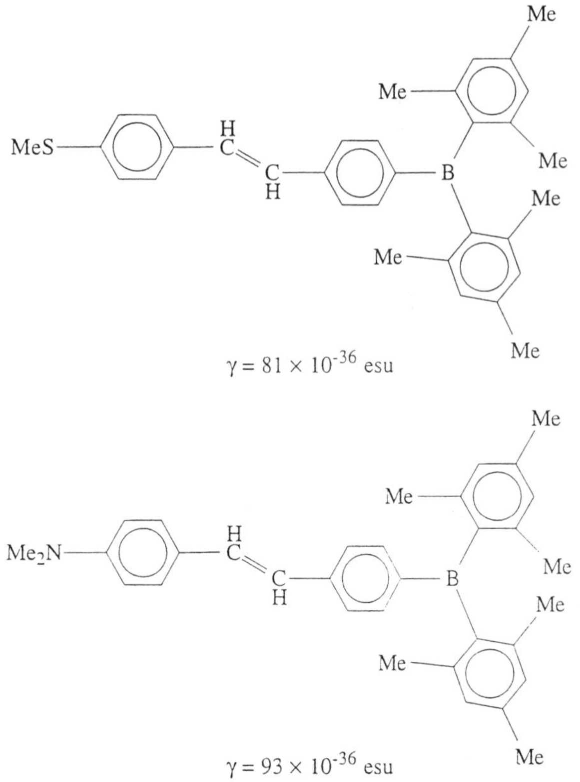

$\gamma = 81 \times 10^{-36}$ esu

$\gamma = 93 \times 10^{-36}$ esu

FIG. 14.    Boron complexes with large $\gamma$.

alkyl, $[Pd_2Me_2(\mu\text{-dppm})_2]$, has also been measured; given the lack of a $\pi$-system for efficient electron delocalization, the nonlinearity is not surprisingly low. The only (isonitrile)ruthenium complex examined has a planar ruthenaphthalocyanine axially ligated by 1,4-diisocyanobenzene to form an oligomer. THG measurements of $\chi^{(3)}$ ($-3\omega;\omega,\omega,\omega$) of a film of this complex ($3.7 \times 10^{-12}$ esu; Table XIV) are significantly less than $\chi^{(3)}(-\omega;\omega,\omega,-\omega)$ obtained by DFWM (Table XIII). A difference of four orders of magnitude was also found with the cubic molecular nonlinearities of the mixed cobalt–iron cluster $[CoFe_2(\mu_3\text{-S})(\mu_3\text{-Se})(CO)_6(\eta^5\text{-}C_5H_5)]$, a significant frequency dependence; as $\lambda_{max}$ is at ~550 nm, the larger value is likely to be significantly resonance enhanced.

A range of organometallic films (either polysilanes or polygermanes, or organotransition metal complexes doped into organic polymer hosts) have been examined for their bulk NLO response; results are listed in Table

## TABLE XII
### THIRD-ORDER NLO MEASUREMENTS OF FULLERENE DERIVATIVES

| Complex | Re($\chi^{(3)}$) | Im($\chi^{(3)}$) | Re $\gamma$ (esu) | Im $\gamma$ (esu) | Conc. (g/liter) | Technique | Polarization | $\tau$ (ps) | Fund. ($\mu$m) | Ref. |
|---|---|---|---|---|---|---|---|---|---|---|
| [($\eta^2$-C$_{60}$)Pt(PPh$_3$)$_2$][a] | — | — | $1.4 \times 10^{-30}$ | $3.4 \times 10^{-30}$ | 0.35 | Z-scan | Linear | 10 | 0.53 | 50 |
| | $1.7 \times 10^{-12}$ esu | $7.8 \times 10^{-13}$ esu | $6.9 \times 10^{-31b}$ | $1.3 \times 10^{-30}$ | 0.35 | Z-scan | Linear | 10 | 0.53 | 50 |
| | $3.2 \times 10^{-12}$ esu | $2.0 \times 10^{-12}$ esu | $1.4 \times 10^{-30b}$ | $3.4 \times 10^{-30}$ | 0.35 | Z-scan | Circular | 10 | 0.53 | 50 |
| [($\eta^2$-C$_{60}$)Pd(PPh$_3$)$_2$][a] | — | $1.32 \times 10^{-16}$ m$^2$ V$^{-2}$ | — | — | 2 | IDA | — | 500 | 0.59 | 51 |
| [($\eta^2$-C$_{70}$)Pd(PPh$_3$)$_2$][a] | — | $1.55 \times 10^{-17}$ m$^2$ V$^{-2}$ | — | — | 2 | IDA | — | 500 | 0.59 | 51 |
| [($\eta^2$-C$_{60}$)Ir(CO)(PMePh$_2$)$_2$][a] | — | $5.78 \times 10^{-17}$ m$^2$ V$^{-2}$ | — | — | 2 | IDA | — | 500 | 0.59 | 51 |
| [($\eta^2$-C$_{70}$)Ir(CO)(PMePh$_2$)$_2$][a] | — | $4.14 \times 10^{-17}$ m$^2$ V$^{-2}$ | — | — | 2 | IDA | — | 500 | 0.59 | 51 |
| | — | $2.64 \times 10^{-17}$ m$^2$ V$^{-2}$ | — | — | 2 | IDA | — | 500 | 0.59 | 51 |

[a] As a solution in toluene.

[b] Authors note that values are less than or comparable to the total experimental error and may not be reliable.

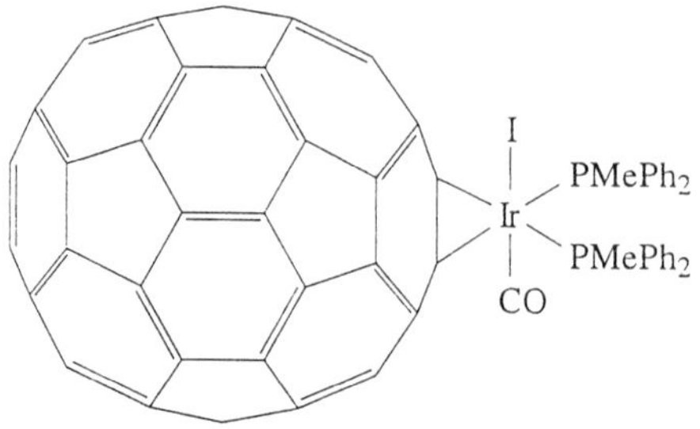

$$Im(\chi^{(3)}) = 4.14 \times 10^{-17} \text{ m}^2 \text{ V}^{-2}$$

FIG. 15.   $[(\eta^2\text{-}C_{60})IrI(CO)(PMePh_2)_2]$.

XIV. Third-harmonic nonlinearities $\chi^{(3)}(-3\omega;\omega,\omega,\omega)$ for some (cyclopenta-dienyl)bis(phosphine)ruthenium complexes containing $p$-substituted benzonitrile ligands in a polymethylmethacrylate host polymer were determined by THG. These give moderate nonlinearities that are not ascribed to three-photon resonance. Comparison of the results for $[Ru(N\equiv CC_6H_4\text{-}4\text{-}C_6H_4\text{-}4\text{-}NO_2)(dppe)(\eta^5\text{-}C_5H_5)]^+$ with those for oligomers of poly(thio-phenevinylene) of similar size using the same technique suggests that $\gamma$ is four times larger for the organometallic; this is explained in terms of reduced bond alternation in $[Ru(N\equiv CC_6H_4\text{-}4\text{-}C_6H_4\text{-}4\text{-}NO_2)(dppe)(\eta^5\text{-}C_5H_5)]^+$, for which $\pi$-back donation enhances the delocalized $\pi$-electron system. Variations in the counterion ($[BPh_4]^-$, $[PF_6]^-$, $[CF_3SO_3]^-$, $[MeC_6H_4\text{-}4\text{-}SO_3]^-$) or phosphine [dppe, (+)-diop] had minimal effects on the nonlinearity for these complexes (Fig. 16).

$\chi^{(3)}$ for varying thickness of polysilane thin films were determined by THG at 1.06 $\mu$m. These materials may have potential in device applications as they readily form thin films and are transparent in the visible range (above 400 nm). The large values of $\chi^{(3)}$ were ascribed to three-photon resonance, the exact nature of which is unknown.[57] Subsequent measurements on a series of polysilanes and polygermane thin films at 1.91 $\mu$m by THG showed them to have the largest values of $\chi^{(3)}$ for transparent polymers. As no attempt has been made to optimize the nonlinearity, it was suggested that $\chi^{(3)}$ might be enhanced by incorporation of conjugated side chains and orienting the polymer chains in a more favorable way.[57]

## B. *Third-Order Calculations*

Very few calculations of cubic nonlinearities have been carried out. Table XV contains computationally derived static $\gamma$ values for ferrocene and some ferrocenyl derivatives.

TABLE XIII

THIRD-ORDER NLO MEASUREMENTS OF OTHER COMPLEXES

| Complex | Solvent | Quantity Measured | Value | $\lambda_{max}$ | Technique | Fund. ($\mu$m) | Ref. |
|---|---|---|---|---|---|---|---|
| [RuCl(PPh$_3$)$_2$($\eta^5$-C$_5$H$_5$)] | thf | Re $\gamma$ | $150 \pm 100 \times 10^{-36}$ esu | 348 | Z-scan | 0.80 | 33 |
| | thf | Re $\gamma$ | $50 \pm 20 \times 10^{-36}$ esu | 348 | DFWM | 0.80 | 33 |
| [RuCl(PMe$_3$)$_2$($\eta^5$-C$_5$H$_5$)] | thf | Re $\gamma$ | $\leq 80 \times 10^{-36}$ esu | 346 | Z-scan | 0.80 | 33 |
| | thf | Re $\gamma$ | $80 \pm 30 \times 10^{-36}$ esu | 346 | DFWM | 0.80 | 33 |
| [Pd$_2$Me$_2$($\mu$-dppm)$_2$] | CHCl$_3$ | Re $\gamma$ | $-1 \times 10^{-46}$ m$^5$ V$^{-2}$ | — | DFWM | 1.06 | 43 |
| | CHCl$_3$ | Im $\gamma$ | $2 \times 10^{-46}$ m$^5$ V$^{-2}$ | — | DFWM | 1.06 | 43 |
| | CHCl$_3$ | $\gamma$ | $2 \times 10^{-46}$ m$^5$ V$^{-2}$ | — | DFWM | 1.06 | 43 |
| [($^t$Bu$_4$-phthalocyaninato)Ru(4-CNC$_6$H$_4$NC)]$_n$ | CHCl$_3$ | $\chi^{(3)}(-\omega;\omega,\omega,-\omega)$ | $0$–$11.5 \times 10^{-8}$ esu | ~550 | DFWM | 0.50–0.80 | 52 |
| [CoFe$_2$($\mu_3$-S)($\mu_3$-Se)(CO)$_6$($\eta^5$-C$_5$H$_5$)] | — | $\gamma$ | $2.7 \times 10^{-14}$ m$^2$ W$^{-1}$ | ~550 | Z-scan | 0.53 | 53 |
| | — | $\gamma$ | $5.8 \times 10^{-18}$ m$^2$ W$^{-1}$ | ~550 | Z-scan | 1.06 | 53 |

## TABLE XIV
### Third-Order NLO Measurements on Organometallic Films and Composites

| Material | Film Thickness (nm) | $\chi^{(3)}$ ($10^{-12}$ esu) | $\gamma$ ($10^{-36}$ esu) | Technique | Fund. ($\mu$m) | Ref. |
|---|---|---|---|---|---|---|
| $[Fe(\eta^5\text{-}C_5H_3(C_6H_{13})\text{-}2\text{-}(E)\text{-}CH=CH\text{-})_2)]_n$ | — | 1–4 | — | THG | 1.0–2.4 | 54 |
| $[Ru(N\equiv CC_6H_4\text{-}4\text{-}NO_2)(dppe)(\eta^5\text{-}C_5H_5)][BPh_4]^a$ | 200–400 | 0.07 | 230 | THG | 1.06 | 55 |
| $[Ru(N\equiv CC_6H_4\text{-}4\text{-}NO_2)(dppe)(\eta^5\text{-}C_5H_5)][PF_6]^a$ | 200–400 | 0.17 | 470 | THG | 1.06 | 55 |
| $[Ru(N\equiv CC_6H_4\text{-}4\text{-}NO_2)(dppe)(\eta^5\text{-}C_5H_5)][CF_3SO_3]^a$ | 200–400 | 0.23 | 510 | THG | 1.06 | 55 |
| $[Ru(N\equiv CC_6H_4\text{-}4\text{-}NO_2)(dppe)(\eta^5\text{-}C_5H_5)][MeC_6H_4\text{-}4\text{-}SO_3]^a$ | 200–400 | 0.23 | 630 | THG | 1.06 | 55 |
| $[Ru(N\equiv CC_6H_4\text{-}4\text{-}NO_2)(dppe)(\eta^5\text{-}C_5H_5)][BF_4]^a$ | 200–400 | 0.27 | 690 | THG | 1.06 | 55 |
| $[Ru(N\equiv CC_6H_4\text{-}4\text{-}NO_2)((+)\text{-}(diop)(\eta^5\text{-}C_5H_5)][PF_6]^a$ | 200–400 | 0.24 | 720 | THG | 1.06 | 55 |
| $[Ru(N\equiv CC_6H_4\text{-}4\text{-}NO_2)((+)\text{-}(diop)(\eta^5\text{-}C_5H_5)][CF_3SO_3]^a$ | 200–400 | 0.21 | 650 | THG | 1.06 | 55 |
| $[Ru(N\equiv CC_6H_4\text{-}4\text{-}NO_2)((+)\text{-}(diop)(\eta^5\text{-}C_5H_5)][MeC_6H_4\text{-}4\text{-}SO_3]^a$ | 200–400 | 0.10 | 320 | THG | 1.06 | 55 |
| $[Ru(N\equiv CC_6H_4\text{-}4\text{-}NMe_2)(dppe)(\eta^5\text{-}C_5H_5)][PF_6]^a$ | 200–400 | 0.21 | 590 | THG | 1.06 | 55 |
| $[Ru(N\equiv CC_6H_4\text{-}4\text{-}Ph)(dppe)(\eta^5\text{-}C_5H_5)][PF_6]^a$ | 200–400 | 0.24 | 660 | THG | 1.06 | 55 |
| $[Ru(N\equiv C\text{-}(E)\text{-}CH=CHC_6H_4\text{-}4\text{-}NO_2)(dppe)(\eta^5\text{-}C_5H_5)][PF_6]^a$ | 200–400 | 0.36 | 1020 | THG | 1.06 | 55 |
| $[Ru(N\equiv CC_6H_4\text{-}4\text{-}C_6H_4\text{-}4\text{-}NO_2)(dppe)(\eta^5\text{-}C_5H_5)][PF_6]^a$ | 200–400 | 0.76 | 2280 | THG | 1.06 | 55 |
| $[Fc(E)\text{-}CH=CH)_3Fc]^b$ | 1000–2000 | 0.029 | | DFWM | 0.53 | 56 |
| $[SiMePh]_n{}^c$ | 150–450 | 1.5 | — | THG | 1.06 | 57 |
| | 120 | 7.2 | — | THG | 1.06 | 56 |
| | 120 | 4.2 | — | THG | 1.91 | 58 |
| $[Si(C_6H_{13}{}^n)_2]_n{}^c$ | 1200 | 1.9 | — | THG | 1.91 | 58 |
| | 50 | 11 | — | THG | 1.06 | 58 |
| | 120 | 5.5 | — | THG | 1.06 | 58 |
| | 240 | 4.6 | — | THG | 1.06 | 58 |
| | 240 | 1.3 | — | THG | 1.91 | 58 |
| $[Ge(C_6H_{13}{}^n)_2]_n{}^c$ | 295 | 3.3 | — | THG | 1.06 | 58 |
| | 295 | 1.1 | — | THG | 1.91 | 58 |
| $[Bu_4(phthalocyaninato)Ru(4\text{-}CNC_6H_4NC)]_n{}^d$ | 160 | 3.7 ± 1.5 | — | THG | 1.06 | 52 |

[a] As a polymethyl methacrylate composite, ca. 10% by mass.

[b] As a polystyrene composite, ca. 2.5% by mass.

[c] At 23°C. Measurements were also performed at lower and higher temperatures.

[d] As a spin-cast film.

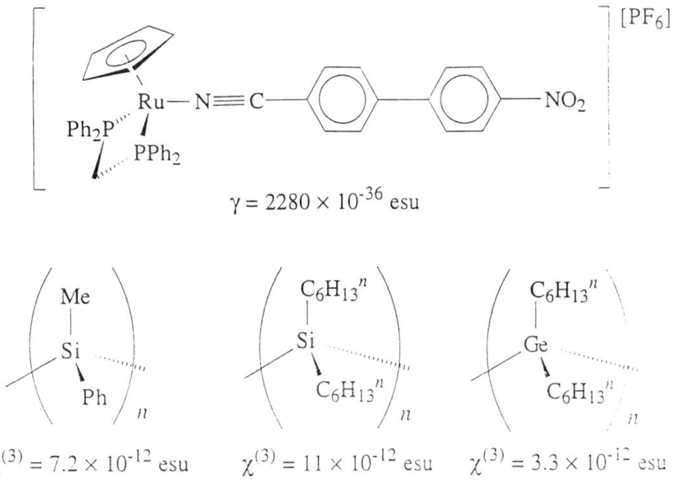
[PF6]
Ru—N≡C
Ph2P
PPh2
NO2
γ = 2280 × 10-36 esu
Me
Si
Ph
n
C6H13n
Si
C6H13n
n
C6H13n
Ge
C6H13n
n
χ(3) = 7.2 × 10-12 esu
χ(3) = 11 × 10-12 esu
χ(3) = 3.3 × 10-12 esu

FIG. 16. Organometallics in films possessing high third-order nonlinearities.

Waite and Papadopoulos have used the method of CHF-PT-EB CNDO (described in Section II,G) for these calculations.[14,15] The average second hyperpolarizability is given by $\langle\gamma\rangle = \frac{1}{5}(\gamma_{xxxx} + \gamma_{yyyy} + \gamma_{zzzz} + 2\gamma_{xxyy} + 2\gamma_{xxzz} + 2\gamma_{yyzz})$ but, to reduce computational expense, Waite and Papadopoulos performed calculations employing the approximation $\langle\gamma\rangle \cong \frac{1}{3}(\gamma_{xxxx} + \gamma_{yyyy} + \gamma_{zzzz})$, which is claimed to reduce computation time by about 50%; using this approximation affords $\gamma$ values differing by about 4% from those calculated using all six tensorial components.[15]

Calculated values of $\gamma$ across these complexes are not particularly sensitive to variation of the exponents in the 3*d*, 4*s*, and 4*p* STOs in the iron basis set. Bonding between the cyclopentadienyl groups and the metal atom is mainly due to *d*–$\pi$ interaction; as a consequence, removing the 3*d* iron orbitals from the calculation results in a dramatic change in $\gamma$ values, whereas the effect of removing the 4*p* orbitals is minimal. Comparison of the computationally derived result for ferrocene with experimental data reveals that the difference is less than an order of magnitude. The experimental value is somewhat resonance enhanced, restricting further comment. Computational analysis of [Fc-(*E*)-CH=CHC$_6$H$_4$-4-NO$_2$] suggests that both the ligand–ferrocenyl $\pi$ interaction and the MLCT are important factors affecting $\gamma$. Introduction of polarizable halogens in this system leads, not surprisingly, to an increase in $\gamma$. The effect of changing geometry was investigated by varying the position of these halo substituents, with up to about 20% variation in computed nonlinearity observed.

The results of ab initio calculations of $\gamma_{zzzz}$ for transition metal–

TABLE XV

COMPUTATIONAL THIRD-ORDER DATA[a]

| Complex | Calculated | | Experimental | | Ref. |
|---|---|---|---|---|---|
| | $\gamma^b$ (10$^{-36}$ esu) | $\gamma^c$ (10$^{-36}$ esu) | $\gamma$ (10$^{-36}$ esu) | Freq. (μm) | |
| FcH | 24.6 | 23.6 | 96.7 | 0.60 | 14,20 |
| | 24.8[d] | | | | 15 |
| FcNH$_2$ | 30.0 | 28.7 | | | 15 |
| FcCN | 24.8 | 23.5 | | | 15 |
| | 24.7[d] | | | | 15 |
| FcCH$_3$ | 27.0 | 25.5 | | | 15 |
| FcCl | 35.5 | 38.4 | | | 15 |
| | 35.1[d] | | | | 15 |
| FcCHO | 26.7 | 27.9 | | | 15 |
| | 26.7[d] | | | | 15 |
| FcC$_2$H | 39.5 | 39.7 | | | 15 |
| | 32.8[d] | | | | 15 |
| FcCHCH$_2$ | 36.8 | 37.1 | | | 15 |
| | 35.3[d] | | | | 15 |

| | | | |
|---|---|---|---|
| FcCOCH$_3$ | 27.7 | 26.9 | 15 |
| FcCH$_2$Cl[e] | 36.2 | 37.1 | 15 |
| FcCH$_2$Cl[f] | 36.9 | 35.9 | 15 |
| FcCHCl$_2$[e] | 65.0 | 34.0 | 15 |
| FcCHCl$_2$[f] | 69.5 | 68.5 | 15 |
| FcI | 42.5 | 37.8 | 15 |
| FcCH$_2$I | 46.9 | 42.7 | 15 |
| FcCHI$_2$[e] | 83.6 | 84.6 | 15 |
| FcCHI$_2$[f] | 102.8 | 105.3 | 15 |
| Fc-(E)-CH$=$CHC$_6$H$_4$-4-NO$_2$ | 69.0 | 75.1 | 15 |
| Fe($\eta^5$-C$_5$H$_4$CHO)$_2$ | 27.3 | 26.2 | 15 |

[a] Converted from literature values given in a.u. by employing the conversion factor 1 a.u. $\approx 0.503717 \times 10^{-39}$ esu.[15]

[b] Calculated using $\gamma = \frac{1}{5}(\gamma_{xxxx} + \gamma_{yyyy} + \gamma_{zzzz} + 2\gamma_{xxyy} + 2\gamma_{xxzz} + 2\gamma_{yyzz})$.

[c] Calculated using $\gamma \cong \frac{1}{3}(\gamma_{xxxx} + \gamma_{yyyy} + \gamma_{zzzz})$.

[d] Calculated using no $4p$ functions of Fe.

[e] Halogen(s) is as near as possible to Fe.

[f] Halogen(s) is as far as possible from Fe.

methylene cations $[MCH_2]^+$ (M = Ti, V, Cr, Mn, Fe, Co) using various methods of electron-correlation correction are shown in Fig. 17.[59] These results show that the calculated values of $\gamma_{zzzz}$ are quite dependent on the electron-correlation method used; the HF method yields values that differ in sign from many of the values obtained using electron-correlation correction and the methods of MP2, MP3, and MP4(SDTQ) appear to overestimate the electron-correlation correction. Shigemoto *et al.*[59] suggest that CCSD(T) and QCISD(T) methods with split-valence plus polarization *p,d,* and *f* basis sets (MIDI + *pdf*)[60] are required to calculate semiquantitative $\gamma$ values of these organometallic fragments. They also state, not surprisingly, that the calculated $\gamma$ values are dependent on the orbital electronic configurations.

## IV

### OPTICAL LIMITING PROPERTIES OF ORGANOMETALLIC COMPLEXES

The optical limiting properties of metal cluster compounds and some fullerene complexes have been studied by RSA, with some of the results

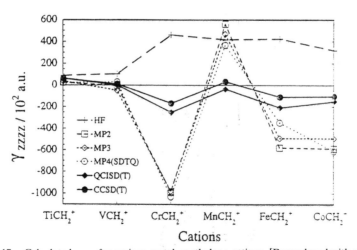

FIG. 17.    Calculated $\gamma_{zzzz}$ for various metal–methylene cations. [Reproduced with permission from I. Shigemoto, M. Nakano, S. Yamada, S. Kiribayshi, and K. Yamaguchi, *Mol. Cryst. Liq. Cryst.* **1996,** *286,* 159. Copyright © 1996 Overseas Publishing Association (Gordon and Breach Science Publishers), The Netherlands.]

reviewed.[61] Table XVI contains optical limiting data for the cluster and fullerene complexes investigated.

The cyclopentadienyliron carbonyl tetramer ($[Fe(CO)(\eta^5\text{-}C_5H_5)]_4$, King's complex) has been shown to exhibit RSA on picosecond time scales.[62] The cross section of the singlet-excited state is about twice that of the ground state, which is much too small to explain the degree of RSA observed on nanosecond time scales.[62] Subsequently, cross sections for ground and first excited state absorptions, first excited state lifetime, and an upper bound on the second excited state lifetime were determined.[63] The contribution of intersystem crossing to the triplet state has been assessed; from the picosecond measurements, it seems that the majority of molecules relax directly from the first excited singlet state to the ground state with minimal population in the triplet-excited state.

Parameters gained from the picosecond measurements have been used in conjunction with the five-level model to attempt to predict the response observed on the nanosecond time scale; however, the nanosecond response was too strong to be modeled by the excited state parameter extracted from the picosecond experiments. In order to explain the difference between picosecond and nanosecond responses, further experiments with varying laser fluences (where fluence is the energy of the laser per unit of area) and delay times between pump and probe beams were performed, which suggested that the large nanosecond responses had significant thermally induced scattering (this thermal effect is too slow to affect the picosecond measurements at fluences used to obtain previous data). In order to eliminate the thermally induced scattering, nanosecond measurements were performed on King's complex embedded in a solid host, resulting in much better agreement with the five-level model using picosecond parameters but even better agreement with a three-level model (i.e., ignoring triplet absorption).[63] It seems then that triplet absorption plays a very minor role in RSA of King's complex.

The ground to first excited state transition has been ascribed to electron transfer from a $d$-orbital of the metal to an antibonding orbital.[64] To determine the origin of the excited state absorption responsible for RSA in King's complex, several derivatives of this complex were prepared and RSA measurements performed on a picosecond time scale,[64] with $[Fe(CO)(\eta^5\text{-}C_5H_4Me)]_4$, $[Fe(COAlEt_3)(\eta^5\text{-}C_5H_5)]_4$, and King's complex affording virtually the same response.[64] Both LMCT and MLCT transitions have therefore been rejected as the source of the excited state absorption, as these cluster modifications impact on charge transfer transitions involving carbonyl ($[Fe(COAlEt_3)(\eta^5\text{-}C_5H_5)]_4$) and cyclopentadienyl ($[Fe(CO)(\eta^5\text{-}C_5H_4Me)]_4$) ligands. The measured excited state cross section in King's complex and its methylcyclopentadienyl analogue showed no solvent dependence for solvents dichloromethane, thf and toluene; the excited state

TABLE XVI

Optical Limiting Data

| Complex | Solvent | $\lambda$ (nm) | Pulse Width | $\sigma_{01}$ ($\times 10^{-18}$ cm$^2$) | $\sigma_{12}$ ($\times 10^{-18}$ cm$^2$) | $\tau_{10}$ | $\tau_{21}$ (ps) | Ref. |
|---|---|---|---|---|---|---|---|---|
| [Fe(CO)($\eta^5$-C$_5$H$_5$)]$_4$ | CH$_2$Cl$_2$ | 532 | 25 ps | 4.1 | 7.8 ± 0.8 | 120 ± 5 ps | <1 | 62–64 |
| [Fe(Me)(CO)($\eta^5$-C$_5$H$_4$Me)]$_4$ | CH$_2$Cl$_2$ | 532 | 25 ps | 4.7 | 9.1 ± 0.9 | 120 ± 5 ps | — | 64 |
| [Fe(COAlEt$_3$)($\eta^5$-C$_5$H$_5$)]$_4$ | CH$_2$Cl$_2$ | 532 | 25 ps | 4.9 | 7.3 ± 1.4 | 120 ± 5 ps | — | 64 |
| [HFeCo$_3$(CO)$_{12}$] | CH$_2$Cl$_2$ | 532 | 8 ns | | | | | 65,66 |
| [NEt$_4$][FeCo$_3$(CO)$_{12}$] | CH$_2$Cl$_2$ | 532 | 8 ns | | | | | 65,66 |
| [HFeCo$_3$(CO)$_{10}$(PMe$_3$)$_2$] | CH$_2$Cl$_2$ | 532 | 8 ns | | | ~115 ns | | 65,66 |
| [HFeCo$_3$(CO)$_{10}$(PPh$_3$)$_2$] | CH$_2$Cl$_2$ | 532 | 8 ns | | | ~115 ns | | 65,66 |
| [($\eta^2$-C$_{60}$)Pt(PPh$_3$)$_2$] | Toluene | 588 | 500 ps | 11.4 | 16.4 | | | 51 |
| [($\eta^2$-C$_{60}$)Pd(PPh$_3$)$_2$] | Toluene | 588 | 500 ps | 5.61 | 9.09 | | | 51 |
| [($\eta^2$-C$_{70}$)Pd(PPh$_3$)$_2$] | Toluene | 588 | 500 ps | 11.6 | 14.2 | | | 51 |
| [($\eta^2$-C$_{60}$)IrI(CO)(PMePh$_2$)$_2$] | Toluene | 588 | 500 ps | 2.56 | 7.11 | | | 51 |
| [($\eta^2$-C$_{70}$)IrI(CO)(PMePh$_2$)$_2$] | Toluene | 588 | 500 ps | 3.46 | 8.93 | | | 51 |

transition responsible has therefore been tentatively assigned to a second
$d$-$d$ transition within the metal core.

A series of iron–cobalt mixed-metal clusters were examined to probe
the effect of ligand variation (CO to $PMe_3$ to $PPh_3$) and removal of hydride
ligand ([cluster]H versus [cluster]$^-$) on the RSA response.[65,66] It was found
that the RSA response varied little on hydride removal, but that ligand
substitution strongly affects optical limiting performance[65]; significantly,
although replacing $PMe_3$ by $PPh_3$ produces large changes in the optical limiting response, no change was observed in the ground state UV-
visible spectrum.

Fullerene derivatives have also been investigated. Complexation of both
$C_{60}$ and $C_{70}$ modifies the absorption spectra significantly, with large shifts
(50–100 nm) in absorption maxima. The first excited state cross sections
$\sigma_{01}$ of the complexes are all larger than those of the nonligated fullerenes
($C_{60}$: $1.45 \times 10^{-18}$ cm$^2$; $C_{70}$: $2.94 \times 10^{-18}$ cm$^2$), and the $\sigma_{12}/\sigma_{01}$ ratios for
the nonligated fullerenes ($C_{60}$: 4.16; $C_{70}$: 2.74) are uniformly larger than
those for the fullerene complexes; the metal complexes are thus without
exception poorer optical limiters than the free fullerenes.

## V

## CONCLUDING REMARKS

The studies summarized in this review have began to establish structure/
NLO property relationships for organometallic systems, but considerably
less is known at present about the design of an efficient third-order NLO
organometallic material than is known of the design of a second-order
NLO complex. Not surprisingly, then a great deal of work investigating
variation in metal, oxidation state, coligands, and coordination geometry
remains to be done. The molecular nonlinearities obtained for some of
the complexes are large, suggesting that the potential for application of
organometallics in photonic devices remains. Thus far, little has been done
toward integrating organometallic compounds into processable polymeric
materials and measuring their properties; such studies should reveal the
true potential of organometallics in nonlinear optics.

## VI

## APPENDIX: ABBREVIATIONS

AC          alternating current
Bu$^n$          normal butyl

| | |
|---|---|
| Bu$^t$ | tertiary butyl |
| CHF | coupled Hartree–Fock |
| CNDO | complete neglect of differential overlap |
| dppm | bis(diphenylphosphino)methane |
| DFWM | degenerate four-wave mixing |
| diop | (1,2-dimethyl-1,3-dioxolane-4,5-diyl)bis(methylene)bis(diphenylphosphine) |
| EB | extended basis |
| EFISH | electric field-induced second harmonic generation |
| Et | ethyl |
| ESA | excited state absorption |
| Fc | ferrocenyl |
| FF | finite field |
| FWM | four-wave mixing |
| HOMO | highest occupied molecular orbital |
| IDS | intensity-dependent absorption |
| LMCT | ligand-to-metal charge transfer |
| LUMO | lowest unoccupied molecular orbital |
| Me | methyl |
| mes | mesityl |
| MLCT | metal-to-ligand charge transfer |
| NLO | nonlinear optical |
| OKG | optical Kerr gate |
| OPL | optical power limiting |
| Ph | phenyl |
| PT | perturbation theory |
| RSA | reverse saturable absorption |
| SA | saturable absorption |
| STO | Slater-type orbitals |
| thf | tetrahydrofuran |
| THG | third harmonic generation |
| TPA | two-photon absorption |

ACKNOWLEDGMENT

We thank the Australian Research Council for support. A.M.M. is the recipient of an Australian Postgraduate Award, and M.G.H. holds an ARC Australian Research Fellowship.

REFERENCES

(1) Whittall, I. R.; McDonagh, A. M.; Humphrey, M. G.; Samoc, M. *Adv. Organomet. Chem.* **1998**, *42*, 291.
(2) Nie, W. *Adv. Mater.* **1993**, *5*, 520.

 (3) Allen, S. *New Scientist* **1989**, *1 July*, 31.
 (4) Nalwa, H. S. *Adv. Mater.* **1993**, *5*, 341.
 (5) Marder, S. R.; Perry, J. W.; Bourhill, G.; Gorman, C. B.; Tiemann, B. G.; Mansour, K. *Science* **1993**, *261*, 186.
 (6) Sutherland, R. L. *Handbook of Nonlinear Optics;* Marcel Dekker: New York, 1996.
 (7) Butcher, P. N.; Cotter, D. *The Elements of Nonlinear Optics;* Cambridge Univ. Press: New York, 1990.
 (8) Sheik-Bahae, M.; Said, A. A.; Wei, T.; Hagan, D. J.; van Stryland, E. W. *IEEE J. Quant. Electr.* **1990**, *26*, 760.
 (9) Boyd, R. W. *Nonlinear Optics;* Academic Press: New York, 1992.
(10) Levine, I. N. *Quantum Chemistry;* Prentice Hall: Englewood Cliffs, NJ, 1991.
(11) McWeeny, R. *Phys. Rev.* **1962**, *126*, 1028.
(12) Diercksen, G.; McWeeny, R. *J. Chem. Phys.* **1996**, *44*, 3554.
(13) Dodds, J. L.; McWeeny, R.; Raynes, W. T.; Riley, J. P. *Mol. Phys.* **1977**, *33*, 611.
(14) Waite, J.; Papadopoulos, M. G. *Z Naturforsch. A* **1987**, *42*, 749.
(15) Waite, J.; Papadopoulos, M. G. *J. Phys. Chem.* **1991**, *93*, 5426.
(16) Pople, J. A.; Beveridge, D. L. *Approximate Molecular Orbital Theory;* McGraw-Hill: New York, 1970.
(17) Winter, C. S.; Oliver, S. N.; Rush, J. D. *Optics Commun.* **1988**, *69*, 45.
(18) Winter, C. S.; Oliver, S. N.; Rush, J. D. in *Organic Materials for Nonlinear Optics;* Hann, R. A.; Bloor, D., Eds.; Royal Society of Chemistry: London, 1989; p. 232.
(19) Winter, C. S.; Oliver, S. N.; Rush, J. D. in *Nonlinear Optical Effects in Organic Polymers;* Messier, J.; Kajzar, F.; Prasad, P.; Ulrich, D., Eds.; Kluwer: Dordrecht, 1989, p. 247.
(20) Ghosal, S.; Samoc, M.; Prasad, P. N.; Tufariello, J. T. *J. Phys. Chem.* **1990**, *94*, 2847.
(21) Calabrese, J. C.; Tam, W. *Chem. Phys. Lett.* **1987**, *133*, 244.
(22) Cheng, L. T.; Tam, W.; Meredith, G. R.; Marder, S. R. *Mol. Cryst. Liq. Cryst.* **1990**, *189*, 137.
(23) Myers, L. K.; Langhoff, C.; Thompson, M. E. *J. Am. Chem. Soc.* **1992**, *114*, 7560.
(24) Yuan, Z.; Stringer, G.; Jobe, I. R.; Kreller, D.; Scott, K.; Koch, L.; Taylor, N. J.; Marder, T. B. *J. Organomet. Chem.* **1993**, *452*, 115.
(25) Calabrese, J. C.; Cheng, L.-T.; Green, J. C.; Marder, S. R.; Tam, W. *J. Am. Chem. Soc.* **1991**, *113*, 7227.
(26) Myers, L. K.; Ho, D. M.; Thompson, M. E.; Langhoff, C. *Polyhedron* **1995**, *14*, 57.
(27) Thompson, M. E.; Chiang, W.; Myers, L. K.; Langhoff, C. *Proc. SPIE-Int. Soc. Opt. Eng.* **1991**, *1497*, 423.
(28) Winter, C. S.; Oliver, S. N.; Rush, J. D.; Manning, R. J.; Hill, C.; Underhill, A. in *Materials for Nonlinear Optics: Chemical Perspectives;* Marder, S. R.; Sohn, J. E.; Stucky, G. D., Eds.; American Chemical Society: Washington, DC, 1991; p. 616.
(29) Oliver, S. N.; Winter, C. S.; Rush, J. D.; Underhill, A. E.; Hill, C. *Proc. SPIE-Int. Soc. Opt. Eng.* **1990**, *1337*, 81.
(30) Blau, W. J.; Cardin, C. J.; Cardin, D. J.; Davey, A. P. in *Organic Materials for Nonlinear Optics III;* Ashwell, G. J.; Bloor, D., Eds.; Royal Society of Chemistry: Cambridge, 1993; p. 124.
(31) Davey, A. P.; Page, H.; Blau, W. *Synth. Met.* **1993**, *55*, 3980.
(32) Davey, A. P.; Byrne, H. J.; Page, H.; Blau, W.; Cardin, D. J. *Synth. Met.* **1993**, *58*, 161.
(33) Whittall, I. R.; Humphrey, M. G.; Samoc, M.; Swiatkiewicz, J.; Luther-Davies, B. *Organometallics* **1995**, *14*, 5493.
(34) Whittall, I. R.; Humphrey, M. G.; Cifuentes, M. P.; Samoc, M.; Luther-Davies, B.; Persoons, A.; Houbrechts, S.; Heath, G. A.; Hockless, D. C. R. *J. Organomet. Chem.* **1997**, *549*, 127.

(35) McDonagh, A. M.; Cifuentes, M. P.; Whittall, I. R.; Humphrey, M. G.; Samoc, M.; Luther-Davies, B.; Hockless, D. C. R. *J. Organomet. Chem.* **1996**, *526*, 99.

(36) Whittall, I. R.; Humphrey, M. G.; Samoc, M.; Luther-Davies., B. *Angew. Chem. Int. Ed. Engl.* **1997**, *36*, 370.

(37) Whittall, I. R.; Humphrey, M. G.; Cifuentes, M. P.; Samoc, M.; Luther-Davies, B.; Persoons, A.; Houbrechts, S. *Organometallics* **1997**, *16*, 2631.

(38) Frazier, C. C.; Chauchard, E. A.; Cockerham, M. P.; Porter, P. L. *Mater. Res. Soc. Symp. Proc.* **1988**, *109*, 323.

(39) Porter, P. L.; Guha, S.; Kang, K.; Frazier, C. C. *Polymer* **1991**, *32*, 1756.

(40) Blau, W. J.; Byrne, H. J.; Cardin, D. J.; Davey, A. P. *J. Mater. Chem.* **1991**, *1*, 245.

(41) Davey, A. P.; Cardin, D. J.; Byrne, H. J.; Blau, W. in *Organic Molecules for Nonlinear Optics and Photonics;* Messier, J.; Kajzar, F.; Prasad, P., Eds.; Kluwer: Dordrecht, 1991; p. 391.

(42) Haub, J.; Johnson, M.; Orr, B.; Woodruff, M.; Crisp, G., presented at *CLEO/QUELS '91;* Baltimore, May 1991.

(43) Page, H.; Blau, W.; Davey, A. P.; Lou, X.; Cardin, D. J. *Synth. Met.* **1994**, *63*, 179.

(44) Guha, S.; Kang, K.; Porter, P.; Roach, J. F.; Remy, D. E.; Aranda, F. J.; Rao, D. V. G. L. N. *Opt. Lett.* **1992**, *17*, 264.

(45) Frazier, C. C.; Guha, S.; Chen, W. P.; Cockerham, M. P.; Porter, P. L.; Chauchard, E. A.; Lee, C. H. *Polymer* **1987**, *28*, 553.

(46) Guha, S.; Frazier, C. C.; Porter, P. L.; Kang, K.; Finberg, S. E. *Opt. Lett.* **1989**, *14*, 952.

(47) Guha, S.; Frazier, C. C.; Chen, W. P.; Kang, K.; Finberg, S. E. *Proc. SPIE-Int. Soc. Opt. Eng.* **1989**, *1105*, 14.

(48) Yuan, Z.; Taylor, N. J.; Marder, T. B.; Williams, I. D.; Kurtz, S. K.; Cheng, L. T. in *Organic Materials for Nonlinear Optics II;* Hann, R. A.; Bloor, D., Eds.; Royal Society of Chemistry: London, 1991; p. 190.

(49) Yuan, Z.; Taylor, N. J.; Ramachandran, R.; Marder, T. B. *Appl. Organomet. Chem.* **1996**, *10*, 305.

(50) Ergorov, A. N.; Mavritsky, O. B.; Petrovsky, A. N.; Yakubovsky, K. V. *Laser Phys.* **1995**, *5*, 1006.

(51) Callaghan, J.; Weldon, D. N.; Henari, F. Z.; Blau, W.; Cardin, D. J. in *Electronic Properties of Fullerenes;* Kuzmany, H.; Fink, J.; Mehing, M.: Roth, S., Eds.; Springer-Verlag: Berlin, 1993; p. 307.

(52) Grund, A.; Kaltbeitzel, A.; Mathy, A.; Schwarz, R.; Bubeck, C. *J. Phys. Chem.* **1992**, *96*, 7450.

(53) Banerjee, S.; Kumar, G. R.; Mathur, P.; Sekar, P. *J. Chem. Soc. Chem. Commun.* **1997**, 299.

(54) Itoh, T.; Saitoh, H.; Iwatsuki, S. *J. Polym. Sci. A Polym. Chem.* **1995**, *33*, 1589.

(55) Dias, A. R.; Garcia, M. H.; Rodrigues, J. C.; Petersen, J. C.; Bjornholm, T.; Geisler, T. *J. Mater. Chem.* **1995**, *5*, 1861.

(56) Sachtleben, M. L.; Spangler, C. W.; Tang, N.; Hellwarth, R.; Dalton, L. in *Organic Materials for Nonlinear Optics III;* Ashwell, G. J.; Bloor, D., Eds.; Royal Society of Chemistry: London, 1993; p. 231.

(57) Kajzar, F.; Messier, J.; Rosilio, C. *J. Appl. Phys.* **1986**, *60*, 3040.

(58) Baumert, J. C.; Bjorklund, G. C.; Jundt, D. H.; Jurich, M. C.; Looser, H.; Miller, R. D.; Rabolt, J.; Sooriyakumaran, R.; Swalen, J. D.; Twieg, R. J. *Appl. Phys. Lett.* **1988**, *53*, 1147.

(59) Shigemoto, I.; Nakano, M.; Yamada, S.; Kiribayashi, S.; Yamaguchi, K. *Mol. Cryst. Liq. Cryst.* **1996**, *286*, 159.

(60) *Gaussian Basis Sets for Molecular Calculations;* Huzinaga, S., Ed.; Elsevier: Amsterdam, 1984.

(61) Tutt, L. W.; Boggess, T. F. *Prog. Quant. Electr.* **1993**, *17*, 299.

(62) Allan, G. R.; Labergerie, D. R.; Rychnovsky, S. J.; Boggess, T. F.; Smirl, A. L.; Tutt, L. *J. Phys. Chem.* **1992**, *96*, 6313.

(63) Boggess, T. F.; Allan, G. R.; Rychnovsky, S. J.; Labergerie, D. R.; Venzke, C. H.; Smirl, A. L.; Tutt, L. W.; Kost, A. R.; McCahon, S. W.; Klein, M. B. *Opt. Eng.* **1993**, *32*, 1063.

(64) Allan, G. R.; Rychnovsky, S. J.; Venzke, C. H.; Boggess, T. F. *J. Phys. Chem.* **1994**, *98*, 216.

(65) Tutt, L. W.; McCahon, S. W. *Opt. Lett.* **1990**, *15*, 700.

(66) Tutt, L. W.; McCahon, S. W.; Klein, M. B. *Proc. SPIE-Int. Soc. Opt. Eng.* **1990**, *1307*, 315.

# Index

# Cumulative List of Contributors
# for Volumes 1–36

# Cumulative Index
## for Volumes 37–43

ISBN 0-12-031143-7

9 780120 311439

90065